Engineering Materials

More information about this series at http://www.springer.com/series/4288

K.M. Gupta · Nishu Gupta

Advanced Semiconducting Materials and Devices

Springer

K.M. Gupta
Department of Applied Mechanics
Motilal Nehru National Institute
of Technology
Allahabad, Uttar Pradesh
India

Nishu Gupta
Department of Electronics and
Communication Engineering
Motilal Nehru National Institute
of Technology
Allahabad, Uttar Pradesh
India

ISSN 1612-1317 ISSN 1868-1212 (electronic)
Engineering Materials
ISBN 978-3-319-19757-9 ISBN 978-3-319-19758-6 (eBook)
DOI 10.1007/978-3-319-19758-6

Library of Congress Control Number: 2015944749

Springer Cham Heidelberg New York Dordrecht London

Printed on acid-free paper

Springer International Publishing AG Switzerland is part of Springer Science+Business Media
(www.springer.com)

Dedicated
to
our respected

Rita Ji

(Wife of Dr. K.M. Gupta and mother of Nishu Gupta)

who

always has been the inspiration and source of the family happiness

Preface

This book is intended to cover the vast and fast growing field of semiconducting materials and devices in accordance with the modern trends. The level of the book covers the syllabi being taught at the undergraduate standard of engineering institutes in India and abroad for students of Electronics and Telecommunication, Electrical Engineering, Materials Science, and Applied Physics. It also covers the syllabi of various competitive and other national level examinations across the globe. The contents of the book will be fully useful to the electronics students of postgraduate studies. The book may be used as a reference book by scientists, engineers, and research scholars for basic, advanced and up-to-date information.

This book is organized into six parts, containing 16 chapters. In Part I, the first two Chaps. 1 and 2 provide an elementary viewpoint on properties and applications of semiconductor materials; some recent advances like spintronic materials, ferromagnetic semiconductors, left-handed materials, DMS, photocatalytic semiconductors, etc.; materials science of crystal structures and imperfections, atomic models and bonding in them. Basic information on semiconductor devices, narrow and wide bandgap semiconductors are given in brief. These chapters are essentially needed to develop the understanding of the subject. Chapters 3 and 4 describe the carrier transport and excess carriers in semiconductors, respectively. These chapters provide detailed mathematical information about the phenomena related to various semiconductor behaviour.

Part II consisting of Chaps. 5 and 6, deals with junctions and interfaces. Chapter 5 describes p–n junctions, diodes and their breakdown mechanism; while Chap. 6 explains a large variety of specific diodes. The Part III comprises Chaps. 7–9. Chapter 7 deals with majority carrier diodes, their operation and response under different biasing conditions. Chapter 8 provides information on construction, characteristics, performance and application of microwave diodes, varacter diode, photodiodes; IMPATT, TRAPATT, BARITT and Gunn diodes, etc. In Chap. 9, the optoelectronic devices describe the solar cells, photodetectors, LEDs and semiconductor lasers, etc.

Part IV, comprising Chaps. 10–12, deals with BJT, FETs and power semiconductor devices. In Chap. 10, various aspects of bipolar junction transistors are dealt in, Chap. 11 provides a thorough discussion on MESFETs, MOSFETs, IGFETs, etc. Chapter 12 focuses on p-n-p-n diode, thyristors, silicon-controlled rectifiers, bilateral devices, etc.

Part V, comprises Chaps. 13 which deals with the brief details of various aspects of semiconductor fabrication. It elaborates the method of production of silicon, semiconductor crystal growth, zone refining, construction of microelectronic circuit. It also deals with manufacturing of wafers, lithography, molecular beam epitaxy, masking, fabrication technique of p-n junction and transistor manufacturing processes.

The last Part VI, comprises of Chaps. 14–16. Chapter 14 incorporates specialized semiconductors in vivid fields like solar cells, thermoelectrics, photocatalytic, plasmonics, photonic, photovoltaic semiconductors, etc. Chapters 15 and 16 presents a comprehensive detail of modern trends in the research and development of semiconductor materials and devices. In Chap. 15, nanostructured semiconducting materials and devices are described; while in Chap. 16, the recent advances and emerging trends in semiconducting materials and devices are given. Both these chapters present up-to-date information about the subject.

Basic and prerequisite information has also been included for easy transition to newer topics. Latest developments in the fields of semiconducting materials, their sciences, processes and applications have been accommodated. Latest topics in optoelectronic devices, metal-semiconductor junctions, heterojunctions, MISFET, LEDs, semiconductor lasers, photodiodes, switching diodes, tunnel diodes, Gunn diode, solar cells, varactor diode, IMPATT diode, advanced semiconductors, etc., have been included.

Topics like *electron theories, high-field effects, Hall effect, transit-time effects, drift and diffusion, breakdown mechanisms, equilibrium and transient conditions, switching, biasing* have been explained. A variety of semiconducting materials and devices, microelectronics, memory devices, advance and futuristic materials are described. Information on ideal diode, real diode, backward diode, etc., are provided to include the contents on more advanced semiconductor devices.

Review and objective type questions based on concepts, design, construction, applications and practical orientations, are included. Wherever required, the mathematical equations have been incorporated to understand the contents easily and in-full, however, they have been kept to a minimum, throughout the text. A large number of numerical and theoretical examples have been worked-out for eaiser understanding to the readers. Numerical exercises for practice and self-valuation are also given with their answers, in order to develop confidence to users. SI units, in general, have been incorporated throughout the text but for familiarity, the MKS units have also been used. Review, objective and numerical problems are enriched with questions of different examinations of many universities and national level institutes. The book is substantiated by a large number of diagrams, tables, equations and glossary of terms.

We gratefully acknowledge the authors and publishers of the books quoted in references which have provided guidelines in preparing this book. We also thank all the authors and publishers of research papers published in different journals whose work has been included. Every effort has been made to seek permission from the copyright holders to reproduce the matter in this book. However, the authors of this book desire excuse from those contributors who could not be contacted, or from those whose acknowledgement is missed undeliberately.

I (K.M.Gupta, and Nishu Gupta with appropriate relations) acknowledge the inspiration and blessings of my respected mother Smt. Bela Devi, brother-in-law Sri Jawahar Lal, sister Smt. Savitri Lal, brothers Sri Gopal Das Gupta, Brij Mohan, Hari Mohan, Madan Mohan, and other family members. I am full of gratitude to my wife Rita, daughter Nidhi, son-in-law Ritesh, son Nishu, daughter-in-law Anamika, grandson Akarsh (Ram) and Ayaansh (Aditya), and granddaughter Anshika (Gauri) for the patience shown and encouragement given to complete this venture. I also pay homage to my loving nephew Jayant (Babul) who left us for his heavenly abode at an early age.

At this time Nishu Gupta recalls his mentor, Professor Rajeev Tripathi, and Dr. Arun Prakash, Assistant Professor and Ph.D. guide for patronage and useful guidance. I (K.M. Gupta) recall with deep sense of respect and honour my distinguished early age friends (Late) Mohd. Naseem, Sri Fazlur Rahman Khan, and Sri Tarak Nath Batabyal, who have always been very positive thinking human beings to me and to the society by their ever helping attitude. Our heartfelt thanks are due to my friends Er. K.R.D. Tewari, Chartered Civil Engineer and Consultant, Allahabad; Er. Ranjeet Singh Virmani, AGM (retd.) Punjab National Bank; my nephew Hemant Jawahar Lal (IRS), Commissioner, Income Tax, Mumbai; to Sri S.C. Sant and Mr. Anurag Sant, the publishers of Umesh Publications for their continued support, guidance and cooperation in preparing this book. We thank Mr. Vishal Kumar Gupta 'Bishu' for his untiring efforts in typing the various versions of the manuscript.

The authors have attempted to contact the copyright holders of all material reproduced in this publication and apologize to copyright holders if permission to publish in this form has not been obtained. If any copyright material has not been acknowledged, please write and let us know so that we may rectify in future reprint. Enormous effort has been made to avoid errors and mistakes, however, their presence cannot be ruled out. Any suggestion to improve the standard of this book, indication towards errors, omissions and mistakes will be highly appreciated.

Allahabad, India
July 2015

K.M. Gupta
Nishu Gupta

Acknowledgements

The authors of this book acknowledge with heartfelt gratitude to Mr. S.C. Sant and Mr. Anurag Sant, the publishers of Umesh Publications (4230/1 Ansari Road, Daryaganj, Delhi-110002) India, for being kind enough to provide substantial literature support and valuable information that proved very useful in preparing this book. Authors (Dr. K.M. Gupta and Nishu Gupta) thank especially to Mr. Anurag Sant for his courteous gesture.

K.M. Gupta
Nishu Gupta

Contents

Abbreviations

APD	Avalanche Photodiode
BARITT	Barrier Injection Transit Time
BCC	Body Centred Cube
BHJ	Bulk Hetero Junction
BIFET	Bipolar Field Effect Transistor
BJT	Bipolar Junction Transistor
BNDC	Bulk Negative Differential Conductivity
CCD	Charge Coupled Device
CFL	Compact Fluorescent Light
CMOS	Complementary Metal Oxide Semiconductor
CVD	Chemical Vapour Deposition
COMFET	Conductivity-Modulated FET
DC	Diamond Cubic
DIAC	Diode Alternating Current
DIBL	Drain-Induced Barrier Lowering
DMOS	Double-Diffused MOS
DMS	Dilute Magnetic Semiconductor
DNG	Double Negative Metamaterials
DOVETT	Double Velocity Transit Time
DSC	Dye-sensitized Solar Cell
DVD	Digital Versatile Disc
EGS	Electronic Grade Silicon
EHP	Electron-Hole Pair
ERF	Electro-Rheological Fluid
FCC	Face Centred Cube
FET	Field Effect Transistor
FLL	Frequency-Locked Loop (circuit)
FS	Ferromagnetic Semiconductor
GEMFET	Grain-Enhanced (MOSFET)
GMR	Giant Magnetoresistance

HCP	Hexagonally Closed Packed
HEMT	High-Electron Mobility Transistor
IC	Integrated Circuit
IGBT	Insulated Gate Bipolar Transistor
IGFET	Insulated Gate Field-Effect Transistor
IGR	Insulated Gate Rectifier
IGT	Insulated Gate Transistor
IMPATT	Impact Avalanche Transit Time
IOC	Integrated Optical Circuit
IR	Infrared
ITO	Indium Tin Oxide
JFET	Junction Field Effect Transistor
LASCR	Light-Activated Silicon Controlled Rectifier
LDR	Light Dependent Resistor
LED	Light Emitting Diode
LHM	Left-Handed Material
LSA	Limited Space-Charge Accumulation
MBD	Magnetic Bipolar Diode
MBE	Molecular Beam Epitaxy
MBT	Magnetic Bipolar Transistor
MESFET	Metal-Semiconductor FET
MGS	Metallurgical Grade Silicon
MISFET	Metal-Insulator-Semiconductor FET
MODFET	Modulation-Doped FET
MOSFET	Metal-Oxide Semiconductor FET
MR	Magneto-Resistance
MTJ	Magnetic Tunnel Junction
NFSOM	Near Field Scanning Optical Microscope
NIM	Negative Index Metamaterial
NLO	Non-Linearly Optical
NM	Nano Membrane
OLED	Organic Light Emitting Diode
OPV	Organic Photovoltaic
PCD	Photo Chemical Deposition
PLL	Phase-Locked Loop (circuit)
PLZT	Lead Lanthanum Zirconate Titanate
POSFET	Piezoelectric Oxide Semiconductor Field Effect Transistor
PSC	Polymer Solar Cell
PSC	Plasmonic Solar Cell
QD	Quantum Dot
QW	Quantum Well
RF	Radio Frequency
SCR	Silicon Controlled Rectifier
SCS	Silicon Controlled Switch
SDL	Semiconductor Disk Laser

SET	Single Electron Transistor
SHIP	Silicon Heterointerface Photodetector
SMD	Surface Mount Device
SNG	Single Negative (Metamaterials)
SOI	Silicon on Insulator
SRAM	Semiconductor Random Access Memory
SRD	Step Recovery Diode
SRR	Split-Ring Resonator
TED	Transferred Electron Device
TFT	Thin Film Transistor
TRAPATT	Trapped Plasma Avalanche Transit Time (diode)
TRIAC	Triode Alternating Current
TTFT	Transparent Thin-Film Transistor
TVS	Transient Voltage Suppression Diode
UJT	Uni-Junction Transistor
UPT	Unipolar Transistor
VCSEL	Vertical Cavity Surface Emitting Laser
VLSI	Very Large Scale Integration
VPE	Vapour Phase Epitaxy
VVR	Voltage Variable Resistor
WBG	Wide Band Gap

Symbols with Units

a	Unit cell dimension (Å)
B	Magnetic flux density (Wb/cm^2)
C_{junc}	Junction capacitance (F/cm^2)
D_{xe}, D_{xh}	Diffusion coefficient for electrons, holes (cm^2/s)
E	Energy (J, eV)
E_F	Equilibrium Fermi level (J, eV)
E_g	Band gap energy (J, eV)
E_r, E_t	Recombination, trapping energy (J, eV)
$f(E)$	Fermi-Dirac distribution function
H_C	Hall coefficient (cm^3/C)
hf	Photon energy (J, eV)
(hkl)	Miller indices
i_B, i_C, i_E	Base, collector, emitter current in a BJT (A)
J	Current density (A/cm^2)
$K = 2\pi/\lambda$	Wave number (cm^{-1})
m_e^*, m_h^*	Effective mass of electrons, holes (kg)
m_0	Rest mass of electron (kg)
N_c, N_V	Effective density of states at the edge of the conduction band, valence band (cm^{-3})
n^+	Heavily doped n-type material
n_e, n_h	Equilibrium concentration of electrons in n-type, p-type material (cm^{-3})
n_i	Intrinsic concentration of electrons (cm^{-3})
n_0	Equilibrium concentration of electrons (cm^{-3})
R	Resistance (Ω)
t_{sd}	Storage delay time (s)
V_{CB}, V_{EB}	Voltage from collector to base, emitter to base in a BJT (V)
V_0	Constant potential (V)
V_d	Drift velocity (cm/s)
w	Sample width, depletion region width (cm)

$\alpha_{absorption}$	Optical absorption coefficient (cm^{-1})
α_r	Recombination coefficient (cm^3/s)
γ	Emitter injection efficiency (%)
δn, δp	Excess electron, hole concentration (cm^{-3})
ε_r	Dielectric constant (F/cm)
λ	Wavelength of light (μm, Å)
μ	Mobility (cm^2/V-s)
σ	Conductivity (Ω-cm)$^{-1}$
τ_n, τ_p	Recombination lifetime for electrons, holes (s)

SI Prefixes of Multiples and Submultiples

Factor	Symbol	Prefix	Factor	Symbol	Prefix
10^{-1}	d	deci	10^{-12}	p	pico
10^{+1}	da	deka	10^{+12}	T	tera
10^{-2}	c	centi	10^{-15}	f	femto
10^{+2}	h	hecto	10^{+15}	P	peta
10^{-3}	m	milli	10^{-18}	a	atto
10^{+3}	k	kilo	10^{+18}	E	exa
10^{-6}	μ	micro	10^{-21}	z	zepto
10^{+6}	M	mega	10^{+21}	Z	zetta
10^{-9}	n	nano	10^{-24}	y	yocto
10^{+9}	G	giga	10^{+24}	Y	yotta

Greek Alphabets

Name	Symbol	Analogous English sound	Name	Symbol	Analogous English sound
Alpha	α	a	Xi	Ξ	X
Beta	β	b		ξ	x
Gamma	Γ	G	Omicron	o	o
	γ	g	Pi	Π	P
Delta	Δ	D		π	p
	δ	d	Rho	ρ	r
Epsilon	ε	e	Sigma	Σ	S
Zeta	ζ	z		σ	s
Eta	η	e	Tau	τ	t
Theta	θ	Th	Upsilon	υ	u
Iota	ι	i	Phi	Φ	Ph
Kappa	κ	k		φ	ph
Lambda	Λ	L	Khi	χ	kh
	λ	l	Psi	Ψ	Ps
Mu	μ	m		ψ	ps
Nu	ν	n	Omega	Ω	O
				ω	o

About the Authors

Dr. K.M. Gupta is Professor (Retd.) in the Department of Applied Mechanics, Motilal Nehru National Institute of Technology, Allahabad. He has over 39 years of teaching, research and consultancy experience. He obtained Diploma (with Honours) in Mechanical Engineering, Bachelor of Engineering (Gr. I.E., AMIE) in Mechanical Engineering, Postgraduation (M.E. with Honours) in 1977, and completed his Doctorate (Ph.D.) degree from University of Allahabad. Although a Mechanical Engineer he has also specialised in Automobile Engineering discipline. He has authored 31 books including the books published by CRC Press, Taylor & Francis Group, USA; Scrivener-Wiley Publishing, USA; edited two books, and written a chapter on 'Handbook of Bioplastics and Biocomposites Engineering Applications' published by Scrivener-Wiley Publishing, USA. He has also authored 123 research papers in reputed International and National Journals and Conferences to his credit. Professor K.M. Gupta has presented his research papers in 17 International conferences abroad at USA, UK, Japan, China, France, Muscat, Bangkok, South Africa, Hongkong, etc. He has also chaired nine international conferences in China, Singapore, Dubai, Bangkok, etc. He has acted as Editor-in-Chief of 'The International Journal of Materials, Mechanics and Manufacturing (IJMMM) Singapore', and has edited many international journals and conferences. He has worked as reviewer for various international and national journals, and has acted as member of several editorial boards. In recognition of his academic contributions at international level, Marquis Publication (USA) has included him in the list of 'World Who's Who in Science and Engineering 2007' and 'Who's Who in the World 2008'. The International Biographical Centre, a leading research institute (Great Britain) selected him as one of the '2000 Outstanding Scientists 2009' from across the world; and Rifacimento International Publisher has included his biographical-note in 'Reference Asia: Asia's Who's Who of Men and Women of Achievement'.

Recipient of many Gold Medals and Prizes for his outstanding career from Diploma to Doctorate (a rare achievement); he has served as Head of Automobile Engineering Department at the Institute of Engineering and Rural Technology, Allahabad. He masterminded the development of several laboratories *viz.* Automobile related Labs, Materials Science Lab., Strength of Materials Lab., Hydraulics Lab., etc., at different Institutes/colleges. He was the trail brazier in establishing an Auto Garage and a Repairing Workshop.

Dr. K.M. Gupta has undergone several industrial trainings at many reputed industries and workshops. He is endowed with vast experience in Curriculum Development activities and consultancy. Earlier, he served as Dean of Research & Consultancy, Head of the Applied Mechanics Department at Motilal Nehru National Institute of Technology Allahabad. He has acted as Chairman of various research selection committees, of research project monitoring committees and other administrative committees of his institute and other universities. He has also served as Chairperson, Community Development Cell (CDC) of MNNIT for several years.

Presently, Dr. Gupta is teaching Materials Science, Engineering Mechanics, Electrical and Electronic Materials, Solid Mechanics, etc. His research interests are in the fields of materials science, composite materials, stress analysis, solid mechanics, etc.

Nishu Gupta is a Research Scholar in the Department of Electronics and Communication Engineering, Motilal Nehru National Institute of Technology Allahabad, India. He received his B. Tech. degree in Electronics and Communication Engineering from Uttar Pradesh Technical University and M. Tech. degree in Nanoscience and Technology from Delhi Technological University (Formerely Delhi College of Engineering), Delhi, India.

Earlier, he served as Software Engineer at Infosys Technologies Limited. He has published 18 research papers to his credit which were published in various international and national journals and conferences in India and abroad, including five papers in international journals. He has attended six international conferences abroad. He is the recipient of honours and awards that include "Top 2000 Intellectuals of the 21st century by International Biographical Centre, Cambridge, England". Marquis Publication (USA) has included him in the list of 'World Who's Who in the world in the year 2009 and 2013 editions. He had been shortlisted by the Ministry of HRD, Government of India for the Commonwealth Scholarship/Fellowship Plan, 2012 offered by the Government of United Kingdom.

His research interests are in the field of semiconductor devices, hybrid solar cells, nano-communication networks and related fields.

Part I
Review of Fundamentals of Semiconductors

Chapter 1
Semiconductor Materials: Their Properties, Applications, and Recent Advances

Abstract In this preliminary chapter discussing the fundamentals of semiconductor, the electrical and electronic materials are classified. Importance of semiconducting materials and their scope are elaborated. Conductors, semiconductors and dielectrics are differentiated from each other on the basis of energy bandgap. Different types of semiconducting materials, their merits and characteristics are listed. Various elemental form, compound and alloyed semiconductors are described. Basic semiconducting devices along with their working principles are briefed. Latest developments in the field of semiconducting materials have been included; the main among them are emerging and futuristic spintronic materials, ferromagnetic semiconductors, emerging wide bandgap semiconductors, left-handed materials, photocatalytic semiconductors, integrated circuit purpose semiconductors etc. Minute insight into the various topics are given through solved numerical and theoretical examples. Review questions, numerical problems and objective type questions are also given with their answers.

Keywords Scope of applications · Energy gap · Different types of Si, Ge, Se · Formulated semiconductors · Spintronic semiconducting materials · Ferromagnetic semiconductors · Wide bandgap semiconductors · Left-handed (LH) materials · Single negative LH materials · Double negative LH materials · Manganese semiconductors · Diluted magnetic semiconductors · Si for ICs · GaAs · Photocatalytic semiconductors · LED stumps in cricket

1.1 Importance of Electronic and Semiconducting Materials

Advancement of any engineering discipline is not possible without the development of materials suitable for appropriate uses. The development necessitates progress in science: physics and chemistry, engineering and technology of the materials. Rapid advancement in electron-based computers, revolutionary changes in electronics engineering from vacuum valves to very large scale integration (VLSI); developments of optoelectronic and junctions and interfaces, new generation AlGaAs chips for hyper high-speed computers, and different kinds of semiconducting diodes

K.M. Gupta and N. Gupta, *Advanced Semiconducting Materials and Devices*,
Engineering Materials, DOI 10.1007/978-3-319-19758-6_1

and transistors are some illustrations which are the outcome of developments in electronics engineering and semiconducting materials technology.

Further advancement in electronics, computers, communication and instrumentation fields are likely to be in the form of whisker-based fibre optics, light-based computers, high temperature superconductors etc. But dreams of these futuristic advances will become a reality only after achieving a breakthrough in certain materials properties. Most likely the 21st century will see the high voltage transmission through hair-sized conducting wires, the hybrid magnets of more than 100 T (tesla) capacities, the magnetic refrigerators operating well above 100 K (kelvin), and hybrid crystals (e.g. Hg-Cd-Te) serving as sensor elements.

1.2 Classification of Electrical and Electronic Materials, and Status of Semiconducting Materials

Semiconducting materials belong to the category of electrical and electronic materials. The Electrical and Electronic engineering materials can be classified into following types [1].

1. Conductors
   - (i) High voltage and low voltage conductors
   - (ii) High temperature and low temperature conductors
   - (iii) Bared and insulated conductors
2. Semiconductors
   - (i) Intrinsic (or element) type
   - (ii) Extrinsic (compound and alloy) type
     - n type
     - p type
3. Dielectrics (or insulators)
   - (i) Solid type
   - (ii) Liquid type
   - (iii) Gaseous type
   - (iv) Ceramic type
   - (v) Polymeric type
   - (vi) Fibrous type
4. Superconductors
   - (i) Metallic type
   - (ii) Ceramic type
   - (iii) Ideal and hard types
   - (iv) Low and high temperature types
   - (v) Magnetic and non-magnetic types

5. Magnetic materials
   (i) Diamagnetic
   (ii) Paramagnetic
   (iii) Ferromagnetic
   (iv) Antiferromagnetic
   (v) Ferrimagnetic (or ferrites)
6. Ferroelectrics
   (i) Zirconates
   (ii) Hafnates
   (iii) Titanates
   (iv) PLZT
7. Piezoelectrics
   (i) Natural (as rochelle salt)
   (ii) Artificial (as tourmaline, metaniobate)
8. Perovskites (or mixed oxides)
9. Spinels, Garnets, and Magnetoplumbites
   (i) Normal spinel (as $ZnFe_2O_4$)
   (ii) Inverse spinel (as magnetite)
   (iii) Metallic garnet
   (iv) Rare earth garnet

An introductory description of few of these materials is given below to appraise the status of semiconducting materials among them.

1.2.1 Conductors

Conductors have plenty of free electrons in their valence orbit which are responsible for electrical and heat conduction. The valence and conduction band energies overlap in them. The mechanism of electron conduction through conductors occurs in accordance with free electron theory. Gold (Au), Silver (Ag), Copper (Cu), Aluminium (Al) are good electrical conductors. Considering the cost factor, copper and aluminium are widely used as transmission wires and cables, and windings of rotating machines (motors, alternators etc.). Electrical resistivity of conductors is of the order of 10^{-9} Ω-m. It is influenced by temperature, impurities, alloying elements, and plastic deformation.

1.2.2 Semiconductors

These are solids having energy gap *Eg* lying in between the conductor and insulator. Their conductivity is more than that of the dielectrics but less than that of conductors. They are basically electronic materials viz. silicon (Si) and germanium (Ge) in element form. The element form remains in pure state and is called *intrinsic semiconductor*. Intrinsic semiconductors are of no practical utility since they desire an extremely high voltage ($\approx 10^8$ V) for conduction across the energy gap. Hence, they are doped to make extrinsic semiconductors.

Extrinsic semiconductors are primarily of *n*-type and *p*-type. They may be in compound form such as GaAs, CdTe etc. and in alloy form such as $HgCd_xTe_{1-x}$. A number of semiconducting compounds are available in oxide, halide and sulphide forms also. The *n* and *p*-types are used to make *n-p* junction for diode, *n-p-n* and *p-n-p* for transistor, etc.

Semiconductors are generally hard and brittle, and possess negative coefficient of temperature resistance. In contrast with conductors whose conductivity increases with purification, the conductivity of semiconductors decreases with purification. Semiconductors are widely used as rectifiers, amplifiers, photocells etc. Their properties are of greater importance in telecommunication, power electronics, computer hardware etc. Complete details about them are given in later chapters.

1.2.3 Dielectrics

These are insulating materials in which the valence electrons are tightly bound to their parent atoms. Therefore, they require a very large electric field to remove these electrons from the atoms. Dielectrics are characterized by (i) a full valence band (ii) an empty conduction band, and (iii) a large energy gap ($Eg > 3$ eV).

Dielectrics are available in solid, liquid, and gaseous states. They may be of natural or synthetic types, polymeric or ceramic in nature, and fibrous or flaked in structure. Mica, bakelite, elastomeric fibre, and paper are some examples of solid dielectrics. Transformer oil, silicon oil, varnishes are liquid dielectrics while air, nitrogen, ozone are gaseous dielectrics. Dielectrics are used as capacitors, as insulation, in strain gauges and sonar devices.

Main properties of dielectrics are their dielectric constant ε_r, dielectric strength, and dielectric loss. Dielectric constant is influenced by temperature and frequency. A good dielectric should have high dielectric constant and high dielectric strength but a lower dielectric loss. The dielectric loss is the least in transformer oil and fused silica ($\tan \delta = 0.0001$). Dielectrics are subjected to different kinds of polarization under the influence of applied electric field. These polarization processes are affected by time, frequency and temperature. The time effect is expressed in terms of *relaxation time*.

1.2.4 Superconductors

These materials exhibit abrupt and sudden changes in their resistance and other behaviour, at extremely low temperatures. It is because of a phenomenon called *superconductivity* which is noticed below a certain critical temperature. Superconductors show extraordinary properties like (i) resistance R of the material ≈ 0, (ii) magnetic flux density $B \approx 0$, and (iii) power (copper) loss $I^2R \approx 0$.

Superconducting materials mainly fall in the category of metals and ceramics, and may be in element, compound, or alloy forms. They may be of magnetic or non-magnetic nature, type I (ideal) or type II (hard), and low temperature or high temperature superconductors. Since the peculiar behaviour of superconductivity in materials is presently available at low temperatures only, there is a need to develop these properties at high temperatures (say room temperature).

Ceramic superconductors are more promising development in this field. Nb_3Sn, $La_{80}Au_{20}$, Nb-Zr-Ti are some examples of superconducting compounds and alloys. Yattrium barium copper oxide ($YBa_2Cu_3C7{-}x$), bismuth strontium calcium copper oxide (Bi Sr Ca CuO) and thallium barium calcium copper oxide (Th Ba Ca CuO) are some modern ceramic superconductors.

1.3 Scope of Application of Semiconducting Materials

Scope of application of electrical and semiconductor materials is very vast. These materials find utilities in not only electrical machines, equipments, devices etc. but are also used as components, circuits and other auxiliaries related to electronics, computers, and instrumentation fields. Their importance is also realized in cable networking, wireless networking, satellites, optical devices etc. They find very useful applications even in medical, mechanical, nuclear, biotechnological fields. The details given below focuses upon the scope of electrical and semiconductor materials in numerous applications [1].

In consumer items	
Heaters Remote control devices Telephone Radio Microwave ovens	Iron press Television Inverters Taperecord Electronic watches
In Electrical Engineering	
Switching contacts Alternators Motors	Bus bars Transformers Voltage dividers

(continued)

(continued)

In Electronics	
Amplifiers Integrated circuits Antenna Broadcasting systems Printed circuits Rectifiers	Filters Regulators Satellite Photoconductive cell Transistors Modulators
In Robotics	
Sensors Controls Manipulators Grippers	Actuators Processors Encoders Pendants
In Computer Engineering	
Hardware Monitors Peripherals Floppy Printers	Memory devices Ports Display devices Hard disk CDs etc.
In Instrumentation	
Transducers Signal generators Microprocessors Strain gauges	Cathode ray oscilloscopes Recorders Thermistors Energy meters, etc.
In Information Technology	
Networking cables Web cameras Cam corders	Multimedia devices Routers Optical fibres, etc.
In Mechanical Engineering	
Furnace thermocouples Arc welding sets	Electrically heated ovens X-ray systems, etc.
In Biotechnology and Medical Sciences	
Electro-cardiography Fibre-optic endoscopy	Electronic photography Magnetic resonance imaging, etc.

Most of above systems/devices are inter-disciplinary, and are indicative only. Besides above, the electronic and semiconductor materials are also used in the fields of aeronautics, marine, defence, chemical, metallurgical, and automobile engineering applications.

Some systems/devices that make use of semiconducting materials for their working are shown in Fig. 1.1 for a ready reference.

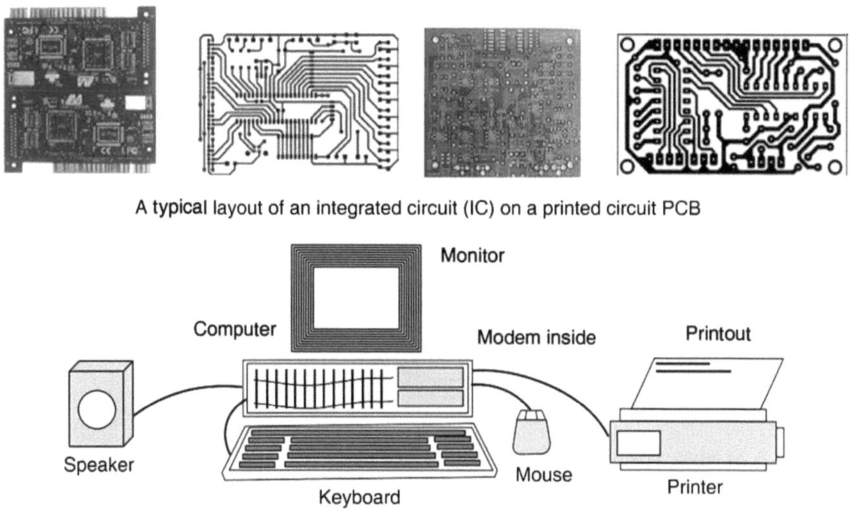

Fig. 1.1 Some systems/devices that make use of semiconducting materials for their working

1.4 Electrons and Their Role in Semiconductivity

Solid substances are composed of atoms, which are discrete indivisible particles having diameter of the order of 1 Å (10^{-10} m). Atoms of a substance are identical but are quite different from the atoms of another substance. They are composed of fundamental (or basic) particles viz. protons, neutrons and electrons. These particles in different substances are similar.

Electron is the most important particle of an atom. The electrical nature of atoms is attributed to a highly systematic configuration of electrons in them. They are universal constituent since all the metals contain them in their structure. Electrons have similar behaviour in conductors, semiconductors, dielectrics, magnetic or superconducting materials. They also have same behaviour in all applications whether a computer, television, radiography, or electrical conduction.

1.4.1 Valence Electrons

Electrons play vital role in determining the electrical and electronic properties of materials. Valence electrons in the outermost orbit (i.e. shell) of an atom decide the manner in which they respond to external effects. For example, due to their free movement within the metal they provide conduction, but on actuation by thermal energy they jump over the energy gap in semiconductors. As a whole the arrangement of electrons in an atom, behaviour of valence electrons, and

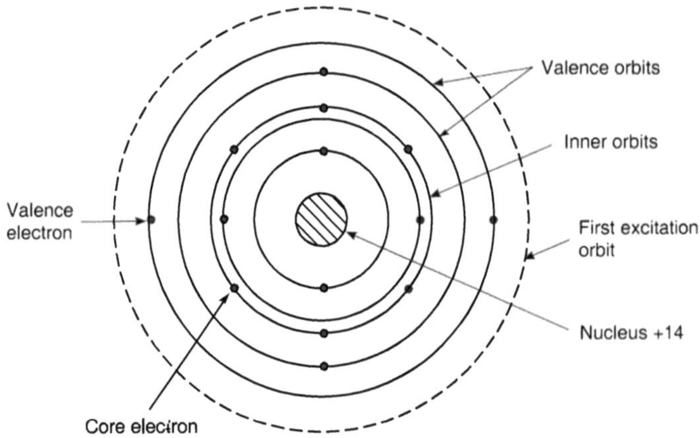

Fig. 1.2 A Si atom showing valence electron which plays an important role in semiconduction

inter-atomic interactions govern the electrical properties (conduction as well) of materials.

Electrons in the outermost shell are called valence electrons. When the outermost shell of an atom is not complete, these electrons are easily detachable. At room temperature, the thermal energy is enough to detach the valence electrons from their parent atoms. Such electrons are called free electrons. Metals have too many free electrons and hence they are good conductors of heat and electricity. On the other hand, insulating materials such as wood, glass, rubber, paper etc. require large energy to detach few electrons from their atoms. The situation in *semiconductors* is intermediatory to those of metals and insulating materials. The electronic structure and orbital model of a Si atom, the most common and important semiconductor is shown in Fig. 1.2.

1.5 Classification of Materials on the Basis of Energy Gap (or Band)

Solids (or materials) are broadly classified into following categories on the basis of their energy gap which is an important electronic behaviour.

1. Conductors
2. Semiconductors and
3. Dielectrics or insulators

Among these, the

- Conductors are mainly metals and alloys. Their electrical resistivity (reciprocal of electrical conductivity) is the least.

Table 1.1 Electrical resistivity of solids at room temperature

Solid	Electrical resistivity (Ω m)	Solid	Electrical resistivity (Ω m)
Metals		***Semiconductors***	
Silver	10^{-9}	Doped germanium	10^{-3}
Gold	10^{-9}	Pure germanium	10^{-1}
Copper	10^{-9}	Pure silicon	10^{-2}–10^{-3}
Aluminium	10^{-8}	***Insulators***	
Nickel	10^{-7}	Glass	10^{5}
Iron	10^{-6}	Fireclay	10^{8}
Antimony	10^{-5}	Bakelite	10^{9}
Bismuth	10^{-4}	Alumina, diamond, rubber, polyethylene	10^{11}
Alloys		Mica, PTFE	10^{13}
Brass	10^{-7}	PVC	10^{15}
Constantan	10^{-3}	Pure silica	10^{17}
Nichrome	10^{-2}		

Reproduced from [1]

- Semiconductors have the properties lying in-between the conductors and dielectrics. They are widely used for making solid state devices.
- Dielectrics have the highest values of resistivity. They are used as insulators and capacitors etc.

The orders of resistivity of solids vary between 10^{-9} Ω-m for good conductors to 10^{17} Ω-m for best quality insulators. The classification of solids on the basis of their electrical resistivities is shown in Table 1.1.

1.5.1 Valence and Conduction Band, and Energy Gap

The outermost filled or partially filled energy band is called valence band. The energy band above the valence band which is empty at 0 K is known as *conduction band*. The electrons available in the valence band are to be pushed into the conduction band for conduction. Thermal energy of electrical voltage is given to the solid for this purpose. The electrons conduct from the top energy level of the valence band to the bottom energy level of the conduction band.

There may or may not be an energy gap between these two bands. Based on the magnitude of energy gap, the solids are classified as conductors, semiconductors or insulators. The energy gap for different solids is shown in Fig. 1.3a–g. Based on the magnitude of *Eg*, various solids are classified as follows.

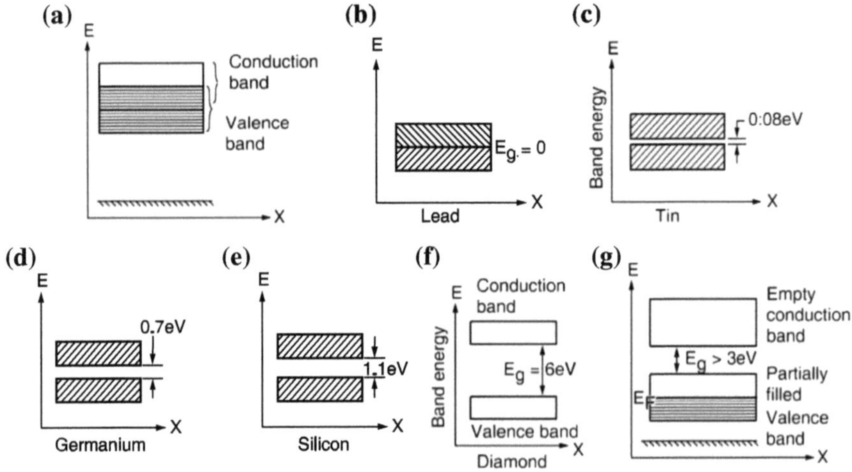

Fig. 1.3 Classification of solids based on energy gap **a** conductor having overlapping energy level, **b** conductor having $Eg = 0$, **c** conductor, **d**, **e** semiconductors, and **f**, **g** insulators. Reproduced from [1]

(ia) Conductors if energy levels overlap, Fig. 1.3a.
(ib) Conductors if $Eg = 0$, Fig. 1.3b.
(ic) Conductors if $Eg < 1$ eV, Fig. 1.3c.
(ii) Semiconductors if 3 eV > Eg > 1 eV, Fig. 1.3d and e. and
(iii) Insulators if $Eg > 3$ eV, Fig. 1.7f and g.

Here eV stands for electron-volts. The symbol x in Fig. 1.3 indicates an arbitrary length of the energy band.

1.5.2 Comparison among Conductors, Semiconductors and Insulators

The properties and characteristics of conductors, semiconductors, and insulators are compared below for a quick remembrance.

Description	*Conductors*	*Semiconductors*	*Insulators*
Conductivity	High to very high	Low to medium	Nil to very low
Resistivity	Low to very low	Medium to low	Very high to high
Temperature coefficient of resistance	Always positive	Always negative	May be positive or negative
Energy band	Overlapping or unfilled	Filled	Filled

(continued)

(continued)

Description	*Conductors*	*Semiconductors*	*Insulators*
Forbidden zone	Nil or very small	Small	Large
Current carriers without external energy	Free electrons	Nil	Nil
With external energy	Free electrons	Electrons and holes	Electrons
Behaviour at low temperatures	Near absolute zero, they become superconductors	In lower range, they become di-electrics	Dielectric constant ε_r drops sharply
Effect of increasing temperature on conductivity	Conductivity decreases	Conductivity increases	Conductivity increases
Effect of high electrical voltage on breakdown	No breakdown	They breakdown	They generally breakdown
Bonding nature of valence electrons with their parent atom	Very loosely (freely) bound	Loosely bound	Very tightly bound
Ionization energy	Much less	Less	Very large

1.6 Introduction to Semiconducting Materials [1]

These are the solids having energy gap *Eg* lying in between the conductor and insulator. Their conductivity is more than that of the dielectrics but less than that of conductors. They are basically electronic materials. Semiconductors are available in following forms.

1. Elements such as silicon (Si) and germanium (Ge).
2. Compounds such as GaAs, InP, AlSb, CdTe, ZnSe etc.
3. Alloys such as $GaAs_x\ P_{1-x}$, $HgCd_xTe_{1-x}$ etc.

Semiconductors are also classified as

(a) Intrinsic, and
(b) Extrinsic.

(i) *n*-type
(ii) *p*-type

The elemental forms of pure Si and pure Ge are intrinsic. In intrinsic form they are not useful. They are, therefore, doped by dopant to make extrinsic semiconductors. Extrinsic forms are directly useful and are widely employed in manufacturing of the solid state devices. They belong to the category of alloys and compounds. Electronics industry requires purity better than $1 : 10^9$ in pure Si and Ge.

Extrinsic semiconductors are primarily of *n*-type and *p*-type. They may be in compound form such as GaAs, CdTe etc. or in alloy form such as $HgCd_xTe_{1-x}$. A number of semiconducuting compounds are available in oxide, halide and sulphide forms also. The *n* and *p*-types are used to make *n-p* junction for diode, *n-p-n* and *p-n-p* for transistor, etc.

Semiconductors are generally hard and brittle, and possess negative coefficient of temperature resistance. In contrast with conductors whose conductivity increases with purification, the conductivity of semiconductors decreases with purification. Semiconductors are widely used as rectifiers, amplifiers, photocells etc. Their properties are of greater importance in telecommunication, power electronics, computer hardware etc. Complete details about them are given in later chapters.

1.6.1 Different Types of Semiconducting Materials

Although the Si and Ge are most commonly used semiconductors, yet there are several other semiconducting materials used in vivid devices. A detailed classification of them is given below in Table 1.2. In elemental form they belong to various columns of the periodic table; in compound form they are binary, ternary and quaternary; and in alloy form they may be homogeneous or heterogeneous types.

Table 1.2 Classification of different kinds of semiconducting materials

Types of semiconductor	Examples
Elemental form	
Column II	Zn, Cd
III	B, Al, Ga, In
IV	C, Si, Ge
V	N, P, As, Sb
VI	S, Se, Te
Binary compounds	
II–VI compounds	ZnS, ZnSe, ZnTe, CdS, CdSe, CdTe
III–V compounds	AlP, AlAs, AlSb, GaN, GaP, GaAs, GaSb, InP, InAs, InSb
IV–IV compounds	SiC, SiGe
Ternary compounds	GaAsP, HgCdTe, AlGaAs
Quaternary compounds	InGaAsP
Ternary alloys	$Al_{0.3}Ga_{0.7}As$ (30 % is Al and 70 % is Ga), $In_{0.53}Ga_{0.47}As$, $In_{0.11}Ga_{0.8}As$, $GaAs_{0.88}Sb_{0.12}$
Quaternary alloys	$In_xGa_{1-x}As_yP_{1-y}$

Reproduced with permission from [1]

1.6.2 Merits of Semiconducting Materials

Semiconductors possess several favourable properties such as listed below.

1. They are much smaller in size. For example, the gauge length of a semiconductor strain gauge may be as small as 0.25 mm or less, and thickness of wafer as less as 0.1 mm or less.
2. They are very light in weight. For example, the weight of a transistor is as low as a fraction of 1 g. And also a calculator, consisting a large number of transistors on its ICs, weighs 146 g only including the cells.
3. They operate on low voltage e.g. 9, 3, 1.5 V etc. For example, the calculator operates on manganese-dry battery of 3 V only.
4. They consume negligible power. For example, the calculator works at an expense of 0.0004 W only for years together.
5. They have a long life. It may be many hundred hours or a few years.
6. They do not develop any creep effect and hardly show any ageing effect.
7. They are least affected by shock loads and are almost shockproof.
8. They operate instantly and do not require any pre-heating.
9. They operate over a practically fruitful wide range of temperatures.

1.6.3 Characteristics of Semiconducting Materials

Main characteristic of semiconducting materials is that their conductivity increases with increase in temperature and also with increase in voltage. This increase is 10–20 times on adding 1 ppm impurity. Some more salient characteristics of semiconductors are given as follows.

1. They have negative temperature coefficient of resistance. It means that

$$\alpha_{semiconductor} \propto -\frac{dR}{dT}, \text{ and also } \propto -\frac{d\rho}{dT}$$

where, ρ is resistivity. The value of α lies in the range of $-4 \times 10^{-6}/^{\circ C}$ to $-6 \times 10^{-6}/^{\circ C}$.
2. The resistivity ρ of semiconductors at very low temperature ($\approx$4 K) matches with those of insulators.
3. They do not follow Ohm's law. The voltage-current relation is nonlinear. The current increases far more rapidly than the voltage. This characteristic is utilized in the working of transistors.
4. The resistance R of semiconductors decreases nonlinearly on adding impurity into it.This characteristic is utilized in the working of rectifiers.

1.6.4 Semiconducting Devices and Their Working Principles

Semiconductors are solid materials of crystalline nature. They possess several properties that are influenced by different factors such as temperature, voltage, impurity, electric field, light illumination, optical behaviour etc. Depending upon the influencing factor, the semiconductors find use in/as following important devices.

1. Thermistors: These utilize the temperature dependency effect of semiconductors.
2. Varistors: These utilize the voltage dependency effect of semiconductors.
3. Rectifiers: These utilize the impurity dependency effect of semiconductors.
4. Strain gauges: These utilize the change in resistance effect of semiconductors.
5. Zener diodes: These utilize the electric field effect of semiconductors.
6. Transistors: These utilize the amplification effects of semiconductors.
7. Photoconductive cells: These utilize the light illumination effect of semiconductors.
8. Photovoltaic cells: These utilize the optical characteristics of semiconductors.
9. Hall effect generators: These utilize the carrier drift effect in semiconductors.

1.7 Element Form Semiconducting Materials

1.7.1 Silicon (Si)

It is the element of IVth column in periodic table. Its structure is diamond cubic (DC) and the atoms are covalent bonded. The bond is directional in nature with a bond angle of 109°. It is most important and widely used material among all known semiconducting elements. This is dull whitish coloured monocrystal possessing the following salient properties [1].

- Atomic mass = 28.09 amu
- Atomic radius = 1.18 Å
- Young's modulus = 103 GPa
- Lattice constant = 5.43 Å
- Electronic configuration = $1s^2 2s^2 2p^6 3s^2 3p^2$

(i) Properties of silicon are sensitive to the presence of interstitial oxygen, which acts as donor and tends to destabilize the material at high temperatures. It is because of creation of neutral complexes due to combination of oxygen with other impurity atoms.

(ii) Its properties are influenced by nuclear radiations which induce such defects that readily combine with oxygen to produce active centres.

(iii) It allows diffusion of metallic impurities such as Fe, Ni, Co, Mn, Cu etc. at moderate to higher temperatures.
(iv) Only 1 atom in about 10^{12} contributes a free electron (and also a hole) to the crystal because of broken covalent bonds.
(v) It is suitable for services over a wide temperature range due to its higher energy gap (more than that of germanium).

1.7.2 Germanium (Ge)

Similar to Si, it is also an element of IVth column in periodic table. Its structure is diamond cubic and atoms are covalent bonded. The bonds are directional in nature with a bond angle of 109°. It is a grey coloured brittle material of metallic look. Its mechanical properties are similar to glass. Germanium possesses the following salient properties.

- Atomic radius = 1.22 Å
- Lattice constant = 5.66 Å
- Atomic mass = 72.59 amu
- Young's modulus = 99 GPa
- Electronic configuration = $1s^2 2s^2 2p^6 3s^2 3p^6 3d^{10} 4s^2 4p^2$

(i) It is capable of creating deep impurity energy levels.
(ii) It has a great sensitivity with elements like Au, Ni, Cu, Li etc., which can move within its crystal with high speeds at low temperatures.
(iii) The doping elements like In, As, Ga, Sb diffuses slowly in it.
(iv) It is prepared by horizontal or vertical pulling from the melt for high purity/quality.
(v) Since it contains the order of 10^{22} atoms/cm^3, but at room temperature (300 K) the $ni \approx 10^{13}$/cm^3; therefore only 1 atom in about 10^9 contributes a free electron (and also a hole) to the crystal due to broken covalent bonds.
(vi) Its carrier mobility is higher (than silicon), so it is useful for high frequency services.

1.7.3 Selenium (Se)

Selenium is a non-metal with properties that are intermediate between those of its periodic table column adjacent elements sulphur and tellurium. Selenium is a polyatomic non-metal sometimes considered as a metalloid. It is an element of group VI in periodic table. It exists in both: crystalline and non-crystalline forms. Its structure is hexagonal. It is covalent bonded with a bond angle of 104°. Its salient properties are as follows.

- Atomic number = 34
- Melting point = 217 °C
- Specific gravity = 4.79
- Young's modulus = 58 GPa

It finds use in the following main applications.

(*i*) xerographic printing	(*ii*) photoelectric light meters	(*iii*) TV cameras
(*iv*) photovoltaic cells	(*v*) photocells in cinematography	

- The largest commercial use of Se accounting for about 50 % of consumption, is for the production of glass. Se compounds confer a red colour to glass. This colour cancels out the green or yellow tints that arise from iron impurities that are typical for most glass. For this purpose, various selenite and selenate salts are added to it. For other applications, the red colour may be desirable in which case, the mixtures of CdSe and CdS are added.

- Copper-indium-gallium-selenide is a material used in the production of solar cells.
- Because of its photovoltaic and photoconductive properties, selenium is used in photocopying photocells, light meters and solar cells. Its use as a photoconductor in plain-paper copiers once was a leading application but in the 1980s, the photoconductor application declined as more and more copiers switched to the use of organic photoconductors.
- It was once widely used in selenium rectifiers. These uses have mostly been replaced by silicon-based devices or are in the process of being replaced. The most notable exception is in power DC surge protection, where the superior energy capabilities of selenium suppressors make them more desirable than metal oxide varistors.
- Zinc selenide was the first material for blue LEDs, but gallium nitride is dominating the market now.
- Cadmium selenide has played an important role in the fabrication of quantum dots.
- Sheets of amorphous selenium convert x-ray images to patterns of charge in xeroradiography and in solid-state, flat-panel x-ray cameras.
- Selenium is a catalyst in some chemical reactions but it is not widely used because of issues of toxicity.
- Selenium is used in the toning of photographic prints. Its use intensifies and extends the tonal range of black-and-white photographic images and improves the permanence of prints.
- Se is used as a gamma source in industrial radiography.

1.7.4 Antimony (Sb)

It is an element of group V in periodic table. Its structure is rhombic and the atoms are covalent bonded. Due to directional nature of bond, the bond angle is 96°. Its other properties are as follows [1].

- Atomic number = 51
- Melting point = 630 °C
- Specific gravity = 6.62
- Ionization energy = 0.039 eV
- Young's modulus = 55 GPa
- Structure is rhombus

It finds use in the following applications.

(i) Photon detection
(ii) Laser

1.7.5 Other Elements

These are used to make various compounds and alloy semiconductors. A summary detail of these elements is given in Table 1.3.

1.7.6 Comparison Between Silicon and Germanium

We have described earlier that the silicon and germanium are covalent bonded. Each atom has four bonds. Now, a comparison of these two extensively used semiconductor elements is given in Table 1.4.

Table 1.3 Properties of various semiconducting elements

Element	Symbol	Atomic number	Specific gravity	Melting point (°C)	Young's modulus (GPa)
Arsenic	As	33	5.72	817	39
Bismuth	Bi	83	9.80	271	34
Boron	B	5	2.34	2030	440
Indium	In	49	7.31	156	10.5
Phosphorus	P	15	1.83	44	4.6
Sulphur	S	16	2.07	119	19.5
Tellurium	Te	52	6.24	450	41
Tin	Sn	50	7.30	232	52

Their structure and other geometrical details are displayed in periodic table given elsewhere in this book. Reproduced with permission from [1]

Table 1.4 Comparison between Silicon and Germanium

Description	Silicon	Germanium
Atomic number	14	32
Type of structure	DC	DC
Valency	4	4
Electrical resistivity at 20 °C (Ω m)	3000	0.47
Energy gap E_g: at 0 K	1.17 eV	0.74 eV
: at 300 K	1.12 eV	0.66 eV
Intrinsic carrier concentration (at 300 K)	$1.1 \times 10^{16}/m^3$	$2.5 \times 10^{19}/m^3$
Mobility μ: of electron (at 300 K)	0.14 m^2/Vs	0.38 m^2/Vs
: of hole (at 300 K)	0.05 m^2/Vs	0.18 m^2/Vs
Dielectric constant ε_r	12	16
Number of atoms	$5.0 \times 10^{22}/cm^3$	$4.4 \times 10^{22}/cm^3$
Number of intrinsic carrier (at 300 K)	$1.5 \times 10^{10}/cm^3$	$2.5 \times 10^{10}/cm^3$
Melting point	1410 °C	937 °C
Specific gravity	2.33	5.32
Coefficient of thermal expansion α	$4.2 \times 10^{-6}/°C$	$6 \times 10^{-6}/°C$
Thermal conductivity (W/cm °C)	0.8	0.55
Heat capacity (cal/gm °C)	0.17	0.08
Limiting operational temperature	≈200 °C	80–85 °C
Cost factor	Costlier	Cheaper

Reproduced with permission from [1]

1.8 Formulated (Compound and Alloyed) Semiconductor Materials

These are made by doping the element semiconductors. Various dopant used for this purpose are P, As, Sb, In, Ga, Al etc. A brief description of some important formulated semiconducting materials is given below.

1.8.1 Gallium Arsenide (GaAs)

This is a compound of gallium (a third column element) and arsenic (a fifth column element) in periodic table. It is made by zone refining technique. Its melting point is about 1250 °C. Production of GaAs is a difficult task due to generation of high vapour pressure in arsenic at about 1200 °C. It has high mobility and large energy gap. Its other properties are as given below [1].

(i) E_g at 300 K = 1.42 eV
(ii) $\varepsilon_r = 13.2$
(iii) Specific gravity = 5.31

(iv) Hole mobility = 0.025 m^2/Vs
(v) Electron mobility = 0.80 m^2/Vs

However, this compound is not competitive with silicon and germanium in respect of purity and structural perfection. It is a useful material for several devices, important amongst them are given as follows

- Switching and parametric diodes
- Tunnel diodes
- Semiconductor lasers
- Hot electron diodes.

For details on semiconductor laser, refer Chap. 9 in this book.

1.8.2 *Indium Antimonite (InSb)*

This is a compound of indium (In) an element of IIIrd column and antimony (Sb) an element of Vth column in periodic table. Electron mobility of InSb at room temperature is highest (≈10 m^2/Vs) among all known semiconductors. It has a low melting point (525 °C) and is easier to produce in single crystal form. Its electrical resistivity at 20 °C is ≈20,000 Ω m. Indium antimonite finds use in the following main applications

- Infrared detectors
- Hall effect devices
- Laser diodes
- Tunnel diodes
- Infrared filter material
- Transistors
- Thermal imaging cameras
- FLIR systems
- Infrared missile guidance systems, and
- Infrared astronomy.

It is a crystalline compound made from the elements indium (In) and antimony (Sb). It is a narrow-gap semiconductor from the III–V group properties are the following

Crystal structure	Zinc Blende
Number of atoms in 1 cm^3	2.94×10^{22}
Density	5.77 g cm^{-3}
Dielectric constant: (Static)	16.8
: High frequency	15.7

(continued)

(continued)

Effective : electron mass	0.014 m_o
: hole mass	0.43 m_o
Electron affinity	4.59 eV
Lattice constant	6.479 Å
Energy gap	0.17 eV
Intrinsic carrier concentration	2×10^{16} cm^{-3}

Other Applications: Indium antimonide (InSb) finds application in following areas also.

- As a terahertz radiation source
- In thermal image detectors using photo-electromagnetic detectors or photodiodes.
- In magnetic field sensors using the Hall effect or magneto-resistance.
- In fast transistors, due to the high carrier mobility of InSb.

1.8.3 Oxides, Sulphides, Hallides, Tellurides and Sellurides

These are extrinsic semiconductors of compound form. Some of them are *n*-type and others are *p*-type. Zinc oxide (ZnO) is a *n*-type and Cuprous oxide (Cu_2O) is a *p*-type semiconductor. Sulphides such as PbS (Gelina), CdS, Cu_2S, ZnS; and halides such as CuBr, CuI, CuCl etc. have deviations from stoichiometric compositions. Tellurides such as Bi_2Te_3, PbTe and sellurides such as PbSe, CdSe and other elements are also the semiconductor compounds. These are used in the following main applications [1].

- BaO: in oxide coated cathodes
- CdS, CdSe, CdTe: in photoconductivity based automatic door opener, street light switching, burglar alarm
- $PbSO_4$, CdS, PbS: in photoconductive devices such as photocells of TV camera and cinematography
- GaP, GaAs, GaSb: in semiconductor lasers
- CsSb: in photomultipliers.

1.8.4 Cadmium Sulphide (CdS)

It is used to make photoconductors of high dissipation capability and excellent sensitivity in visible spectrum, and to prepare cadmium sulphide cell by depositing a layer of CdS. This layer generally contains Ag, Sb, In etc. as impurity. Its energy gap is 2.4 eV. CdS finds use in the following main applications

- As constituent of cathode-ray phosphorus
- To measure a fixed amount of illumination as with light meter
- To record modulating light intensity on sound track
- As ON-OFF light relay in digital and control circuits.

1.8.5 Silicon Carbide (SiC)

It is hard and refractory in nature. Its energy gap is large, about 3 eV. Its melting point is also very high (about 2400 °C). Due to above properties, the SiC is useful as

- High-temperature rectifiers
- High-temperature transistors

1.8.6 Lead Sulphide (PbS)

It is used to make commercial photoconductive cells and finds use in the following main applications

- Infrared detection
- Infrared absorption measurement

1.8.7 Indium Arsenide (InAs)

Indium arsenide (or indium monoarsenide) is a semiconductor composed of indium and arsenic. It has the appearance of grey cubic crystal with a melting point of 942 °C. InAs is well known for its high electron mobility and narrow energy bandgap. It is widely used as terahertz radiation source, as it is a strong photo-emitter.

Crystal structure of InAs is shown in Fig. 1.4. Other parameters at 300 K are the following.

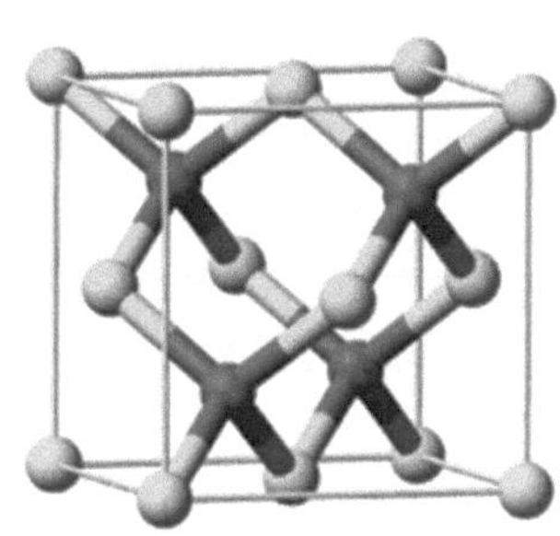

Fig. 1.4 Crystal lattice structure of InAs (Reproduced from [2])

Crystal structure	Zinc Blende
Number of atoms in 1 cm^3	3.59×10^{22}
de Broglie electron wavelength	400 Å
Debye temperature	280 K
Density	5.68 g cm^{-3}
Dielectric constant (static)	15.15
(high frequency)	12.3
Effective electron mass	$0.023m_o$
Effective hole masses (m_h)	$0.41m_o$
Electron affinity	4.9 eV
Lattice constant	6.0583 Å

Its electrical parameters are the following.

Breakdown field	$\approx 4 \cdot 10^4$ V cm^{-1}
Mobility: of electrons	$\leq 4 \cdot 10^4$ $cm^2V^{-1}s^{-1}$
: of holes	$\leq 5 \cdot 10^2$ cm^2 $V^{-1}s^{-1}$
Diffusion coefficient: of electrons	$\leq 10^3$ cm^2s^{-1}
: of holes	≤ 13 cm^2 s^{-1}

1.9 Choicest Materials for Different Semiconductor Devices

Various semiconducting materials mentioned above possess varying properties, and hence are suitable for application in vivid devices. Main among these materials and their suitability for different devices are given below [1].

Devices/product/services	Suitable semiconductors
Diodes	Si, Ge, GaAs
CPU, microprocessor	AlGaAs chips
Transistors	Si, Ge
Photocells	Se, CdS, PbS (i.e. Galena or galenite)
Rectifiers	Si, Se, CuO
ICs	Si chip, GaAs and Si hybrid chips
Light detectors	InSb, CdSe, PbTe, HgCdTe

(continued)

(continued)

High-frequency devices	Ge
Fluorescent screens such as T.V. screens	ZnS, Phosphor coating of oxides and sulphides of Zn, Cd, Be
Infrared detectors	Si, Ge
Nuclear radiation detectors	Si, Ge
Gunn diode (a microwave device)	GaAs, InP
Semiconductor lasers	GaAs, AlGaAs, GaP, GaSb
Solar cells/batteries	Photovoltaic action materials such as Se
Cinematography	Photocell effect based materials such as Se, CdS, PbSO4
Xerox-type photocopier	Se
Hyper high-speed computers	New generation AlGaAs chips
Sensor elements for guided missiles	Hg-Cd-Te crystals
Automatic door opener	Photoconductivity based materials such as CdSe, CdS, CdTe
Stroboscope disk (flashing light called stobotron)	Opto-electronic polymers
Red phosphor for TV tubes	Yttrium (Y)
Diamond transistors	Phosphorus-doped diamond film *n*-type semiconductor
Semiconductor lenses and mirrors for high power lenses	Synthetic diamond
Light absorption and optics	Single crystal diamond
Light emitting diode (LED)	
For visible green color light	Gap, CdS
For visible red color light	GaAsP, CdSe
For visible yellow light	GaP
For visible blue light	SiC
For invisible infrared light	GaAs, InSb
For ultraviolet region of light	ZnS
Avalanche photodiode	InAlAs, InGaAs, GaAsSb
Photon detector	InP, InAs, InSb
Photoconductors	
in green light	CdS
in red light	CdSe
in infrared region	CdTi

1.10 Spintronics and Spintronic Materials [1]

The development of electronics has spread in many specialize branches such as bio-electronics, molecular electronics, spintronics, rubber electronics, polymer electronics, nano electronics etc. Amongst these the spintronics (i.e. spin electronics) is one of the fast emerging fields. It relates to spin-dependent phenomena which are manipulated to achieve a desired electronic outcome, such as *quantum-computing*. The spintronics devices store information into spins of up and down orientations, which are then attached to the mobile electrons to carry them along a wire, to be read at a terminal.

Several kinds of spintronic devices are in the stage of development and in concept. They are/may be bipolar devices such as spin diodes, spin transistors, spin-polarized solar cells, magnetic diodes etc. These devices are attractive for the purpose of magnetic sensor and memory storage applications. Other aspects of spintronics are in the use of NSOM (Near Field Scanning Optical Microscopy) to detect electrons in semiconductor quantum dots, in quantum computing, and in SET (Single Electron Transistor) as a single electron spin detector.

1.10.1 Major Fields of Spintronic Research

Modern researches on spintronics concentrate on development of advanced materials. Almost all material systems are tried for it. Magnetic metal multilayer, Ga_{1-x}, Mn_xAs type ferromagnetic semiconductor, GaN etc. are prominent among them. However, the ferromagnetic semiconductors are the most favourable material system, which combine the characteristics of ferromagnetism and semiconductor both. Some major fields of current spintronic research are the following:

1. Development of spin-based devices such as *p-n* junctions and amplifiers.
2. Spin relaxation behaviour in metals and semiconductors.
3. Spin polarized transportation through the semiconductor/semiconductor interfaces.
4. Spin-based quantum computation.
5. Quantum computer hardware.

1.10.2 Operational Mechanisms of Spintronic Devices

Spin orientation of conduction electrons survives for nanoseconds (10^{-9} s), whereas the time elapsed in decay of electron momentum is tens of femto seconds (10^{-15} s). Consequently, the spintronic devices are especially useful for quantum computing. Here, the electron spin represents a bit of information, which is known as 'qubit'.

The understanding of spin relaxation and efforts needed to enhance the spin memory of conduction electrons in metals and semiconductors are essential for study of spintronic devices.

1.10.3 Working Principle of Spintronic Devices

On application of magnetic field in normal metals, there is a change in their resistance. This change is normally of the order of 1 % for the magnetic field of 1 tesla (T). In some ferromagnetic metals, the direction of magnetization can be reversed when a magnetic field of the order of 0.0001 T is applied. This result in a phenomenon called *Magneto-resistance* (MR), which is utilized in commercial production of small magnetic 'read heads'. They can sense very small magnetic fields in written information on hard discs. This causes considerable decrease in the space required for storing the bits of information. Thus the storage capacity of hard discs increases.

1.10.4 Emerging and Futuristic Spintronic Materials

The utility of semiconductor spintronics devices depends on the availability of suitable materials which retain ferromagnetic properties above room temperature. Some recent and important materials in this list are the following

- Half-metallic materials such as Fe_3O_4, CrO_2 etc.
- Colossal magneto-resistive materials such as Sr-doped $LaMnO_3$
- Double perovskits
- Heusler alloys having composition of X_2YZ, where X and Y are transition elements and Z is an element of group III, IV or V.
- Half-Heuslar alloys having a composition of XYZ.
- Ferromagnetic semiconductors.

1.11 Ferromagnetic Semiconductor

Magnetic bipolar diodes (MBD) such as *p*-type GaMnAs and *n*-type GaAs are salient applications of ferromagnetic semiconductor. In this diode, the *p*-region is a magnetic semiconductor while the *n*-region is a conventional semiconductor compound. Similarly, a magnetic bipolar transistor (MBT) may be of great use, if produced. Conceptually, it may consist of a conventional semiconductor emitter and collector, and a base of *p*-type magnetic semiconductor. With the advent of such materials, presently existing memory devices e.g. SRAM (semiconductor random access memory) and DRAM (dynamic random access memory) can be combined together to produce MRAM (magnetic random access memory), which combines the advantages of high speed access, high density information storage, and non-volatility.

Ferromagnetic semiconductors such as $CrBr_3$ are highly spin polarized, Co-doped TiO_2 are optically transparent, and Mn-doped III–V compounds have small solubility of magnetic dopant. Thus the ferromagnetic semiconductors having a wider range of spin polarization are desirable materials. In order to be ferromagnetic at room temperature, they are required to have Curie temperature above room temperature.

1.12 Emerging Wide Bandgap Semiconductors

Other popular/emerging and futuristic materials having wide bandgap semiconductors that retain ferromagnetic properties above room temperature are given below [1].

InMnAs. It is a metal-insulator semiconductor which is suitable for working at low temperature. It may be used as magnetic sensors.
GaMnAs. It is suitable for high spin-injection. It may be used for infrared LED and lasers.
CoTiO$_2$. It has high coherent length.
NiGaAs. It is suitable for efficient carrier injection. It may be used for high speed digital electronic applications.
ZnGeP$_2$. It is a chalcopyrite material that exhibits unusual nonlinear optical properties. It can be used in optical oscillations.
ZnSnAs$_2$. This is a chalcopyrite material. It shows promise for frequency converters and IR- generation.
ZnSiN$_2$. It is a wide bandgap chalcopyrite having lattice parameter close to GaN and SiC. Achievement of ferromagnetism in it will enable its usefulness to make ultraviolet solar-blind detector and microwave power electronic devices.
EuS. It is a dilute magnetic semiconductor (DMS) material. It can be used as tunnelling barrier, which can effectively function as spin filter.

1.13 Left Handed (LH) Materials [1]

This is a novel metamaterial characterized by simultaneously negative permittivity ε and negative permeability μ, i.e.

$$\varepsilon < 0 \,\text{and}\, \mu < 0$$

LH materials are recently developed interesting materials having novel concepts and applications. The unusual properties of LHMs will lead to

1. Reversal of Doppler effect, and
2. Reversal of Snell's law.

Consequences of these unusual properties are that

(i) The light rays incident on a convex lens will diverge instead of converging, as happens in conventional i.e. right handed (RH) materials.
(ii) It will support waves with antiparallel group and phase velocities, known as backward waves.
(iii) In a left-handed medium, the light propagates in opposite direction to the direction of energy flow.
A left handed material has negative index of refraction.
The name left-handed comes from the fact that the properties of these materials are just opposite to those having $\varepsilon > 0$ and $\mu > 0$ i.e. right-handed materials (RHMs).

Applications. The unusual behaviour of LHMs has lead to novel applications such as:

- Microwave components
- Miniaturized antennas
- Probes
- Waveguides
- Interconnections between nanodevices and external terminations
- Novel filters in cellular phones
- Micro-lenses

1.13.1 Single Negative Left-Handed Materials

Single negative (SNG) metamaterials have either negative relative permittivity (ε_r) or negative relative permeability (μ_r), but not both. They act as metamaterials in combination with a different SNG, jointly acting as a DNG double negative. Epsilon negative media (ENG) display a negative ε_r, while μ_r is positive. Many plasmas exhibit this characteristic. For example, noble metals such as gold or silver exhibit this characteristic in the infrared and visible spectrums. Mu-negative media (MNG) display a positive ε_r, while μ_r is negative. *Gyrotropic* or *gyromagnetic* materials exhibit this characteristic.

Gyrotropic Material. A *gyrotropic* material is one that has been altered by the presence of a quasi-static magnetic field, enabling a *magneto-optic effect*. A magneto-optic effect is a phenomenon in which an electromagnetic wave propagates through such a medium. In such a material, left- and right-rotating elliptical polarizations can propagate at different speeds. Joining a slab of ENG material with the slab of MNG material resultes in properties such as listed below.

- Resonances
- Anomalous tunnelling
- Transparency, and
- Zero reflection.

Like negative index materials, SNGs are innately dispersive; so their ε_r, μ_r and refraction index n alter with changes in frequency.

1.13.2 Double Negative Left-Handed Materials

A LHM possesses many unusual properties such as negative refraction, backwards wave propagation, reversed Doppler effect, etc. Recently, the UCSD group brought this medium to realization with a man-made composite, which consists of metallic split ring resonators (SRR) and wires. This composite was shown to possess a frequency window where the effective permittivity, ε_{eff}, and permeability, μ_{eff}, are simultaneously negative. These types of metamaterials have a refractive index, $n = -\sqrt{\varepsilon\mu}$ and are also commonly referred to as double negative materials (DNG). Left-handed materials and their unusual properties are attracting increasing attention in view of their vast potential for many applications. Many other composite metallic designs that possess these unusual properties have been explored.

1.13.3 Negative Index Metamaterials

In negative index metamaterials (NIM), both permittivity and permeability are negative resulting in a negative index of refraction. These are also known as double negative metamaterials or double negative materials (DNG). Other terms for NIMs include "left-handed media", "media with a negative refractive index", and "backward-wave media". In optical materials, if both permittivity ε and permeability μ are positive, wave propagation travels in the *forward* direction. If both ε and μ are negative, a *backward* wave is produced. If ε and μ have different polarities, waves do not propagate. The materials exhibiting a negative index of refraction have only been demonstrated as artificially constructed materials. In this regard, the depiction of Fig. 1.5 is self-explanatory.

1.13.4 Double Positive Medium

Double positive mediums (DPS) do occur in nature, such as naturally occurring dielectrics. Permittivity and magnetic permeability are both positive and wave

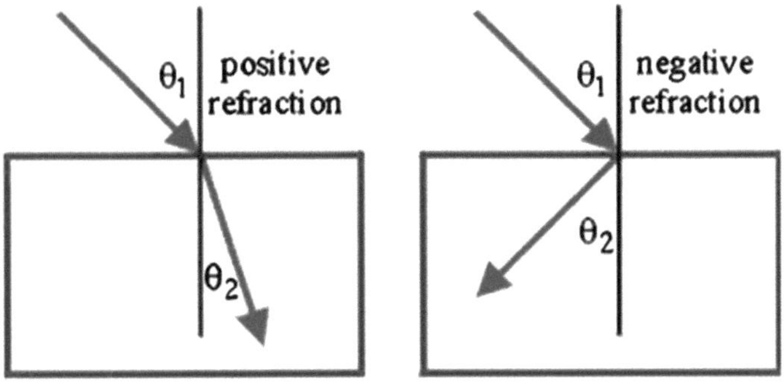

Fig. 1.5 Explanation of negative index metamaterial

propagation is in the forward direction. Artificial materials have been fabricated which combine DPS, ENG and MNG properties.

1.14 Manganese Semiconductor

Manganese sulfide (MnS) is a magnetic semiconductor material with a number of technological applications (optoelectric devices and lithium ion battery cathodes). In its bulk thermodynamic ground state, it adopts an antiferromagnetic (AFM) ordered rock-salt structure. The calcium and manganese chalcogenides offer an opportunity to examine the solubility limits in simple isomorphous systems. The phase relation between them is that a few percent of CaS in solution hardens MnS. CaS shows higher melting temperature than MnS. The bonds between metal and non-metal species in CaS are more ionic than in MnS. CaS is wide-band-gap semiconductor, shows phase transition from B1–B2 phase, crystallizes in the rocksalt, zinc blende, wurtzite and NiAs structures and has technological impact due to various applications in optoelectronics and luminescent devices. This semiconductor is also a potentially very good choice for protective coatings for its hardness, high thermal conductivity, and large bulk modulus.

1.15 Diluted Magnetic Semiconductor

Diluted magnetic semiconductors (DMSs) are the materials in which a fraction of host atoms of a semiconductor are substituted by transition metal atoms such as V, Cr, Mn, Fe, Co and Ni. In the class of II–VI based diluted magnetic semiconductors, much focus has been paid to $Mn_xCa_{1-x}S$ as the material CaS can accommodate a large percentage of Mn atoms and possesses a band gap suitable for optical applications. DMSs have been investigated due to their possibility of having magnetic properties. They can be used in developing spintronics, integrated opto-electronic and nanostructure quantum devices. The II–VI alloys show novel magneto-optical and magneto-transport properties if the cation is randomly and partially substituted by Fe, Co, Ni and Mn. The development of new spintronics devices such as light emitting diodes and magnetic sensors requires the finding of semiconductors with improved ferromagnetic properties. The orientation of research on DMSs was managed towards (II–Mn) VI system.

The ferromagnetic (FM) $Mn_xCa_{1-x}S$ semiconductors in the rock-salt phase using spin polarized method has a magnetic field application such as magnetic switching and recording. The interaction of lattice and spin between the dopant magnetic ions and the lattice can induce ferromagnetic (FM) behaviour, depending on the magnitude of d-d (J) and sp-d (α, β) exchange integrals between the paramagnetic dopant and host lattice. It is easy to dope Mn impurities into II–VI compounds which do not introduce any carriers in this system. The important feature of DMSs is that the energy band gap and other physical parameters can be controlled by varying the composition of magnetic ions in these materials.

1.16 Silicon: The Semiconductor Used as Raw Material in Making ICs

Pure silicon is the basis for most integrated circuits. It provides the base (or substrate) for the entire chip and is chemically doped to provide the N and P regions that make up the integrated circuit components. The silicon of about 1 ppb imurity is used. Typical N-type dopants include phosphorus and arsenic. Boron and gallium are typical P-type dopants. Aluminum is commonly used as a connector between the various IC components. The thin wire leads from the integrated circuit chip to its mounting package may be aluminum or gold. The mounting package itself may be made from ceramic or plastic materials. Figure 1.6 shows an Si IC.

Si is so dominant because of the following reasons.

Fig. 1.6 Silicon integrated circuit (Reproduced from [3])

1. It forms an oxide that is of very high quality, seals the surface with very few pin holes or gaps. This allows gap MOSFET (see Chap. 11) to be more easily made as the SiO_2 forms the insulating layer for the Gate.
2. It forms a very tough Nitride, Si_3N_4. Silicon Nitride forms a very high bandgap insulator which is impermeable. This is used to passivate (seal) the die. This is also used to make hard masks and in other process steps.
3. Si has a reasonable bandgap of ~ 1.12 eV, not too high so that room temperature can not ionize it, and not so low that it has high leakage current.
4. It forms a nice gate material. Most modern FET's used in VLSI have used Si as the gate material. It turns out that it is very easy to deposite non-crystalline Si on surfaces and it is easily etched to great precision.

Basically the success of Si is the success of MOSFET, which with scaling and extreme integration has driven the industry. MOSFET's are not so easily manufactured in other material systems, and the same level of integration can not be achieved in other semicondcutors, because

- GeO_2—is partially soluble
- GaAs—does not form a oxide
- CO_2—is a gas.

1.16.1 Gallium Arsenide (GaAs) for Making Integrated Circuit

Gallium arsenide (GaAs) is a compound of *gallium* and *arsenic*. It is a III–V direct bandgap semiconductor with a zinc blende crystal structure. Gallium arsenide is used in the manufacture of devices such as microwave frequency integrated circuits, monolithic microwave. The most important advantage of gallium arsenide is speed. Electrons travel about five times faster in gallium arsenide than they do in silicon. Gallium arsenide also has a high resistance to electrical current before it is doped with any impurities to form circuit elements, grated circuits, infrared light-emitting diodes, laser diodes, solar cells and optical windows.

The interest in GaAs circuits over silicon has increased since the investigation of GaAs transistors.The advantages are numerous and varied.

- For microwave analog work, GaAs offers lower noise and power consumption than silicon.
- For military and space applications, GaAs integrated circuits offer exceptional radiation hardness.
- For automotive and geological applications, GaAs can work in wide temperature ranges between −200 and +200 °C

- GaAs optical devices offer high speed sources and detectors for optical fibre networks.
- GaAs devices have displayed unequaled gain, noise performance, and bandwidth in analog microwave applications.
- More recently, high speed digital circuits have been developed because of the high switching capability of GaAs devices.

The development of high speed logic gates with modest power dissipation and high density has increased the interest in refinement towards very large scale integration (VLSI). This allows the development of high speed signal processors, computer circuits and memory circuits. It would also be possible to integrate circuits with both the necessary speed and optical properties for intelligent optical fibre links.

1.17 Semiconducting Photocatalytic Materials in the Services of Pollution Free Environment

The worsening environmental pollution has caused awareness about worldwide pollution crisis. For the sustainable development, the development of pollution-free technologies for environmental is an urgent need. Among the wide variety of green earth and renewable energy projects, the semiconductor photocatalysis has emerged as one of the most promising technologies for the potential applications of photocatalysis the following semiconducting materials are mainly used:

- ZnO
- Cu_2O
- Iron oxides semiconductor
- Sulfides semiconductor
- Ternary oxides
- Semiconductor chalcogenide
- Bismuth oxyhalides

Detailed information on these topics are given in Chap. 14.

1.18 LED Stumps: The Advent of Semiconductors in Cricket

Powered by hidden low voltage batteries, once the wicket is hit, the bails instantaneously bring flash red LED lights and send a radio signal to the stumps (Fig. 1.7) which also light up. The particular semiconductors used for LED manufacture are

Fig. 1.7 LED Stumps showing use of semiconductors. (Reproduced from [4])

gallium arsenide (GaAs), gallium phosphide (GaP), or gallium arsenide phosphide (GaAsP). The impurities commonly added are zinc or nitrogen; but silicon, germanium, and tellurium are also used. Gold and silver compounds most commonly used for contact purpose, because they form a chemical bond with the gallium at the surface of the wafer.

1.19 Glimpse of Some Salient Semiconductors

The following illustrations indicate the glimpse of some semiconducting materials.

- Sb-Bi are used in thermocouples to measure a very small temperature difference.
- Red LEDs are made of $GaAs_{0.6}P_{0.4}$.
- Alloys such as $Al_{0.3}Ga_{0.7}As$ and $In_xGa_{1-x}As_yP_{1-y}$ are used for semiconductor laser purpose.
- Hall effect sensors are used as manipulator of robot.
- Input and output control unit of robots are generally made of *npn* and *pnp* switches.
- The feedback of drive system in robots uses the optical encoders
- GaInNAsSb semiconductor is grown by MOVPE process for solar cell applications.
- $Ga_{0.94}In_{0.06}N_{0.02}As_{0.96}Sb_{0.02}$ alloy films have a band energy gap of 1.13 eV room temperature.
- Pen drive uses a flash memory chip which is made of a semiconducting material.
- Remote controlled devices such as for TV, A/C etc. use IR light, which uses semiconducting materials for its operation.
- Bi_2Te_3 is a thermoelectric semiconductor.
- $CoAs_3$ is an example of Skutterudites type TE semiconductor.
- Clathrates, a kind of TE semiconductors may of following structures.

  - $X_2Y_6E_{46}$ (where E represents Si, Ge, or Sn)
  - $Sr_8Ga_{16}Ge_{30}$ and $Ba_8In_{16}Sn_{30}$ are the type I clathrates.

- Half-Heusler(HH) intermetallic alloys are: MgAgAs, TiNiSn,LiMgP.
- β-Zn_4Sb_3 is a *p*-type semiconducting (TE) intermetallic compound, which presses high ZT, where Z represents 'figure of merit'.
- Examples of challogenides TE semiconductors are

  - Te-Ag-Ge-Sb
  - $AgPb_{18}SbTe_{20}$

1.20 Solved Examples

Example1.1 Make a comparison between silicon and gallium arsenide.

Solution. Some electronic properties of gallium arsenide are superior to those of silicon. It has a higher electron velocity and higher electron mobility, allowing gallium arsenide transistors to function at frequencies in excess of 250 GHz. As a wide direct band gap material with resulting resistance to radiation damage, GaAs is an excellent material for space electronics and optical windows in high power applications. A comparison of their properties is shown below.

Property	Silicon	GaAs
Electron mobility (cm^2/Vs)	1500	8000
Hole mobility (cm^2/Vs)	600	400
Energy gap (eV)	1.12 (indirect)	1.43 (direct)

- Higher electron mobility for GaAs shows promise for high speed devices and circuits.
- Direct gap allows for emission of photons in LEDs and LASER devices.

Review Questions

1. Discuss the importance of semiconductor materials and enumerate their scope in different fields.
2. Explain the role of electrons in semiconductors and semiconductivity. What are valence electrons?
3. Compare and distinguish the semiconductors from conductors and insulators.
4. What are different types of semiconductor materials? Enumerate their uses.
5. Describe the merits and characteristics of semiconducting materials.
6. Describe various element forms of semiconducting materials.
7. Enlist various semiconductor devices and state the characteristic property of semiconductor which is utilized in their working.
8. Compare the properties and characteristics of Si and Ge.
9. Discuss various compound and alloy form of semiconductor materials. Write the properties and applications of oxides, sulphides, halides, tellurides and sellurides.
10. Suggest the choicest semiconductors for producing the following devices.

    (a) ICs (b) Automatic door opener (c) Solar cells

    (d) LEDs (e) Photoconductors (f) Rectifiers
11. Write notes on the following:
    (a) Spintronic materials
    (b) Futuristic semiconductor materials
    (c) Ferromagnetic semiconductors
12. Discuss the role of following materials as futuristic electronic/semiconducting materials.
    (a) Left-handed materials
    (b) Wide bandgap semiconductors
13. Why are some LH materials called as single negative and some other as double negative? Write their examples and applications.
14. Discuss the properties and characteristics of manganese semiconductor and dilute magnetic semiconductor. Write their examples and applications.
15. Describe the silicon and GaAs as semiconductors for making ICs. Why is Si, a dominant material for this purpose?
16. Explain the especialities of GaAs and write its applications. Compare GaAs with Si from various viewpoints.

17. Write detailed notes on the following semiconducting materials.
    (a) Negative index metamaterials.
    (b) Gyrotropic materials
    (c) Double Positive materials
    (d) Semiconductors useful for making pollution-free environment.
18. How do the LED stumps of cricket game function with the help of semiconductors?
19. Enlist some salient types and applications of semiconductors.

Objective Questions

1. Given below are some statements.

   A. The phosphorus is a Vth column elemental form of semiconductor.

   B. Unit cell of Si is an agglomerated structure.

   C. Zincblende is the structure selenium.

   D. The number of effective atoms per unit cell in Si is 6.

   Of these, the correct statements are
   (a) A and B (b) B and C
   (c) A, C and D (d) A, B and D

2. Which of the following pairs does *not* match?
   (a) Theseutilize the voltage dependency effect of semiconductors.
   (b) Use of selenium
   (c) Use of GaAs
   (d) Semiconductor used for LED in infrared region of the light spectrum

3. Consider the following statements.
   Extrinsic semiconductors show high electrical conductivity because the impurities are
   1. of high conductivity 2. highly charged
   3. highly mobile
   Which of these statements are *not* correct ?
   (a) 1 and 2 (b) 2 and 3
   (c) 1 and 3 (d) 1, 2 and 3

4. Consider the following statements.
   A. A semiconductor is an amorphous solid.
   B. Semiconductors are usually opaque in appearance.
   C. Semiconductors behave like dielectrics at extremely low temperatures.
   D. Temperature coefficient in semiconductors is negative only.

   Of these, the correct statements are
   (a) A and B (b) B and C
   (c) C and D (d) B, C and D

5. Match list I with list II, and choose the correct answer from the codes given below the lists.

List I	List II
A. InSb	1. 10^{-8} to 10^{-9} ohm m
B. Resistivity of Si	2. 10^{-2} to 10^{-3} ohm m
C. Photoengraving	3. For cutting windows in oxide layer
D. Structure of As	4. Simple orthorhombic
	5. III-V compound
	6. II-VI compound

Codes :

	A	B	C	D		A	B	C	D
(a)	5	1	4	3	(b)	6	2	4	3
(c)	5	2	3	4	(d)	6	1	3	4

6. The correct sequence of increasing order of electrical resistivity of the given materials is
   (a) diamond, doped germanium, silicon, gold.
   (b) gold, silicon, doped germanium, diamond.
   (c) gold, doped germanium, silicon, diamond.
   (d) gold, diamond, silicon, doped germanium

Answers

1. (c) 2. (d) 3. (b) 4. (b)
5. (c) 6. (c)

References

1. Gupta, K.M., Gupta, N.: Advanced Electrical And Electronics Materials, Scrivener Wiley Publishing, USA (2015)
2. Lide, D.R.: Handbook of Chemistry and Physics (87 ed.), Boca Raton, FL: CRC Press, pp. 4–61, ISBN 0-8493-0594-2 through http://en.wikipedia.org/wiki/Indium_arsenide
3. http://www.channel4learning.com/sites/gcsease/engineering/image_engineered_assembly_pinthru_integrated.html
4. Posted by Alen Dsouza in Technology "New Zing LED Bails Glows up ICC World Twenty20 tournament" through http://geekstroke.com/technology/new-zing-led-bails-glows-up-icc-world-twenty20-tournament/

Chapter 2
Overview of Crystals, Bonding, Imperfections, Atomic Models, Narrow and Wide Bandgap Semiconductors and, Semiconductor Devices

Abstract In this chapter, the basic concepts of Materials Science are presented that will help to understand the lattice structure of semiconductors. Monocrystalline and polycrystalline solids, Bravais crystals system; different unit cells along with their geometry and examples are elaborated. Fundamentals and illustrations of Miller indices are presented for the understanding of the planes, family of planes, directions and family of directions through the unit cells. Atomic bonding: bonding forces, types and nature of bonds in semiconductors are presented. Covalent bonds, bond length, bond angle and percentage ionic character in compound semiconductors are explained. Diamond cubic structure of Silicon and Germanium, and zincblende and wurtzite lattice structures are given. Different types of crystal imperfections with emphasis on point imperfections are suggested. Details of Bohr atomic model, electron energy, spectral of hydrogen, modern concept of atom, quantum states etc. are discussed which will be useful for understanding the topics given in later chapters. Important applications and brief descriptions of some semiconductor devices are discussed. Details of Narrow bandgap and wide bandgap semiconductors along with their properties and applications are presented. Minute insight into the various topics are given through solved numerical and theoretical examples. Review questions, numerical problems and objective type questions are also given with their answers.

Keywords Bravais crystal system · Miller indices · Covalent bond in semiconductors · Diamond cubic · Structure of silicon and germanium · Crystal imperfections · Atomic models · Bohr atomic model · Sommerfeld's relativistic atomic model · Quantum states · Semiconductor devices · Narrow bandgap semiconductors · HgCdTe wide bandgap semiconductor · Selenium rectifiers

2.1 Crystal Structure

On the basis of their structure, materials may be classified into two groups viz.

1. Crystalline, and
2. Non-crystalline (or amorphous)

K.M. Gupta and N. Gupta, *Advanced Semiconducting Materials and Devices*, Engineering Materials, DOI 10.1007/978-3-319-19758-6_2

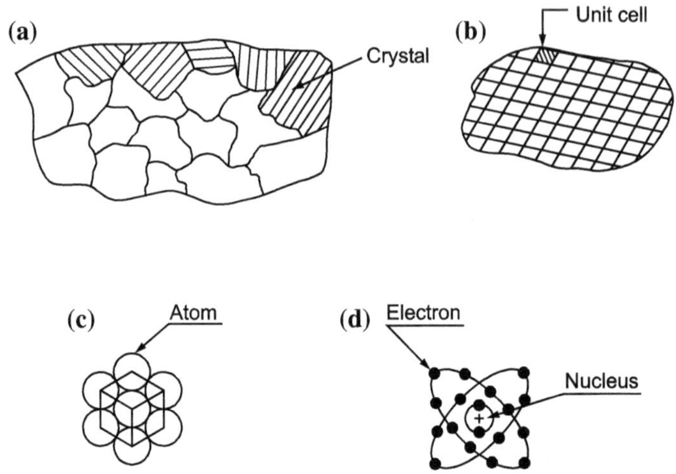

Fig. 2.1 Construction of a solid material. **a** Smallest visible part of polycrystalline solid, **b** Crystal, **c** unit cell, **d** atom. Reproduced from [1]

Generally metals are crystalline, and non-metals are non-crystalline. But this is not a rule. Plastics, a non-metal may be obtained in almost crystalline form. Crystalline solids have periodically repeating arrangement of atoms. Such solids can be further sub-classified as follows.

(i) Monocrystalline, and
(ii) Polycrystalline.

Most of materials in engineering applications are polycrystalline. Monocrystalline material has a single crystal. It finds use in specific applications. As an example, single crystal quartz is employed in generating ultrasonic waves. We are concerned here with the geometry of crystalline materials.

The smallest visible part of a material is made up of large number of *crystals*. These crystals may be of different shapes and sizes. They generally have random orientation. Each crystal is further composed of basic structural item called unit cell. Unit cells are of different types (see Article 2.2). These unit cells contain atoms arranged in a very systematic pattern. The complete system of solid materials may be summarily as shown as in Fig. 2.1.

2.2 Bravais Crystal System

The crystals have unit cells of various geometries. The geometries are defined in terms of their linear dimensions **a**, **b** and **c**, and angular dimensions α, β and γ. Dimensions a, b and c are along the x, y and z axes respectively; and angles α, β and γ between xy, yz and zx axes respectively.

Bravais has classified 14 space lattices into 7 crystal systems. These are displayed in Table 2.1. Examples of common semiconducting materials are shown 'bold'.

2.3 Miller Indices: The Crystallographic Notation of Atomic Planes

Miller is the name of a material scientist. Indices is plural of index. Miller indices are the styles to designate the planes and directions in the unit cells and crystals. Such designations become essential for investigations of various properties in different elements. In the most general way, Miller indices are expressed by (*h k l*) and no comma is used between them. Here, symbols *h, k* and *l* indicate unknown integers. The kinds of brackets have special meanings which are elaborated as below.

$(h\,k\,l)$ denotes a plane.

$[h\,k\,l]$ denotes a direction.

$\{h\,k\,l\}$ denotes family of planes.

$\langle h\,k\,l\rangle$ denotes family of directions.

Miller indices (*h k l*) *are* expressed as a reciprocal of intercepts *p*, *q* and *r* made by the plane on the three rectangular axes *x, y* and *z* respectively. The intercepts are measured in terms of the dimensions of unit cell. These are the unit distances from the origin along the three axes.

Thus,

$$h = 1/p,\ k = 1/q \ \text{and}\ l = 1/r$$

where

p intercept of the plane on x-axis,
q intercept of the plane on y-axis, and
r intercept of the plane on z-axis.

Reciprocal of these intercepts are then converted into whole numbers. This can be done by multiplying each reciprocal by a number obtained after taking LCM (lowest common multiplier) of denominator. This gives the Miller indices of required plane. The Miller indices are expressed by three smallest integers.

Table 2.1 Bravais crystal system

Unit cell (geometry)	Space lattice	Abbreviation	Example
1. Cubic $a = b = c$ $\alpha = \beta = \gamma = 90°$	1. Simple (Points at the eight comers of the unit cell)	SC	Po
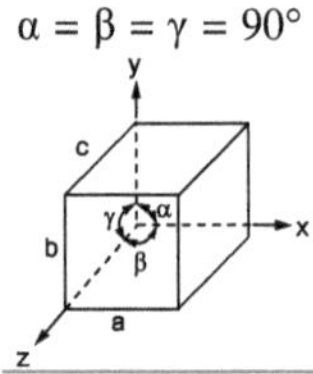	2. Body Centred (Points at the eight corners and at the body centre)	BCC	Li, Na, V, Ta, Cr, Mo, W, Fe, CsCl
	3. Face Centred (Points at the eight corners and at the six face centres)	FCC[a]	Ni, Pt, Cu, Ag, Au, Al, Pb, Ne, Ar, Kr, Xe, NaCl
2. Tetragonal $a = b \neq c$ $\alpha = \beta = \gamma = 90°$	4. Simple (Points at the eight corners of the unit cell)	ST	Pa, In
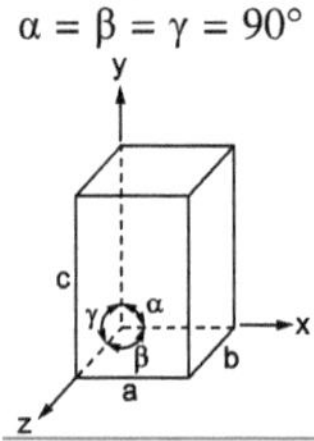	5. Body Centered (Points at the eight corners and at the body centre)	BCT	White Sn, KH_2PO_4, U between 668–774 °C
3. Orthorhombic $a \neq b \neq c$ $\alpha = \beta = \gamma = 90°$	6. Simple (Points at the eight corners of the unit cell)	SO	As, Sb, Bi
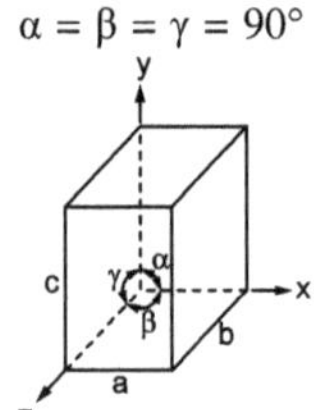	7. End Centred. (Also called side Centred or base Centred) (Points at the eight corners and at the face centres opposite to each other)	ECO	$BaSO_4$, KNO_3, $MgSO_4$
	8. Body Centred (Points at the eight corners and at the body centre)	BCO	
	9. Face Centred (Points at the eight corners and at the six face centres)	FCO	S, Ga
4. Rhombohedral (or trigonal) $a = b = c$ $\alpha = \beta = \gamma \neq 90°$ 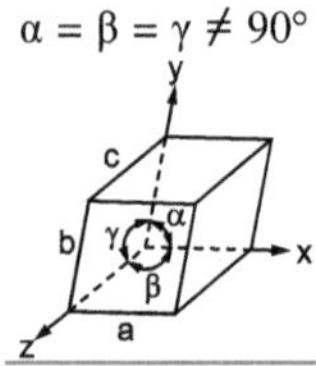	10. Simple (Points at the eight corners of the unit cell)	SR	B, Sb, SiO_2, $CaCO_3$, Lithium niobate ($LiNbO_3$), Lithium tantalate ($LiTaO_3$)

(continued)

Table 2.1 (continued)

Unit cell (geometry)	Space lattice	Abbreviation	Example
5. Hexagonal $a = b \neq c$ $\alpha = \beta = 90°$ $\gamma = 120°$ 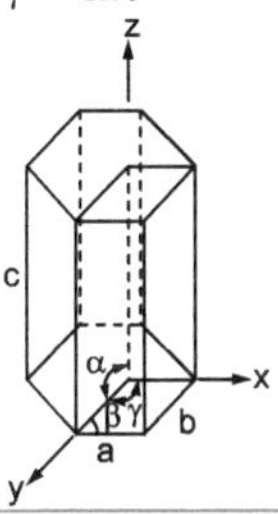	11. Simple [(i) Points at the eight corners of the unit cell outlined by thick lines OR (ii) Points at the twelve corners of the hexagonal prism and at the centres of the two hexagonal faces]	SH	La, Pr, Nd, Am, Se, Te, graphite, AgCl
6. Monoclinic $a \neq b \neq c$ $\alpha = \beta = 90° \neq \gamma$ 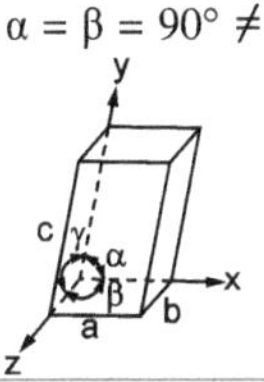	12. Simple (Points at the eight corners of the unit cell)	SM	$NaSO_4$, Li_2SO_4, $CaSO_4$, Rochelle salt, $FeSO_4$
	13. End Centred (Points at the eight corners and at two face centres opposite to each other)	ECM	S
7. Triclinic $a \neq b \neq c$ $\alpha \neq \beta \neq 90° \neq \gamma$ 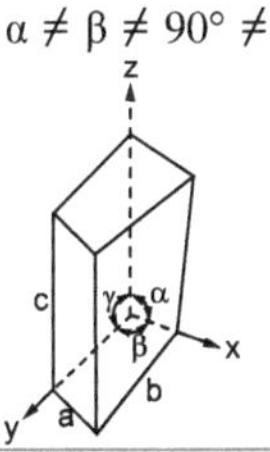	14. Simple (Points at the eight corners of the unit cell)	STC	$K_2Cr_2O_7$, $CuSO_4$

[a]FCC is also known as CCP (Cubic closed packed). Reproduced from [1]

2.3.1 Determining the Miller Indices of a Given Plane [1]

Refer Fig. 2.2 showing a plane *ABC* in a cubical unit cell. We have to determine Miller indices of plane *ABC*. Origin *O* and coordinate axes *x*, *y* and *z* are shown in the unit cell in which $OP = OQ = OR = a$. The plane in question intercepts at *A, B* and *C* on *x, y* and *z* axes in such a way that

$$OA = (2a/3) \text{ along } x\text{-axis,}$$
$$OB = (2a/5) \text{ along } y\text{-axis, and}$$
$$OC = (la/3) \text{ along } z\text{-axis.}$$

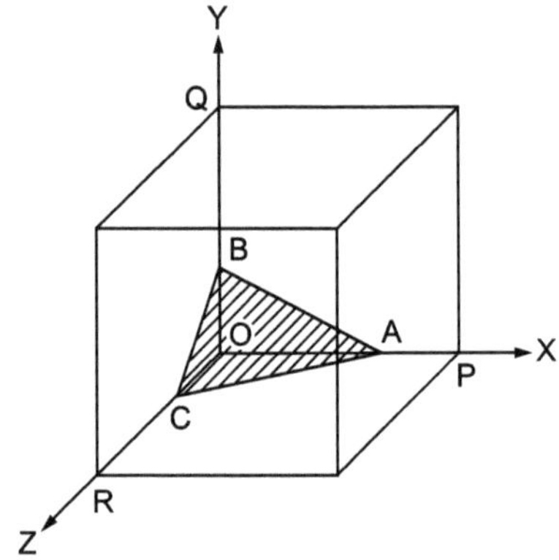

Fig. 2.2 Determination of Miller indices of an unknown alone

To obtain the required Miller indices, we proceed as follows:

(i) The intercepts are $c_1 = OA$, $c_2 = OB$ and $c_3 = OC$, therefore

$$\begin{aligned} p(\text{along x-axis}) &= \text{OA}/\text{OP} = (2/3a)/a = 2/3 \\ q(\text{along y-axis}) &= \text{OB}/\text{OQ} = (2/5)a/a = 2/5 \\ r(\text{along z-axis}) &= \text{OC/OR} = (1/3)a/a = 1/3 \end{aligned}$$

(ii) As $h = 1/p$, $k = 1/q$ and $l = 1/r$; therefore

$$\begin{aligned} h &= 1/(2/3) = 3/2 \\ k &= 1/(2/5) = 5/2 \\ l &= 1/(1/3) = 3/2 \\ \therefore \quad (h\,k\,l) &= (3/2\;5/2\;3) \end{aligned}$$

(iii) As $(h\ k\ l)$ should have smallest possible integers, hence we can write

$$(h\,k\,l) = \tfrac{1}{2}(3\,5\,6)$$

Here the factor 1/2 is the result of conversion of reciprocals to integers, and is usually omitted.

We, therefore, conclude that Miller indices of plane ABC is

$$(h\,k\,l) = (3\,5\,6)$$

Miller indices of a plane in non-cubical unit cells may be determined in the same manner.

2.3.2 *Family of Planes*

Different planes in a cubical unit cell are shown in Fig. 2.3. Their Miller indices are (1 0 0), (0 1 0) and (0 0 1). Miller indices of a plane may also be as $(\bar{1}\ 0\ 0)$. Similarly other planes in a cubic unit cell may be designated as $(0\ \bar{1}\ 0)$ and $(0\ 0\ \bar{1})$. Digits in all these planes are 1, 0 and 0 at different sequence. They belong to the same family and are called family of planes. These planes can be designated by using only a curly bracket. As another illustration, the planes having Miller indices (1 2 3), (2 3 1), (3 2 1), (3 1 2), $(\bar{1}\ 2\ 3)$, $(1\ \bar{2}\ 3)$, (2 1 3) etc. form a family of planes. They can be designated by (1 2 3) which are the lowest integers.

Example 2.1 Obtain the Miller indices of a plane whose intercepts are a, $b/2$, and $3c$ on x, y and z axes respectively in a simple cubic unit cell.

Solution. The intercepts c_1 (along x-axis) = a, c_2 (along y-axis) = $b/2$, and c_3 (along z-axis) = $3c$.

Therefore,

$$p = c_1/a = a/a = 1$$

$$q = c_2/b = (b/2)/b = 1/2$$

and

$$r = c_3/c = 3c/c = 3$$

$$\therefore \quad h = 1/p = 1, \quad k = 1/q = 2$$

And

$$l = 1/r = 1/3$$

Hence

$$(h\,k\,l) = (1\,2 \quad 1/3) = 1/3(3\,6\,1) \text{ or } 3\,6\,1$$

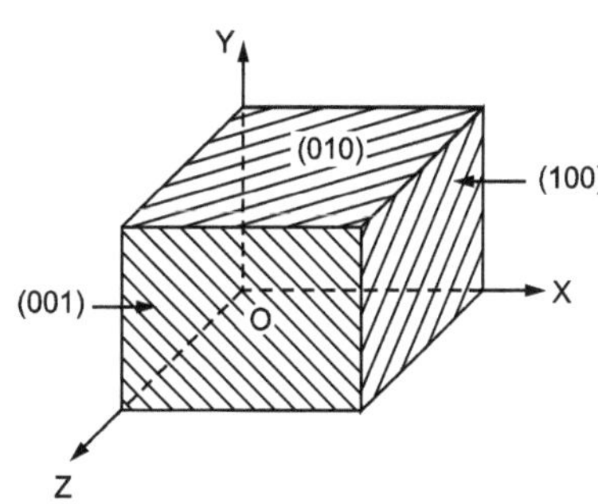

Fig. 2.3 Planes showing Miller indices {1 0 0} in a cubic unit cell. Reproduced from [1]

Example 2.2 The ratio of the intercepts of an orthorhombic unit cell $a : b : c = 0.429 : 1 : 0.379$. What are the Miller indices of faces with the following intercepts?

$$0.214 : 1 : 0.188$$
$$0.858 : 1 : 0.754$$
$$0.429 : 1 : 0.126$$

Solution. The intercepts in unit axial dimensions are

$$\begin{aligned} 0.214/0.429 : 1 : 0.188/0.379 &= 1/2 : 1 : 1/2 \\ 0.858/0.429 : 1 : 0.754/0.379 &= 2 : 1 : 2 \\ 0.429/0.429 : 1 : 0.126/0.379 &= 1 : 1 : 1/3 \end{aligned}$$

Their reciprocals are

$$2 : 1 : 2$$

$$1/2 : 1 : 1/2$$

and

$$1 : 1 : 3$$

Reducing them to a smallest whole number and enclosing in brackets, we get the Miller indices as (2 1 2), (1 2 1), (1 1 3).

2.3.3 *Miller Indices: Crystallographic Notation of Atomic Crystal Directions*

Miller indices are also used to specify directions within a unit cell or a crystal. An unknown direction is designated by $[h\ k\ l]$. Here a square bracket is used to designate direction. To illustrate the procedure of determining crystal direction, let us take a line passing through the origin and parallel to a given direction. A vector r passing through the origin O of the unit cell to a lattice point may be given by

$$r = r_1 x + r_2 y + r_3 z \tag{2.1}$$

where r_1, r_2 and r_3 are integers.

The length of the projections of this line on the coordinate axes x, y and z are noted. These intercepts are reduced to smallest integers to obtain $[h\ k\ l]$. Consider vector $r = OB$ in Fig. 2.4 in which B is a lattice point. Here $r_1 = 1$, $r_2 = 1$ and from the geometry of the cubical unit cell along x, y and z axes. Hence this direction is [1 1 1].

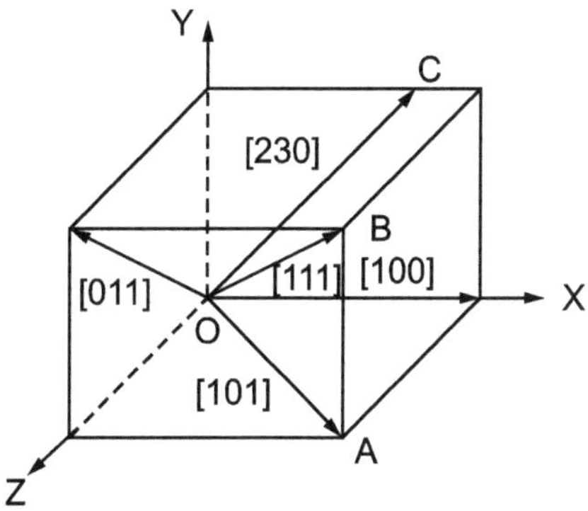

Fig. 2.4 Miller indices of directions [1 1 1], [1 0 0], [1 0 1], [0 1 1]. Reproduced from [1]

2.3.4 Family of Directions

It is analogous to family of planes. Directions [1 3 0], [3 1 0], [0 1 3], [0 3 1], [$\bar{1}$ 3 0], [$\bar{3}$ 1 0] etc. can be represented by ⟨1 3 0⟩.

2.4 Chemical (or Atomic) Bonding

If we try to break a material, it requires application of some force. The magnitude of applied force varies widely for different materials. A Si or Ge crystal may break by applying a small force, but steel necessitates application of substantial external force. One may think as to why is it so? In fact, atoms in the solids are held together by internal forces. These forces are known as bonding forces (or atomic bonding forces or chemical bonding forces). The geometry of materials remains intact due to atomic bonding in them.

Semiconductors, like other materials, have atoms arranged in a systematic manner. Coulomb's forces, both attractive and repulsive, work on them. The net force, out of these two opposite natured forces, is responsible for bonding. Bonding is the nature of a material to hold the atoms together. The mechanical, physical, electrical and other properties are greatly influenced by the bonds. Therefore different solids differ in their properties. In fact, there exists a correlation between the nature of bond and the material's properties.

2.4.1 Type of Bond in Semiconductors

Atomic bonds in solids are classified into two main types. These are

1. Primary or fundamental bonds, and
2. Secondary or molecular bonds.

Primary bonds are stronger and more stable than the secondary bonds. They are further classified into following types.

(i) Ionic bond
(ii) Covalent bond
(iii) Metallic bond

Out of these three bonds, the ionic bonds are strongest while the metallic bonds are weakest. *Semiconductors are covalent bonded.*

2.4.2 Nature of Bond in Semiconductors

Bonds, whether primary or secondary, may have properties which are either

(i) Directional, or
(ii) Non-directional.

Depending upon this behaviour, they are termed as either 'directional bond' or 'non-directional bond'. The bond strength is unequal in different directions in case of *directional bonds*. It is, thus, *direction dependent*. The non-directional bonds possess equal strength in all directions. Hence they are *direction independent*. The ionic and metallic bonds are non-directional, while the covalent bonds are directional. Thus the bonds in semiconductors have directional nature.

2.5 Bonding Forces

The bonding force in a material is provided by the atoms. The forces between two atoms are of attractive and repulsive nature. Both these forces exist simultaneously. When two atoms reach in close proximity, their electrons come much closer to each other as compared to their nuclei. At larger distances, the repulsion between the electrons is balanced by the attractions between the electrons of one atom and nucleus of the other. The repulsion between electrons dominates at closer spacing. Thus the repulsive force rises rapidly as the electron orbits begin to overlap. In metals, the repulsive forces exist between the positively charged ions. The net bonding force is the resultant of attractive and repulsive forces described above. It is obtained by superimposing the two curves of attractive and repulsive forces.

Variation of attractive, repulsive and net (or resultant) forces as a function of atomic distance of separation r is shown in Fig. 2.5. The net bonding force F between the atoms is expressed by,

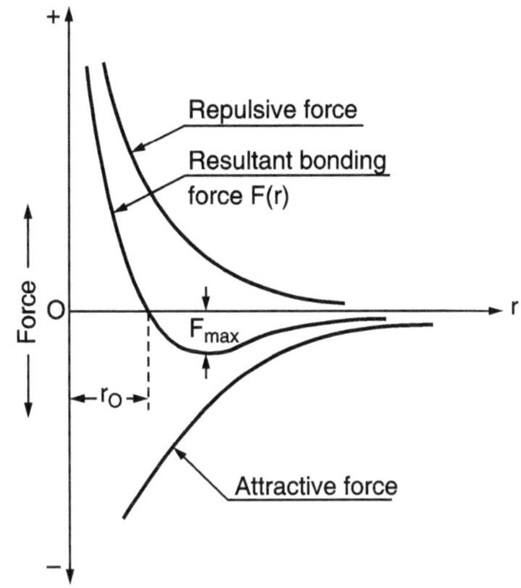

Fig. 2.5 Variation of forces versus distance of atomic distance

$$F(r) = \frac{A}{r^M} - \frac{B}{r^N} \text{ for N > M} \tag{2.2}$$

where r is centre to centre spacing between the atoms, A and B are the materials constants, M and N are the constants depending on the type of bonds. Following are the usual values of these constants.

M = 2 for primary bonds
N = 7–10 for metallic bonds,
= 9–12 for covalent bonds, and
= 10–12 for ionic bonds. In Eq. 2.2, the first and second terms of right hand side represent the attractive and repulsive forces respectively.

2.5.1 Bonding Length and Bond Forces

When the two atoms of a material approach each other under the influence of externally applied force, they are drawn together until equilibrium spacing is reached. This spacing between the two atoms is called bond length. It has been shown by r_0 in Fig. 2.5. Here the force of attraction just balances the force of repulsion, hence F(r) = 0. The bond length is smaller in primary bond than those in secondary bonds. Range of bond lengths is of the order of

$$r_0 = 1 \text{ to } 2\ \text{Å} \text{ for primary bonded materials, and}$$
$$r_0 = 2 \text{ to } 5\ \text{Å} \text{ for secondary bonded materials.}$$

2.6 Covalent Bond and Semiconductors

A covalent bond is formed when the pair of electrons is shared by several atoms. As a result of this sharing, their energies are lowered. Stable covalent bonds are formed between many non- metallic elements as the atoms of these elements usually posses half filled outer electron orbit. Such elements are hydrogen, carbon, nitrogen, oxygen and chlorine etc. Elements like silicon, germanium, arsenic and selenium etc. form partly covalent and partly metallic bond.

2.6.1 Type of Covalent Bond in Semiconductors

Covalent bonds may be further classified into following three kinds.

1. Sigma (σ) bond such as in Se, Te etc.
2. Pi (π) bond such as in O_2.
3. Hybrid bond such as in diamond.

End to end overlap of p suborbits gives rise to σ-bond. Bonding in semiconducting materials like sulphur, selenium and tellurium are the examples. π-bonds are formed when overlap of p suborbits are lateral. Bond in oxygen molecule is the example of this kind. Schematic diagrams are shown in Fig. 2.6 to explain σ-bond and π-bond.

2.6.2 Bond Angle in Semiconductors

Covalent bonds are easily formed between the atoms having unfilled p suborbits. The p suborbits are directional in nature. They allow overlapping of suborbits in the direction of densest electron probability cloud. Hence the covalent bonds are directional. The bond is formed at specific angles in different elements. This angle is called bond angle θ.

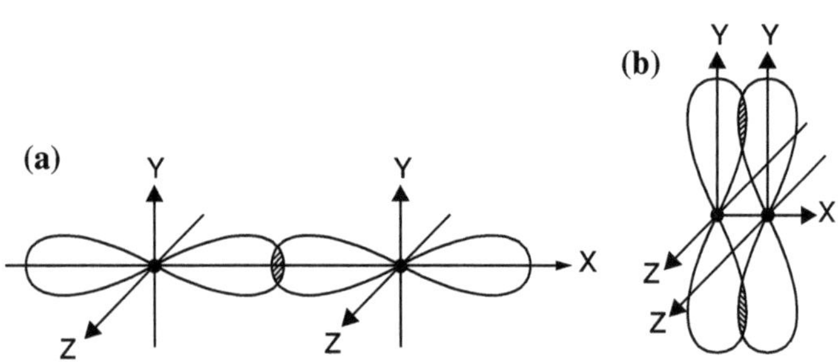

Fig. 2.6 Types of covalent bonds **a** σ-bond which is found in semiconducting materials like S, Se, Te, and **b** π-bond

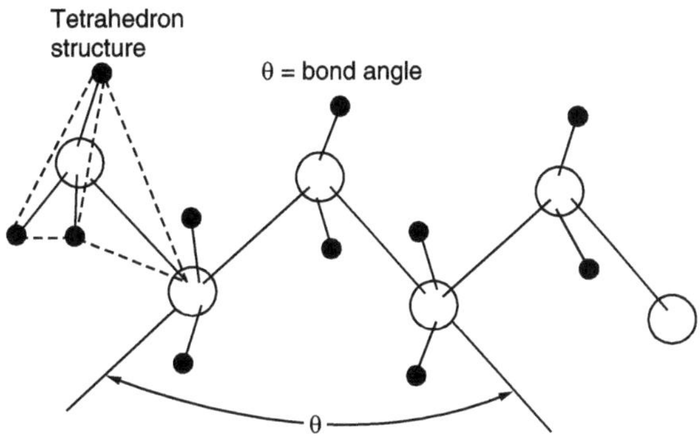

Fig. 2.7 Tetrahedral bond angle

Bond angle is 90° between the unbonded p suborbits. It is true if the p suborbits are bonded without showing any distortion and also in the case when distortion of this suborbit does not take place during formation of bond. Diamond develops tetrahedral bond angle of 109.5°. Figure 2.7 shows the concept of bond angle. Bond angles in some semiconducting elements are given in Table 2.2.

2.6.3 *Mixed Bond in Compound Semiconductors*

Whereas the elemental semiconductors are generally covalent bonded, the compound semiconductors have a mixed nature of bonding in which both: the ionic and covalent bonding forces exist together. Hence the compound semiconductors are generally composed of mixed bonding. Normally the mixed bonding in them is of ionic-covalent type. It results due to the difference in electronegativities of the elements forming the mixed bond. The *ionic* character thus imparted may be determined from 'Pauling equation' given by

$$I_c = \left[1 - e^{-0.25(E_{NP}-E_{NQ})^2}\right] \times 100 \tag{2.3}$$

where Ic is percentage of ionic character, E_{NP} and E_{NQ} are the electronegativities of atoms P and Q forming the mixed bond. Difference in electronegativity of about 1.7–1.8 corresponds to 50 % ionic character.

Table 2.2 Bond angle in some semiconductor elements

Element	Symbol	Bond angle (°)
Antimony	Sb	96
Phosphorous	P	99
Tellurium	Te	104
Selenium	Se	104
Sulphur	S	107

Some examples of ionic-covalent bonded compound semiconductors are the following.

- GaP, GaAs, GaSb
- ZnS, ZnSe, ZnTe
- InP, InAs, InSb
- CdS, CdSe, etc.

Example 2.3 Estimate the percentage ionic character of the compound semiconductors InAs and GaAs using Pauling's equation. The electronegativities are 1.5 for In, 2.2 for As, and 1.8 for Ga.

Solution. Using Pauling's equation given by Eq. 2.3, we have

For *InAs*:

$$\begin{aligned}\%\text{ of Ionic character} &= \left[1 - e^{-0.25(2.2-1.5)^2}\right] \times 100\\ &= (1 - 0.884) \times 100 = 11.53\,\%\end{aligned}$$

For *GaAs*:

$$\begin{aligned}\%\text{ of Ionic character} &= \left[1 - e^{-0.25(2.2-1.8)^2}\right] \times 100\\ &= (1 - 0.96) \times 100 = 4\,\%\end{aligned}$$

2.7 Diamond Cubic (DC) Structure of Silicon and Germanium

The structures of Si and Ge are diamond cubic (DC). The name DC comes from diamond which has hybrid covalent bond. Each of its atoms has four bonds. The bonds are directional in nature. The covalent bonds are primary bonded and extend in a three-dimensional network.

The directional bond angle in diamond is 109.5°. The structure of diamond is known as diamond cubic (DC). Unit cell of a DC is shown in Fig. 2.8. It contains a total of 18 atoms. Of these, 8 are placed on corners of the cube, one on each of the

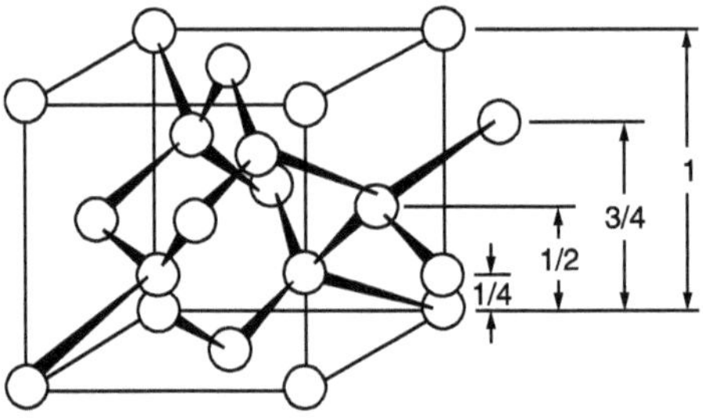

Fig. 2.8 A DC structure

six faces, and four completely inside the cube. The two inside atoms are placed at three-fourth (3/4) and other two at one-fourth (1/4) distance above the base when the height of unit cell is unity i.e. 1. These four atoms are placed on body diagonal, and are not the lattice points.

Diamond is the hardest known solid. Other geometrical details of DC are given below:

Number of atoms per lattice point	2
Basis	2
Distance of separation between two atoms	$a\sqrt{3}/4$
Atomic packing factor	0.34
Effective number of atoms per unit cell	8
Specific gravity	3.5

Besides Si and Ge, the DC structure also exists in diamond, grey tin and compound semiconductors etc.

2.8 Lattice Structures of Some Compound Semiconductors

The basic lattice structure of many semiconductors is diamond cubic (DC), which is a characteristic of Si and Ge. In many compound semiconductors, although the atoms are arranged in DC structure but they are different on alternating sites. Consequently, they differ from DC and are of following kinds.

1. Zinc blende: which is typical of III–V compounds, and
2. Wurtzite: which is typical of II–VI compounds.

The zinc blende and wurtzite lattice structures are shown in Fig. 2.9a, b. In zinc blende structure, for example, if one FCC sub-lattice is composed of Ga atoms and

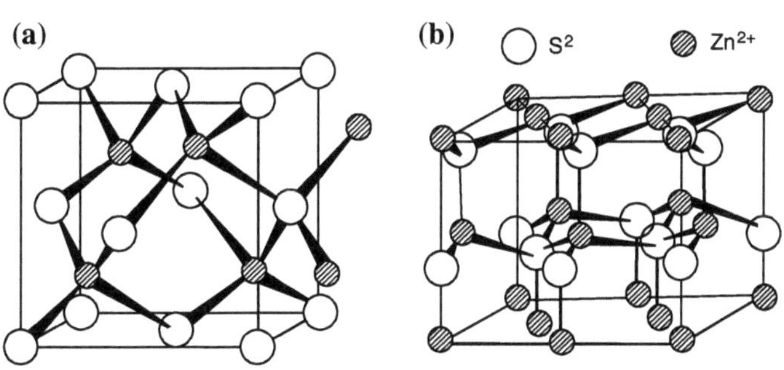

Fig. 2.9 Structures of **a** zinc blende, and **b** wurtzite. Reproduced from [1]

the intermediatory sub-lattice is of As, it will form a GaAs zinc blende. Most of the compound semiconductors have this type of lattice.

Main examples of semiconductors falling under each kind are given below.

- Zinc blende: AlP, AlAs, AlSb, GaP, GaAs, GaSb, InP, InAs, InSb, ZnSe, ZenTe, Cd Te
- Wurtzite: Si(α), CdSe

Some semiconductors possess the structure of both kinds, zinc blende and wurtzite. These are

- ZnS
- GaN
- CdS

2.9 Lattice Structure of Zinc Sulphide

In zinc sulphide (ZnS), each Zn^{2+} ion is tetrahedrally surrounded by four S^{2-} ions and each S^{2-} ion is tetrahedrally surrounded by four Zn^{2+} ions. Coordination number of both ions is 4, therefore it is called a 4:4 arrangement. Zinc sulphide exists in two different forms viz. (i) zinc blende, and (ii) wurtzite. Their structures are shown in Fig. 2.9a, b. Zinc blende is related to FCC structure while wurtzite relates to HCP structure.

2.10 Crystal Imperfections

Till now, we have discussed crystal structures and bonding. We have also described crystalline and non-crystalline solids. Crystals discussed up to now are perfect (ideal) crystals. They are perfectly ordered arrangement of atoms at the points of infinite space lattice. Any departure from this idealized arrangement makes the crystals imperfect. In general, crystalline materials have *real crystals.* They are of finite dimensions. They have broken bonds at the boundaries where bonding forces remain unbalanced. Thus the boundary of a crystal is a defect in itself.

These imperfections lead to several deficiencies in the solids. Solids are generally imperfect. A perfect single crystal is rarely available. Polycrystalline solids are, inevitably, imperfect. Presence of imperfections in crystalline materials is generally undesired. They decrease the mechanical strength of materials.

Imperfections affect structure-sensitive properties of crystals. Their presence is advantageous in certain applications. A part per million (ppm) doping of phosphorus in silicon changes the behaviour of intrinsic semiconductor and makes it suitable for various applications.

2.10.1 Types of Imperfections

Imperfections in crystalline solids may be of following types.

S. No.	Main types	Sub-types
1.	Point imperfections (zero-dimensional defects)	Vacancy, Substitutional impurity, Interstitial impurity, Frenkel's defect, Schottky's defect
2.	Line imperfections (1-dimensional defects)	Edge dislocation, Screw dislocation, Mixed dislocation
3.	Surface or planer imperfections (2-dimensional defects)	Grain boundary, Twin or twinning, low angle boundary, high angle boundary, twist boundary, stacking fault, interphase
4.	Volume imperfections (3-dimensional defects)	Pores, Foreign particle inclusions, Non-compatibility regions, Dissimilar natured regions

Point imperfections, line and surface imperfections may occur together in crystals. These defects are not visible to the naked eye. They can be visualized by using X-rays diffraction techniques and microscopes.

Imperfections are also classified on the basis of their dimensions, such as

(i) Nano-level (10^{-9} m) imperfections
(ii) Angstrom level (10^{-10} m) imperfections
(iii) Micro-level (10^{-6} m) imperfections

2.10.2 Point Imperfections

As the name suggests, they are imperfect point-like regions in the crystal. These defects are of one or two atomic diameters only. Hence these are known as zero-dimensional defects.

- **Vacancy** refers to a vacant atomic site in a crystal. At these sites the atoms are missing. One or more atoms may remain absent from their respective locations.
- **Substitutional Impurity**. This defect refers to a foreign atom that substitutes a parent atom at its site in the crystal. The substituting foreign atoms are called *solute* and the substituted (or dislodged) parent atoms are known as the *solvent.* Solute and solvent of comparable sizes mix randomly to form an alloy. For example boron or antimony doped in germanium is an example of substituted impurities in a crystal.
- **Interstitial Impurity**. When a small sized foreign atom occupies a void space in the parent crystal (or its unit cell), the defect is known as interstitial impurity. Dislodging of parent atoms from their sites does not occur in this case.

2.11 Bohr's Quantum Atomic Model

Bohr conceived-off a new atomic model employing the principle of *quantum theory* suggested by Planck. This model provided adequate explanation for stability of the atom. The model also accounted for the origin of spectral lines in the hydrogen atom. He proposed new ideas which are known as Bohr's postulates. These are:

(i) Electrons revolve in non-radiating stationary orbits. Centripetal force provided by Coulomb's force of attraction between the electron and the nucleus keeps the electron in orbital motion, Fig. 2.10.
Thus,

$$\frac{mv^2}{r} = \frac{1}{4\pi\varepsilon_0}\frac{(Ze)(e)}{r^2} \tag{2.4}$$

where, Z = atomic number of nucleus, m = mass of the electron, v = velocity of electron in the orbit, r = radius of the orbit, e = charge of electron.

(ii) The angular momentum of the moving electron is an integral multiple of $h/2\pi$ where h is *Planck's constant*. Thus,

$$mvr = nh/(2\pi) \tag{2.5}$$

where n = 1, 2, … ∞, and is called *principal quantum number.*

(iii) The electron does not radiate energy while moving in stationary orbit. Energy is emitted when electron falls from higher energy orbit to lower energy orbit. If the electron jumps- up to higher energy orbit from lower energy orbit, absorption of energy takes place. The energy absorbed or emitted is expressed by Bohr's frequency condition

$$\Delta E = E_f - E_i = hf \tag{2.6}$$

where f is the frequency of emitted radiation, E_i and E_f are the energies of initial and final orbits respectively.

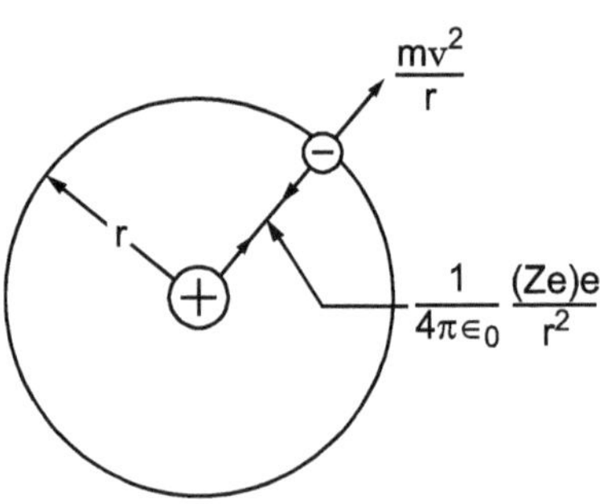

Fig. 2.10 Balance of forces keeps the electron in orbital motion

2.11.1 Radii of Orbits, Velocity and Frequency of Electrons

The radius of *n*th stationary orbit is obtained from Eqs. 2.4 and 2.5 on eliminating *v* from them. It is, then, given as

$$r_n = \frac{n^2 \varepsilon_0 h^2}{\pi m Z e^2} \tag{2.7}$$

Velocity of *n*th orbit electrons is determined by

$$v_n = \frac{Ze^2}{2n\varepsilon_0 h} \tag{2.8}$$

The orbital frequency of an electron in *n*th orbit is given as

$$f_n = \frac{v_n}{2\pi r_n} \tag{2.9}$$

Substituting the values of *rn* and *vn* from Eqs. 2.7 and 2.8 in Eq. 2.9, we get

$$f_n = \frac{mZ^2 e^4}{4\varepsilon_0^2 n^3 h^3} \tag{2.10}$$

As ε_0, *h*, *m* and *e* are constants, hence for Z = 1, Eq. 2.7 shows that $r_n \propto n^2$. Thus for *n* = 1, 2, 3, 4, …; the radii of orbits are proportional to 1^2, 2^2, 3^2, 4^2, …, i.e. in the ratio of 1:4:9:16, …, etc.

Equation 2.8 reveals that $v_N \propto \frac{1}{n}$. It means that the velocity of electrons in outer orbits is lower than those in the inner orbits. Similarly Eq. 2.10 interprets that $f_n \propto \frac{1}{n^3}$. Therefore orbital frequency of electrons, for *Z* = 1, is in the ratio of $\frac{1}{1^3} : \frac{1}{2^3} : \frac{1}{3^3} : \frac{1}{4^3} : \cdots$ etc.

2.11.2 Normal, Excited and Ionized Atoms

Hydrogen atom has a single electron. When this electron lies in the innermost *K* orbit (*n* = 1), the atom is said to be in *normal state.* Normal state is also called *ground state.* The electron in normal state remains stable. The atom is said to be *excited* if its electron is forced into an outer orbit. The electron absorbs energy in this case. The atom is said to be *ionized* when its electron is removed out from the atom completely. The electron emits energy in ionized state.

The radius r_0, velocity v_0 and frequency f_0 of hydrogen atom in ground state (n = 1) are obtained using Eqs. 2.7, 2.8 and 2.10. These values are

$$r_0 = 0.53\,\text{Å},\ v_0 = 2.2 \times 10^6\ \text{m/s}\ \text{ and }\ f_0 = 0.66 \times 10^6\ \text{m/s}^2$$

2.11.3 Electron Energy

Energy of an electron is the sum of kinetic energy E_k due to its motion and potential energy E_n due to its position in the orbit. The value of kinetic energy is given as

$$E_k = \frac{1}{2} m v_n^2 = \frac{mZ^2 e^4}{8\varepsilon_0^2 n^2 h^2} \tag{2.11}$$

and the value of potential energy may be obtained from

$$E_p = \int_{\infty}^{r} F dr = \frac{Ze^2}{4\pi\varepsilon_0} \int_{\infty}^{r} \frac{1}{r^2} dr = -\frac{Ze^2}{4\pi\varepsilon_0 r}$$

In above expression, right hand side term of Eq. 2.4 has been taken for F, which is Coulombian force. On substituting the value of r from Eq. 2.7, we get

$$E_p = -\frac{mZ^2 e^4}{4\varepsilon_0^2 n^2 h^2} \tag{2.12}$$

The potential energy of an electron is equal to the work done in bringing the electron from infinity to the present position. This work is required to be done due to Coulombian force produced by the nucleus. Since this force is attractive, the work done and hence the potential energy E_p has negative value.

On summing-up Eqs. 2.11 and 2.12, the energy level of electron in nth orbit is obtained as

$$E_n = -\frac{mZ^2 e^4}{8\varepsilon_0^2 n^2 h^2} \tag{2.13}$$

For hydrogen atom, the above equation yields a value of

$$E_n = -\frac{13.61}{n^2}\ \text{eV} \tag{2.14}$$

It can be seen from Eq. (2.13) that total energy of electron is inversely proportional to the square of principal quantum number i.e. $E_n = 1/n^2$.

It is also evident from Eq. (2.7) that $r_n \propto n^2$. Thus the energy of electrons and radii of their orbits are quantized depending on principal quantum number.

Example 2.4 Using Bohr's theory, calculate the ionization potential for hydrogen atom.

Solution. The ionization energy for hydrogen ($Z = 1$) is the energy required to remove an electron from the first orbit ($n = 1$) to an infinite distance ($n = \infty$). This energy can be expressed in eV. The *ionization potential* in volts then simply becomes the value of *ionization energy* (in eV).

The values of constants on the right side of Eq. (2.13) are

$$\varepsilon_0 = 8.85 \times 10^{-12}\,\mathrm{C^2\,N^{-1}\,m^{-2}}; \quad m = 9.11 \times 10^{-31}\,\mathrm{kg}$$
$$e = 1.602 \times 10^{-19}\,\mathrm{C}; \quad h = 6.62 \times 10^{-34}\,\mathrm{Js}$$

Therefore, the ionization energy is

$$E = \frac{mZ^2e^4}{8\varepsilon_0^2h^2n^2} = \frac{(9.11 \times 10^{-31})(1^2)(1.602 \times 10^{-19})^4}{8(8.85 \times 10^{-12})^2(6.62 \times 10^{-34})^2(1)^2} = 2.18 \times 10^{-18}\,\mathrm{J}$$
$$= \frac{2.18 \times 10^{-18}}{1.602 \times 10^{-19}} = 13.6\,\mathrm{eV}$$

Thus, the ionization potential for hydrogen atom is 13.6 V.

2.11.4 *Frequency of Radiation and Spectral Series of Hydrogen*

Bohr investigated spectral series emitted by an excited hydrogen atom. When the atom is in unexcited (ground) state, $n = 1$, the electron possesses lowest energy and does not absorb or emit any energy.

If the electron absorbs energy in any way, it is raised to higher energy level corresponding to the value of *n*. The energy can be given to the electron by providing thermal energy, or illumination of the atom, or by strike of an accelerated particle.

When the electron absorbs 13.6 eV of energy, it is raised to $n = \infty$ level. If it absorbs 10.2 eV of energy, it will be raised from $n = 1$ to $n = 2$ level. Electron, if excited to any level, returns back to ground state in different steps. For example, the return from $n = 4$ state to $n = 1$ state may be as $4 \rightarrow 3 \rightarrow 2 \rightarrow 1$ or $4 \rightarrow 3 \rightarrow 1$ or $4 \rightarrow 2 \rightarrow 1$.

During this action, electron radiates energy difference of the two states. The frequency of this radiation is *f*. The emitted energy *hf* is called *photon*.

Let n_i and n_f *be* the principal quantum numbers of initial and final states of electrons; and E_i and E_f be their respective energies. Then the frequency of emitted radiation given by Eq. 2.5 becomes

$$f = \frac{1}{h}(E_f - E_i) \tag{2.15}$$

Considering Eq. (2.12), it can be rewritten as

$$f = \frac{mZ^2e^4}{8\varepsilon_0^2h^3}\left(\frac{1}{n_i^2} - \frac{1}{n_f^2}\right) \tag{2.16}$$

Velocity of light c and the wavelength λ are related to frequency of radiation f by

$$c = f\lambda \tag{2.17}$$

Defining wave number* as reciprocal of wavelength

$$v = 1/\lambda = f/c \tag{2.18}$$

Equation 2.16 may be modified as

$$v = RZ^2\left(\frac{1}{n_i^2} - \frac{1}{n_f^2}\right) \tag{2.19}$$

where, R is Rydberg's constant, and is given by

$$R = \frac{me^4}{8\varepsilon_0^2h^3c} \tag{2.20}$$

Its value is equal to 1.097×10^7/m.

The series of lines in the hydrogen spectrum arise due to transition of electrons from different outer orbits to a fixed inner orbit. Various series in this regard are obtained from Eq. 2.19 for $Z = 1$. These series are

1. Lyman series—for which $n_i = 1$, $n_f = 2, 3, 4, \ldots$
2. Balmer series—for which $n_i = 2$, $n_f = 3, 4, 5, \ldots$
3. Paschen series—for which $n_i = 3$, $n_f = 4, 5, 6, \ldots$
4. Brackett series—for which $n_i = 4$, $n_f = 5, 6, 7, \ldots$
5. Pfund series—for which $n_i = 5$, $n_f = 6, 7, 8, \ldots$

The hydrogen spectrum and different series are shown in Fig. 2.11.

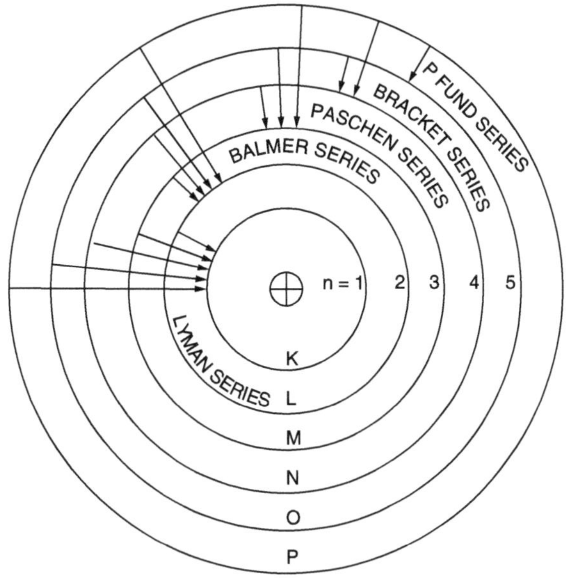

Fig. 2.11 Different spectral series of hydrogen spectrum. *Note* In spectroscopy, frequencies are generally expressed in wave numbers

2.12 Sommerfeld's Relativistic Atomic Model

Sommerfeld in his theory:

(i) Accounted for relativistic variation of mass with the velocity in accordance with theory of relativity given by Einstein.
(ii) Introduced the idea of elliptical orbits to explain the group of fine spectrum.
(iii) Quantised the angular momentum and the radial momentum.

Sommerfeld introduced azimuthal quantum number to account for the elliptical orbits of electrons. He also conceived of suborbits ($s, p, d, f, g, \ldots$) within the orbits ($K, L, M, N, O, \ldots$). The path of suborbits is elliptical whose major axis is a and minor axis is b. The main features of this atomic model are

(i) The azimuthal quantum number l assigned to different suborbits are

$l = 0$	1	2	3	4	for
s	p	d	f	g	respectively

(ii) Maximum number of electrons in any suborbit is obtained by $2(2l + 1)$. Thus maximum number of electrons in different suborbits are

2 in s suborbit, 6 in p suborbit,
10 in d suborbit, 14 in f suborbit.

(iii) The number of suborbits are equal to the principal quantum number of the orbit. Hence,

K orbit (n = 1) has 1 suborbit i.e. s
L orbit (n = 2) has 2 suborbits i.e. s, p
M orbit (n = 3) has 3 suborbits i.e. s, p, d
N orbit (n = 4) has 4 suborbits i.e. s, p, d, f
O orbit (n = 5) has 5 suborbits i.e. s, p, d, f, g

(iv) The geometry of elliptical path may be obtained by

$$\frac{b}{a} = \frac{l+1}{n} \tag{2.20}$$

(v) One suborbit has circular path which means $a = b$.
Thus f suborbit will be circular, and s suborbit will have largest major axis. The result is shown in Fig. 2.12 for N-shell.

Example 2.5 Calculate the dimensions of elliptical locus of different suborbits of N orbit, and show them on diagrams.

Solution. According to Sommerfeld's atomic model, the N orbit (n = 4) will have 4 suborbits viz. s, p, d and f. From Eq. 2.20,

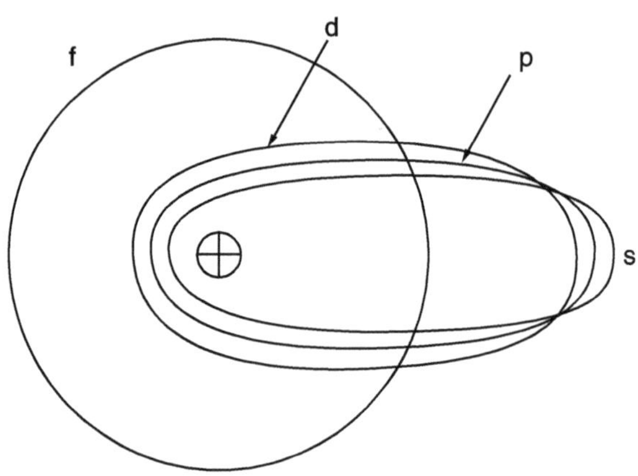

Fig. 2.12 Sommerfeld atomic model showing various suborbits in a N-shell

For s suborbit, $\frac{b}{a} = \frac{0+1}{4}$ $\therefore b = a/4$

For p suborbit, $\frac{b}{a} = \frac{1+1}{4}$ $\therefore b = 2a/4 = a/2$

For d suborbit, $\frac{b}{a} = \frac{2+1}{4}$ $\therefore b = 3a/4$

For f suborbit, $\frac{b}{a} = \frac{3+1}{4}$ $\therefore b = 4a/4 = a$

2.13 Modern Concept of Atomic Model

The present concept of atom is based on quantum mechanical theory. It takes into account the ***Heisenberg's uncertainty principle.*** This principle states that the momentum p and the position x of an electron cannot be predicted precisely. Therefore,

$$(\Delta p)(\Delta x) \geq \frac{h}{2\pi} \tag{2.21}$$

where Δp is the uncertainty in momentum, and Δx is the uncertainty in position. Consequently definite discrete orbits of electrons are not possible. The electron orbitals can at best be interpreted as *probability density cloud* around the nucleus.

The modern concept of atom also considers the concept of wave nature of electron. According to this concept a particle (say electron) of mass m moving with velocity v is associated with a wave propagating in the direction of moving particle. This wave is known as *de Broglie wave* or *matter wave.* Wavelength λ of such a wave associated with the particle is given by

$$\lambda = \frac{h}{mv} \tag{2.22}$$

Velocity of electron accelerated by a potential difference of V volts can be obtained from

$$\mathrm{eV} = \frac{1}{2}mv^2 \quad \text{or} \quad v = \frac{\sqrt{2\,\mathrm{eV}}}{m}$$

$$\therefore \quad mv = \sqrt{2m\,\mathrm{eV}} \tag{2.23}$$

Therefore, wavelength of electron wave in Eq. 2.22 becomes

$$\lambda = \frac{h}{\sqrt{2m\,\mathrm{eV}}}$$

On substituting $h = 6.62 \times 10^{-34}$ Js, $m = 9.1 \times 10^{-31}$ kg, and $e = 1.602 \times 10^{-19}$ C, we get

$$\lambda = \frac{12.25}{\sqrt{V}}\ \text{Å} \tag{2.24}$$

Modern concept of atomic model, also known wave-mechanical model, assumes behaviour electron as a *standing* or *stationary wave.* Such a wave will establish when the length of orbit is a whole number multiple of electron wavelength as shown in Fig. 2.13.

Hence, $2\pi r = n\lambda = \frac{nh}{mv}$

Or

$$mvr = \frac{nh}{2\pi} \quad \text{for} \quad n = 1, 2, 3, \ldots\text{etc.} \tag{2.25}$$

Here r *is* radius of the circular orbit, mvr is angular momentum of electron as a particle.

The stationary waves have their nodes and antinodes. Motion is practically zero at nodes, and there are almost no charge on them at these points. The amount of charge is maximum at antinodes. Whole of electron charge and mass is uniformly distributed around nucleus of the atom.

The wavelength of electron is of the order of interatomic spacing in the crystals. A beam of electron incident on a crystal shows diffraction pattern thus depicting concept of wave nature of electron.

Example 2.6 Calculate the minimum uncertainty in determining the position of a particle when the uncertainty in its momentum does not exceed 10^{-27} kg ms^{-1}.

Solution. From Eq. 2.21,

$$(\Delta p)(\Delta x) \geq \frac{h}{2\pi}$$

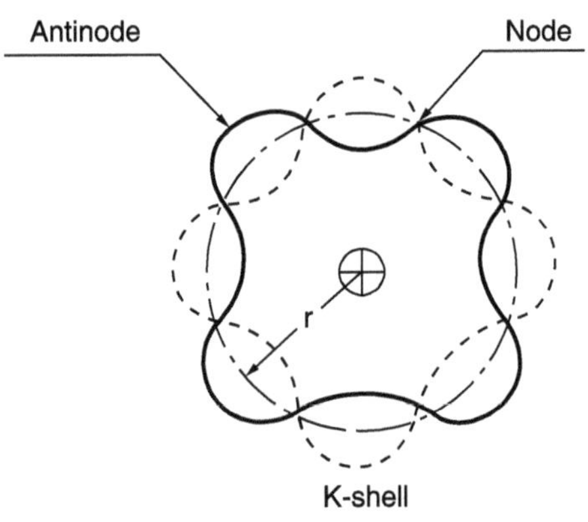

Fig. 2.13 Wave nature of electron in *K*-shell of an atom showing stationary wave

Given is the value of $(\Delta p) \leq 10^{-27}$ kg ms^{-1}, and taking $h = 6.626 \times 10^{-34}$ Js, the uncertainty in position (displacement) of the particle will be obtained from

$$
\begin{aligned}
(\Delta x) &\geq \frac{h}{2\pi(\Delta p)} \\
&\geq \frac{6.626 \times 10^{-34}}{2\pi(10^{-27})} \\
&\geq 1.054 \times 10^{-7}\ \text{m}
\end{aligned}
$$

Thus the minimum uncertainty = 1.054×10^{-7} m.

2.14 Quantum States

Electron orbital is called *quantum state,* and it has the following *quantum numbers.*

(i) principal quantum number n
(ii) orbital or azimuthal quantum number l
(iii) magnetic quantum number m_1
(iv) spin quantum number m_s

Principal quantum number n takes the integer values 1, 2, 3, 4, … ∞. Azimuthal quantum number l takes the integer values 0, 1, 2, …5, …, $n - 1$ and relates to the shape of electron orbital. The suborbitals *sharp, principal, diffuse* and *fundamental* are abbreviated by *s, p, d* and *f* respectively. The values of l for these suborbitals are 0, 1, 2, 3, … respectively.

The three dimensional (spatial) orientation of electron probability density cloud is defined by magnetic quantum number ml. There are (2l + 1) values of ml ranging between $-l, -l + 1, \ldots, -1, 0, 1, \ldots, l - 1, l$. Spin of electron is defined by m_s. Two possible cases of spinup and spindown take-on the values of ms = +½ and −½.

2.14.1 *Pauli's Exclusion Principle*

According to Pauli's exclusion principle, *no two electrons in an atom can have the same quantum state.* Thus electrons cannot have identical values for all the four quantum numbers. In other words, not more than one electron can occupy each quantum state. This principle provides a method for estimating the capacities and structure of different shells and sub-shells. The electronic configuration of different elements of the periodic table can also be studied with the help of this principle.

Table 2.3 Illustration of Pauli's exclusion principle

Orbit	Principle quantum number (n)	Azimuthal quantum number (l)	Magnetic quantum number m_l	Spin quantum number (m_s)
K	1	0	0	+1/2
	1	0	0	+1/2
L	2	0	0	+1/2
	2	0	0	−1/2
	2	1	0	+1/2
	2	1	0	−1/2
	2	1	−1	+1/2
	2	1	−1	−1/2
	2	1	+1	+1/2
	2	1	+1	−1/2
M	3	0	0	+1/2
				−1/2
	3	1	+1	+1/2
				−1/2
			0	+1/2
				−1/2
			−1	+1/2
				−1/2
	3	2	+2	±1/2
			+1	±1/2
			0	±1/2
			−1	±1/2
			−2	±1/2

Thus we find that if quantum numbers of two electrons *n*, *l* and m_l are the same, then value of quantum number m_s will certainly be different. m_s for one electron will be + and for the other −. The arrangement of electrons in an atom may be illustrated as shown in Table 2.3.

2.15 Important Applications of Semiconductor Devices

Compound semiconductors are widely employed to fabricate a number of semiconductor devices. Main among them are listed below.

1. Equilibrium condition junction diodes e.g.

   Rectifying diodes as (i) half-wave rectifier, (ii) full-wave rectifier
   Zener diode (i) for meter protection (ii) as peak clipper

2. Transient condition diodes e.g.
   - Switching diodes
   - Varactor diodes
   - Metal-semiconductor junction Schottky diode
3. Bipolar junction transistors (BJT) e.g.
   - Field effect transistor (FET)
   - Metal-semiconductor field-effect transistor (MESFET)
   - Metal-insulator-semiconductor field effect transistor (MISFET)
   - Insulated gate field effect transistor (IGFET)
4. Optoelectronic devices e.g.
   - Photodiodes
   - Photodetectors
   - Photocells
   - Solar cells
   - Light emitting diodes (LEDs)
   - Optical fibres
   - Semiconductor lasers
5. Negative conductance microwave devices e.g.
   - Tunnel diodes
   - Impatt diode
   - The Gunn diode
6. Power devices e.g.
   - p-n-p-n diode
   - Silicon controlled rectifiers (SCRs)
   - Silicon controlled switch (SCS)
   - Bialateral devices such as (*i*) Diac, (*ii*) Triac
7. Integrated circuits (ICs)
8. Transducers
9. Microprocessors

2.15.1 Brief Description of Some Semiconductor Devices

Brief description of some important devices is given below.

Junction. The boundary between *p* and *n* semiconductors is known as *junction.* The region of small thickness at the junction is called *depletion layer.* The *rectifying action* takes place in this region.

Transistor is a two junction (*n-p* and *p-n*) system exhibiting amplifying action of electron signals. Revolution in the electronics industry may be attributed to miniature, cheap, reliable, instant acting and less power consuming solid state transistors. Metal oxide semiconductor field-effect transistor (MOSFET) is a modern transistor used in street lights and as burglar alarm. CdS, CdSe and CdTe are commonly used photoconducting semiconductors.

Photodiodes. A photodiode is a two terminal device, which responds to photon absorption.

Photoconductors (or photo detectors) are used to detect and measure the quanta of light such as in automatic door opener, in switching the street lights and as burglar alarm. CdS, CdSe and CdTe are commonly used photoconducting semiconductors.

Photocells convert the light energy into electrical energy. They are fabricated from CdS, Se and $PbSO_4$. Photocells are used in cinematography, fire-alarms and television cameras etc.

Solar cells made from semiconducting materials are of immense utility in satellites and space- going vehicles. They are also used in calculators, solar power generation and solar autovehicles.

Light-emitting diode is an incoherent light source that is used as a light source in fibre optic systems and other devices.

Laser is the source of a highly directional, monochromatic coherent light which is used as a light source for various optical and electronic devices.

Optical fibre is a means of transmitting optical signals from a source to a detector.

Thermistors are used to determine the temperature of various systems in process industries, ovens and furnaces etc. They are made from sintered manganese oxide that contains dissolved germanium or lithium.

Integrated circuits are manufactured on a single chip that contains diodes, transistors, resistors and capacitors etc. They are generally monolithically constructed using either unipolar or bipolar techniques.

If we open our calculators, we will see I.Cs in it. Fabrication of I.Cs. is discussed under the section on microelectronics.

The energy gap, effective mass, and mobility of various compound semiconductors are tabulated in Table 2.4.

2.16 Narrow Bandgap Semiconductor Materials

Narrow band gap semiconductors are those semiconducting materials which have a small band gap as compared to silicon. Generally, the materials with forbidden band width less than 0.5 eV, or equivalent to those with forbidden bandwidth corresponding to infrared absorption cut-off wavelength over 2 μm, are treated as narrow band gap semiconductors. The characteristics of energy band of narrow gap materials include strongly non-parabolic conduction bands and spin-orbit splittings even larger than the fundamental band gaps. The developments of narrow band gap semiconductors have played a vital role in the field of infrared radiation detectors and emitters, and other high speed devices. With the synthesis of artificially fabricated alloy semiconductor $Hg_{1-x}Cd_{(x)}Te$ having forbidden band width about

Table 2.4 Electronic properties of compound semiconductors at room temperatures. Reproduced from [1]

Compoud system of III–V columns	Symbol	Energy gap E_g (eV)	Effective mass		Mobility (m^2/Vs)		Application
			Electron (m_e^a/m_o)	Hole (m_h^a/m_o)	Electron (μ_e)	Hole (μ_h)	
Indium phosphide	InP	19	0.07	0.4	0.46	0.015	Photon detection
Indium arsenide	InAs	0.33	0.02	0.41	3.30	0.04	–
Indium antimonide	InSb	0.16	0.013	0.6	7.80	0.075	Photon detection
Gallium phosphide	GaP	2.24	0.82[a]	0.60[a]	0.05	0.002	Laser
Gallium arsenide	GaAs	1.4	0.072	0.54	0.80	0.025	Diode, laser
Gallium antimonide	GaSb	0.77	0.047	0.5	0.40	0.14	Laser
Cadmium sulphide	CdS	2.42	–	–	–	–	Photoconductor in green light
Cadmium selluride	CdSe	1.74	0.13[a]	0.45[a]	0.08	–	Photoconductor in red light
Cadmium telluride	CdTi	1.45	–	–	–	–	Photoconductor in infrared region
Lead sulphide	PbS	0.38	–	–	–	–	Photoconductors
Lead selluride	PbSe	0.32	–	–	–	–	Photoconductors
Lead telluride	PbTe	0.33	–	–	–	–	Photoconductors
Silicon carbide	SiC	2.70	–	–	0.04	0.005	–

[a]At 4 K

0.1 eV, an ideal intrinsic infrared radiation detecting material was found. Besides its many detector applications, this alloy system has enabled detailed studies of band gap variation of transport processes, optical properties, magneto-optic effect, lattice vibration characteristics etc.

Later, more importance was given to research of HgCdTe thin film materials. These were prepared by liquid phase epitaxy(LPE), metal organic chemical vapour deposition (MOCVD) and molecular beam epitaxy(MBE) in addition to the bulk material growth methods. The position of HgTe and CdTe is shown in Fig. 2.14 depicting the band gap versus lattice constant.

2.16.1 HgCdTe as Narrow Bandgap Semiconductor [2]

HgCdTe (i.e. mercury cadmium telluride) is a direct narrow bandgap zincblende, II–VI ternary alloy of CdTe and HgTe with a tunable bandgap spanning the shortwave infrared to the very long wave infrared regions. HgCdTe was the material that revolutionised the infrared detection industry. Its unique properties which could be created by varying the concentration of cadmium and mercury proved quite useful in the semiconductor arena. The properties of HgCdTe are displayed in Table 2.5.

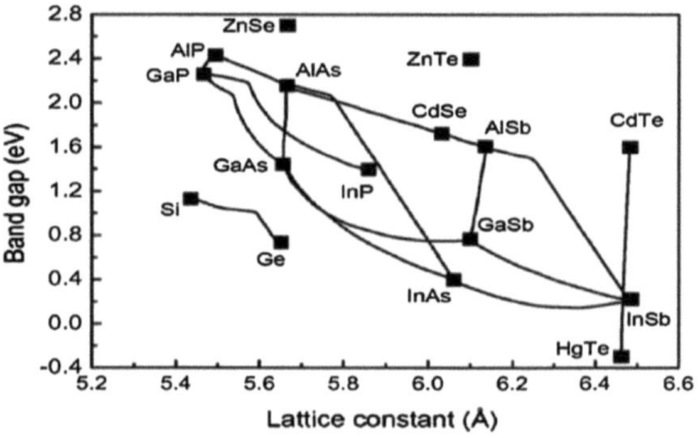

Fig. 2.14 Energy band gap versus lattice constant for various narrow bandgap semiconductors, depicting the position of HgTe and CdTe. Reproduced from [2]

2.16.2 *Applications as Infrared Detectors*

HgCdTe ternary alloy is a nearly ideal IR detector material system. Its position is conditioned by three key features; namely

- Tolerable energy band gap over the 1–30 μm range,
- Large optical coefficients that enable high quantum efficiency, and
- Favourable inherent recombination mechanisms that lead to high operating temperature.

These properties are direct consequence of the energy band structure of the blende semiconductor. Moreover, the specific advantages of HgCdTe are the ability to obtain both low and high carrier concentrations, high mobility of electrons, and low dielectric constant. The extremely small change of lattice constant with composition makes it possible to grow high quality layered and graded gap structures. As a result, HgCdTe can be used for detectors operated at various modes, as given below.

- Photoconductor
- Photodiode, or
- Metal–insulator–semiconductor (MIS) detector.

2.16.3 *Ternary Stannide Phase Narrow Bandgap Semiconductors: Na_2MgSn [3]*

In experiments conducted by researchers in April 2014 in China, the electronic structures, chemical bonding, elastic and optical properties of the ternary stannide phase Na_2MgSn were investigated by using density-functional theory. The calculated energy band structures show that the Na_2MgSn is an indirect semiconductor material with a narrow band gap 0.126 eV. The density of state (DOS) and the partial density of state (PDOS) calculations show that the DOS near the Fermi level is mainly from the Na 2p, Mg 3p and Sn 5p states. Population analysis suggests that

Table 2.5 Properties of HgCdTe semiconductors having different amounts of cadmium and mercury

Property	HgTe	$Hg_{1-x}Cd_xTe$						CdTe
x	0	0.194	0.205	0.225	0.31	0.44	0.62	1.0
a (Å)	6.461	6.464	6.464	6.464	6.465	6.468	6.472	6.481
T	77 K	77 K	77 K	77 K	140 K	200 K	250 K	300 K
E_g (eV)	−0.261	0.073	0.091	0.123	0.272	0.474	0.749	1.490
λ_c (μm)	–	16.9	13.6	10.1	4.6	2.6	1.7	0.8
η_i (cm^{-3})	–	1.9×10^{14}	5.8×10^{13}	6.3×10^{12}	3.7×10^{12}	7.1×10^{11}	3.1×10^{10}	4.1×10^{5}
m_c/m_0	–	0.006	0.007	0.010	0.021	0.035	0.053	0.102

Reproduced from [1] and modified

there are strongly bonded Mg-Sn honeycomb layers in Na_2MgSn. The Na_2MgSn is mechanically stable soft material and behaves in a brittle manner. Detailed analysis of all optical functions reveals that Na_2MgSn is a better dielectric material. Its reflectivity spectra shows that Na_2MgSn promise as good coating materials in the energy regions of 6.24–10.49 eV.

2.17 Wide Bandgap Semiconductor [4]

Wide bandgap semiconductors are the semiconductor materials with electronic band gaps significantly larger than 2.5 electronvolt (eV). The exact threshold of "wideness" often depends on the context; but for common usage, "wide" bandgap typically refers to material with a band gap significant greater than that of the commonly used semiconductors, silicon (1.1 eV) or gallium arsenide (1.4 eV).

Semiconductors including II–VI compounds such as ZnSe, ZnS, and ZnO have been the wide-band-gap semiconductors; however, the nitride compounds with column IIIA elements (AlN, GaN, and InN) have emerged as exceptionally versatile semiconductors. Compared to GaAs with an energy gap corresponding to infrared emission (890 nm), the AlGaInN family covers the entire visible spectrum from infrared (InN) to blue (InGaN), and extending into deep ultraviolet (AlGaN), creating opportunities in display, illumination, high-density optical storage, and biological and medical photonics. Transistors and diodes made from wide-band-gap GaN create opportunities beyond low-power digital electronics; including applications in automobiles, aircraft, utility distribution, and wireless communications.

A list of wide bandgap semiconducting materials are displayed in Table 2.6 along with various related quantities.

2.17.1 Advances in Wide Bandgap Materials for Semiconductor Spintronics [5]

Existing semiconductor electronic and photonic devices utilize the charge on electrons and holes in order to perform their specific functionality such as signal processing or light emission. The relatively new field of semiconductor spintronics seeks to exploit the spin of charge carriers in new generations of transistors, lasers and integrated magnetic sensors. The ability to control of spin injection, transport and detection leads to the potential for new classes of ultra-low power, high speed memory, logic and photonic devices. The utility of such devices depends on the availability of materials with practical (>300 K) magnetic ordering temperatures. In this section, the recent progress in dilute magnetic semiconductors (DMS) such as (Ga, Mn)N, (Ga, Mn)P, (Zn, Mn)O and (Zn, Mn)$SiGeN_2$ exhibiting room

Table 2.6 Wide bandgap semiconductors along with band gaps, and dielectric constants

Semiconductor	Band gap (eV)	Dielectric constant
Si	1.1	12
SiO_2	9	2.25
Si_3N_4	5.3	3.8
$SrTiO_3$	3.3	6.1
$PbTiO_3$	3.4	6.25
$BaZrO_3$	5.3	4
$PbZrO_3$	3.7	4.8
Ta_2O_5	4.4	4.84
$SrBi_2TaO_9$	4.1	5.3
TiO_2	3.05	7.8
ZrO_2	5.8	4.8
HfO_2	6	4
Al_2O_3	8.8	3.4
Y_2O_3	6	4.4
La_2O_3	6	4
$ZrSiO_4$	6	3.8
$HfSiO_4$	6	3.8

Modified and reproduced from [4]

temperature ferromagnetism, the origins of the magnetism and its potential applications in novel devices such as spin-polarized light emitters and spin field effect transistors, is presented.

2.17.2 Future Aspect of Wide Bandgap Materials Power Generating Window [6]

Wide bandgap chalcopyrite compound films ($CuInGaS_2$ i.e. CIGS) reveal the semi-transparent and bifacial properties that are beneficial for power generating window applications. Importantly, solar cell devices with these films demonstrate a synergistic effect for bifacial illumination resulting in a 5.4–16.3 % increase of the apparent power conversion efficiency compared to the simple sum of the efficiencies of the front and rear side illumination only.

This extra output power acquisition due to bifacial irradiation is apparently not influenced by the light intensity of the rear side illumination, which implies that weak light (e.g., indoor light) can be efficiently utilized to improve the overall solar cell efficiency of bifacial devices.

Wide band gap materials may be utilized to generate electricity on see-through glass for commercial application in homes, offices, and commercial buildings. Further

- They do not require expensive high-temperature or high-vacuum production methods, rather, can be sprayed on to glass at room temperature.
- They generate electricity from both natural and artificial light sources, outperforming today's commercial solar and thin-film technologies by as much as tenfold.
- They measure less than 1/10th the thickness of 'thin' films, (which is about 1/1000th the thickness of human hair).

2.17.3 Recent Advances in Wide Bandgap Materials

UV sensitive organic semiconductors attracted increasing attention because of the advantages like easy processing as well as low fabrication cost compared to inorganic semiconductors including diamond, SiC, GaN, ZnO. For more and vivid details, read Chaps. 14, 15 and 16.

2.18 Rectifiers

Rectifiers convert ac into dc. It is mainly of two types viz. (i) Copper oxide rectifier, and (ii) Selenium rectifier. Copper oxide rectifier is a plate of 99.98 % pure copper on which a film of cuprous oxide is produced by special purpose. One side of plate is cleaned with cuprous oxide and an electrode is soldered directly to copper. The second electrode is soldered to cuprous oxide film. When positive potential is applied to the oxide layer and negative of the copper, and it corresponds to forward biasing of a P-N junction. These rectifiers, hence, have low permissible current density.

2.18.1 Selenium Rectifier

Selenium rectifier uses more than 99.9 % pure selenium. Crystalline selenium, which has a melting point of 220 °C, is used for making rectifiers. It is constructed by using Nickel-plated steel plate. Selenium under pressure at 130 °C is raised to 180–215 °C and put on the nickel plate making a layer of 1 mm on it. Crystalline selenium will change to grey crystalline selenium. Over this electrode of tin, an alloy is sprayed. This makes the active disc of the rectifiers.

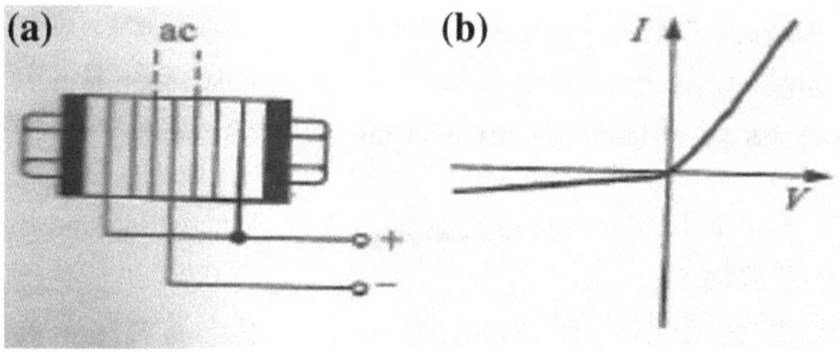

Fig. 2.15 A selenium rectifier showing its **a** construction, and **b** behaviour

Large-area rectifiers employ selenium, copper sulfide, and cuprous oxide as P-type semiconductor bonded to a metal base. Low resistance direction occurs when the semiconductor is positive. The rectifier consists of many such disks stacked in series for higher voltage use and are bridge connected as shown in Fig. 2.15.

2.19 Solved Examples

Example 2.7 Compare the non-directional and directional bonds.

Solution. The differences between non-directional and directional bonds are discussed below:

S. No.	Non-directional bond	Directional bond
1.	Ionic, metallic and Van der Waals bonds are the examples of this kind	Covalent and hydrogen bonds are the examples in this category
2.	The overlapping electrons orbitals are spherically symmetric	The overlapping electrons orbitals are directionally oriented
3.	The bond does not form at any specific angle	The bond forms at some angle called the bond angle
4.	The electrons are in common pool, and in the form of free electron gas	Electrons are localised. Two neighbouring atoms bind together and stay

Example 2.8 Bring out the differences between dipole and dispersion bonds.

Solution. The differences between dipole and dispersion bonds are given as below:

Sl no.	Dipole bond	Dispersion bond
1.	Bonds form on account of electrical dipoles produced due to non-coincidence of centers of positive and negative charges	Bonds form due to attraction of fluctuating charges of adjoining atoms

(continued)

(continued)

Sl no.	Dipole bond	Dispersion bond
2.	Secondary forces of attraction arise during bonding	Momentary polarization occurs in the molecules
3.	It is a type of molecular bonding	Fluctuating and oscillating dipoles develop during bonding
4.	Hydrogen fluoride and water are such examples	Inert gas crystals are the examples
5.	These are stronger than the dispersion bonds	These are feeble and weakest amongst all bonds

Example 2.9 Can we produce a material having a bonding force of infinite value?

Solution. No, this is not possible. We can produce a material having bonding forces equal to its theoretical strength only. Theoretical strength σ_{th} of a material is the maximum bonding force per unit cross-sectional area ($\sigma_{th} = F_{max}/A$). According to calculations as per Coulomb's theory, the theoretical strength may reach almost to the value of Young's modulus E of the material. Thus for steel, $\sigma_{th} = E = 200$ GPa.

Example 2.10 What are 'electron affinity', 'electronegativity' and 'ionization energy' in respect of atomic bonding?

Solution. *Electron affinity* is the work done to attract an extra electron from infinity to the outermost orbit of a neutral atom. *Electronegativity* is the tendency of an atom to attract electrons during bond formation with other atoms. *Ionization energy* is the energy required to remove a loosely bound electron from any orbit of an atom.

The concept of electron affinity (expressed in volt/atom) influences upon the atomic behaviour in bonding. The electronegativity demarcates between metals and non-metals, and ionization energy has an effect on the properties of materials.

Example 2.11 Explain the Frenkel's defect and write its features.

Solution. An ion, displaced from a regular location to an interstitial location, in an ionic solid is called Frenkel's defect, Fig. 2.16a. The ions of two different kinds are known as *cations* and *anions*. Cations are the smaller ions while anions are the larger ones. Cations may easily get displaced into the void. Anions, on account of their larger size, do not displace in small sized voids. The presence of this defect does not change the overall electrical neutrality of the crystal. Imperfections in CaF_2 and silver halides are the examples of this kind.

Example 2.12 What is Schottky defect? How is it different from Frenkel's defect?

Solution. When a pair of one cation and one anion are absent from an ionic crystal, Fig. 2.16b, the defect is called Schottky's defect.

The valancy of missing pair of ions maintains electrical neutrality in the crystal. Such imperfections are dominant in alkali halides such as LiCl, LiBr, LiI etc.

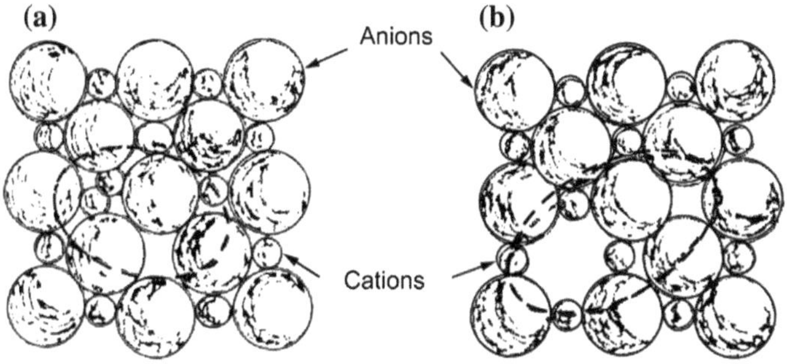

Fig. 2.16 **a** Frenkel's defect in an ionic crystal shows presence of a cation elsewhere within the crystal, and **b** Schottky's defect in an ionic crystal shows a missing pair of anion and cation

Review Questions

1. Explain composition of a solid illustrating crystals, unit cells, atoms and electrons in it.
2. Differentiate between the following.
   (a) Crystalline and amorphous solids
   (b) Monocrystalline and polycrystalline structures
   (c) Unit cell and a crystal
3. Describe the Bravais crystal system, particularly mentioning the details of semiconducting elements.
4. What is the importance of Miller indices? How does it help in the study of crystallography?
5. Detail out the procedure to obtain Miller index in a crystal.
6. Draw the planes whose Miller Indices are $(11\bar{1})$ and $\{100\}$.
7. Discuss different types of bonds in semiconductors. How is directional bond different from non-directional bond ?
8. Draw the bonding force F vs r (the distance of atomic separation) curve, and explain the bond length and bond energy.
9. What is covalent bond? What are its different types ? Also explain the band angle in semiconductors.
10. Discuss the mixed bond and quote some examples of ionic-covalent bonded compound semiconductors.
11. Sketch and explain the DC structure of Si and write its various geometrical details.

12. What are zinc blende and wurtzite structures? Quote their examples. Explain them inrespect of the structure of ZnS.
13. Sketch the lattice structure of any three compound semiconductors.
14. Classify imperfections as exhaustively as you can, and briefly discuss them.
15. Discuss different types of point imperfections.
16. Discuss Bohr atomic modal and explain as how the electron energy is calculated from it.
17. How is the wave number related to Rydberg's constant? Explain various spectrum series in the hydrogen atom.
18. Explain the improvement inSommerfelt's atomic model.
19. Describe modern concept of atom. How is the arrangement of subshells in different shells?
20. Write notes on the following.

    (a) Heisenberg's uncertainty principle (b) Probability density cloud

    (c) de Broglie wave (d) Quantum numbers and their significance.
21. What do you mean by electronic configuration? Write electron configuration of silicon, copper, and germanium.
22. Enlist the various application of semiconductor devices, discuss them briefly.
23. Which semiconductors are called 'narrow bandgap' and which are called 'wide bandgap'? compare their properties, characterstics, and applications. Quote their examples also.
24. Describe the especislities, properties and characterstics of HgCdTe as narrow bandgap semiconductor. Discuss its application as IR detector.
25. Give a detailed account of wide bandgap semiconductors. Quote its examples and applications. Write its future scope for power generating window.
26. What are rectifiers? Sketch and explain its construction, working and applications of selenium rectifier.
27. Write notes on the following.

    (a) Ternary stannide phase narrow bandgap semiconductors

    (b) Recent advances in wide bandgap materials.

Numerical Problems

1. The dimensions of an orthorhombic unit cell are 2.3 Å, 4.6 Å and 6.9 Å. Obtain the intercepts of a plane of Miller indices (120).
   [**Ans.** 2.3 Å, 2.3 Å,∞]

2. Determine the Miller indices of a plane that makes an intercept of 2 Å, 3 Å and 4 Å onthe coordinate axes of an orthorhombic crystal with $a : b : c = 4 : 3 : 2$. Draw it also.

3. Calculate the Miller indices of a plane passing througho rthorhombic gallium crystal,and intersecting the three coordinate axes along a, b and c at 1.763, 8.040 and 1.915 Å respectively. The lattice constants for gallium are a = 3.526 Å,b= 4.020 Å, and c = 7.660 Å.
   [**Ans.** (418)]

4. For a hydrogen atom in the ground state, determine in electron volts: (*a*) the kinetic energy, (*b*) the potential energy, (*c*) the total energy, (*d*) the energy required to remove the electron completely.
   [**Ans.** (*a*) 13.6 eV (*b*) –27.2 eV (*c*)–13.6 eV (*d*) 13.6 eV]

5. At what speed must the electron revolve around the nucleus of a hydrogen atom in order that it may not be pulled into the nucleus by electrostatic attraction? Also calculate itsorbital energy and frequency.
   [**Ans.** $v = 2.25 \times 10^6$ m/s, $E_n = -21.7 \times 10^{-19}$ J, $f = 6.7 \times 10^{15}$ Hz = 6700 T Hz]

6. Find the number of electrons in the atoms which have the following levels filled in the ground state (*a*) *K* and *L* shells, the 3*s* subshell and one half of the 3*p* subshell (*b*) The *K*, *L*, and *M* shells and the 4*s*, 4*p* and 4*d* subshells.
   [**Ans.** (*a*) 15 Phosphorus (*b*) 46 Palladium]

7. Calculate for the hydrogen atom (*a*) velocity of electron in the ground state (*b*) radius of the orbit in the ground state (*c*) time taken by the electron to traverse the first Bohr'sorbit.
   [**Ans.** (*a*) 2.19×10^6 m/s (*b*) 0.0529 m (*c*)1.519×10^{-16} s]

8. The ultraviolet wavelength of 800 Å and 700 Å is allowed to fall on hydrogen atom inthe ground state. It is found to liberate electrons with kinetic energy 1.8 eV and 4.0 eV respectively. Find the value of Planck constant.
   [**Ans.** 6.57×10^{-34} Js]

9. The energy of an electron in the second and third Bohr's orbits of the hydrogen atom is–5.42×10^{-19} joule and–2.41×10^{-19} joule respectively. Calculate the wavelength of the emitted radiation when the electron drops from the third to the second orbit.
   [**Ans.** 6600 Å]

10. How many revolutions does an electron make in the $n = 2$ state of hydrogen atom before dropping to $n = 1$ state. The average life time of an excited state is about 10^{-8}s. Assume the required constants.
    [**Ans.** 4.03×10^8]

11. Calculate the frequency of radiation and the wave number when the electron jumps from third orbit to the second orbit in a hydrogen atom.
    [**Ans.** 4.56×10^{14}/s, 1.52×10^6/m]

12. Obtain the difference in wavelengths between the first line of Balmer series of hydrogen and of deuterium (heavy hydrogen). The mass of electron, proton and deuterium nucleiare in the ratio of 1 : 1836 :3669. Rydberg constant for hydrogen is 10 967 700 per metre.
    [**Ans.** 1.79×10^{-10}/m]

Objective Questions

1. An unit cell is defined as
   (a) the smallest group of atoms, which when regularly repeated, forms a crystal.
   (b) the basic of a crystal.
   (c) a tetragon containing the largest number of atoms.
   (d) an agglomerated structure.
2. In a simple cubic structure, are the direction [100] and plane (100) perpendicular?
   (a) No (b) yes
   (c) may be yes (d) can't be predicted
3. The crystal structure of Boron and SiO2 are
   (a) cubic (b) rhombohedral
   (c) tetragonal (d) triclinic
4. The Miller indices of a plane are proportional to
   (a) the reciprocal of numerical parameters of the intercepts.
   (b) the under root of unit cell dimensions.
   (c) the intercepts of the planes on the coordinate axes.
   (d) interplaner distance.
5. A crystalline solid has all similar atoms relative to neighbouring atoms in
   (a) periodical positions (b) reversed positions
   (c) random fluctuating position (d) similar positions
6. The [110] direction in a cubic unit cell is parallel to the
   (a) face diagonal of unit cell (b) none
   (c) body diagonal of the cube (d) edge of the cube
7. Miller indices of the diagonal plane of a cube are
   (a) (200) (b) (010) (c) (111) (d) $(\bar{1}10)$
8. If the Bohr radius of hydrogen atom is r, the radius of the third Bohr orbit for the electron in an ionized helium atom is
   (a) $\frac{2r}{3}$ (b) $\frac{2r}{9}$ (c) $\frac{9r}{2}$ (d) $\frac{3r}{2}$
9. If Z is the atomic number of a multi-electron atom, the energy of an electron in the Bohr'satomic model is proportional to
   (a) Z^2 (b) $1/Z$ (c) Z (d) $1/Z^2$
10. Consider the spectral line resulting from the transition $n = 2 \rightarrow n = 1$ in the atoms and ions given below. The shortest wavelength is produced by
    (a) deuterium atom (b) hydrogen atom
    (c) doubly ionized lithium (d) singly ionized helium

11. In Bohr model of hydrogen atom,
    (a) the radius of the nth orbit is proportional to n
    (b) the total energy of electron in nth orbit is proportional to n
    (c) the angular momentum of the electron in an orbit is an integral multiple of Planck's constant
    (d) the magnitude of the potential energy of an electron in any orbit is greater than its kinetic energy
12. The maximum number of electrons in the nth main energy level of an atom is
    (a) $2\sqrt{n}$ (b) 2^2 (c) 2n (d) 2 / n
13. If l is the number of subshell in the main shell, the maximum number of electrons in this subshell is
    (a) 2l(2l + 1) (b) 2l (c) 2(2l + 1) (d) 4l
14. If the atomic number of an element is Z, and its atomic mass number is A, the number of protons in its nucleus is
    (a) A (b) Z (c) A – Z (d) A/Z
15. Energy is emitted when the electron
    (a) falls from L orbit to K orbit
    (b) jumps from K orbit to L orbit
    (c) falls or jumps from K orbit to L orbit
    (d) none of the above
16. When n_f is the quantum number of final state, then hydrogen spectrum series for $n_f = 3$ is given by
    (a) Paschen (b) Pfund (c) Lyman (d) Plank
17. The azimuthal quantum number in O shell (n = 5) will be
    (a) 1, 2, 3, 4, 5 (b) 0, 1, 2, 3, 4 (c) 5 (d) 4
18. The principal quantum number in ionized state may be assumed to be
    (a) 1 (b) ∞ (c) 7 (d) 0
19. The ratio of major axis to minor axis in the elliptical path will be least in case of
    (a) psuborbit (b) s suborbit (c) f suborbit (d) d suborbit
20. Which of the following atomic orbital do not follow quantum theory?
    (a) 4f (b) 3p (c) 6s (d) 3f
21. The interplaner distance in a cubic crystal is equal to the length of the cube.Miller indices of the related plane will be
    (a) (100) (b) $(\bar{1}1\bar{0})$ (c) $(\bar{1}0\bar{1})$ (d) (200)
22. The [110] direction in a cubic unit cell is parallel to
    (a) face diagonal of unit cell (b) edge of the cube
    (c) body diagonal of the cube (d) none of the above

Answers

1. (a)	2. (b)	3. (b)	4. (a)	5. (d)	6. (d)
7. (c)	8. (c)	9. (c)	10. (c)	11. (d)	12. (c)
13. (c)	14. (b)	15. (a)	16. (a)	17. (b)	18. (b)
19. (c)	20. (b)	21. (a)	22. (b)		

References

1. Gupta, K.M., Gupta, N.: Advanced Electrical And Electronics Materials. Scrivener Wiley Publishing, USA (2015)
2. Rogalski, A.: HgCdTe infrared detector material: history, status and outlook. Rep. Prog. Phys. **68**, 2267–2336 (2005). doi:10.1088/0034-4885/68/10/R01
3. Wang, Y.-F., Xia, Q.-L., Pan, L.-X., Yu, Y.: First principle calculation on ternary stannide phase narrow band gap semiconductors Na_2MgSn. Trans. Nonferrous Metals Soc. China **24**(6), 1853–1858 (2014)
4. http://en.wikipedia.org/wiki/Wide-bandgap_semiconductor
5. Pearton, S.J., Abernathy, C.R., Norton, D.P., Hebard, A.F., Park, Y.D., Boatner, L.A., Budai, J.D.: Advances in wide bandgap materials for semiconductor spintronics. Mater. Sci. Eng. **R 40** 137–168 (2003)
6. Moon, S.H., Park, S.J., Hwang, Y.J., Lee, D.-K., Cho, Y., Kim, D.-W., Min, B.K.: Printable, wide band-gap chalcopyrite thin films for power generating window applications. Sci. Rep. **4408**, 1–5 (2014)

Chapter 3
Carrier Transport in Semiconductors

Abstract Role of valence electrons and free electrons in carrier transport through semiconductors is presented. Various electrons theories such as Free election theory, Energy band theory, and Brillouim zone theory are described. Mechanism of conduction by free electrons is elaborated. Fermi energy level, intrinsic and extrinsic semiconductors, energy diagrams of n-type and p-type semiconductors are given. Direct and indirect energy band semiconductors and effect of alloying on them are discussed. Concepts of effective mass, density of state, temperature dependence of carrier concentrations are developed. Drift of carriers under electric and magnetic fields, collision time, effects of temperature on mobility of carriers, effects of doping on mobility are explained. Degenerate semiconductors and effect of heavy doping are incorporated. The phenomenon of Hall Effect and its significance; Hall voltage, Hall angle, Hall coefficient etc. are described. Relation between Density of states and Fermi energy, momentum space, Fermi sphere, Fermi radius, Fermi velocity, Fermi momentum, Fermi temperature etc. are derived. Minute insight into the various topics are given through solved numerical and theoretical examples. Review questions, numerical problems and objective type questions are also given with their answers.

Keywords Electron theories · Energy band theory · Brillouin zone theory · Direct band semiconductors · Indirect band semiconductors · Charge carriers in semiconductors · Fermi energy · Intrinsic semiconductors · Extrinsic semiconductors · Effective mass · Degenerate semiconductors · Hall effect · Fermi sphere

3.1 Electrons and Their Role in Conductivity

The solid substances are composed of atoms, which are discrete indivisible particles having diameter of the order of 1 Å (10^{-10} m). Atoms of a substance are identical but are quite different from the atoms of another substance. They are composed of

Articles 3.1 to 3.4.4 have been reproduced from [1], and articles 3.5 to 3.5.1 have been reproduced with permission from [1].

K.M. Gupta and N. Gupta, *Advanced Semiconducting Materials and Devices*, Engineering Materials, DOI 10.1007/978-3-319-19758-6_3

fundamental (or basic) particles viz. protons, neutrons and electrons. These particles in different substances are similar.

Electron is the most important particle of an atom. The electrical nature of atoms is attributed to a highly systematic configuration of electrons in them. They are universal constituent since all the metals contain them in their structure. Electrons have similar behaviour in conductors, semiconductors, dielectrics, magnetic or superconducting materials. They also have same behaviour in all applications whether a computer, television, radiography, or electrical conduction.

3.1.1 Valence Electrons

Electrons play vital role in determining the electrical properties of metals. Valence electrons in the outermost orbit of an atom decide the manner in which they respond to external effects. For example, due to their free movement within the metal they provide conduction, but on actuation by thermal energy they jump over the energy gap in semiconductors. As a whole the arrangement of electrons in an atom, behaviour of valence electrons, and inter-atomic interactions govern the electrical properties (conduction as well) of materials.

3.1.2 Free Electrons

An atom is completely stable when its outermost shell contains 8 electrons. Chemically inert Neon, Argon, Krypton etc. are thus highly stable. Removal of an electron from stable inert gas configuration requires a relatively large energy. Electrons in the outermost shell are called valence electrons. When the outermost shell of atom is not complete, these electrons are easily detachable. At room temperature, the thermal energy is enough to detach the valence electrons from their parent atoms. Such electrons are called free electrons. Metals have too many free electrons and hence they are good conductors of heat and electricity. On the other hand, insulating materials such as wood, glass, rubber, paper etc. require large energy to detach few electrons from their atoms. The situation in semiconductor is intermediatory to those of metals and insulating materials.

3.2 Electron Theories of Solids

Electrical properties of solids are determined by electrons in outermost orbit of an atom. The valence electrons in an atom decide the manner in which they will respond to external effects. The cohesive and repulsive forces, in the formation of atomic bond are not sufficient to describe many properties of solids.

Various electron theories have been propagated to study the behaviour of solids. The acceptable theories in this regard are the following.

1. Free electron theory,
2. Energy band theory, and
3. Brillouin zone theory.

Whereas the free electron theory explains the mechanism of conduction in metals, the energy band and Brillouin zone theories explain the mechanism of semiconduction in semiconductors.

3.2.1 *Free Electron Theory*

We have already studied that the electrons in outermost orbit are not bound to its atom, and are free to move throughout the solid. These free electrons are known as Fermi gas or electron cloud, and their potential field remains uniform throughout the solid due to the ion-cores. The free electron theory is based on the assumption that the average potential energy E_p is a constant throughout the solid, and its energy difference $dE_p = 0$. So the total energy E is equal to the kinetic energy E_k only. This energy is substantially lower than the energy of bound electrons in an isolated atom.

de-Broglie's wavelength. Due to dual nature of particle (i.e. electron), the de-Broglie's wavelength is given by

$$\lambda = \frac{h}{mv} \tag{3.1a}$$

where h is Planck's constant, v is velocity of the free electron and m its mass. Defining wave number k by

$$k = \frac{2\pi}{\lambda} \tag{3.1b}$$

and neglecting relativistic effect[1] as the velocity of electrons is much lower than the velocity of light ($v_{electron} \ll v_{light}$), the total energy may be expressed by

$$E = E_k = \frac{1}{2}mv^2 \tag{3.2a}$$

Kinetic energy in terms of wave number. Substitution of Eqs. (3.1a) and (3.1b) in Eq. (3.2a) a parabolic relation of continuous energy. It is shown in Fig. 3.1 as kinetic energy versus wave number curve. The kinetic energy is given by

---

[1]Einstein has given that mass m at velocity v will be $m = m_0\sqrt{(1-(v/c)^2)}$ where m_0 is rest mass and c is velocity of light.

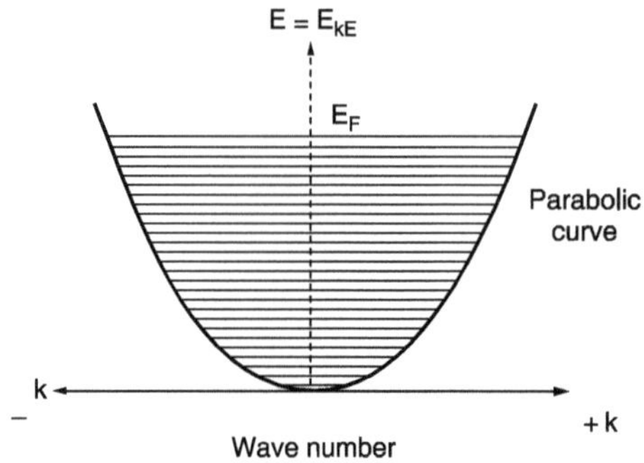

Fig. 3.1 Relation between the kinetic energy $E_k = E$ and wave number k of a free electron, in free electron theory where potential energy is assumed constant

$$E = \frac{h^2 k^2}{8\pi^2 m} \tag{3.2b}$$

As the electrons in an atom move in the opposite directions at same speed, so the wave number may take both: the positive and negative values. In critical conditions, these values are $k = \pm n\pi/l$ where n is principal quantum number and l is the length of solid. Substituting this value in Eq. (3.2b), we get

$$E = \frac{h^2 n^2}{8ml^2} \tag{3.3a}$$

Energy equation for 3-dimensional solid. This equation is true for the case of unidirectional flow of electrons. The free electrons can move in any direction in a three-dimensional solid. Hence, we can write $n^2 = n_x^2 + n_y^2 + n_z^2$, and now Eq. (3.3a) gets modified to

$$E = h^2 \left(\frac{n_x^2 + n_y^2 + n_z^2}{8ml^2} \right) \tag{3.3b}$$

The electron waves are the travelling waves by nature, and have constant velocity throughout the solid. Hence, the time averaged probability of finding an electron is constant throughout the solid.

3.2.2 *Mechanism of Conduction by Free Electrons*

The electrical and thermal conduction through a metal bar is a common feature noticed by us. The electrical conductivity in an aluminium rod or in a copper wire, and the thermal conductivity through a steel bar is possible due to movement of free electrons.

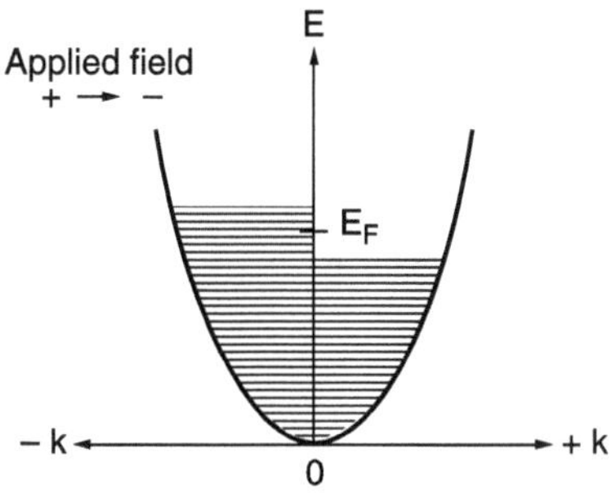

Fig. 3.2 Redistributed E versus k profile on application of external field

For every electron moving in a direction with some speed, there is another electron moving in the opposite direction with the same speed. So the net velocity becomes unidirectional on application of an external field, and hence the solid conducts.

Redistribution of energy field. As shown in Fig. 3.2, the negatively charged electrons accelerate towards the positive end of the applied field. In this action, some electrons acquire additional velocity while the electrons moving in opposite direction lose some velocity. Due to this, E versus k curve of Fig. 3.1 gets modified. Such occurrence is the characteristics of conductors, and is possible only when empty energy levels are available above the Fermi energy level E_F. In this figure, it is shown that some electrons acquire $E > E_F$ while others have $E < E_F$.

Some electrons gain extra velocity at the cost of electrons moving in opposite direction

3.3 Energy Band Theory

Assumption. The assumption that a constant potential field exists in the free electron theory was questioned by many researchers. Potential energy of an electron is a function of its position with respect to the ion-cores. Considering the Heisenberg's uncertainty principle and Bragg's diffraction pattern of electrons, the potential energy of an electron cannot be neglected as compared to its dimension. This is due to the fact that the 'standing wave' of an electron gives rise to a periodic variation in its amplitude. The probability of finding an electron remains maximum at the crest of the waveform. The two possible waveforms are sine wave and cosine wave, formed due to superimposition of 'travelling' de-Broglie waves. These waveforms satisfy Bragg's law given as

$$2d\sin\theta = n\lambda \tag{3.4}$$

and are shown in Fig. 3.3. Thus the difference in potential energy cannot be assumed to be zero, and the energy will be given by

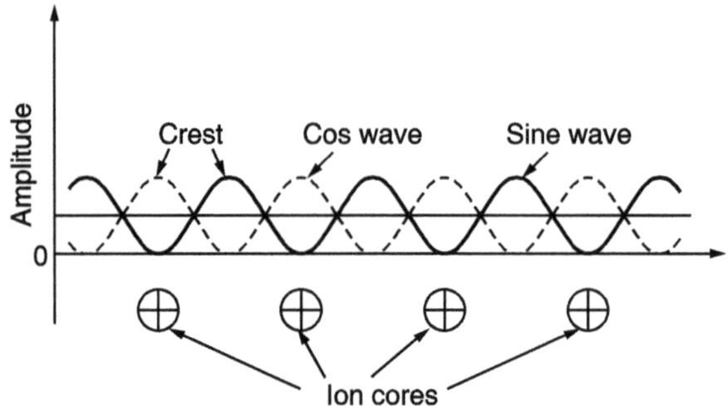

Fig. 3.3 Sine and cosine standing waveforms are obtained due to superimposition of travelling waves. The electron is likely to be located at the crest

$$E_e = E_k + E_P \tag{3.5}$$

Critical conditions. At critical conditions of wave number $k = \pm n\pi/d$ for n = 1, 2, 3, … etc., the electron is described by a standing wave. When the waveforms are sine or cosine type, the potential energy shows deviations. This deviation results in break of E versus k curve (Fig. 3.4a), giving rise to an energy gap. (According to free electron theory, the E vs. k curve is continuous). This energy gap E_g is between different orbits K, L, M, … of an electron, and is shown in Fig. 3.4b.

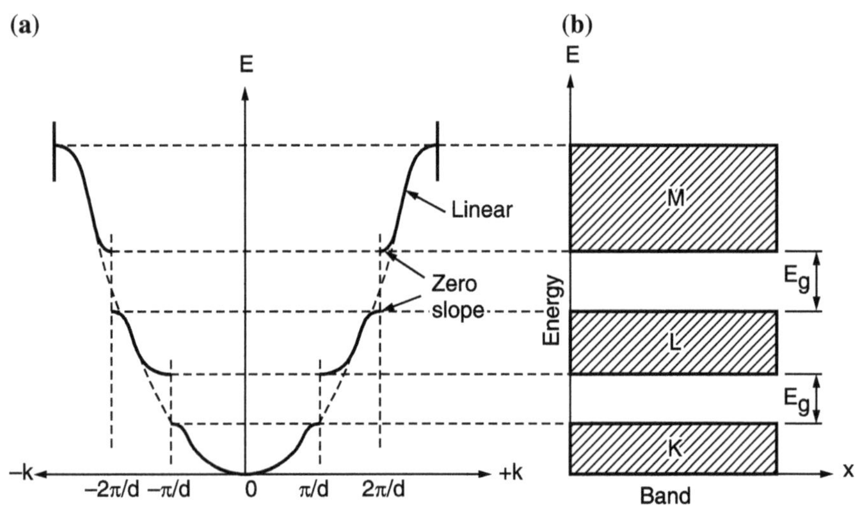

Fig. 3.4 **a** Consideration of potential energy of electron yields a break in E versus k *curve* at critical values, and **b** energy gap E_g

Table 3.1 Energy gap in some semiconducting materials

Material	Energy gap E_g (eV)		Material	Energy gap E_g (eV)	
	at 0 K	at 300 K		at 0 K	at 300 K
Sn(gray)	0.08	–	InSb	0.23	0.17
Ge	0.74	0.66	InAs	0.42	0.36
Si	1.17	1.12	GaAs	1.52	1.42
C (diamond)	5.48	5.47	Gap	2.34	2.26
CdSe	3.42	3.35	SiC (α)	3.03	2.99
ZeO	1.85	1.70	AlP	–	2.45

Magnitude of energy gap. Magnitude of this energy gap is an indication of the difference in potential energy for electron locations of two different waveforms. The two closely spaced energy levels are known as energy bands. The slope of E versus k curve is linear except at the critical values of k where this slope becomes zero.

Energy gap in some materials is given in Table 3.1.

3.4 Brillouin Zone Theory

The discontinuous plot of E versus k in Fig. 3.4a depicts only unidirectional motion of an electron. Brillouin zone theory considers the motion of electron in all possible directions such as [100], [110] etc., and all possible reflections in Bragg's Eq. 3.4, where the value of glancing angle θ can vary between 0° and 90°. Critical values of k correspond to $\pm n\pi/(d \sin \theta)$ instead of $\pm n\pi/d$ which was used for unidirectional case. Here d is interplaner spacing.

Effective zones on different planes and directions. Now consider a set of parallel planes (100) in a cubic crystal. For the first order reflection (n = 1), the critical condition will be $k = \pm \pi/(d \sin \theta)$. As the direction [100] is perpendicular to plane (100), hence movement of an electron along the direction [100] will have critical condition for $k = \pm\pi/(d \sin 90°) = \pm\pi/d$.

Now if direction of electron motion is changed to [110], then for deflection from planes (100), the critical condition will be obtained at $k = \pi/(d \sin 45°) = \sqrt{2}\pi/d$. In this case, the Bragg angle is 45°. Figure 3.5a–b show E-k curve for electron moving along [100] and [110] directions. The two curves are for the first order reflection from set of (100) planes. The gaps occur at $k = \pi/d$ for [100] motion and at $k = \sqrt{2}\pi/d$ for [110] motion. The effective energy gap is depicted in above figures. This effective gap is known as forbidden gap (or zone), and are those energy levels that an electron cannot take.

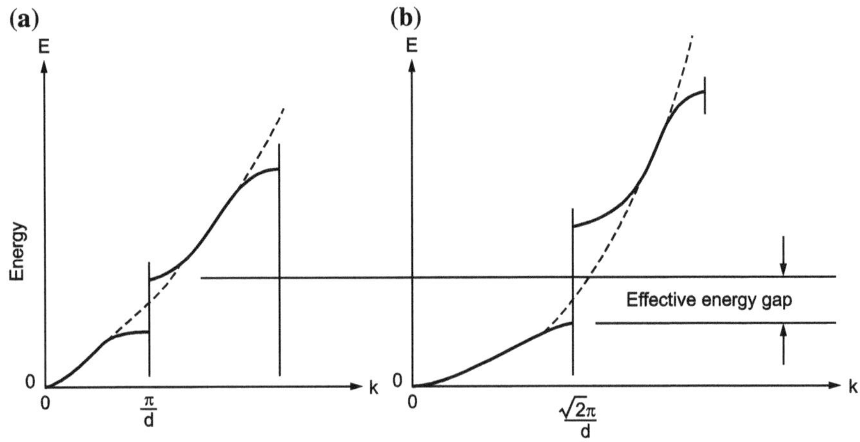

Fig. 3.5 Effective energy gap for electron moving along direction [100] (**a**), and [110] direction (**b**)

3.4.1 Meaning of Brillouin Zones

This theory demarcates different Brillouin zones as follows.

(i) First Brillouin zone, and
(ii) Second Brillouin zone, etc.

Brillouin zones are the ranges of k lying between the forbidden gaps. A Brillouin zone may be obtained by plotting the values of k for different sets of planes. First and second Brillouin zones, for a two-dimensional crystal, are shown in Fig. 3.6a–b respectively. The first Brillouin zone contains all energy levels up to first discontinuity in E-k curve. Range of k for this discontinuity is from $-\pi/d$ to $+\pi/d$. Similarly, second Brillouin zone contains all energy levels between first and second discontinuities lying between $\pm 2\pi/d$. Similar explanation may be given for third Brillouin zone and so-on.

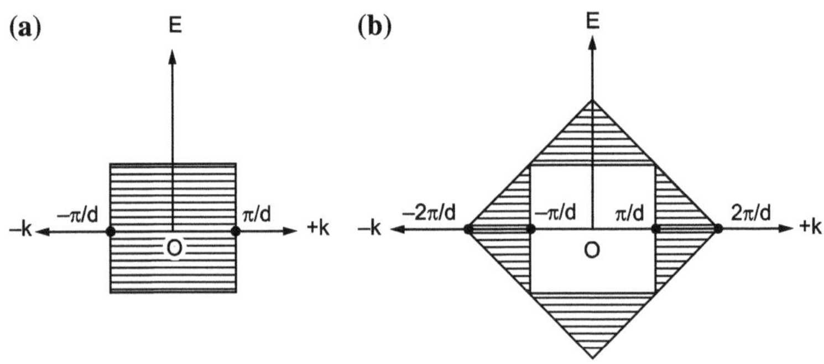

Fig. 3.6 **a** First Brillouin zone, and **b** second Brillouin zone, in a two dimensional crystal

3.4.2 First and Second Brillouin Zones

Brillouin zone is a 3-dimensional representation of the allowable values of k. In a 3-dimensional crystal, the critical value of k depends on the relative direction of the moving electron to the crystal lattice, due to the changing value of θ, and also because different sets of planes may diffract the electron. For 1-dimensional lattice, the critical values of k in Fig. 3.7 would be

$$k_n = \frac{n\pi}{a} \tag{3.6}$$

where n = 1, 2, 3, … and a is the distance between atoms. The region between k_1 and k_2 is called the first Brillouin zone. The second one is the region from k_2 to k_3. Figure 3.8 shows the two zones. In a 2-dimensional square lattice, the components

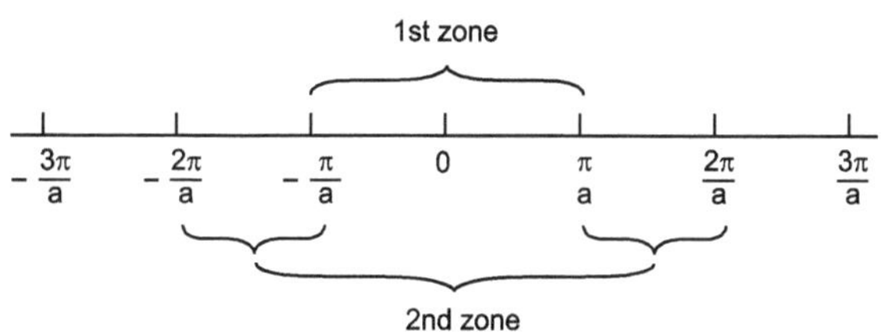

Fig. 3.7 The first two Brillouin zones for a 1-dimensional lattice

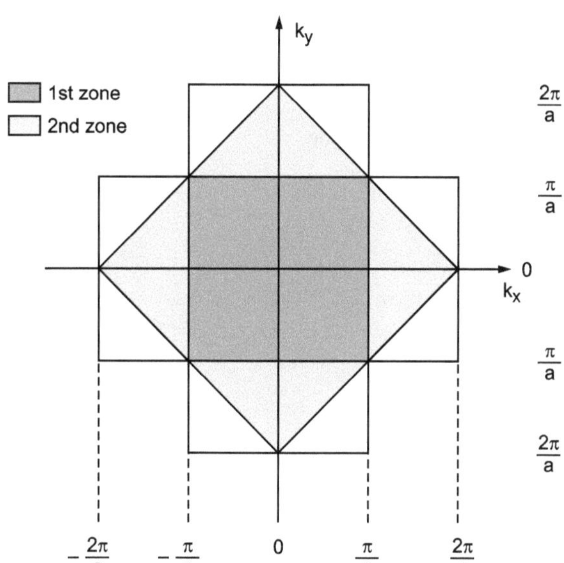

Fig. 3.8 The first two Brillouin zone for a square lattice

of k are k_x and k_y. Diffraction will occur whenever the k satisfies Eq. 3.6. If we apply Eq. 3.6 and remember that both the vertical and horizontal planes can diffract, we arrive at

$$k_x n_1 + k_y n_2 = (\pi/a)\left(n_1^2 + n_2^2\right) \tag{3.7a}$$

where n_1 is the integer for diffraction by the vertical planes, and n_2 for the horizontal planes. For the first zone, one integer is ±1 and the other is zero. This is the minimum value of Eq. 3.7a. For the second zone, each integer is ±1. The first two zones of the square lattice are shown in Fig. 3.8. The higher zones correspond to higher values of n_1 and n_2.

3.4.3 Brillouin Zones for Simple Cubic Lattice

The Brillouin zones of a simple cubic lattice in 3-dimensions can be calculated using an equation of the form as given below.

$$k_x n_1 + k_y n_2 + k_z n_3 = (\pi/a)\left(n_1^2 + n_2^2 + n_3^2\right) \tag{3.7b}$$

From this equation, it follows that the first zone for a simple cubic lattice is a cube (Fig. 3.9) whose walls intersect the k_x, k_y and k_z axes at the points $+\pi/a$ and $-\pi/a$. The second zone in 3-dimensions is made up of pyramids which rest on each

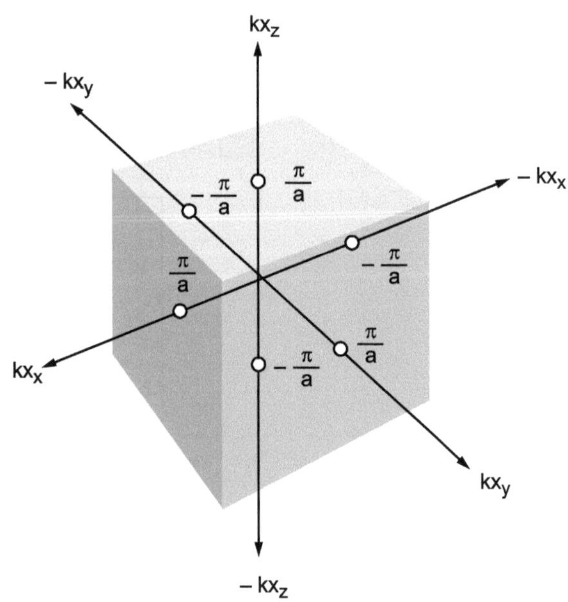

Fig. 3.9 The first Brillouin zone for a cubic lattice. The second zone consists of altitude π/a and square bases 2 π/a wide, resting on the cubic faces of the first zone

face of the first zone cube in a manner similar to the triangles of the second zone in Fig. 3.8.

3.4.4 Brillouin Zones for BCC, FCC and HCP Lattices

In a similar manner, the first and second Brillouin zones for the BCC, FCC and HCP lattices can be calculated to obtain the polyhedra shown in Fig. 3.10. The equations which define the boundaries of these zones are all based on the Bragg equation for the reflection of a wave by a periodic lattice. Thus, the Brillouin zones in metals are polyhedra whose plane surfaces are parallel to the reflecting planes, which are responsible for X-ray diffraction.

In both BCC and FCC metals, each Brillouin zone holds as many quantum states (without considering electron spin) as there are atoms or primitive unit cells in a crystal. For a crystal of N unit cells, there are N states is the first zone. The quantum rule applies to the filling of quantum states in a Brillouin zone. Electrons occupy the states of lowest energy first. Two electrons of opposite spin occupy each state. If there are N atoms in the crystal, then for mono-valent metals the N/2 quantum states of lowest energy are filled. The first Brillouin zone would therefore be half filled.

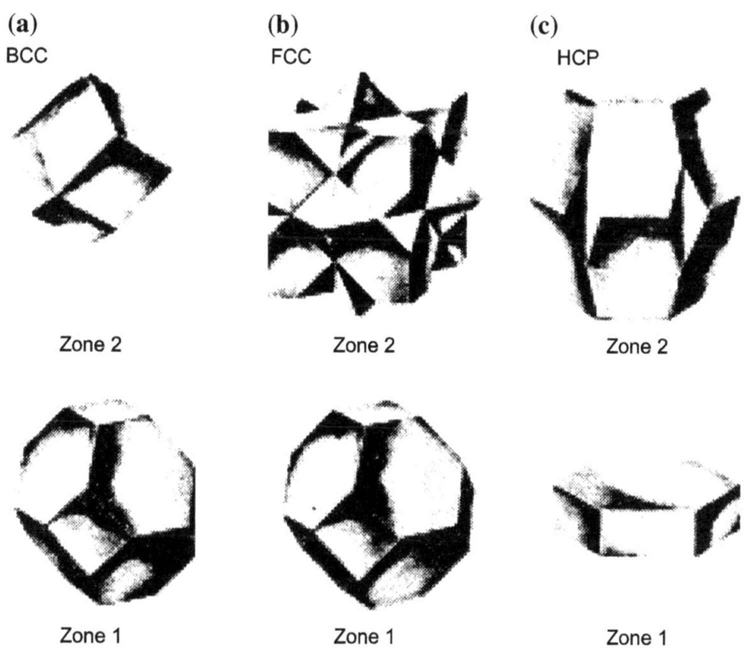

Fig. 3.10 The first two Brillouin zones for the **a** BCC lattice, **b** the FCC, and **c** the HCP lattice

For multi-valent metals the situation becomes more complicated because the zone may overlap. When the zones overlap, it is impossible to fill up one zone without starting to fill the next one.

3.5 Direct and Indirect Energy Band Semiconductors

The E versus k curve shown in Fig. 3.4a represents the wave function of the electron in x-direction only. It is typically true for a single electron travelling through a perfectly periodic lattice. However, since the periodicity of most lattices is different in different directions, the E versus k curve is also different for various crystal directions. For illustration, we cite the examples of GaAs and Si according to which

- the band structure of GaAs for $k = 0$ has a maximum in valence band and a minimum in conduction band. Therefore, an electron making a transition from conduction band to valence band can do so without any change in the value of k. However,
- the band structure for Si, has a maximum in valence band at a different value of k than a minimum in conduction band. Therefore, a transition from conduction band to valence band requires some change in the value of k.

Accordingly, the energy bands in semiconductors have two different classes viz.

(i) direct energy band, and
(ii) indirect energy band

Thus the semiconductors falling in above categories are also known as

1. Direct semiconductors, and
2. Indirect semiconductors

Different semiconducting materials falling in each of these categories is listed below.

Direct energy band semiconductors		**Indirect energy band semiconductors**	
GaAs	ZnS	Si	GaP
GaN	ZnSe	Ge	PbS
GaSb	ZnTe	SiC (a)	PbSe
InP	CdS	AlP	PbTe
InAs	CdSe	AlAs	
InSb	CdTe	AlSb	

3.5.1 *Differences Between Direct and Indirect Semiconductors*

These are given as below.

S.No.	*Description*	Direct semiconductor	Indirect semiconductor
1.	When an electron in conduction band falls to an empty state in valence band	The energy difference E_g is given-off as a photon of light as shown in Fig. 3.11	The electron cannot fall directly to valence band. However, it undergoes a change in its • Momentum, and • Energy
2.	During transition	There is no change in the value of k ($k = 0$) as shown in Fig. 3.11	There is a change in the value of k as shown in Fig. 3.12. Also the energy is given up as heat to the lattice
3.	Suitability of applications	Semiconductor light emitters and lasers for materials capable of *direct* band-to-band transitions	Semiconductor light emitters and lasers for materials capable of *vertical* transitions between defect states

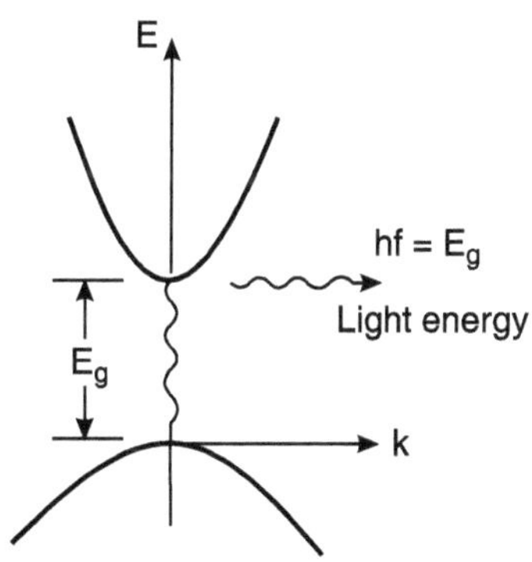

Fig. 3.11 Direct electron transition accompanied with photon emission

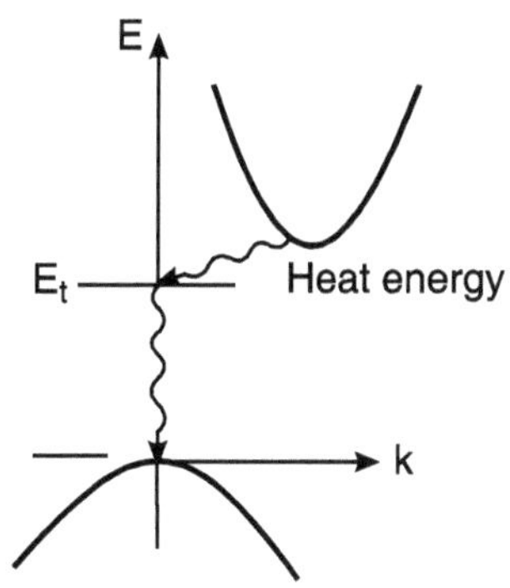

Fig. 3.12 Indirect electron transition via a defect level, accompanied with heat emission and a change in the value of k

3.5.2 *Variation of E_g with Alloy Composition*

Often we require a change in the value of E_g for certain semiconductor, to make it suitable for a particular application. For that the composition of semiconductor has to be varied. It is done by alloying the semiconductor in varying proportion. In the same way the band structure of ternary and quaternary alloys is also altered by varying their composition. Its mechanism is explained through Fig. 3.13a–c, which illustrate the band structures of binary compound GaAs and AlAs, and effect of changing the fraction x in ternary compound $Al_xGa_{1-x}As$.

Figure 3.13a shows that for GaAs the direct minimum is lower than the indirect minima, but this is just opposite (see Fig. 3.13b) for AlAs. Figure 2.9c depicts the variation of Al fraction x in conduction bands of AlGaAs. The range of x for compositions GaAs (x = 0) to AlAs (x = 1) shows the smallest band gap E_g at x = 0. 38. Thus the ternary AlGaAs is

- direct semiconductor up to x = 0.38, and
- indirect semiconductor for x > 0.38

3.5.3 *Effect of Alloying on $GaAs_{1-x}P_x$*

In another illustration, the variation in energy band for a ternary alloy $GaAs_{1-x}P_x$ as a function of alloy composition is shown in Fig. 3.14. It is illustrated that the GaAsP behaves as

- direct semiconductor up to x = 0.45, and
- indirect semiconductor for x > 0.45.

The ternary alloy is a direct semiconductor due to GaAs and is indirect semiconductor due to composition of GaP. It is about $GaAs_{0.55}P_{0.45}$.

Applications. As the alloy composition decides whether the ternary compound will be a direct semiconductor or indirect, the percentage composition becomes the criteria for selection of materials. For example, since the light emission is most effective for direct materials in which electrons drop from conduction band to valence band without changing k and momentum, hence

- LEDs are made of $GaAs_{1-x}P_x$ having x < 0.45, and
- Red LEDs are made for x ≈ 0.4.

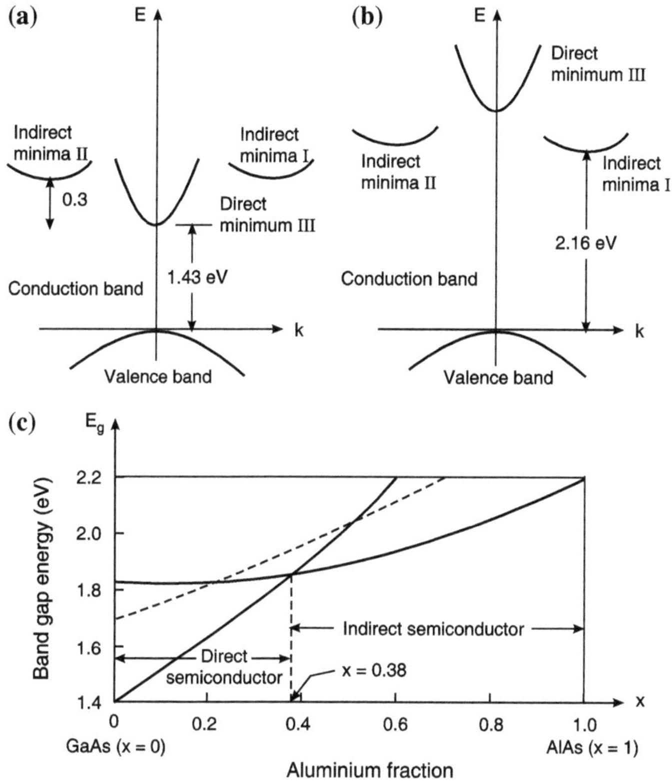

Fig. 3.13 Illustration of variation in E_g by changing alloy composition **a** GaAs shows direct minimum is lower than indirect minima, **b** AlAs band diagram, and **c** effect of variation of Al fraction x in conduction band of ternary alloy $Al_xGa_{1-x}As$

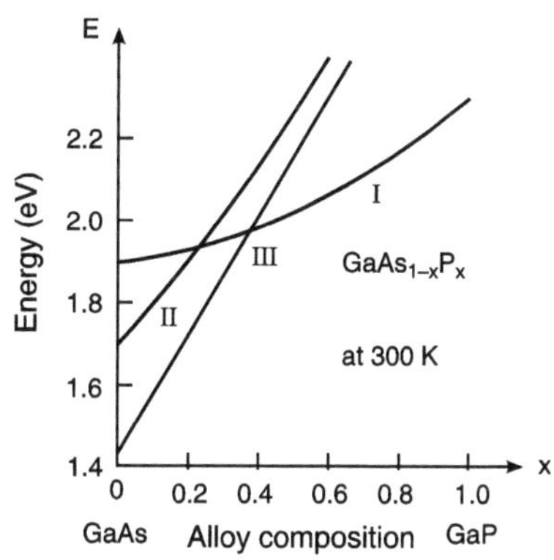

Fig. 3.14 Conduction band energy of ternary alloy $GaAs_{1-x}P_x$ as a function of variation in fraction x of alloy composition

Table 3.2 Number of free electrons in some elements

Metal	Number of free electrons (per m^3)	Metal	Number of free electrons (per m^3)
Al	1.67×10^{29}	Au	5.65×10^{28}
Cu	6.30×10^{28}	Zn	3.0×10^{28}
Ag	6.10×10^{28}	Si	2.0×10^{29}

3.5.4 Charge Carriers in Semiconductors

Semiconduction in semiconductors occur by charge carriers. Following two types of charge carriers take part in semiconduction.

1. Electron (charge) carrier i.e. negative (or n-type) carrier, and
2. Hole (charge) carrier i.e. positive (or p-type) carrier

The electrons lying in valence band, conduct when they receive enough thermal energy to be excited across the band gap. Thus the electrons occupy empty conduction band and conduct. An empty state in valence band is known as a hole. When the conduction band electron and the hole are created by excitation of a valence band electron to the conduction band, they are referred to as electron-hole pair (EHP). Their behaviour in intrinsic and extrinsic (n and p types) semiconductors is different.

3.5.5 Fermi Energy Level

Definition. The Fermi energy level E_F, corresponds to the highest filled energy level at 0 K. The free electrons occupy energies up to E_F at 0 K. When the temperature goes up above 0 K, some lower energy electrons at energy $E(E < E_F)$ move up to higher energy levels $E > E_F$. This motion of electrons under an applied electric field occurs only when the electrons are located in partially filled energy bands. Fermi energy level of an element can be determined by

$$E_F = \left(\frac{3n}{2\rho_E}\right)^{2/3} \tag{3.8}$$

where n is number of free electrons per unit volume of a metal, and ρE is an energy density constant whose value is $6.82 \times 10^{27}/m^3\ (eV)^{3/2}$. Number of free electrons in some elements are given in Table 3.2.

3.5.6 *Fermi-Dirac Probability Distribution*

The thermal behaviour of electrons in an atom can be explained by Fermi-Dirac probability function p(E) given by

$$p(E) = \frac{1}{1 + e^{\frac{(E-E_F)}{kT}}} \tag{3.9}$$

where k is Boltzmann's constant, and T the absolute temperature. The Fermi-Dirac distribution is shown in Fig. 3.15. It illustrates that p(E) = 0 or 1, at T = 0 K. Variation in p(E) with increasing temperature is also shown. At higher temperatures, more and more electrons occupy energy greater than Fermi energy. The kinetic energy of electrons in a metal having maximum energy is equal to its Fermi level. Only some electrons may attain this level.

Alternate definition of Fermi energy. The Fermi energy level of some elements is given in Table 3.3. The value of p(E) is 0.5 at cross-over point. This point is invariant at all temperatures. Thus the Fermi energy level can also be defined as the level at which probability of occupation by an electron is 50 %. For $E > E_F + kT$, in Eq. 3.9, the exponential term in denominator becomes too large than unity. Therefore, Eq. 3.9 may be rewritten as

$$p(E) = e^{-\frac{(E-E_F)}{kT}} \tag{3.10}$$

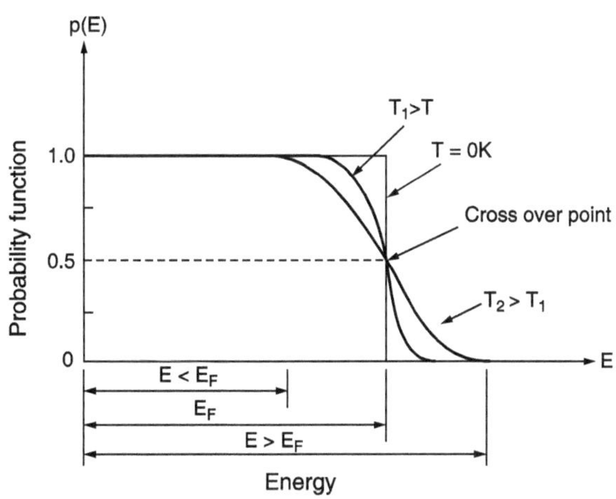

Fig. 3.15 The Fermi-Dirac distribution of free electrons as a function of temperatures

Table 3.3 Fermi energy level of some elements

Element	Be	Al	Zn	Cu	Au	Ag	Cs	K	Na	Li
Fermi energy E_F (eV)	12	11.8	11	7.04	5.51	5.51	1.6	2.1	3.2	4.7

Example 3.1 There is 5 % probability for an electron to occupy an energy state which is 0.4 eV above the Fermi energy. Estimate the temperature at which this can happen.

Solution. Using Eq. 3.9 for Fermi-Dirac probability distribution

$$p(E) = \frac{1}{1 + e^{\frac{(E - E_F)}{kT}}} \tag{i}$$

in which p(E) = 5 % = 0.05, E = E_F + 0.4, and T is to be determined. On putting the value of Boltzmann constant k = 8.614 × 10^{-5} eV/K in Eq. (i), we find

$$0.05 = \frac{1}{1 + e^{\frac{0.4}{8.614 \times 10^{-5} T}}}$$

$$\therefore \quad 1 + e^{\frac{0.0464}{(T \times 10^{-5})}} = 20$$

or

$$e^{\frac{0.0464}{(T \times 10^{-5})}} = 19$$

or

$$\frac{0.0464}{T} \times 10^{-5} = \ln 19$$

$$= 2.9444$$

$$T = \frac{0.0464 \times 10^{-5}}{2.9444}$$

$$= \mathbf{1575.8}\,\mathrm{K}$$

3.6 Intrinsic Semiconductors

Reason of unsuitability of intrinsic semiconductor for use. We noticed that the conduction of electrons from valence band to conduction band needs crossing-over the forbidden gap. The applied electric field required for this conduction is extremely high in intrinsic semiconductors. For example, the forbidden gap in germanium is 0.7 eV and in silicon 1.1 eV. The distance between locations of electrons near an ion-core and away from it is about 1 Å. Thus the semiconduction of an electron from valence band to conduction band will take place when a field gradient of about 0.7 V/1 Å in Ge and 1.1 V/1 Å in Si is applied. This comes out to be an impracticable value of about 0.7 × 10^{10} V/m and 1.1 × 10^{10} V/m respectively.

On the other hand, thermal energy at room temperature can excite a limited number of electrons across the energy gap. The number of electrons crossing-over the forbidden gap may be calculated by Fermi-Dirac probability distribution,

$$p(E) = \frac{1}{1 + e^{\frac{(E-E_F)}{kT}}} \tag{3.11}$$

3.6.1 Energy Diagram

The energy diagram for an intrinsic semiconductor is shown in Fig. 3.16. It shows that the Fermi level E_F lies in the middle of forbidden energy gap E_g. This Fermi level is also referred to as intrinsic energy level E_i. The energy gap E_g is equal to the difference in the energies of valence band E_v and conduction band E_c. The term (E − E_F) in Eq. 3.11 is therefore equal to $E_g/2$. For pure germanium, (E − E_F) is $E_g/2 = 0.35$ eV, which is (13.5) times higher than kT at room temperature (=0.026 eV). Therefore, as compared to the exponential term in Eq. 3.11, the value unity (i.e. 1) is neglected, and Eq. 3.11 can be rewritten as

$$p(E) = \frac{1}{1 + e^{\frac{(E-E_F)}{kT}}} \tag{3.12}$$

Number of electrons crossing-over to conduction band. If the total number of electrons available in valence band for semiconduction is N, and out of this only n numbers cross-over to conduction band, then n/N = p(E). Thus Eq. 3.12 modifies to

$$n = Ne^{-\left(\frac{E_g}{2kT}\right)} \tag{3.13}$$

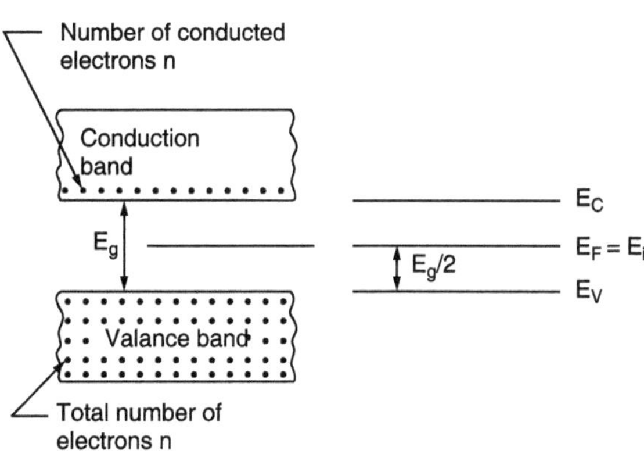

Fig. 3.16 Energy diagram of an intrinsic semiconductor showing energy levels of conduction band, valence band, energy gap and intrinsic energy

3.6.2 Holes, Mobility and Conductivity

If n number of electrons cross the gap, n sites become vacant in the valence band. These vacant sites are called holes. Thus the number of electrons n_e and number of holes n_h are equal ($n_e = n_h$). Both the electrons and the holes take part in semi-conduction. Electrons conduct in the conduction band and the holes in the valence band. They move in opposite directions with certain drift velocity under an applied field gradient ζ. This movement of electrons and holes is known as mobility. Mobility of an electron and of a hole is designated by μ_e and μ_h respectively, and is defined as

$$\mu = \frac{v_d}{\zeta} \tag{3.14}$$

Effect of temperature. Mobility is a temperature sensitive property, and decreases with increasing temperature. It varies with temperature as $\mu_e \propto T^{-2.4}$ and $\mu_h \propto T^{-2.2}$. Its value is more for electrons than for holes ($\mu_e > \mu_h$). Since both these charge carriers (electrons and holes) contribute towards conduction, hence conductivity σ of an intrinsic semiconductor is obtained from

$$\sigma = n_e e_e \mu_e + n_h e_h \mu_h \tag{3.15}$$

where $e_e = e_h = e$ is electronic charge, n_e and n_h are carrier concentrations per unit volume of electrons and holes respectively. Intrinsic carrier concentration of an electron or a hole in germanium at 300 K is $2.5 \times 10^{19}/m^3$.

Example 3.2 Determine the intrinsic carrier density of pure silicon whose resistivity at room temperature is 3000 Ω m. The mobilities of electrons and holes in silicon at room temperature are 0.14 and 0.05 m^2/V-s. Electron charge $e = 1.602 \times 10^{-9}$ C.

Solution. We know that the resistivity $\rho = 1/\rho$ and $n_e = n_h = n$ in an intrinsic semiconductor.Hence using Eq. 3.15, in which

$$\mu_e = 0.14, \mu_h = 0.05, \textit{and}\, e_e = e_h = e$$

we get

$$\frac{1}{3000} = n \times 1.602 \times 10^{-19} \times (0.14 + 0.05)$$
$$n = 1.095 \times 10^{16}/\text{m}^3$$

Example 3.3 The resistivity of Ga-Sb at 300 K is found to be 2×10^{-3} Ω-m. The electron and hole mobilities are 0.3 and 0.1 m^2/V-s respectively. Calculate the carrier density. Given $e = 1.6 \times 10^{-19}$ Coulomb.

Solution. Since

$$Conductivity = \frac{1}{Resistivity\ \rho} = \frac{1}{2 \times 10^{-3}} = 500/\Omega - \mathrm{m}$$

As

$$\sigma = n_e e_e \mu_e + n_h e_h \mu_h$$

where $e_e = e_h = 1.6 \times 10^{-19}$ C, μ_e and $\mu_h = 0.3$ and 0.1 m^2/V-s resistivity, and taking $n_e = n_h$, we get

$$500 = n_e(\mu_e + \mu_h)$$

Or

$$n \times 1.6 \times 10^{-19}(0.3 + 0.1) = 500$$
$$n = 7.8125 \times 10^{21}/\mathrm{m}^3$$

Example 3.4 The mobilities of silicon are μ_e=$0.17\,\mathrm{m}^2/\mathrm{V{-}s}$ and $\mu_h = 0.035\,\mathrm{m}^2/\mathrm{V{-}s}$ at room temperature. If the carrier density in the material is known to be $1.1 \times 10^{16}/m^3$, calculate the resistivity of silicon.

Solution. Given are: $e_e = e_h = 1.6 \times 10^{-19}$ C

$$\sigma = n_e e_e \mu_e + n_h e_h \mu_h \qquad \text{(i)}$$

$$n_e = n_h = 1.1 \times 10^{16}/\mathrm{m}^3 \quad \text{(for intrinsic semiconductor)}$$

and μ_e and μ_h are 0.17 and 0.035 m^2/V-s respectivelyTherefore, on substituting the given values in Eq. (i), we get

$$\sigma = n_e e_e(\mu_e + n_h)$$
$$= \left(1.1 + 10^{16}\right) \times \left(1.6 + 10^{19}\right)(0.17 + 0.035)$$

Hence,

$$resistivity\ \rho = \frac{1}{Conductivity\ \sigma} = \frac{1}{3.608 \times 10^{-9}} = \mathbf{2{,}771.618}\,\Omega{-}\mathrm{m}$$

3.7 Extrinsic Semiconductors

In the last article we discussed that conduction in intrinsic semiconductor is due to inherent crystal properties. Contrary to this, conduction in extrinsic semiconductors occurs due to the presence of foreign impurities. Deliberate impuring of intrinsic semiconductors is done by adding impurities to obtain extrinsic semiconductors. The impuring agents are called dopants, and the process is known as doping. The dopants are the elements of third column and fifth column of the periodic table. They increase the conductivity of semiconductors considerably.

Based on the type of dopants used, extrinsic semiconductors are classified as

(i) n-type, and
(ii) p-type.

- The n-type semiconductor is also called negative or electron carrier type, and p-type as positive or hole carrier type.

3.7.1 *n-Type Semiconductors and Their Energy Diagram*

In this case, the fifth column elements such as As, Sb or P are used as dopants. Figure 3.17a shows a germanium crystal doped with phosphorus. Four of the five electrons in the outermost orbit of phosphorus atom form a tetrahedral bond with four germanium neighbours. The fifth electron remains loosely bound to its parent atom, and moves in the electric field of germanium crystal with an electron orbit of large radius. The energy of fifth electron is close to the conduction band, and is

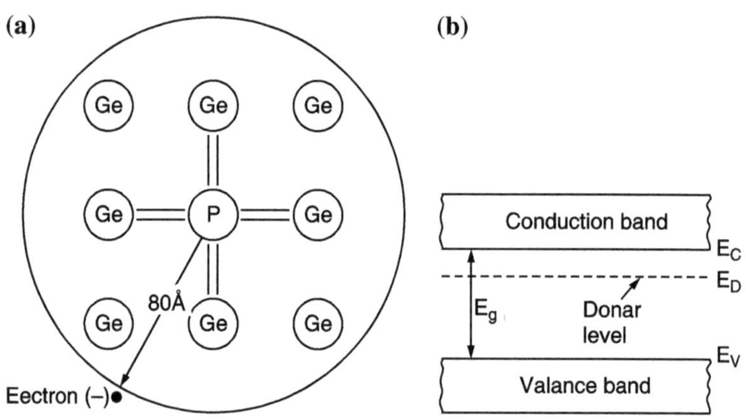

Fig. 3.17 A n-type semiconductor showing **a** germanium doped with fifth column phosphorus, and **b** energy diagram showing donor energy level

shown in Fig. 3.17b. This energy level is known as donor energy level E_D as the dopant has donated one of its electrons to the semiconductor.

Law of mass action. Now the cross-over of fifth electron into the conduction band is much easier than it would have been from the valence band. The energy needed to excite the fifth electron into the conduction band is known as ionization energy E_I, and is given in Table 3.4. The number of donor level electrons excited into the conduction band at room temperature is large, as $E_I \ll E_g$. The number of electrons n_e in the conduction band are more than the number of holes n_h in the valence band (i.e. $n_e > n_h$). The electrons remain the majority charge carriers. However, the law of mass action ($n_e \times n_h$ = constant) holds good.

3.7.2 p-Type Semiconductors and Their Energy Diagram

An intrinsic crystal can also be doped by elements of third column such as Ga, In or Al. Doping of a silicon crystal by an aluminium atom is illustrated in Fig. 3.18a.

Table 3.4 Ionization energies of dopants in intrinsic semiconductors

Type	Dopant	Ionization energy (eV)	
		Silicon	Germanium
n	Sb	0.039	0.010
	P	0.044	0.012
	As	0.049	0.013
p	B	0.045	0.010
	Al	0.057	0.010
	Ga	0.065	0.011
	In	0.160	0.011

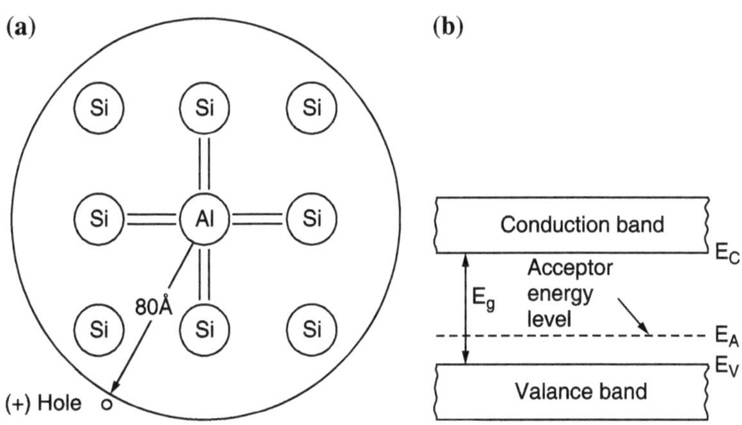

Fig. 3.18 A p-type semiconductor showing **a** silicon doped by third column aluminium, and **b** energy diagram showing acceptor energy level

Aluminium requires a fourth electron in addition to three electrons in its outer orbit to form a tetrahedral bond. The fourth electron comes from the neighbouring silicon atom. Migration of electron from silicon atom creates a hole in it. The hole thus created (in silicon) moves around aluminium atom with a positive charge.

Acceptor energy level. Calculations reveal that the holes orbit at a radius of about 80 Å, and are close to the valence band. The energy level possessed by them is known as acceptor energy level E_A, and is shown in Fig. 3.18b. The ionization energy needed to excite the holes into the valence band is given in Table 3.4. The holes are majority charge carriers. In this case ($n_h > n_e$), and the law of mass action holds goods here also.

- The conductivity of extrinsic semiconductors, both n-type and p-type, may be found from Eq. 3.15 knowing that $n_e \neq n_h$.

Example 3.5 A current of density 1000 A/m^2 flows through a n-type germanium crystal. This crystal has a resistivity of 0.05 Ω m and the electron mobility of 0.4 m^2/Vs. Calculate the drift velocity and time taken by the electrons to travel 100 μm in the crystal.

Solution. The given data are

$$\mathrm{I_d} = 1000\,\mathrm{A/m^2}, \rho = 0.05\,\Omega\,\mathrm{m}$$
$$\text{travel length} \quad 1 = 100\,\mu\mathrm{m} = 100 \times 10^{-6}\mathrm{m}$$
$$\text{and} \quad \mu_\mathrm{e} = 0.4\,\mathrm{m^2/Vs}.$$

Let t be the time taken by electrons to travel a distance of l. For a n-type semiconductor we know that $n_e \gg n$. Therefore neglecting the term $n_h\, e_h\, \mu_h$, Eq. 3.15 may be written as

$$\sigma = n_e e \mu_e$$
$$\frac{1}{0.05} = n_e \times 1.602 \times 10^{-19} \times 0.4$$

or

$$n_e = \frac{1}{0.05 \times 1.602 \times 10^{-19} \times 0.4} = 3.13 \times 10^{20}/\mathrm{m}^3$$

we also know that the drift current I_d and drift velocity v_d are related as

$$I_d = nev_d$$
$$\therefore \quad v_d = \frac{I_d}{ne} = \frac{1000}{3.13 \times 10^{20} \times 1.602 \times 10^{-19}} = \mathbf{20\,m/s}$$

The distance l travelled in time t at drift velocity v_d, by an electron may be obtained from

$$
\begin{aligned}
l &= tv_d \\
\therefore \qquad t &= \frac{l}{v_d} = \frac{100 \times 10^{-6}}{20} \\
&= 5 \times 10^{-6}\,\text{s} = \mathbf{5\,\mu s}
\end{aligned}
$$

Example 3.6 A p-type germanium rod, 10 mm long and 1 mm in diameter, has a resistance of 100 Ω. Determine the impurity concentration in the rod. The mobility of holes in germanium = 0.19 m^2/Vs.

Solution. The area of cross-section of rod is found as

$$
a = \frac{\pi d^2}{4} = \frac{\pi (1 \times 10^{-3})^2}{4} = 7.85 \times 10^{-7}\,\text{m}^2
$$

The resistivity of the rod is obtained from

$$
\begin{aligned}
\rho &= \frac{aR}{l} = \frac{7.85 \times 10^{-7} \times 100}{10 \times 10^{-3}} \qquad \text{(i)} \\
&= 7.85 \times 10^{-3}\,\Omega\,\text{m}
\end{aligned}
$$

In a p-type semiconductor, the holes are majority carriers as $n_h \gg n_e$. Therefore the term $n_e.e.\mu_e$ can be neglected from Eq. 3.15. Knowing that the resistivity $\rho = \frac{1}{\sigma}$, the Eq. (i) can be written as

$$
\begin{aligned}
\rho &= \frac{1}{\sigma} = \frac{1}{n_h \,.\, e \,.\, \mu_h} \\
\therefore \qquad 7.85 \times 10^{-3} &= \frac{1}{n_h \times 1.602 \times 10^{-19} \times 0.19} \\
\therefore \qquad n_h &= \frac{1}{7.85 \times 10^{-3} \times 1.602 \times 10^{-19} \times 0.19} \\
&= 4.18 \times 10^{21}/\text{m}^3
\end{aligned}
$$

Intrinsic concentration is neglected, and assuming that each impurity atom produces one electron in the rod, the impurity concentration may be approximated as

$$
N_h = n_h = 4.18 \times \frac{10^{21}}{\text{m}^3}
$$

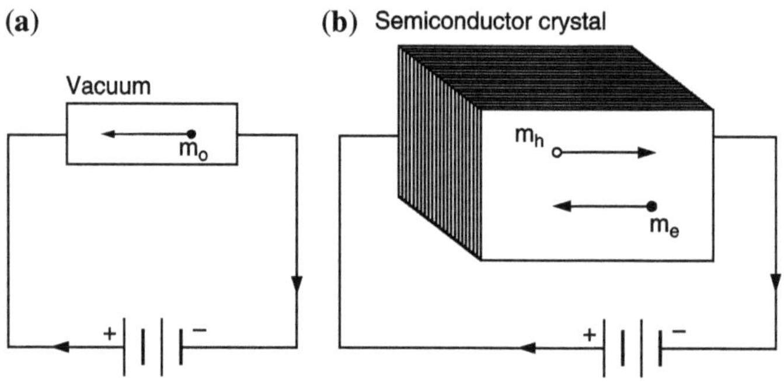

Fig. 3.19 **a** Rest mass of electron carrier in vacuum, and **b** effective mass of electron and hole carriers in a semiconductor crystal

Table 3.5 Rest mass m_0 and effective mass m*

Intrinsic semiconductor	Effective mass of electron m_e*	Effective mass of hole m_h*
Silicon	0.99 m_0	0.59 m_0
Germanium	0.55 m_0	0.37 m_0

3.8 Effective Mass

Rest mass. The mass of an electron and a hole is one of the basic properties in design of semiconductor devices. The mass of these carriers in vacuum is called the rest mass m_0, Fig. 3.19a. In real applications, the electrons or hole carriers move within a solid and not in vacuum. While moving, they collide with the atoms of semiconductor, which causes their periodic deceleration. These carriers are also subjected to a complex crystalline field in addition to the applied field. Hence, there mass in movement is different from their mass at rest, and is known as effective mass m*. The effective mass of an electron and a hole is denoted by m_e* and m_h* respectively, Fig. 3.19b. This mass is taken into consideration during design calculations for semiconductor devices.

The density of state effective mass and the rest mass for intrinsic semiconductors is shown in Table 3.5.

3.9 Carrier Concentrations

Design of semiconductor devices requires calculation of electronic properties of materials. The electronic properties generally required are

- Fermi energy E_F
- Probability distribution function p(E)

- Density of state N(E)dE in the energy range dE
- Concentration of electrons ne and hole nh in conduction band
- Effective density of states N_c and N_v in conduction and valence band, etc.

Study of carrier concentration is necessary because it is essential to know the number of charge carriers per cm^3 in semiconducting material. Consideration of majority carrier concentration is generally required for this purpose.

Amongst the above properties, we have already studied the Fermi energy and Fermi-Dirac probability distribution in art. 3.6.1 and 3.6.2. The other properties will be studied now.

3.9.1 Density of State, and Electron and Hole Concentration at Equilibrium

Meaning. Density of state means the population density of electrons in a metal. It has relevance to Fermi-Dirac distribution. In the last article we discussed that the Fermi probability function p(E) determines the probability of energy level E occupied by an electron. It tells us about the energy level but not about the number of electrons in those levels. The density of state N(E) indicates the number of electrons across the energy band. This number is not uniform across the energy band, rather it is greatest at the centre of the band. The product of p(E), N(E) and the number of electrons ne for metals are related by

$$n_e = \int_0^{E_F} p(E)N(E)dE \tag{3.16}$$

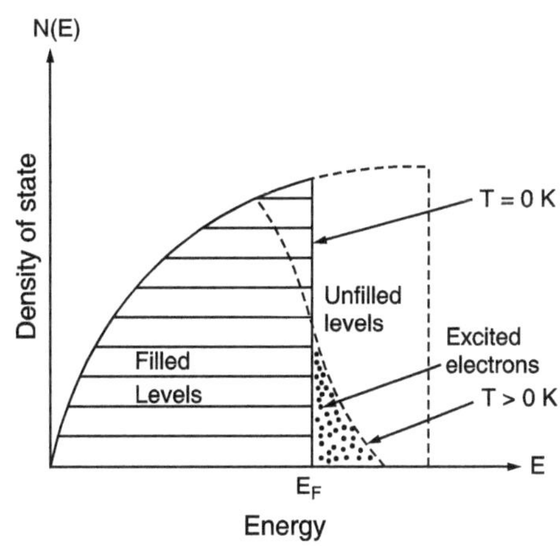

Fig. 3.20 Illustration of density of state showing a small fraction of excited electrons possessing $E > E_F$ at $T > 0$ K

Effective density of state. This relation is shown in Fig. 3.20 over a range of band energy at 0 K and temperature above 0 K. It illustrates that only a small fraction of electrons within the energy range of kT can be excited above Fermi level. Here k is Boltzmann constant and T is absolute temperature. The effective density of energy states can be found by employing quantum mechanics. If the effective density of states at the conduction and the valence bands are N_C and N_V respectively, then

$$N_C = 2\left(\frac{2\pi m_e^* kT}{h^2}\right)^{\frac{3}{2}} \quad (3.17a)$$

$$N_V = 2\left(\frac{2\pi m_h^* kT}{h^2}\right)^{\frac{3}{2}} \quad (3.17b)$$

where m_e* and m_h* are the effective mass of an electron and a hole respectively, and h is the Planck's constant.

The number of negative and positive charge carriers in their respective bands may be found from

$$N_e = Nc^{e^{\frac{|E_C - E_F|}{kT}}} \quad (3.18a)$$

$$N_e = Nv^{e^{\frac{|E_F - E_V|}{kT}}} \quad (3.18b)$$

where E_F is the Fermi energy. Now the product of positive and negative charge carriers is

$$n_e \times n_h = n_i^2 \quad (3.19)$$

and the term $\sqrt{N_C N_V}$ depends on the band structure of the semiconductor. For a specific material, the product

$$\sqrt{N_C N_V} = constant \quad (3.20)$$

Example 3.7 Calculate the conduction electron and hole densities at room temperature in Si-doped with 10^{19} donors/m^3, if the intrinsic carrier concentration is 1.5×10^{16}/m^3.

Solution. Using product of mass relation between the number of conduction electrons n and number of holes p as

$$np = n_i^2$$

we write

$$p = \frac{n_i^2}{n} = \frac{(1.5 \times 10^{16})^2}{10^{19}} = 2.25 \times 10^{13}/\text{m}^3$$

Example 3.8 A silicon sample is doped with 10^{17} arsenic atoms per cubic centimeter. Calculate the hole concentration at 300 K. What will be the location of Fermi energy level relative to intrinsic energy level? Take intrinsic carrier concentration as $1.5 \times 10^{10}/\text{cm}^3$ at room temperature.

Solution. Given data are: total number of donor atoms $N_D = 10^{17}/\text{cm}^3$ and $n_i = 1.5 \times 10^{10}/\text{cm}^3$ Since $N_D \gg n_i$, we can approximate

$$n_e = N_D$$

Using Eq. 3.19, the hole concentration n_h is obtained as

$$n_h = \frac{n_i^2}{n_e} = \frac{(1.5 \times 10^{10})^2}{10^{17}}$$
$$= \frac{2.25 \times 10^{20}}{10^{17}} = \mathbf{2.25 \times 10^3}/cm^3$$

The relative locations of E_F and E_i are found from

$$E_F - E_i = kT \ln\left(\frac{n_e}{n_i}\right) \tag{i}$$

Since the value of kT at room temperature (300 K) = 0.0259 eV, therefore Eq. (i) becomes

$$E_F - E_i = 0.0259 \ln\left(\frac{10^{17}}{1.5 \times 10^{10}}\right)$$
$$= \mathbf{0.407}\, eV$$

Thus the E_F is located at 0.407 eV away from E_i, towards E_c side as given below.

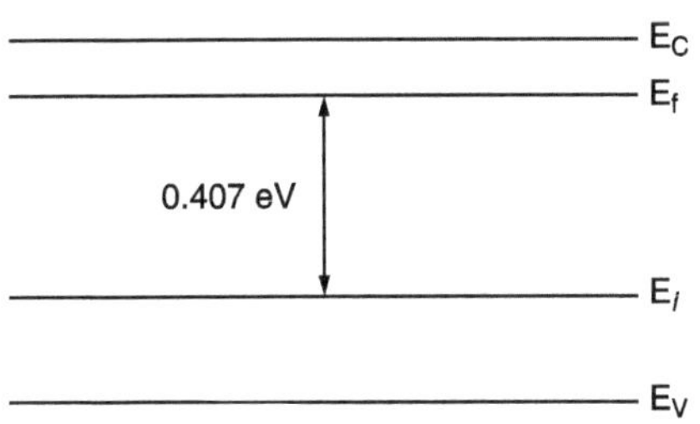

3.10 Temperature Dependency of Carrier Concentrations

The carrier concentration is a function of temperature. It is evident from the following relations.

$$n_o = n_i e^{\frac{E_F - E_i}{kT}} \tag{3.21a}$$

$$p_o = n_i e^{\frac{E_i - E_F}{kT}} \tag{3.21b}$$

Here n_o and p_o are electron and hole concentration respectively in conduction band. Figure 3.21 depicts the variation of electron concentration as a function of reciprocal of temperature for a n-type extrinsic Si. It has a donor concentration of $N_d = 10^{15}/cm^3$. The curve shows three distinct regions.

1 Ionization region. It occurs at large values of 1000/T i.e. at low temperatures where donor electrons are bound to donor atoms. At about 1000/T = 10 i.e. 100 K, all donor atoms are ionized and $n_o \approx N_d = 10^{15}/cm^3$.
2 Extrinsic region. It occurs when every available extrinsic electron is transferred to the conduction band, hence $n_i \approx N_d$ and n_o becomes almost constant.
3 Intrinsic region. It occurs at smaller values of 1000/T i.e. at higher temperatures when $n_i \gg N_d$.

Amongst the above three regions, the extrinsic region is desirable for the operation of semiconductor devices. It is due to its constant characteristic. The extrinsic range can also be extended beyond the highest temperature at which the device is to operate. This is accomplished by either

- generating the thermal electron-hole pair, or
- doping

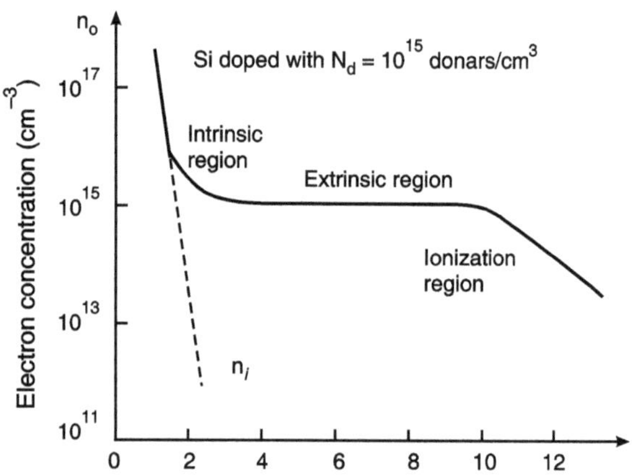

Fig. 3.21 Depiction of variation in intrinsic carrier concentration on a semi-log *curve*, as a function of reciprocal of temperature

In most semiconductor devices, the control of the carrier concentration by doping is preferred.

Example 3.9 If the conductivity of an intrinsic semiconductor is to be increased by 30 %, to what temperature should it be heated above the room temperature?

Solution. For an intrinsic semiconductor, the conductivity is given by

$$\sigma = \sigma_0 exp\left(\frac{-E_g}{2kT}\right) \tag{i}$$

For increase in conductivity by 30 %, let it is heated to a temperature of T + ΔT, where T is room temperature = 300 K

∴ From Eq. (i)

$$1.3 = 1 \times \frac{exp\left(\frac{-E_g}{2k(T+\Delta T)}\right)}{\exp\left(-\frac{E_g}{2kT}\right)}$$

or

$$1.3 = exp\left[\frac{-E_g}{2k}\left(\frac{1}{T+\Delta T} - \frac{1}{T}\right)\right]$$

or

$$T + \Delta T = \frac{1}{\frac{1}{T} - \frac{2k}{E_g}\log_e 1.3}$$

$$\therefore \quad T + \Delta T = 305.9\,\text{K}$$

It means that a temperature rise of ΔT = 305.9 – 300 = 5.9 K will lead to a rise of 30 % in conductivity.

3.10.1 Temperature Dependency of n_i

Intrinsic carrier concentration n_i has a strong temperature dependence. It can be visualized from the following relations.

$$n_i = \sqrt{N_C N_V}\,.\,e^{-\frac{E_g}{2kT}} \tag{3.22}$$

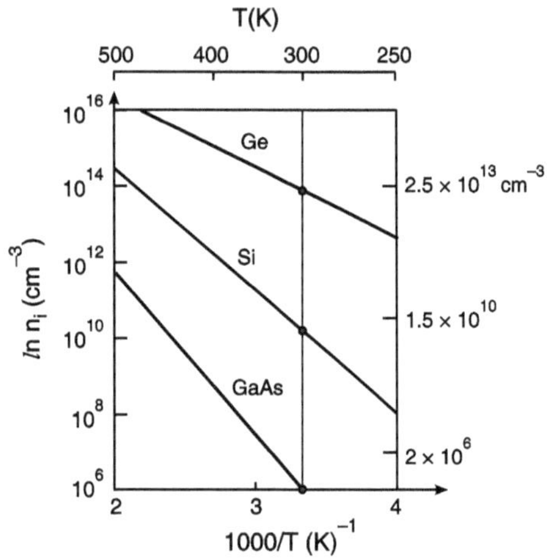

Fig. 3.22 Depiction of dependency of electron concentration as a function of reciprocal of temperature

$$N_C = 2\left(\frac{2\pi m_e^* kT}{h^2}\right)^{\frac{3}{2}}, \quad and \quad N_V = 2\left(\frac{2\pi m_h^* kT}{h^2}\right)^{\frac{3}{2}} \tag{3.22}$$

On combining above equations, we obtain

$$n_i = 2\left(\frac{2\pi kT}{h^2}\right)^{\frac{3}{2}} \cdot \left(m_e^* m_h^*\right)^{\frac{3}{4}} e^{-\frac{E_g}{2kT}} \tag{3.23}$$

It shows an exponential dependence of ni on T. Figure 3.22 depicts this variation on a semi- log curve. The ln n_i versus 1000/T plot. It appears to be linear on the pretext that the effects of $T^{3/2}$ dependency and E_g are neglected.

3.11 Drift of Carriers in Electric and Magnetic Fields

For understanding the mechanism of operation of semiconductor devices and their design, it is essential to understand drift behaviour of carriers in presence of electric and magnetic fields. It basically involves an understanding of collisions of charge carriers with lattice and impurities, and their mobility.

3.11.1 Drift Velocity and Collision Time

The accelerated movement of electrons results in a gain of extra velocity which is above their normal velocity v_n. This gained velocity is called the drift velocity v_d. When accelerated, these electrons collide with obstacles present within the solid. The drift velocity in a real situation is shown in Fig. 3.23a whose idealized

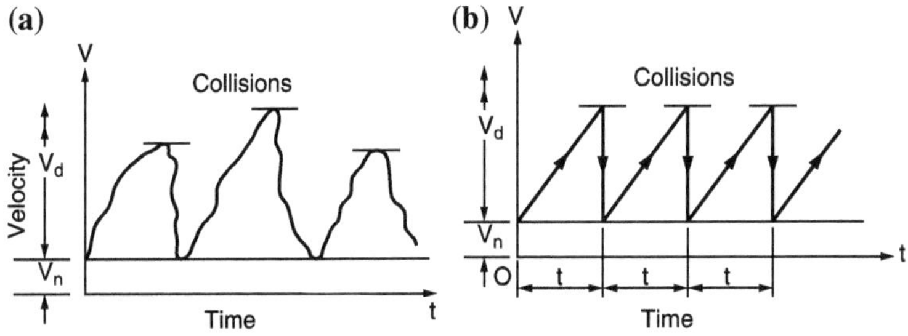

Fig. 3.23 Accelerated movement of electrons under an applied field shows normal and drift velocities v_n and v_d in **a** real situation, and **b** an idealized case

representation is depicted in Fig. 3.23b. If an electron of mass m and charge e undergoes an acceleration a due to the applied field gradient ζ, then

$$\zeta e = ma \tag{3.24}$$

The accelerated electrons collide with obstacles at successive interval of time t called collision time. As $a = \frac{v_d - 0}{t}$, hence Eq. 3.24 may be rewritten as

$$\zeta e = \frac{mv_d}{t} \tag{3.25}$$

or

$$v_d = \frac{\zeta e t}{m} \tag{3.26}$$

Current density. The current density Id is the flux caused due to the flow of n number of free electrons, and is expressed by

$$I_d = nev_d \frac{ne^2 t\zeta}{m} \tag{3.27}$$

Conductivity. The flux per unit potential gradient is called conductivity σ which is determined from

$$\sigma = \frac{ne^2 t}{m} \tag{3.28a}$$

Resistivity. The electrical resistivity ρ is given by

$$\rho = \frac{1}{\sigma} \tag{3.28b}$$

Example 3.10 How much is electric field between a pair of conducting plates, 5 mm apart, when the potential difference between them is 230 V?

Solution. The electric field E between two plates is equal to the potential gradient, which may be obtained as follows.

$$\text{As} \quad E = -\frac{dV}{dx} = -\frac{V}{d}$$

$$\therefore \quad E = -\frac{230\,\text{V}}{5\,\text{mm}} = -\frac{230\,\text{V}}{0.005\,\text{m}} = \mathbf{-4.6 \times 10^4 \frac{V}{m}}$$

Example 3.11 In a certain copper conductor, the current density is 2.4 A/mm^2 and electron density is 5×10^{28} free electrons per m3 of the copper. Determine the drift velocity of the electrons. Given e = 1.6×10^{-19} coulombs.

Solution. Given are: $I_d = 2.4$ A/mm^2 = 2.4×10^6A/m^2, n = 5×10^{28} electron/m^3, and e = 1.6×10^{-19} coulombs.

$$\because \quad \text{I}_\text{d} = \text{nev}_\text{d}$$

$$\therefore \quad v_d = \frac{I_d}{ne}$$

$$= \frac{2.4 \times 10^6}{(5 \times 10^{28}) \times (1.6 \times 10^{-19})} = 3 \times 10^{-4}\frac{\text{m}}{\text{s}} = \frac{0.3\,\text{mm}}{\text{s}}$$

Example 3.12 The carrier concentration in a material of conductivity 0.018/Ω m is 10^{19} electrons/m^3. A voltage of 0.16 volts is applied across the 0.29 mm thick material. Determine the drift velocity of the carriers. Take standard values of mass of electron and electronic charge.

Solution. The given data and the standard values for electron are

$$\text{n} = 10^{19}\text{electrons/m}^3, \quad \text{e} = 1.602 \times 10^{-19}\,\text{C},$$
$$\sigma = 0.018/\Omega\,\text{m}, \quad \text{m} = 9.1 \times 10^{-31}\,\text{kg},$$
$$\text{V} = 0.16\,\text{V}, \quad \text{and} \quad \text{t} = 0.29\,\text{mm}$$

where V is voltage and t is thickness of the material. As the electric field gradient ζ is equal to voltage per unit thickness, therefore

$$\zeta = \frac{\text{V}}{t} = \frac{0.16}{0.29 \times 10^{-3}} = 551.72\frac{\text{V}}{\text{m}}$$

We can write Eq. 3.26 as

$$\frac{v_d}{\zeta} = \frac{et}{m} \tag{i}$$

Substitution of Eq. (i) in Eq. 3.28a, yields

$$\sigma = \frac{nev_d}{\zeta} \tag{ii}$$

Therefore,

$$0.018 = \frac{10^{19} \times 1.602 \times 10^{-19} \times v_d}{551.72} = 0.0029 v_d$$
$$v_d = \frac{0.018}{0.0029} = \mathbf{6.19}\frac{m}{s}$$

3.11.2 Mean Free Path (or Mean Free Length) and Conductivity

The average distance travelled by an electron between two successive collisions is called mean free path l. It is expressed by

$$l = \frac{l_1 + l_2 + \cdots l_P}{p} \tag{3.29}$$

where l_1, l_2, …, l_p are the distances travelled between any two random collisions, and p is the number of collisions. The effects of mean free path on conductivity are:

- More the mean free path in a crystal, more will be its conductivity.
- In an ideal crystal at 0 K, $l \to \infty$ therefore $\sigma \to \infty$,
- But in a real crystal $l \neq \infty$ therefore σ is finite.

It happens because there are no collisions in an ideal crystal as it is free from imperfections, whereas the real crystals have too many imperfections. The effect of high temperature is to decrease the mean free path and conductivity of the material.

3.12 Effects of Temperature on Mobility of Carriers

Mechanism. The mobility of electron and hole carriers are governed by following scattering mechanisms, which themselves result from temperature.

1. Lattice scattering. In this mechanism, a carrier moving through a crystal is scattered by inherent vibration of the lattice, caused by temperature. The frequency of such scattering increases with increase in temperature, therefore the thermal agitation of the lattice also becomes greater. Hence the mobility decreases, as shown in Fig. 3.24.
2. Impurity scattering. In this mechanism, the scattering dominates at low temperatures.

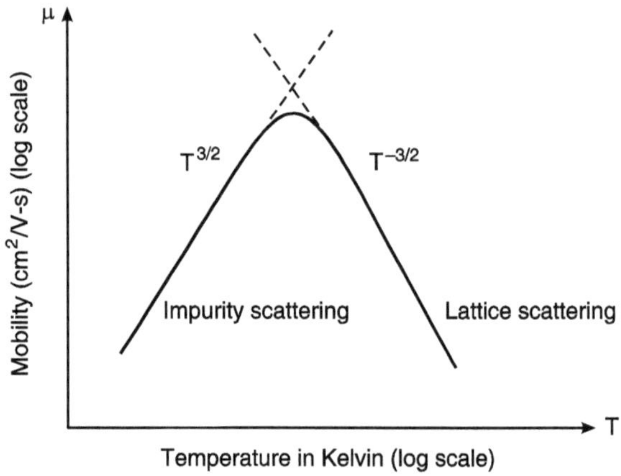

Fig. 3.24 Effect of temperature on mobility of carriers

At lower temperatures, as the atoms are less agitated, therefore the thermal motion of the carriers is slow. Hence, there is an increase in mobility with increasing temperature, Fig. 3.24.

Mobility-temperature relation. The approximate temperature dependencies for both the above mechanisms are as follows.

i. for lattice scattering $T^{-1.5}$, and (3.30a)

ii. for impurity scattering $T^{+1.5}$, and (3.30b)

Since the total mobility of a carrier is the contribution of mobilities due to two or more scattering mechanisms, it is expressed as

$$\frac{1}{\mu_{total}} = \frac{1}{\mu_{scattering1}} + \frac{1}{\mu_{scattering2}} + \cdots. \tag{3.31a}$$

Considering Eqs. (3.30a) and (3.30b), the expression of mobility-temperature relation can be written as

$$\mu = \frac{1}{\frac{1}{a}T^{\frac{3}{2}} + \frac{1}{b}T^{-\frac{3}{2}}} \tag{3.31b}$$

3.12.1 *Effects of Doping on Mobility*

The effect of doping (i.e. impurity concentration) on mobility of carriers is shown in Fig. 3.25. It shows a decrease in mobility of electrons and holes for Si, Ge, and GaAs with increasing doping concentration. The reason of decrease in mobility is attributed to a rise in impurity scattering due to the presence of larger dopants.

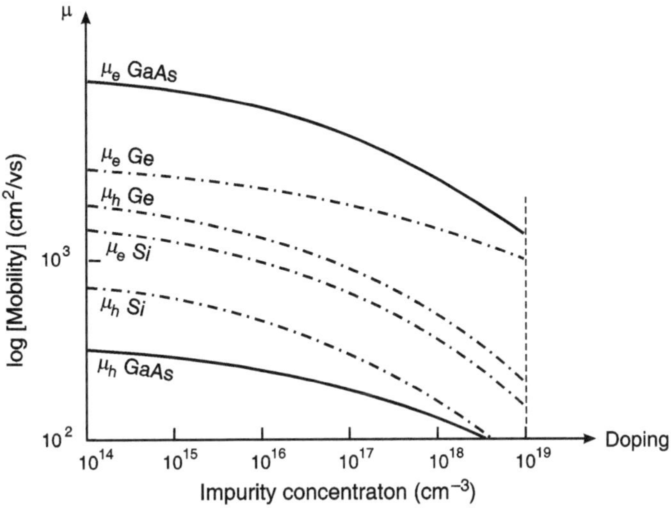

Fig. 3.25 Effect of doping on mobility of Si, Ge and GaAs

3.13 Degenerate Semiconductors

We know that the amount of impurities doped in semiconductor is in a very small fraction. It is generally of the order of $1{:}10^9$ to 10^{12}. Although the ppm[2] addition of dopants has a tremendous effect on conductivity and other properties of semiconductors, but it has almost no effect on the following.

(i) on total atomic density of the material.
(ii) on charge transport within the donor or acceptor levels because of a very few impurity atoms, and that too very widely spaced throughout the material.
(iii) on interactions between the impurities.

In fact, the donor or acceptor energy levels appear to be composed of discrete and non-interacting energy states.

3.13.1 *Effect of Heavy Doping*

If doping in semiconductor is continued uninterrupted, a limit will reach at which the impurities will be closely packed within the material lattice and will interact with each other. In this condition when the concentration is high (say, about 10^{20} donors/cm^3), following occurrences may be noticed within the material.

---

[2] 1 ppm = 0.0001 %.

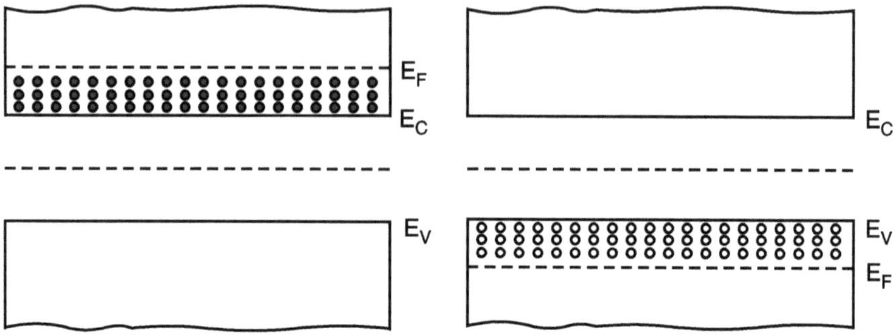

Fig. 3.26 The Fermi level lies **a** within the conduction band in degenerate n-type semiconductor, and **b** within the valence band in degenerate p-type semiconductor

- Some effect on total atomic density of material.
- Charge transport within the donor or acceptor levels.
- Considerable interactions between the impurities.
- Non-discrete (i.e. continuous) donor or acceptor energy levels.

In such condition, the energy levels form a band which may overlap the bottom of conduction or valence band, as the case may be.

3.13.2 Degenerate Types

Depending upon the nature of heavy doping, the semiconductors may be called as either

1. degenerate n-type, or
2. degenerate p-type.

It is known as degenerate n-type when the electron concentration ne in conduction band exceeds the effective density of states $N(E)_{eff}$ i.e. $n_e > N(E)_{eff}$. In this case the Fermi level does not lie within the energy band gap, rather lies within the conduction band as shown in Fig. 3.26a. The semiconductor is known as degenerate p-type when the acceptor concentration is higher than the effective density of states i.e. $n_p > N(E)_{eff}$. In this case the Fermi energy level lies in the valence band as shown in Fig. 3.26b.

3.13.3 Filled and Empty Energy States in Conventional and Degenerate Semiconductors

The filled or empty energy states in conventional and degenerate semiconductors are also different. These are mentioned below.

Energy states	In conventional semiconductors	In degenerate n-type semiconductor	In degenerate p-type semiconductor
Below E_F	Mostly filled	–	–
Above E_F	Mostly empty except for a small distribution	–	–
Region between E_c and E_F	–	For most part, filled with electrons	–
Region between E_v and E_F	–	–	Almost completely filled with holes

Example 3.13 A semiconductor is said to be degenerate when the Fermi level is located at an energy of 3 kT or less from the conduction band, in the case of a donor doped semiconductor. Calculate the minimum doping density for silicon at 300 K to become degenerate if n_i is $1.4 \times 10^{16}/m^3$, Take Boltzmann constant $k = 1.38 \times 10^{-23}$ J/K.

Solution. Using $E_F - E_i = kTln\left(\frac{n}{n_i}\right)$ and approximating $E_i = \left(\frac{E_C+E_V}{2}\right)$, we get

$$E_F - \left(\frac{E_C - E_V}{2}\right) = kTln\left(\frac{n}{n_i}\right)$$

As

$$E_F = E_C - 3kT$$

$$\therefore \quad E_C - 3kT - \left(\frac{E_C - E_V}{2}\right) = kTln\left(\frac{n}{n_i}\right)$$

or

$$\left(\frac{E_C - E_V}{2}\right) - 3kt = kTln\left(\frac{n}{n_i}\right)$$

But the energy gap is $(E_C - E_V)$, therefore

$$= \frac{(1.1 \times 1.6 \times 10^{-19})}{2} - 3 \times 1.38 \times 10^{-23} \times 300$$

$$\rho_{doping} = 1.38 \times 10^{-23} \times 300 \ln\left(\frac{n}{1.4 \times 10^{16}}\right)$$

$$\therefore \quad = \mathbf{1.2 \times 10^{24}/m^3}$$

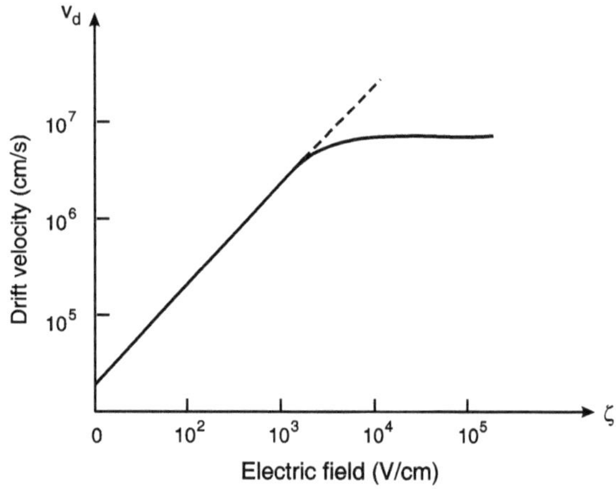

Fig. 3.27 Effect of high-electric field on drift velocity

3.14 High-Field Effects

In deriving the expression for carrier drift process, it is assumed that the drift current is proportional to the applied electric field. Although this assumption is valid over a wide range of applied field ζ, but it does not hold good for larger fields (generally $\zeta > 10^3$ V/cm). It is because the high fields cause a sub-linear dependence to the drift velocity vd. Such dependence is shown in Fig. 3.27.

This behaviour is noticed in Si, Ge, and some other semiconductors. There are other effects also of high-fields. These are

- negative conductivity such as in GaAs, and
- avalanche multiplication

These phenomena are described in later chapters of this book.

3.15 The Hall Effect

Meaning. The electrons in a n-type semiconductor and the holes in a p-type semiconductor usually drift in the direction x of an applied electric field. When a magnetic field is applied in z-direction which is perpendicular to the direction of drift of holes or electrons, they tend to deflect. This is an undesirable effect and needs to be prevented. Establishment of an electric field along y-direction to prevent deflection of holes or electrons is known as Hall effect.

Illustration of the phenomenon. To illustrate this phenomenon, let us take a p-type semiconductor bar of length l, width b and thickness t lying in the field of an electric current I_x and a magnetic flux density β_z as shown in Fig. 3.28.

A single hole in the bar experiences a force F_y along y-direction whose magnitude is given by

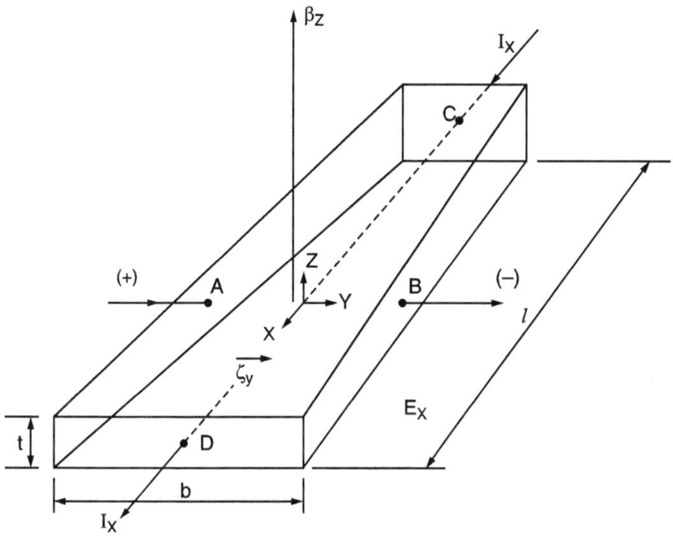

Fig. 3.28 Establishment of electric field along y-direction to prevent deflection of holes along x-direction, is called Hall effect

$$F_y = e_h(\zeta_y - u_x\beta_z) \tag{3.32}$$

where eh is charge on the hole, ζ_y is electric field strength along y-direction, and v_x is velocity of drifting holes.

Hall Voltage. To maintain steady state flow of holes down the length of bar, it is desired that the holes do not deflect. Therefore, F_y in Eq. 3.32 must be zero, and hence

$$\zeta_y = u_x\beta_z \tag{3.33}$$

Establishment of ζ_y along the width of the bar is known as Hall effect. The resulting voltage V_{AB} along y-direction is called Hall voltage and is given by

$$V_{AB} = \zeta_y \, . \, b \tag{3.34}$$

3.15.1 Significance of Hall Effect

The idea of Hall effect is utilized to determine the following quantities.

1. Nature of an unknown semiconductor; whether n-type or p-type,
2. Mobility of the semiconductor,

3. Conductivity, and
4. Resistivity of the semiconductor.

Hall Coefficient. An unknown piece of semiconductor whether n-type or p-type can be investigated by measuring Hall coefficient H_c. It is found from

$$H_C = \frac{1}{e_h h_0} \tag{3.35}$$

where h_0 is hole concentration. A negative value of H_C indicates n-type and a positive value confirms p-type semiconductor.

Relation between Hall voltage and carrier concentration. The electric field gradient ζ_y can also be expressed as

$$\zeta_y = I_{dx}\beta_z H_c = \frac{I_{dx}\beta_z}{e_h h_0}$$

Therefore

$$\mathrm{h}_0 = \frac{I_{dx}\beta_z}{e_h h_0} \tag{3.36}$$

where I_{dx} is drift current density along x-direction and is equal to I_x/bt. We can also write Eq. 3.36 as

$$h_0 = \frac{\left(\frac{I_x}{bt}\right)\beta_z}{e_h\left(\frac{V_{AB}}{b}\right)} = \frac{I_x\beta_z}{e_h t V_{AB}} \tag{3.37}$$

Determining the resistivity and mobility. All the quantities of R.H.S. of Eq. 3.37 can be measured experimentally. Then the resistivity ρ may be obtained from

$$\rho = \frac{Rbt}{l} \tag{3.38}$$

where R is resistance of semiconductor. As $R = (V_{CD}/I_X)$ mobility of hole may be determined from I x g and conductivity $\sigma = 1/\rho$, hence

$$\mu_h = \frac{\sigma}{eh_0} = \frac{\frac{1}{\rho}}{e\left(\frac{1}{eH_c}\right)} = \frac{H_c}{\rho} \tag{3.39a}$$

$$= \frac{l}{rbtH_c} = \frac{I_x l}{V_{CD}btH_c} \tag{3.39b}$$

Example 3.14 The magnetic flux density in a Hall model is 0.5 tesla. The Hall coefficient has been observed to be 3.66 × 10^{-4} m^3/C. If the resistivity of semiconductor at room temperature is 0.00893 Ω-m, determine the Hall angle.

Solution. Given are β_z = 0.5 tesla = 0.5 Wb/m^2, Hall coefficient $H_C = 3.66 \times 10^{-4}$ m^3/C, and $\rho = 0.00893$ Ω m. We have to determine the Hall angle θ_H. As

$$\begin{aligned} \tan\theta_H &= \mu\beta_z \\ &= (\sigma H_c)\beta_z \\ &= \left(\frac{H_c}{\rho}\right)\beta_z \end{aligned}$$

$$\begin{aligned} \tan\theta_H &= \frac{3.66 \times \frac{10^{-4}\,\mathrm{m}^3}{\mathrm{C}}}{0.00893\,\Omega\,\mathrm{m}} \times 0.5\frac{Wb}{\mathrm{m}^2} \\ &= 2.049 \times 10^{-2} \end{aligned}$$

or

$$\theta_H = \mathbf{1.17^o}$$

Example 3.15 An aluminium piece 15 mm thick, 60 mm wide and 180 mm long is placed under a magnetic field of 0.6 tesla. It carries a current of 25 ampere. If the hole mobility and electrical conductivity are 0.0012 m^2/Vs and 3.8 × 10^7 per Ω m respectively, determine the (a) Hall coefficient, (b) Hall voltage, and (c) resistance of metal.

Solution. Given are

$$\begin{aligned} l &= 180\,\mathrm{mm} = 180/1000\ \mathrm{m} = 0.18\,\mathrm{m}, \\ \mathrm{b} &= 60\,\mathrm{mm} = 0.06\,\mathrm{m},\ \mathrm{t} = 15\,\mathrm{mm} = 0.015\,\mathrm{m} \\ \beta_z &= 0.6\mathrm{T} = 0.6\,\mathrm{Wb/m^2},\ \mathrm{I_X} = 25\ \mathrm{A} \\ \mu_h &= 0.0012\ \mathrm{m^2/Vs},\ \rho = 1/\sigma = \frac{1}{3.8 \times 10^7} \end{aligned}$$

(a) The Hall coefficient is found as

$$\begin{aligned} H_C &= \mu_h \cdot \rho = \frac{0.0012}{3.8 \times 10^7} \\ &= \mathbf{3.166 \times 10^{-11}\frac{Vm}{AT}} \end{aligned}$$

(b) The Hall voltage is found from

$$\begin{aligned} V_{AB} &= \frac{H_C I_x \beta_z}{t} \\ &= \frac{3.66 \times 10^{-11} \times 25 \times 0.6}{0.015} \\ &= 3166 \times 10^{-11} = \mathbf{3.166 \times 10^{-8}}\, V \end{aligned}$$

(c) The resistance is found from Eq. 3.38 as

$$R = \frac{\rho l}{bt} \text{ i.e. } R = \frac{l}{\sigma bt}$$

$$\therefore \quad R = \frac{0.18}{3.8 \times 10^{7} \times 0.06 \times 0.015} = \mathbf{5.263 \times 10^{-6}\,\Omega}$$

Example 3.16 Calculate (a) the Hall coefficient and (b) conductivity of a semiconducting sample whose electron mobility is 1.065×10^{-3} m^2/Vs and relaxation time is 6×10^{-15} s. We take n $=10^{19}$/m^3 for this sample.

Solution. Given are $\mu_e = 1.065 \times 10^{-3}$ m^2/Vs, $\tau = 6 \times 10^{-15}$ s

$$\because \quad \mu_e = \sigma H_c \text{ and } \sigma = \frac{ne^2\tau}{m_e}$$

$$\therefore \quad \mu_e = \frac{ne^2\tau H_C}{m_e}$$

$$\therefore \quad 1.065 \times 10^{-3} = \frac{1 \times (1.6 \times 10^{-19})^2 \times (6 \times 10^{-15}) H_c}{9.1 \times 10^{-31}}$$

or

$$\begin{aligned} H_c &= \frac{(1.065 \times 10^{-3}) \times (9.1 \times 10^{-31})}{(1.6 \times 10^{-19}) \times (6 \times 10^{-15})} \frac{Vm^3}{AWb} \\ &= 6.309 \times 10^{18} \end{aligned}$$

$$\begin{aligned} \therefore \quad \sigma &= \frac{\mu_e}{H_c} = \frac{1.065 \times 10^{-3}}{6.309 \times 10^{18}} \\ &= \mathbf{5.924 \times 10^{-22} (\Omega\, m)^{-1}} \end{aligned}$$

3.16 Relation Between Density of States and Fermi Energy

After discussing the Fermi energy level and density of state, we shall now develop expressions to relate them. We shall also derive the expressions for Fermi radius, Fermi velocity, Fermi momentum, Fermi temperature etc. For all these the use of Schrödinger equation is essential, as this is the starting point of wave mechanics. Schrödinger equation is a differential equation capable of describing the motion of an electron. Its solution may be found for different cases such as particles in a box, particles in a potential well, hydrogen atom etc. The case of particles in a box is analogous to the case of electrons in metals.

Hence, we consider a particle (i.e. electron) of mass m which is free to move in a 1-dimensional box of length l as shown in Fig. 3.29a. Potential energy P.E. of the electron at the bottom of the box is constant and for convenience may be taken as zero. Let the potential energy is infinite for $x < 0$ and $x > l$; therefore the wave-function ψ changes along l as shown in (3.27). Based on the above assumptions, the total energy possessed by an electron is found in accordance with the free electron theory described earlier in art. 3.2.1. It has been expressed there by Eq. (3.3a), which is reproduced below.

$$E = \frac{h^2 n^2}{8ml^2} \tag{3.40a}$$

where $n = 1, 2, 3, \ldots \infty$.

The above equation indicates that a particle within the box can have energy values in a discrete set only i.e. the energy is quantized.

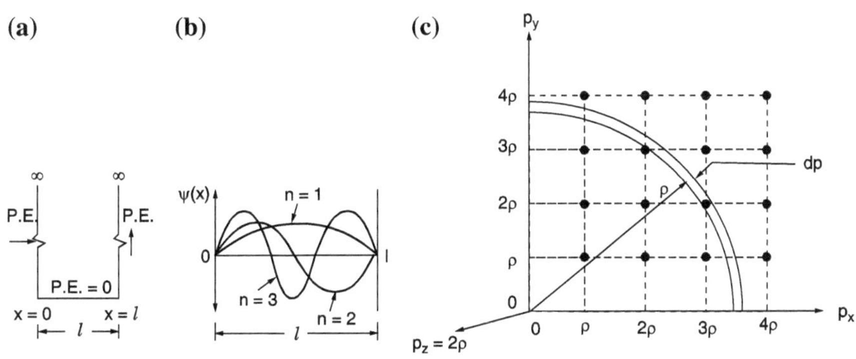

Fig. 3.29 **a** Potential energy P.E. of an electron at the bottom of potential well is assumed zero, **b** variation of wavefunction ψ along length l, and **c** the p_x-p_y plane in momentum space. Each dot represents three quantum numbers

3.16.1 Quantization of Energy i.e. 3-dimensionalization

If the constant potential region is treated as a 3-dimensional box of very large dimensions, then we have to solve the 1-dimensional Schrödinger equation three times. The energy levels can be calculated from Eq. (3.3b) which is reproduced here as

$$E = \frac{h^2\left(n_x^2 + n_y^2 + n_z^2\right)}{8ml^2} \tag{3.40b}$$

where n = 1, 2, 3, … ∞

The energy levels corresponding to each set of quantum numbers n_x, n_y, and n_z can be filled- up by putting two electrons of opposite spin in each level in accordance with the Pauli's Exclusion principle. The energy of the highest filled state is known as Fermi energy E_F.

3.16.2 Momentum Space

For an electron lying in the cube of a metal, each component of momentum is quantized. Thus

$$p_x = n_x\rho, p_y = n_y\rho, \quad p_z = n_z\rho \tag{3.41}$$

where ρ = h/2 l. A diagrammatical representation of components of momentum is shown in Fig. 3.29c) on three mutually perpendicular axes. This 'volume' is called momentum space. The electrons will occupy only those points in momentum space which are expressed by Eq. 3.41. These are marked by dots in the figure.

3.16.3 Fermi Sphere

The Schrödinger equation in 3-dimensional case may be expressed as

$$\nabla^2\psi + \frac{3mE}{h^2}\psi = 0 \tag{3.42}$$

where $\nabla^2 = \frac{\partial^2}{\partial x^2} + \frac{\partial^2}{\partial y^2} + \frac{\partial^2}{\partial z^2}$ is Laplacian operator and ψ is wave function which is generally complex.

Let we denote $\frac{2mE}{h^2} = r_c^2$ where r_c is a constant, then

$$E = \frac{h^2 r_c^2}{2m} \quad (3.43a)$$

Therefore for $E = E_F$, this equation can be rewritten as

$$E = \frac{h^2 r_F^2}{2m} \quad (3.43b)$$

Since the shape of Fermi surface in momentum space for free electron is spherical (see Fig. 3.27c), therefore K_F is Fermi radius.

3.16.4 Derivation of Different Fermi Parameters

Fermi states $\mathbf{n_F}$. If we consider a parallelepiped box of $l_x \times l_y \times l_z$ size, then in momentum space, the volume occupied by each solid will be equal to $\frac{2\pi}{l_x} \cdot \frac{2\pi}{l_y} \cdot \frac{2\pi}{l_z} = \frac{8\pi^3}{l_x l_y l_z}$. Therefore, the number of Fermi states inside the Fermi sphere will be

$$\begin{aligned} n_F &= \frac{\mathit{Volume\ of\ Fermi\ space}}{\mathit{Volume\ of\ solid}} \frac{\frac{4}{3}\pi(r_F)^3}{\frac{8\pi^3}{l_x l_y l_z}} \\ &= \frac{l_x l_y l_z (r_F)^3}{6\pi^2} \end{aligned} \quad (3.44)$$

Fermi radius $\mathbf{r_F}$. If there are a total number of N electron in n_F Fermi states, and each state occupies two electrons (one having spin-up and the other spin-down), then

$$n_F = \frac{N}{2}$$

which on considering Eq. (3.44), becomes

$$\frac{l_x l_y l_z (r_F)^3}{6\pi^2} = \frac{N}{2}$$

or

$$r_F^3 = \frac{3\pi^2 N}{l_x l_y l_z}, i.e. \quad r_F = \left(\frac{3\pi^2 N}{l_x l_y l_z}\right)^{\frac{1}{3}} \quad (3.45)$$

Fermi energy $\mathbf{E_F}$. It can be found by substituting Eq. (3.44) into Eq. (3.43b), and is given by

$$E_F = \frac{h^2\left[\left(\frac{3\pi^2 N}{l_x l_y l_z}\right)^{\frac{1}{3}}\right]^2}{2m}$$
$$\frac{h^2}{2m} = \left(\frac{3\pi^2 N}{l_x l_y l_z}\right)^{\frac{2}{3}} \tag{3.46}$$

Fermi velocity $\mathbf{V_F}$. Since $E_F = (1/2)mv_F^2$, hence using Eq. (3.46), we find

$$v_F^2 = \frac{2E_F}{m} = \frac{2}{m}\frac{h^2}{2m}\left(\frac{3\pi^2 N}{l_x l_y l_z}\right)^{\frac{2}{3}}$$
$$v_F = \left[\frac{2}{m}\frac{h^2}{2m}\left(\frac{3\pi^2 N}{l_x l_y l_z}\right)^{\frac{2}{3}}\right]^{1/2} \tag{3.47}$$
$$\frac{h}{2m} = \left(\frac{3\pi^2 N}{l_x l_y l_z}\right)^{\frac{1}{3}}$$

Fermi momentum $\mathbf{P_F}$. Since momentum = mass × velocity, therefore on considering Eq. (3.47), we find

$$p_F = m \times \frac{h}{m}\left(\frac{3\pi^2 N}{l_x l_y l_z}\right)^{\frac{1}{3}}$$
$$= h\left(\frac{3\pi^2 N}{l_x l_y l_z}\right)^{\frac{1}{3}} \tag{3.48}$$

Fermi temperature $\mathbf{T_F}$. As F_F = (Boltzmann constant) × T_F, therefore using Eq. (3.47), we find

$$T_F = \frac{h^2}{2mk}\left(\frac{3\pi^2 N}{l_x l_y l_z}\right)^{\frac{2}{3}} \tag{3.49}$$

3.16.5 Relation Among Density of Fermi States (E_F), E_F and N

As

$$\rho(E) = \frac{dN}{dE}, \quad so \quad \rho(E_F) = \frac{dN}{dE_F} \tag{3.50a}$$

∵ From Eq. (3.46),

$$N = \frac{l_x l_y l_z}{3\pi^2}\left(\frac{2m}{h^2}\right)^{\frac{3}{2}} \times E_F^{\frac{3}{2}}$$

$$\therefore \quad \frac{dN}{dE_F} = \frac{l_x l_y l_z}{3\pi^2}\left(\frac{2m}{h^2}\right)^{\frac{3}{2}} \times \frac{3}{2} E_F^{\frac{1}{2}}$$

or

$$\rho(E_F) = \frac{l_x l_y l_z}{2\pi^2}\left(\frac{2m}{h^2}\right)^{\frac{3}{2}} \sqrt{E_F} \tag{3.50b}$$

This equation shows that the density of state is proportional to the square root of Fermi energy.

A relation between E_F and N can also be established by multiplying and dividing the above equation by 3. Then we have

$$\begin{aligned}\rho(E_F) &= \frac{3 \times l_x l_y l_z}{3 \times 2\pi^2}\left(\frac{2m}{h^2}\right)^{\frac{3}{2}} \sqrt{E_F} \\ &= \frac{3}{2}\frac{N}{E_F}\end{aligned} \tag{3.50c}$$

Example 3.17 Calculate the Fermi energy in aluminium whose density is 2700 kg/m^3 and atomic mass is 26.98 amu. Take electron rest mass = 9.109×10^{-31} kg, h = 6.626×10^{-34} Js, and e = 1.602×10^{-19} C.

Solution. The number of free electrons n per unit volume of a metal (i.e. electron density) may be found from

$$\begin{aligned} n &= \frac{\textit{Avogadro's no.} \times \textit{density of metal}}{\textit{atomic mass unit}} = \frac{N_A \rho}{amu} \\ &= \frac{(6.023 \times 10^{26}) \times 2700}{26.98} = 6.027 \times 10^{28} \frac{atoms}{m^3}\end{aligned}$$

Therefore from Eq. (3.46), we observe

$$\begin{aligned} E_F &= \frac{h^2}{2m}\left(3\pi^2 n\right)^{\frac{2}{3}} = \frac{(6.626 \times 10^{-34})^2}{2 \times 9.109 \times 10^{-31}} \times \left(3\pi^2 \times 6.027 \times 10^{28}\right)^{\frac{2}{3}} \\ &= \left(2.409 \times 10^{-37}\right) \times \left(31845.23 \times 10^{56}\right)^{\frac{1}{3}} \\ &= \left(2.409 \times 10^{-37}\right) \times \left(1.471 \times 10^{20}\right) = 3.543 \times 10^{-17}\,\text{J} \\ &= \frac{3.543 \times 10^{-17}}{1.602 \times 10^{-19}} = \mathbf{221.16\,eV}\end{aligned}$$

Example 3.18 Determine the density of states of 1 cm^3 of silver at the Fermi level. Take E_F for silver as 5.51 eV. Assume the effective mass m* to be equal to the rest mass m_0. Take $m_0 = 9.1 \times 10^{-31}$ kg and $h = 6.626 \times 10^{-34}$ Js.

Solution. The density of state ρ(E) may be determined from Eq. (3.16). At Fermi level, it may be found from the equation as given below.

$$N(E_F) = \frac{volume}{2\pi^2}\left(\frac{2m*}{h^2}\right)^{\frac{3}{2}}\sqrt{E_F} \quad \text{(i)}$$

Here volume = 1 cm^3 = 1×10^{-6} m^3. On substituting the appropriate values in Eq. (i), we find

$$\begin{aligned} N(E_F) &= \frac{1 \times 10^{-6}}{2\pi^2}\left[\frac{2 \times 9.1 \times 10^{-31}}{(1.05 \times 10^{-34})^2}\right] \times \sqrt{5.51 \times 1.6 \times 10^{-19}} \\ &= \left(5.066 \times 10^{-8}\right) \times \left(16.507 \times 10^{37}\right) \times \left(9.389 \times 10^{-10}\right) \\ &= \mathbf{785.17 \times 10^{19} energy\ state/Joule} \end{aligned}$$

Example 3.19 Fermi velocity of the electron of a certain metal from among the list given below is 0.85×10^6 m/s. Identify this metal. Take rest mass of electron = 9.109×10^{-31} kg.

Metal Al	Al	Cu	Ag	Na	K	Cs
E_F(eV)	11.8	7.04	5.51	3.2	2.1	1.6

Consider (a) rest mass, and (b) variable mass of electron.

Solution. (a) Since the KE $= \frac{1}{2}$ mv^2, the Fermi energy may be calculated as below.

$$\begin{aligned} E_F &= \frac{1}{2}m_0V_F^2 \\ &= \frac{1}{2} \times \left(9.109 \times 10^{-31}\right) \times \left(0.85 \times 10^6\right)^2 \\ &= 3.290 \times 10^{-19}\,\text{J} \\ &= \frac{3.290 \times 10^{-19}}{1.602 \times 10^{-19}} = 2.05\,\text{eV} \end{aligned}$$

As this value is close to 2.1, hence the metal in question is potassium (K).

(b) Considering the mass varying with velocity, it is found from Einstein equation given as

$$
\begin{aligned}
m &= m_0\sqrt{1-\left(\frac{v}{c}\right)^2} \\
&= \left(9.109\times 10^{-31}\right) = \sqrt{1-\left(\frac{0.85\times 10^6}{3\times 10^8}\right)^2} \\
&= \left(9.109\times 10^{-31}\right)(0.99999) = 9.108\times 10^{-31}\,\text{kg} \\
\therefore \quad E_F &= \frac{1}{2}mv_F^2 \\
&= \frac{1}{2}\times\left(9.108\times 10^{-31}\right)\left(0.85\times 10^6\right)^2 \\
&= 3.290\times 10^{-19}\,\text{J} \\
&= \frac{3.290\times 10^{-19}}{1.602\times 10^{-19}} = 2.05\,\text{eV}
\end{aligned}
$$

Hence, the metal in question is Potassium.

3.17 Solved Examples

Example 3.20 Explain as to why the ionization energy of impurity carrier is higher in silicon than in germanium.

Solution. The ionization energy is the amount of energy needed to detach the valence electron from the impurity atom. The total energy of an orbiting electron is the sum of its kinetic energy and potential energy given as

$$
E_{total} = E_{KE} + E_P = \frac{1}{2}mv^2 - \frac{q^2}{4p\varepsilon_0\varepsilon_r r} \tag{3.51}
$$

Above equation shows that for other factors being constant, the ionization energy is inversely proportional to dielectric constant ε_r. Since $\varepsilon_r = 16$ for germanium and 12 for silicon, hence ionization energy is higher in silicon than in germanium.

Example 3.21 What is electron scattering? How does it influence the behaviour of electron and the semiconducting crystal?

Solution. Collision of electrons with obstacles within a conducting material is an inherent phenomenon. Depending upon various factors, they scatter over certain angle θ, after collision. This is called electron scattering.

The phenomenon of electron scattering causes loss in its velocity after collision. It also causes transfer of heat energy from electron to crystal lattice. Probability of scattering remains a constant in case the scattering is spherically symmetric.

Example 3.22 Explain the relaxation process and relaxation time.

Solution. Relaxation is a time dependent phenomenon that occurs due to collision of electrons with the obstacles present within the crystal (material). Jumping of atoms by diffusion, atomic vibration, flow of grain boundaries, oscillation and damping are some relaxation processes.

Time interval of applied forces/fields and the time taken for completion of these relaxation processes may be comparable in some cases and quite different in other cases. A relaxation process may be too slow or to fast as compared to force/field (application) time interval.

Relaxation processes are specified by relaxation time t_r, which is expressed as

$$p = 1 - e^{\left(-\frac{t}{t_r}\right)} \tag{3.52}$$

where p is the fraction of relaxation process and t is any arbitrary time. When $t = t_r$ then $p = 0.63$ i.e. the relaxation time is the time taken to complete 63 % of a process. For different processes, t_r varies from 10^{-13} to 10^7 s.

Example 3.23 What are the sources of obstacles present in conducting materials? Why do we consider an average value of collision time in electrical designing?

Solution. The obstacles in materials remain present in different forms. They may be present in natural form such as cementite in steel and iron, oxides in copper and aluminium; or may be present in manmade (unnatural) form such as grain boundaries. These obstacles are located randomly in different orientations within the (conducting) materials. That is why the collision of electrons with them occurs at different intervals of time. Therefore, the collision time is determined as an average value, and is defined as the average time elapsed between two successive collisions.

Example 3.24 Determine the drift velocity for an electron in Cu at the Fermi surface, when the applied electric field is 100 V/m. Also determine the velocity of electron having Fermi energy $F_F = 7.1$ eV. Take mobility $\mu_e = 0.0032$ m^2/Vs Assume any other quantity and constant, if required.

Solution. Given are $\zeta = 100\,\text{V/m}$. We take $e = 1.6 \times 10^{-19}$ C and $m = 9.11 \times 10^{-31}$ kg.

$$\text{As} \quad F_F = \frac{1}{2} m v_F^2$$

$$\therefore \quad v_F = \sqrt{\frac{2E_F}{m}} = \sqrt{\frac{2 \times 7.1 \times 1.6 \times 10^{-19}}{9.11 \times 10^{-31}}}$$

$$= \mathbf{1.58 \times 10^6 m/s}$$

And the mean drift velocity is given by

$$\begin{aligned} v_{dmean} &= \mu_e \zeta = 0.0032 \times 100 \\ &= \mathbf{0.32 m/s} \end{aligned}$$

Example 3.25 The energy band gap of binary compounds GaAs and GaP are about 1.43 eV and 2.3 eV respectively. Discuss their suitability as electronic devices.

Solution. The 1.43 eV band gap corresponds to near infrared region of the light spectrum, while

2.3 eV band gap corresponds to wavelengths in green portion of the light spectrum. Hence GaAs is suitable to make microwave devices like the Gunn diodes, while GaP is suitable for light emitting diodes (LEDs) purposes.

Example 3.26 High-frequency transistors are generally made of germanium and not of silicon, but silicon is used in electronic devices operating over a wider temperature range—why?

Solution. The carrier mobility of silicon is lower than that of germanium. Therefore, it is inferior to germanium for high frequency services. But the energy gap in silicon is higher than that of germanium, so it offers a greater temperature range service for electronic devices.

Review Questions

1. Define the following terms.
   (a) Homopolar bond (b) Bipolar solid (c) Electronegativity
   (d) Wave number (e) Hall angle (f) Energy band
2. Name various electron theories in reference to energy bands. How is free electron theory diferent from energy band theory?
3. Explain energy band theory stating the assumptions made therein.
4. Why is the energy versus wave number curve discontinuous in energy band theory? Plot E vs k curve and show the forbidden gap on it.
5. Explain the Brillouin zone theory stating the assumptions made therein. How is this theory different from energy band theory?
6. Discuss the meaning of first and second Brillouin zones. Show and explain the Brillouin zones in BCC, FCC and HCl crystals.
7. What is meant by direct and indirect energy bands? Differentiate between direct and indirect semiconductors. Enlist their examples also.
8. What is the necessity of varying Eg with varying alloy composition? Explain as to how it is done in a ternary alloy.
9. Differentiate between `Fermi energy level' and `Fermi-Dirac probability function'. What is their importance in the study of electronic behaviour of solids?
10. Enumerate different types of semiconductors. Intrinsic semiconductors are not suitable for applications in electronic devices—why? Show that the Fermi level for a pure germanium lies in the middle of its forbidden gap.
11. Explain how electrons and holes both conduct in a pure silicon crystal. How is conductivity influenced by mobility of electrons and holes?
12. Differentiate between n-type and p-type semiconductors. Name various dopants used to prepare them from intrinsic elements. Draw energy diagrams for both the above types, show salient levels on them, and explain the mechanism of conduction in them.
13. Draw energy band diagrams for Arsenic-doped-Silicon, and Indium-doped-Germanium. Explain important differences between them.
14. Describe 'effective mass' and 'density of states', and discuss their importance in designing solid state devices.
15. Derive expressions to relate density of states with 'Fermi energy'.
16. Define and explain the following.
    (a) Momentum space (b) Fermi sphere (c) Fermi states
    (d) Fermi radius (e) Fermi momentum
17. Explain the phenomenon of Hall effect. Discuss the method of identifying an unknown piece as n-type or -type semiconductor.
18. Write notes on the following.
    (a) Drift velocity (b) Electron-hole pair
    (c) Density of state (d) Relaxation process and relaxation time

Numerical Problems

1. The velocity of electrons in a solid at Fermi level is 9×10^{15} m/s. Obtain the Fermi level of the solid. What this solid is likely to be ? Assume data suitably, if required. [**Ans:** 2.3 eV, potassium]

2. Determine the resistivity of an intrinsic germanium sample at 300 K when intrinsic concentration is 25×10^{19} per cubic metre. Take electron and hole mobilities at room temperature as 0.39 m^2/V s and 0.19 m^2/V s respectively. [**Ans :** 0.43 ohm m]

3. The intrinsic carrier density of silicon at 30°C is 11×10^{16} per m^3. The energy gap is 1.17 eV, and the electron and hole mobilities are 0.14 m^2/V and 0.05 m^2/V s respectively at the same temperature. Calculate the (a) resistivity of silicon, and (b) number of electrons available in valence band for thermal excitation.

   [**Ans :** 3000 ohm m (b) $6 \times 10^{25}/m^3$]

4. A silicon sample is doped with 10^{23} arsenic atoms per cubic meter. Calculate the hole concentration at room temperature. Determine the location of Fermi energy level with respect to intrinsic energy level. The number of intrinsic concentration is 15×10^{16} per cubic meter.

   [**Ans :** $225 \times 10^9/m^3$, EF is located at 0.407 eV above E_i towards E_c]

5. 7 gm of p-type germanium doped with aluminium is melted with 7 gm n-type germanium doped with arsenic. In both the types the conductivity is 10/ohm m. Calculate the conductivity of the resultingmaterial. Assume the required data suitably. [**Ans :** 2.563/ohm m]

   [Hint: Obtain concentration of donor and acceptor level. As theacceptor impurity isgreater, hence the resulting material is of p-type]

6. In the above problem, if p-type germanium is 7 gm and n-type is 6 gm, calculate the conductivity of the resulting material. What type of semi-conductor will it be? [**Ans :** 2.0/ohm m, p-type]

7. In a silicon doped with antimony, 54.35% donor electrons are promoted to the conduction band at 320 K. Calculate the donor ionization energy in the semiconductor.

[**Ans :** 0.015 eV]

8. The wavelength of the light from a GaP laser is 5490 Å. Compute the energy gap of it.

[**Ans :** 2.26 eV]

9. A silicon sample is doped with 10^{17} phosphorus atoms per cubic centimeter. A current of 1 mA passes through 100 m thick sample which is placed in a magnetic field of 10^5 Wb/cm^2. Taking mobility as 700 cm^2/V s, calculate (a) resistivity, (b) Hall coefficient, and (c) Hall voltage.

[**Ans :** (a) 8.93×10^{-4} ohm m (b) -625×10^{-6}m^3/C (c) $-62.5\ \mu$ V]

10. In the above problem, predict the type of silicon sample. [**Ans** : n-type]
11. The hole and electron mobilities in a silicon single crystal at 300 K are 0.025 m^2/Vs and 0.17 m^2/Vs respectively. Determine the diffusion coefficient of holes and electrons at this temperature.

[**Ans :** $D_h = 0.00091$ m^2/s, $D_e = 0.0044$ m^2/s]

Objective Type Questions

1. When a semiconductor is doped with a p-type impurity, each impurity atom will

   (a) acquire negative charge

   (b) acquire positive charge

   (c) remain electrically natural

   (d) give away one electron

2. An electron rising through a potential of 500 V, will acquire an energy of

   (a) 800 x 10^{-22} joules (b) 800 eV

   (c) 500 joules (d) 500 eV

3. For pure germanium, the density of the charge carriers is 2.5×10^{13} /cm^3 and electron and hole mobilities at 300 K are 3600 cm^2 / V-sec and 1700 cm^2/ V-sec respectively. Its conductivity would be

   (a) 0.02 mho/cm (b) 0.5 mho/cm

   (c) 0.05 ohm-cm (d) 0.2 ohm-cm

4. A specimen of intrinsic germanium, with the density of charge carriers of 2.5 $10^{13}/cm^3$ is doped with impurity atoms such that there is one impurity atom for every 10^6 germanium atoms. The density would be

   (a) 4.4 $10^{13}/cm^3$ (b) 1.4 $10^{10}/cm^3$

   (c) 4.4 $10^{10}/cm^3$ (d) 1.4 10^4/cm/sec

5. The current passed through a semiconductor placed in a magnetic field is proportional to x. The magnetic field intensity is proportional to y. Then the voltage developed across the semiconductor will be proportional to

   (a) x–y (c) x/y

   (c) x + y (d) xy

6. The first critical condition at which free electrons are diffracted in an FCC crystal would occur at which one of following values of the wave number k? (a is lattice parameter)

   (a) $\frac{2}{a}$ (b) $\frac{\pi a}{\sqrt{3}}$

   (c) $\frac{\pi}{a}$ (d) $\frac{\sqrt{3}}{a}$

7. The slope of discontinuous E vs k curve at k ±2/d is ________.

8. The acceptor atom in a p-type semiconductor carries a _________ charge at normal temperature.

9. Hall effect is most significant in

   (a) Silver (b) Sodium

   (c) lithium (d) gold

Answers

1. (a) 2. (d) 3. (a) 4. (b) 5. (d) 6. (d) 7. (zero) 8. negative 9. (b)

Reference

1. Gupta, K.M., Gupta, N.: Advanced Electrical And Electronics Materials. Scrivener Wiley Publishing, USA (2015). (Most of the matter of this chapter has been reproduced with the permission of the publisher of this book)

Chapter 4
Excess Carriers in Semiconductors

Abstract Most semiconductor devices operate by creation of charge carriers which are in excess of the charge carriers available at thermal equilibrium. Conduction in semiconductors are dominated by these excess carriers. Therefore, in this chapter optical absorption and its mechanism, absorption coefficient and factors affecting it, are explained. Luminescence, phosphorescence, electro-luminescence, mechanism of excitation and recombination are discussed. Carrier lifetime and its derivation, indirect recombination, steady state carrier generation and quasi-Fermi levels are described. Photoconductivity and photoconductive devices, photoconductive cell, photo-multiplier tube, and factors affecting the selection of semiconductors for photo-conducting purpose are explained. Concepts of diffusion of carriers, analysis of drift and diffusion carriers, Einstein relation, continuity equation for diffusion and recombination are presented. Equations have been derived for diffusion and diffusion length. Solution has been provided for charge transport and impurity distribution by error function method and Gaussian distribution method. Long diode and short diode are also discussed. Quantum efficiency, trapping centre, dark resistance of photoconductor, and derivation of junction-voltage etc. are briefed through solved solutions. Minute insight into the various topics are given through solved numerical and theoretical examples. Review questions, numerical problems and objective type questions are also given with their answers.

Keywords Optical absorption · Luminescence · Photconductivity · Diffusion of carriers · Einstein relation · Continuity equation · Error function method · Gaussian distribution · Long diode · Short diode

4.1 Introduction

Most semiconductor devices operate by creation of charge carriers which are in excess of the charge carriers available at thermal equilibrium. The conduction processes in semiconductor materials are dominated by these excess carriers. The excess charge carriers are created by different means such as

K.M. Gupta and N. Gupta, *Advanced Semiconducting Materials and Devices*,
Engineering Materials, DOI 10.1007/978-3-319-19758-6_4

- Optical excitation
- Electron bombardment
- Injection across a forward-biased p-n junction

In optical excitation, the discrete energy packets are released when electromagnetic radiations/light interact with the surface of a material. These packets are called 'quanta' or 'photons'. When photons are absorbed, the electrons from lower energy states are excited to higher energy state.

Photons are basic constituents of any radiation. They possess properties like energy E_{photon}, mass m, and momentum p. These quantities are expressed as

$$E = hf, m = \frac{E}{c^2}, \quad and \quad p = mc = \frac{hf}{c}$$

In this chapter, we shall mainly study the photons and the phenomenon related to them.

4.2 Optical Absorption

By optical absorption we mean the absorption of photons by materials. In context of this subject, the material means semiconductors. Hence in this chapter, we shall mean the optical absorption as the absorption of photons by semiconductors. An important use of this phenomenon is made in

- measuring the energy of band gap.

Measurement of band gap energy of semiconductors is often required for the analysis and proper design of semiconductor devices. It is generally accomplished by a technique involving absorption of incident photons by the material.

All incident photons on a semiconductor are not absorbed. While some of them are absorbed, the rest are transmitted. The absorption or transmission depends on the following condition.

1. Photons are absorbed if $E_{Photon} \geq E_g$, and
2. Photons are transmitted if $E_{Photon} < E_g$.

Here, $E_{photon} = hf$ is the energy of photon where h is Planck's constant and f is frequency of radiation. The absorption at energies greater than band gap energy is due to the fact stated below.

While the valence band contains several electrons, the conduction band has many empty states. Therefore, the electrons can be easily excited from valence band and hence the probability of photon absorption is high. However, a photon having energy less than E_g cannot excite electrons from the valence band to the conduction band. That is why the absorption of photons is negligible in pure (intrinsic) semiconductors like Si and Ge.

4.2.1 Mechanism

Optical absorption of a photon is shown in Fig. 4.1. It shows that an electron excited from location A in valence band to location B in conduction band by optical absorption may have energy more than E_c.

The excited electron leaves a hole at A. It is common fact that almost all electrons have energy equal to or near to E_c, if the semiconductor sample is not heavily doped. Therefore in case of scattering, the excited electron loses energy to the lattice, until its velocity reaches the velocity of other conduction band electrons at thermal equilibrium. The electron and hole created by the above absorption process are referred to as 'excess carriers'. Although these carriers exist in their respective bands, but they freely contribute to the conductivity of semiconductor. The phenomenon of optical absorption of photon shown in Fig. 4.1, can be illustrated as follows.

(i) *Step AB shows creation of an electron-hole pair.*
(ii) *Step BC shows giving-up of energy by excited electron to the lattice due to scattering.*
(iii) *Step CD shows recombination of electron with the hole in valence band.*

4.2.2 Absorption Coefficient

When a beam of photons falls on a semiconductor; a part of it is reflected, some fraction is transmitted and the rest is absorbed. Therefore for a given wavelength; if the absorbed, reflected and transmitted fractions are expressed by α_A, α_R, and α_T respectively; then we have

$$\alpha_A + \alpha_R + \alpha_T = 1 \tag{4.1}$$

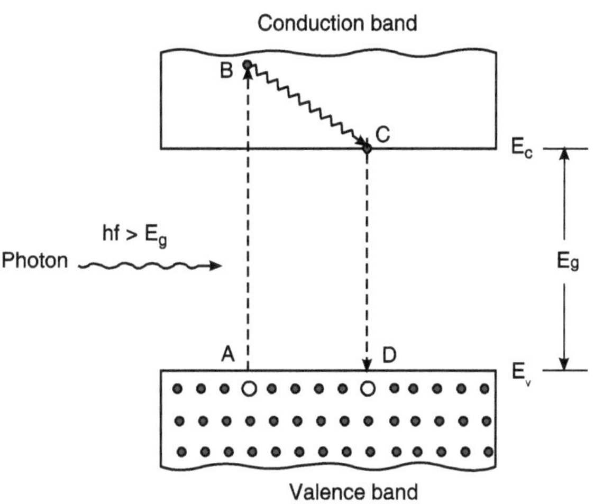

Fig. 4.1 Depiction of mechanism of optical absorption

The fraction of incident light absorbed, depends on the wavelength of photon and the sample thickness on which it falls.

In order to determine the absorptivity, we assume a beam of photon of intensity I_0 (photons/cm^2-s) and wavelength λ that falls on a sample of thickness t. This intensity decreases with increase in distance x along its thickness as shown in Fig. 4.2. Since the degradation of intensity $-\frac{dI}{dx}$ is proportional to remaining intensity at distance x, therefore we can write

$$-\frac{dI}{dx} \propto I \tag{4.2a}$$

or

$$-\frac{dI}{dx} = \alpha I \tag{4.2b}$$

where α is proportionality constant and is known as absorption coefficient.

Solution of this first order differential equation may be obtained as

$$I = I_0 e^{-\alpha x} \tag{4.3}$$

It depicts an exponential decrease as shown in Fig. 4.2. Its unit is cm^{-1}. This solution can be used to calculate the intensity of transmitted light I_t through the sample of thickness t. It is given by

$$I_t = I_0 e^{-\alpha x} \tag{4.4}$$

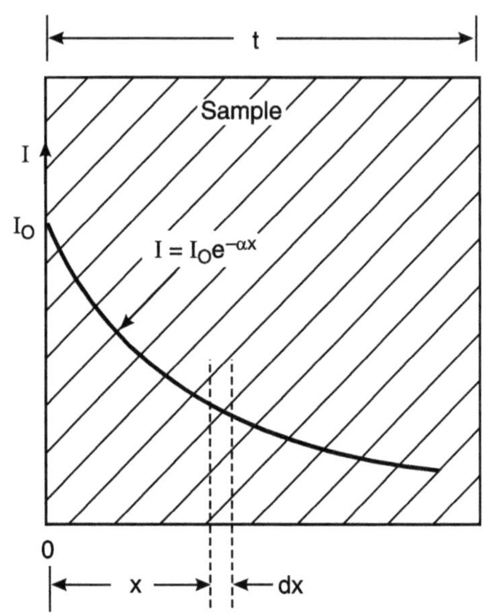

Fig. 4.2 A decrease in intensity of photon along the thickness of specimen is due to absorption

4.2.3 *Factors Affecting the Absorption Coefficient*

The absorption coefficient mainly depends on the following two factors.

1. Photon wavelength λ, and
2. Type of semiconducting material.

The effect of wavelength on optical absorption is shown in Fig. 4.3. It shows too little absorption for longer wavelengths (i.e. smaller hf) and substantial absorption for smaller wavelengths (i.e. larger hf). Since the photon energy $E_{phonton}$ and wavelength are related as $E_{photon} = hc/\lambda$, where c is velocity of the light; hence on substituting $h = 6.626 \times 10^{-34}$ Js and $c = 2.998 \times 10^{8}$ m/s, we get

$$E_{photon} = \frac{hc}{\lambda} = \frac{1.24}{\lambda}\ \text{eV} \tag{4.5}$$

in which λ is in μm.

The effect of type of semiconductor is shown in Fig. 4.4. It depicts a spectrum of light for different band gap energies E_g. The salient observations of it are given as follows.

(i) Semiconductors lying in infrared region are

- InSb
- Ge
- Si
- GaAs

(ii) Semiconductors lying in visible region are

- CdSe (with red visibility)
- GaP (with yellow visibility)
- CdS (with green visibility)
- SiC (with blue visibility)

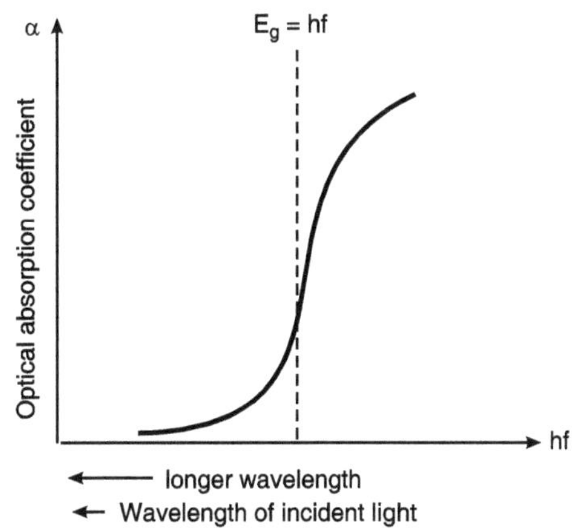

Fig. 4.3 Dependence of optical absorption coefficient a for a semiconductor as a function of wavelength of incident light

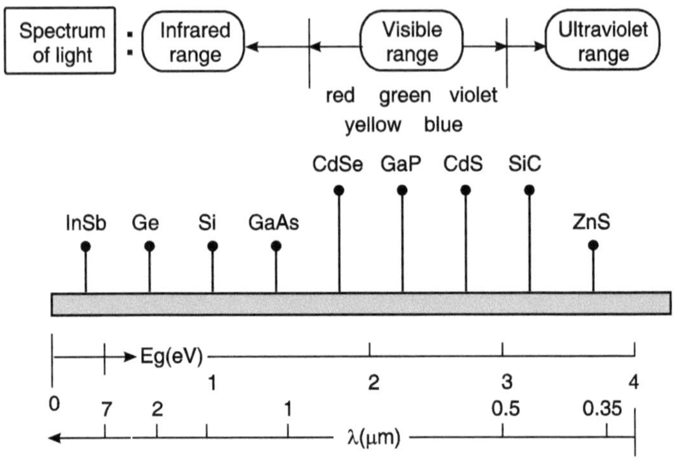

Fig. 4.4 Some semiconductors under different visibility range of light spectrum

(iii) Semiconductors lying in ultraviolet region are

- ZnS

Values of E_g and λ are not to the scale

Example 4.1 Calculate the energy of a photon of sodium light having wavelength of 5.893×10^{-7} m (a) in joule, and (b) in electron-volt. Take $h = 6.62 \times 10^{-34}$ Js and velocity of light $c = 3 \times 10^8$ m/s.

Solution. Given are: $\lambda = 5.893 \times 10^{-7}$ m

(a)
$$\because \quad E_{photon} = hf = \frac{hc}{\lambda}$$
$$\therefore \quad = \frac{(6.62 \times 10^{-34}) \times (3 \times 10^8)}{5.893 \times 10^{-7}} \text{ J}$$
$$= \mathbf{3.375 \times 10^{-19}\ J}$$

(b)
$$\because \quad 1\,\text{eV} = 1.6 \times 10^{-19}\,\text{J}$$
$$\therefore \quad _{photon} = \frac{3.375 \times 10^{-19}}{1.6 \times 10^{-19}} = \mathbf{2.11\,eV}$$

Example 4.2 A 0.45 μm thick sample of GaAs is illuminated by a monochromatic light of photon energy 3 eV. The incident power on the sample is 15 mW. If the absorption coefficient is 50,000/cm, determine (a) the total energy absorbed by the sample per second, the rate of excess thermal energy given up by the electrons to the lattice before recombination, and (c) the number of photons given off from recombination. Assume perfect quantum efficiency. The outgoing radiation is of 2.35 eV.

Solution. Given are:

$$
\begin{aligned}
t &= 0.45\ \mu\text{m} = 0.45 \times 10^{-6}\text{m} = 0.45 \times 10^{-4}\text{cm} \\
E_{\text{photon}} &= 3\,\text{eV},\ \alpha = 50,000/\text{cm} = 5 \times 10^{4}\text{cm}^{-1}, \\
I_0 &= 15\,\text{mW} = 15 \times 10^{-3}\text{W}
\end{aligned}
$$

(a) Using Eq. 4.4, we write

$$
\begin{aligned}
\text{I}_\text{t} &= \text{I}_0\text{e}^{-\alpha t} = 15 \times 10^{-3}.e^{(-5\times10^{4}\times0.45\times10^{-4})} \\
&= 0.015 \times 0.105 = 1.575 \times 10^{-3}\text{W} \\
&= 1.575\,\text{mW}
\end{aligned}
$$

Thus the total power absorbed is

$$
\begin{aligned}
\text{I}_{\text{absorbed}} &= \text{I}_0 - \text{I}_\text{t} = 15 - 1.575 \\
&= 13.425\,\text{mW} = 13.425 \times 10^{-3}\text{J/s}
\end{aligned}
$$

(b)

$$
\begin{pmatrix} \text{The rate of excess thermal} \\ \text{energy given by the} \\ \text{electrons to the lattice} \\ \text{before recombination} \end{pmatrix} = \begin{pmatrix} \text{The amount of} \\ \text{energy converted} \\ \text{to heat per second} \end{pmatrix}
$$

As the fraction of each photon energy unit which is converted to heat is

$$
\frac{hf_1 - hf_2}{hf_1} = \frac{3 - 2.35}{3} = 0.216
$$

∴ Total amount of energy converted to heat per second is

$$
\begin{aligned}
&= 0.216 \times (13.425 \times 10^{-3}) \\
&= 2.899 \times 10^{-3}\text{J}/s
\end{aligned}
$$

(c) As the quantum efficiency is perfect, therefore there is a single (i.e. one) emitted photon for each absorbed photon. Hence the number of photons given off from recombination per second is

$$
\begin{aligned}
n_{photon} &= \frac{I_{absorbed}}{e \times E_{photon}} = \frac{13.425 \times 10^{-3}\text{J/s}}{1.6 \times 10^{-19}\frac{\text{J}}{\text{eV}} \times \frac{3\text{eV}}{photon}} \\
&= 8.39 \times 10^{16}\text{photons/s}
\end{aligned}
$$

4.2.4 Capability of a Material to Absorb Light

The capability of a material to absorb light depends on the type of charges contained in it. The absorption takes place due to the interaction of electromagnetic radiation of light and the charges in the material. In general, following three types of charge distribution exist in materials.

(i) Tightly bound charges such as the inner shell electrons of the lattice ions or the lattice ions themselves.
(ii) Loosely bound charges such as outer shell electrons of the lattice ions.
(iii) Free charges like conduction electrons in metals and semiconductors.

Tightly bound charges provide the phenomenon of optical dispersion. In such materials, the excitation of light near the frequency of resonance leads to a pronounced absorption. The presence of both: the bound electrons and ions thus lead to absorption in the UV and infrared regions of the spectrum.

The loosely bound charges can be best described by band theory of electrons. The absorption in such charges can result from inter band excitation. The absorption will occur only if the photon energy is more than the band gap energy.

The free electron materials are very polarizable and interact readily with an incident electromagnetic wave. They do not move independently of the lattice structure, but during their motion may interact with a photon. A loss of energy may result before the charge re-radiates the light and hence absorption occurs. In such materials, the absorption increases with the number of electrons, and the absorption coefficient also varies proportionately to the square of the wavelength.

4.3 Luminescence

Luminescence is the property by virtue of which a material emits the light. In semiconducting materials, the light is emitted under certain conditions such as given below.

(i) when electron-hole pairs (EHP) are generated, or
(ii) when the carriers fall to their equilibrium state after being excited to higher impurity levels.

Several semiconductors exhibit the property of luminescence, specially the compound semiconductors with direct band gaps.

Types of luminescence. Depending upon the excitation mechanism involved, the luminescence can be classified as follows.

1 Photo-luminescence in which the carriers are excited by photon absorption.
2 Cathode-luminescence in which the carriers are created by high-energy bombardment of electrons on the material.

3 Electro-luminescence in which the carriers are excited by introducing current in the material.
4 Chemi-luminescence in which the luminescence is created by chemical action. Amongst above types, the first three types of excitation are most important mechanisms for applications in vivid electronic devices.

4.3.1 Photo-Luminescence

It is the phenomenon of emission of light from a semiconductor on account of recombination of excited electron-hole pair (EHP). The recombination may occur directly or indirectly. If excitation of EHP occurs under steady state condition, the rate of recombination remains the same as the rate of generation. In this case, one photon is emitted for each photon absorbed. Recombination in semiconductors takes place at varying rates: fast and slow. Accordingly, the photo-luminescence may be of following two types [1].

1. **Fluorescence**. It is a fast process property of material in which the emission of photon stops in about 10^{-8} s after the excitation is removed. This time period matches with the order of the value of mean lifetime (see Article 4.4) of EHP. The materials belonging to this category are the following.
   - Glass surface coated with tungstates or silicates such as in fluorescent lamps.
   - Television screen coated with sulphides, oxides, tungstates etc.
2. **Phosphorescence**. It is a slow process property of material in which the emission of photon continues for a longer duration, lasting for seconds and minutes after removal of excitation. Materials falling in this category are termed as phosphors, and their examples are the following.
   - ZnS coated with Cu as impurity
   - CdS coated with Ag as impurity
   - KCl coated with Tl as impurity
   - NaI coated with Tl as impurity.

4.4 Phosphorescence

Phosphorescence is a specific type of photoluminescence related to fluorescence. Unlike fluorescence, a phosphorescent material does not immediately re-emit the radiation it absorbs. The slower time scales of the re-emission are associated with "forbidden" energy state transitions in quantum mechanics. As these transitions occur very slowly in certain materials, absorbed radiation may be re-emitted at a lower intensity for up to several hours after the original excitation.

Example. Commonly seen examples of phosphorescent materials are the glow-in-the-dark toys, paint, and clock dials that glow for some time after being charged with a bright light such as in any normal reading or room light. Typically the glowing then, slowly fades out within minutes (or up to a few hours) in a dark room. Ironically, white phosphorus (from which phosphorescence takes its name) does not actually exhibit this property, but rather chemiluminescence.

4.4.1 Phosphorescence Materials

Common pigments used in phosphorescent materials include zinc sulfide and strontium aluminate. However, the development of strontium aluminate $SrAl_2O$, with a luminance approximately 10 times greater than zinc sulfide, has relegated most zinc sulfide based products to the novelty category. Strontium aluminate based pigments are now used in exit signs, pathway marking, and other safety related signage.

Most photoluminescent events in which a chemical substrate absorbs and then re-emits a photon of light, are fast, of the order of 10 ns. Light is absorbed and emitted at these fast time scales in cases where the energy of the photons involved matches the available energy states and allowed transitions of the substrate. In the special case of phosphorescence, the absorbed photon energy undergoes an unusual intersystem crossing into an energy state of higher spin multiplicity, usually a triplet state. As a result, the energy can become trapped in the triplet state with only classically “forbidden” transitions available to return to the lower energy state. Most phosphorescent compounds are relatively fast emitters, with triplet lifetimes on the order of milliseconds. However, some compounds have triplet lifetimes up to minutes or even hours, allowing these substances to effectively store light energy in the form of very slowly degrading excited electron states. If the phosphorescent quantum yield is high, these substances will release significant amounts of light over long time scales, creating so-called “glow-in-the-dark” materials.

4.4.2 Mechanism of Excitation and Recombination in Photo-Luminescence

An example depicting the slow process (i.e. phosphorescence) is shown in Fig. 4.5. The semiconductor shown here has an energy level E_t in the energy band gap, which might be due to an impurity or a defect. This energy level has a strong tendency to trap electrons from the conduction band. Different occurrences of this mechanism are elaborated as follows.

- ***AB path***. It shows the absorption of an incoming photon of energy $hf_1 > E_g$, thereby creating an electron-hole pair.

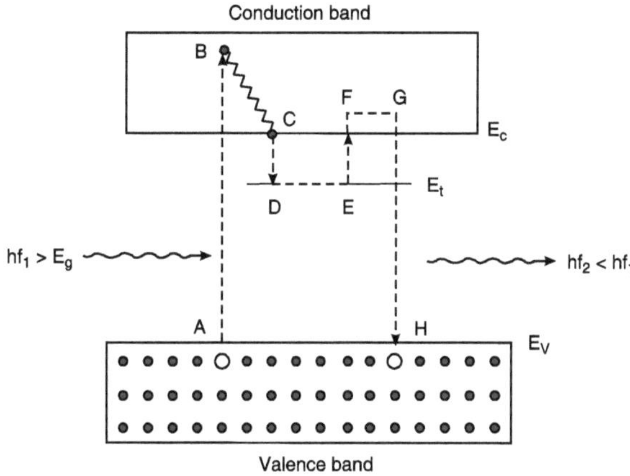

Fig. 4.5 Mechanism showing the excitation and recombination in the phenomenon of photo-luminescence

- ***BC path***. It shows the giving-up of energy by the excited electron to the lattice due to scattering. The process continues till the energy level reaches to a value of energy which is at the bottom energy value of conduction band.
- ***CD path***. It indicates the trapping of electron at impurity energy level E_t.
- ***DEF path***. The DE path depicts that the electron remains trapped at E_t level until it is thermally re-excited to conduction band as shown by EF path.
- ***FGH path***. It shows direct recombination as the electron falls to empty state in valence band, thereby giving-off a photon of hf_2 of the value of about E_g.

The delay time between excitation and recombination may be relatively long, if the probability of thermal re-excitation from the trap FG is small.

4.5 Electro-Luminescence

Electroluminescence (EL) is an optical and electrical phenomenon in which a material emits light in response to the passage of an electric current or to a strong electric field. This is distinct from black body light emission resulting from heat (incandescence), from a chemical reaction (chemiluminescence), sound (sonoluminescence), or other mechanical action (mechanoluminescence). Electroluminescence is the result of radiative recombination of electrons and holes in a material, usually a semiconductor. The excited electrons release their energy as photons—light. Prior to recombination, electrons and holes may be separated either by doping the material to form a p-n junction (in semiconductor electroluminescent devices such as light-emitting diodes) or through excitation by impact of high-energy electrons accelerated by a strong electric field (as with the phosphors in electroluminescent displays).

We have studied earlier that the electro-luminescene effect can be created by introducing the electric current into a semiconductor. This electrical current can be used in different ways to generate the photon emission from semiconductors. One such way is 'injection'. Therefore, the name of the process is injection electro-luminescence whose use is made in light-emitting diodes (LEDs). In them the minority carriers are injected by electric current, into the regions of a crystal where they can recombine with majority carriers. It results in emission of recombination radiation.

The effect of electro-luminiscence can be seen in devices incorporating the phosphor powder (such as of ZnS) in a plastic binder. The phosphor gives-off the light when an alternating current (a.c.) field is applied on it. Such device is known as 'electro-luminescence cell', which can be used as lighting panel.

Destriau effect. The above explanation stating that the emission of photons in certain phosphors occurs when they are subjected to alternating electric field was observed for the first time by Destriau. Hence this phenomenon is known as 'Destriau effect'.

4.5.1 Examples of Electroluminescent Materials

Electroluminescent devices are fabricated using either organic or inorganic electroluminescent materials. The active materials are generally wide bandwidth semiconductors to allow exit of the light. The most typical inorganic thin-film EL is ZnS:Mn with yellow-orange emission. Examples of the range of electroluminescent material include:

- Powdered zinc sulfide doped with copper (producing greenish light) or silver (producing bright blue light) or thin-film zinc sulfide doped with manganese (producing orange-red colour)
- Naturally blue diamond, which includes a trace of boron that acts as a dopant.
- Semiconductors containing Group III and Group V elements, such as indium phosphide (InP), gallium arsenide (GaAs), and gallium nitride (GaN).
- Certain organic semiconductors, such as $[Ru(bpy)3]^{2+}(PF6\text{-})^{2}$, where bpy is 2,2′-bipyridine

4.5.2 Practical Implementations

1. An electroluminescent nightlight in operation uses 0.08 W at 230 V.
2. The most common electroluminescent devices are composed of either powder primarily used in lighting applications or thin films for information displays.
3. Light-emitting capacitor (or LEC) is used to describe the electroluminescent panels. Flat electroluminescent panels are made as night lights and backlights for instrument panel displays. The drawbacks of conventional

electroluminescent lights are low efficiency and lifetimes, limited to hundreds of hours. Electroluminescent panels are a capacitor where the dielectric between the outside plates is a phosphor that gives-off photons when the capacitor is charged. By making one of the contacts transparent, the large area exposed emits light.
4. Electroluminescent automotive instrument panel backlighting.
5. Powder phosphor-based electroluminescent panels are frequently used as backlights to liquid crystal displays. They readily provide a gentle illumination to the entire display while consuming relatively little electric power. This makes them convenient for

   - battery-operated devices such as pagers,
   - wristwatches,
   - computer-controlled thermostats, and
   - gentle green-cyan glow

6. Thin film phosphor electroluminescence provides bright, long-life light emission in yellow-emitting manganese-doped zinc sulfide material. Displays using this technology were manufactured for medical and vehicle applications where ruggedness and wide viewing angles were crucial, and liquid crystal displays were not well developed.
7. Recently, blue-, red-, and green-emitting thin film electroluminescent materials that offer the potential for long life and full colour electroluminescent displays have been developed.

   In either case, the EL material must be enclosed between two electrodes and at least one electrode must be transparent to allow the escape of the produced light. Glass coated with indium-tin-oxide is commonly used as the front (transparent) electrode while the back electrode is coated with reflective metal. Additionally, other transparent conducting materials, such as carbon nanotube coatings can be used as the front electrode.

4.5.3 Advantageous Features

Electroluminescent technologies have low power consumption compared to lighting technologies, such as neon or fluorescent lamps. This, together with the thinness of the material, has made EL technology valuable to the advertising industry. Relevant advertising applications include electroluminescent billboards and signs. EL manufacturers are able to control precisely which areas of an electroluminescent sheet illuminate, and when. This has given advertisers the ability to create more dynamic advertising.

In principle, EL lamps can be made in any colour. However, the commonly used greenish colour closely matches the peak sensitivity of human vision, producing the greatest apparent light output for the least electrical power input. Unlike neon and fluorescent lamps, EL lamps are not negative resistance devices, therefore no extra circuitry is needed to regulate the amount of current flowing through them. Electroluminescent lighting is now used as an application for public safety identification involving alphanumeric characters on the roof of vehicles for clear visibility from an aerial perspective. Electroluminescent lighting has also made its way into clothing as many designers have brought this technology to the entertainment and night life industry.

4.6 Carrier Lifetime

Carriers, whether electron carrier or hole carrier, decay during the process of electron-hole pair recombination. The decay is generally of exponential nature. This decay is alternatively referred to as lifetime, and is known as 'carrier lifetime'. It can be either electron decay lifetime or hole decay lifetime. Since the process of recombination may occur directly or indirectly, the carrier lifetime is also obtained accordingly. Hence carrier lifetime is known as

1. Direct recombination lifetime, and
2. Indirect recombination lifetime.

The carrier recombination lifetime is generally of the order of 10^{-8}s. Its determination for direct recombination is simple, but determining carrier lifetime resulting from indirect recombination is more complicated. Their mechanism and recombination kinetics are discussed in subsequent sections.

4.7 Derivation of Carrier Lifetime in Direct Recombination Mechanism

When an electron in conduction band transits to valence band for combining with a hole, an excess population of electrons and holes are decayed. At the same time, the energy lost by the electron (in making transition) is given up as photon. Since in case of direct recombination, the process occurs spontaneously; therefore the probability of recombination of electron and hole is constant with time. Hence the rate of decay of electrons $\frac{dn}{dt}$ at any time t is proportional to the number of electrons n and number of holes p remaining at time t. This can be expressed as

$$\begin{aligned} \frac{dn}{dt} &\propto -np \\ \frac{dn}{dt} &\propto -\alpha_r np \end{aligned} \tag{4.6}$$

Here α_r is recombination proportionality constant. A minus sign indicates a loss (due to decay) of carriers.

Now considering the rate of thermal generation of electrons in conduction band, the net rate of change in electron concentration in conduction band may be expressed as

$$\frac{dn}{dt} = \alpha_r n_i^2 - \alpha_r np \tag{4.7}$$

where n_i is intrinsic concentration of electrons per cm^3.

4.7.1 Assumptions and Simplifications

If we assume that

(i) an excess electron-hole population is created at t = 0 by some means such as a flash of light, and
(ii) the initial concentration of excess electron Δn and excess holes Δp are equal (i.e. Δn = Δp), then the instantaneous concentration of excess electrons δn and excess holes δp will also be equal. It is because the electrons and holes recombine in pairs. Hence Eq. 4.7 may be rewritten in terms of equilibrium values n_0 and p_0 as

$$\frac{d(\delta n)}{dt} = -\alpha_r n_i^2 - \alpha_r[(n_0 + \delta n)(p_0 + \delta p)] \tag{4.8a}$$

As δn = δp, therefore Eq. 4.8a simplifies to

$$\frac{d(\delta n)}{dt} = -\alpha_r\left[(n_0 + p_0)\delta n + \delta n^2\right] \tag{4.8b}$$

On assuming that the excess carrier concentrations are small, the term δn^2 can be neglected. Furthermore if the semiconductor material is extrinsic, the equilibrium minority carrier term can be neglected. For example, in case of n-type ($n_0 \gg p_0$) and in case of p-type ($p_0 \gg n_0$). Therefore on taking p-type semiconductor, Eq. 4.8b modifies as

$$\frac{d(\delta n)}{dt} = -\alpha_r p_0 \delta n \tag{4.9}$$

4.7.2 *Solution of Equation to Determine Lifetime*

Solution of Eq. 4.9 yields an expression given by

$$(\delta n) = \Delta n e^{(-\alpha_r \rho_0 t)} = \Delta n e^{\left(-\frac{t}{\tau_{rln}}\right)} \tag{4.10}$$

It shows an exponential decay of carrier from the original excess carrier concentration Δn. In above expression τ_{rln} is recombination lifetime whose value is

$$\tau_{rln} = \frac{1}{\alpha_r p_0} \tag{4.11a}$$

This lifetime is also known as minority carrier lifetime since the above analysis is based on consideration of minority carriers. By similar reasoning, the decay of excess holes in n-type extrinsic semiconducting material occurs when

$$\tau_{rln} = \frac{1}{\alpha_r n_0} \tag{4.11b}$$

4.7.3 *Generalization of Expression*

A more general expression for the carrier lifetime τ_{cln} may be obtained by combining Eqs. 4.11a and 4.11b. It is given as

$$\tau_{cln} = \frac{1}{\alpha_r (n_0 + p_0)} \tag{4.12}$$

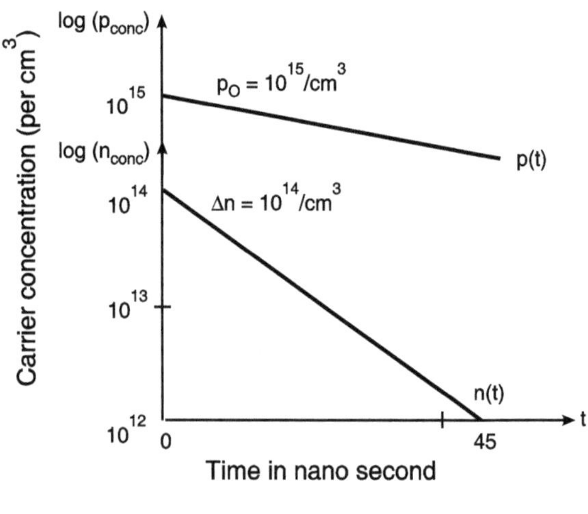

Fig. 4.6 Decay of excess electrons and holes by recombination on semi-logarithmic graph

This is, however, valid for a low injection level. Decay of carrier concentration as a function of time is shown on semi-log curve in Fig. 4.6. It depicts that if 10^{14} EHP/cm^3 are created at t = 0, the decay time of these carriers can be determined. The exponential decay of n as a function of t is linear on this semi- logarithmic graph.

4.8 Indirect Recombination (i.e. Capture or Trapping Process)

In indirect semiconductors such as AlP, SiC, GaP, PbS etc., the recombination occurs within the band gap via recombination energy levels. It occurs by any such lattice defect or impurity which serves as recombination centre and is capable of annihilating a pair. Annihilation is possible by receiving a carrier of one type (say n-type) and capturing the carrier of opposite type (say p-type). A typical recombination energy level E_r is shown in Fig. 4.7. It is below the Fermi energy level E_F at equilibrium. This energy level E_r is substantially filled up with electrons.

On creation of excess electrons and holes in semiconductor, each electron-hole pair recombines at E_r in following manner.

(i) Hole capture (shown by AB in figure). The first step in recombination process is the capture of hole as the recombination centres are filled at equilibrium. This situation is equivalent to an electron at energy level E_r falling to valence band, thereby leaving an empty state in E_r at B. In hole capture, the energy is given up in the form of heat to the lattice.

(ii) Electron capture (shown by CD in figure). Similarly when an electron from conduction band falls to an empty state in energy level E_r, the energy is again given up.

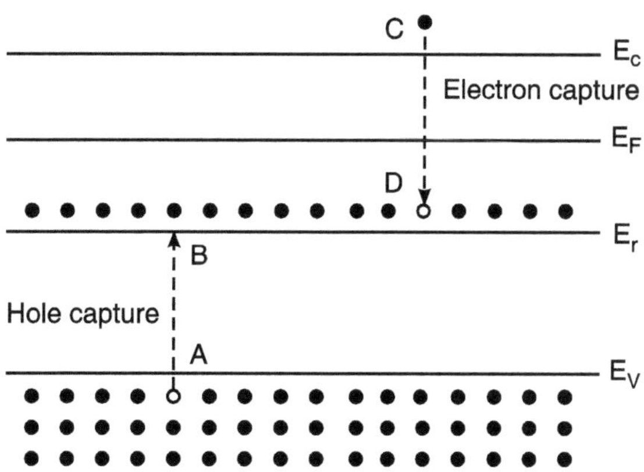

Fig. 4.7 Illustration of hole capture processes at a filled recombination centre and electron capture process at an empty centre

After completion of above events, the recombination centre attains its original state and is filled with an electron, but with a missing EHP. It indicates the recombination of one EHP. Now the recombination centre gets ready for repetition of another such event.

Example 4.3 Calculate the carrier lifetime in (a) Si, and (b) GaAs, when electrons as minority carriers are injected into p-type region having a hole concentration of $10^{18}/\text{cm}^3$. The injected electron density is small as compared to the density of majority carrier. Take recombination coefficient of Si as $1.79 \times 10^{-15}\ \text{cm}^3/\text{s}$ and that for GaAs as $7.21 \times 10^{-10}\ \text{cm}^3/\text{s}$.

Solution. Given are: $\alpha r_{Si} = 1.79 \times 10^{-15}\ \text{cm}^3\text{s}$, $\alpha r_{GaAs} = 7.21 \times 10^{-10}\ \text{cm}^3\text{s}$, and $p_0 = 10^{18}/\text{cm}^3$.

Hence using Eq. 4.12, we write

$$t_{cln} = \frac{1}{\alpha_r(n_0 + p_0)}$$

$$\approx \frac{1}{\alpha_r(p_0)} \quad (\text{as } n_0 \ll p_0)$$

(a) for Si $\quad t_{cln} = \frac{1}{(1.79 \times 10^{-15}) \times 10^{18}} = \mathbf{0.56\,ms}$

(b) for GaAs $\quad t_{cln} = \frac{1}{(7.21 \times 10^{-10}) \times 10^{18}} = \mathbf{1.39\,ns}$

4.9 Steady State Carrier Generation

Various recombination mechanisms are important in a simple at thermal equilibrium or with a steady state EHP generation-recombination balance. For example, a semiconductor at equilibrium experiences thermal generation of EHPs at a rate given by $\alpha_r n_0 p_0 = \alpha_r n_i^2$. This generation is balanced by the recombination rate so that the equilibrium concentrations of carriers n_0 and p_0 are maintained:

$$g(T) = \alpha_r n_i^2 = \alpha_r n_0 p_0 \tag{4.13}$$

This equilibrium rate balance can include generation from defect centers as well as band-to- band generation.

If a steady light is thrown on the sample, an optical generation rate g_{op} will be added to the thermal generation, and the carrier concentrations n and p will increase to new steady state values. The balance between generation and recombination in terms of the equilibrium carrier concentrations and the departures from equilibrium δn and δp, may now be written as

$$g(T) = g_{op} = \alpha_r np = \alpha_r (n_0 + \delta n)(p_0 + \delta p) \tag{4.14}$$

As δn = δp; therefore steady state recombination and no trapping, Eq. (4.14) becomes

$$g(T) = g_{op} = \alpha_r n_0 p_0 + \alpha_r \left[(n_0 + p_0)\delta n + \delta n^2 \right] \tag{4.15}$$

The term $\alpha_r n_0 p_0$ is just equal to the thermal generation rate g(T). Thus, neglecting the δn^2 term for low-level excitation, we can rewrite Eq. (4.15) as

$$g_{op} = \alpha_r (n_0 + p_0)\delta n + \frac{\delta n}{\tau_n} \tag{4.16}$$

The excess carrier concentration can be written as

$$\delta n = \delta p = g_{op}\tau_n \tag{4.17}$$

4.9.1 *Quasi-Fermi Levels*

It is often desirable to refer to the steady state electron and hole concentrations in terms of Fermi levels, which can be included in band diagrams for various devices. The Fermi level E_F used earlier is meaningful only when n_o excess carriers are present. However, we can write the expressions for steady state concentrations in the same form as the equilibrium expressions by defining separate quasi-Fermi levels F_n and F_p for electrons and holes. Then the resulting carrier concentration equations will be

$$\begin{aligned} n &= n_i e^{\frac{(F_n - E_i)}{kT}} \\ p &= n_i e^{\frac{(E_i - F_p)}{kT}} \end{aligned} \tag{4.18}$$

These can be considered as defining relations for the quasi-Fermi levels.

The quasi-Fermi levels of Fig. 4.8 illustrate the deviation from equilibrium caused by the optical excitation. The steady state F_n is only slightly above the equilibrium E_F, whereas F_p is greatly displaced below E_F. From the figure it is obvious that the excitation causes a large percentage change in minority carrier hole concentration and a relatively small change in the electron concentration.

Thus, the quasi-Fermi levels F_n and F_p are the steady state analogues of the equilibrium Fermi level E_F. When the excess carriers are present, the deviations of F_n and F_p from E_F indicate how far the electron and hole- populations are from the equilibrium values n_0 and p_0. A F_n and F_p for a Si sample given concentration of excess EHPs causes a large shift in the minority carrier quasi-Fermi level compared

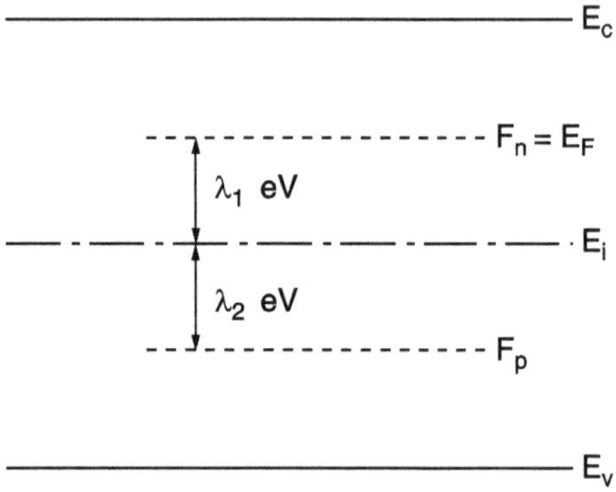

Fig. 4.8 A typical Quasi-Fermi levels

with that for the majority carriers. The separation of the quasi-Fermi levels $F_n - F_p$ is a direct measure of the deviation from equilibrium (at equilibrium $F_n = F_p = E_F$). The concept of quasi-Fermi levels is very useful in visualizing minority and majority carrier concentrations in devices where these quantities vary with position.

4.10 Photoconductivity

On creation of excess electrons and holes in a semiconductor, the conductivity of material increases. This has been described by the equation $\sigma = n_e\, e_e\, \mu_e + n_h\, e_h\, \mu_h$ in Chap. 2 also. Although the excess electrons and holes can be created by different means; but if these arise from optical luminescence, the increase in conductivity is called as 'photoconductivity'. This is an important phenomenon, which is utilised in the analysis and operation of semiconductor devices. Some of the practical applications are found in following devices.

- Diodes
- Transistors
- Photocells
- Lasers

4.10.1 Applications of Photoconductive Devices

There are numerous applications of photoconductive devices which change their resistance when exposed to light. Main among them are the following.

1 Light detectors such as used in

   (i) automatic controlling of night lights at home, streets, industries etc. These lights turn-on at dusk and turn-off at dawn, automatically.

(ii) exposure meters of cameras to measure the illumination level.

2 Moving object counters such as used in

(i) detecting a submarine, ship etc.
(ii) detecting an aircraft
(iii) low speed counting

3 Burglar alarms
4 Optical signaling systems
5 Television cameras
6 Photographic light meters
7 Infrared detectors
8 Automatic door opener
9 Fire alarms
10 Cinematography etc.

4.10.2 Photoconductive Materials and Factors Affecting Their Selection

A number of photoconductive materials are used in various semiconductor devices. Main among them are the following.

S.No	Example	Symbol	E_g (eV)	Main application
1	Germanium	Ge	0.67	In infrared portion of spectrum
2	Telurium	Te	0.34	–
3	Cadminum sulphide	CdS	2.42	Photoconductor in green light (visible range).
4	Cadminum selluride	CdSe	1.74	Photoconductor in red light
5	Cadmium telluride	CdTe	1.45	Photo conductor in infrared region
6	Lead sulphide	PbS	≈0.38	Photoconductors
7	Lead selluride	PbSe	≈0.32	Photoconductors
8	Lead telluride	PbTe	≈0.33	Photoconductors
9	Zinc sulphide	ZnS	3.82	Photo-luminescence
10	Zinc oxide	ZnO	3.3	–
11	Gallium selluride	GaSe	2.09	–
12	Gallium arsenide	GsAs	1.37	Diode, laser
13	Gallium antimonide	GaSb	0.82	Laser

(continued)

(continued)

S.No	Example	Symbol	E_g (eV)	Main application
14	Indium antimonide	InSb	0.18	In infrared portion of spectrum
15	Indium phosphide	InP	1.37	Photon detection
16	Alloy	CdS_xSe_{1-x}	1.80–2.50	–
17	Alloy	$GaAs_xP_{1-x}$	1.41–1.95	–

Reproduced from [1]

A semiconductor suitable for certain application, may not be suitable for other application. Similarly all semiconductors are not suitable for making the same device. Their selection is based on different factors. These are described below.

4.10.3 Factors Affecting the Selection of Semiconductor

Selection of an appropriate photo conducting material for a specific application depends on following factors.

- Sensitive wavelength range
- Time response, and
- Optical sensitivity of the material

In this respect, the under mentioned characteristics form the basis for selection of semiconductors.

1. Semiconductors are most sensitive to those photons whose energies are equal to or a bit more than the band gap E_g. Therefore,
   (i) less energetic photons are not absorbed.
   (ii) highly energetic photons ($h_f \gg E_g$) are absorbed at the surface, but they contribute only a little towards bulk conductivity.
2. Some semiconductors are sensitive to photons of those energies which are lesser than that of the band gap energy (i.e. $h_f < E_g$). It is because they respond to carrier excitations from those impurity levels which lie within the band gap.
3. Semiconductors used to make photoconductive cells have limited time response due to trapping of carriers, recombination time, and the time needed for carriers to drift through the device in an electric field. However, with proper choice of materials and device geometry, the time response properties can be improved.

4. Semiconductors meant to be used as photoconductors should have high optical sensitivity. For maximum photoconductive response, the material should have high mobilities and long lifetime. That is why, InSb is a good choice for making photoconductive devices. Its electron mobility is about 10^5 cm^2/Vs, and it is used as

   - sensitive infrared detector.

4.11 Photoconductive Cell [1]

Its operation is based on the principle that resistivity of semiconductor materials like Se, CdS, PbS, and thalmium-sulphide (TlS) etc. is decreased when irradiated. In means that, such materials have 'dark' high resistance and irradiated low resistance.

Construction. The simplest form of such a cell using selenium is shown in Fig. 4.9a, b. It consists of a semiconductor material with two electrodes attached to it. The resistance of the cell, when unilluminated (dark) is always high so that the current through the circuit shown in Fig. 4.9a is low. When the cell is illuminated, the cell resistance is decreased and the circuit current becomes large. The shape of the semiconductor material is so made as to obtain a large ratio of 'dark to light' resistance.

The commonly-used CdS cell is shown in Fig. 4.9b. It has a very high dark/light ratio and gives maximum response at about 5000 atomic unit. In order to increase the contact area with the sensitive material, the two electrodes are extended in an inter-digital pattern.

Uses. Photoconductive cells employing TIS and PbS are used for detecting ships and aircrafts with the help of radiations given out by their exhausts or funnels, and for telephony by modulated infrared light.

A circuit employing such a cell which may be used for low-speed counting or as a fire or burglar alarm is shown in Fig. 4.10a. Figure 4.10b shows a similar circuit using a thyratron and an a.c. supply. In each case when sufficient light falls on the cell, the valve conducts and energises the relay. Resistor R can be changed to adjust the light level for relay operation.

4.11.1 Photo-Multiplier Tube

In a phototube, the current is amplified by electron multiplication through secondary emission. Cross-section of one such multiplier tube is shown in Fig. 4.11. It consists of 9 electrodes (called dynodes) which are maintained at increasing potentials in sequence from the photocathode 0 to the anode marked 10 in the diagram. When light falls on the cathode, photoelectrons are emitted which are

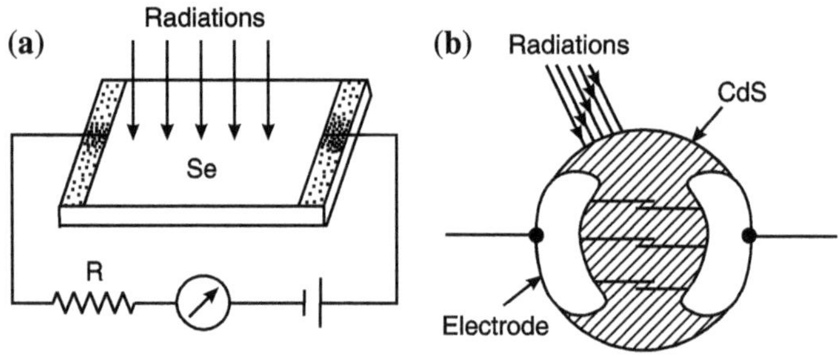

Fig. 4.9 Photoconductive cell showing **a** Se in circuit, and **b** CdS

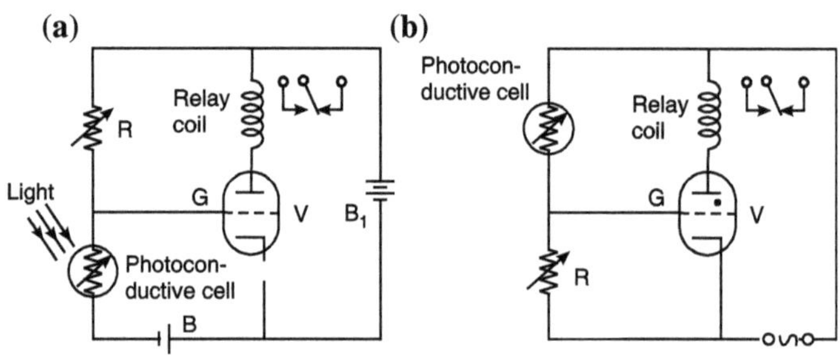

Fig. 4.10 A fire alarm circuit incorporating a photoconductive cell with a **a** battery, and **b** an a.c. supply

accelerated towards dynode no. 1, since it is at a higher potential. On colliding with this dynode, they liberate secondary electrons which are attracted by dynode no. 2 since it is at a relatively still higher potential. These electrons release further secondary electrons from it which are attracted by dynode no. 3. This process of secondary emission from different dynodes continues till the electrons are finally collected by the anode. It is obvious that at each dynode, the number of secondary electrons keeps on multiplying.

4.12 Diffusion of Carriers

Diffusion is a flow process that governs the movement of carriers within the material. It so happens when the excess carriers are created non-uniformly in a semiconductor. As a result, the electrons and holes change their position under the influence of electric field gradient and/or concentration gradient. Then a net motion of carriers from the region of high concentration to a region of low concentration starts. Thus the diffusion is a charge transport process.

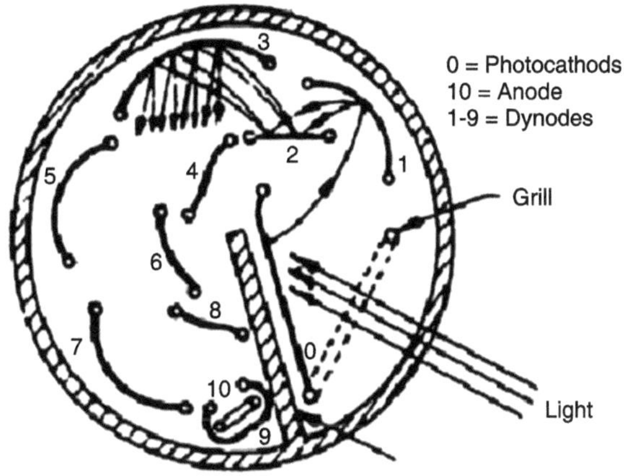

Fig. 4.11 Cross-section of a photo-multiplier tube

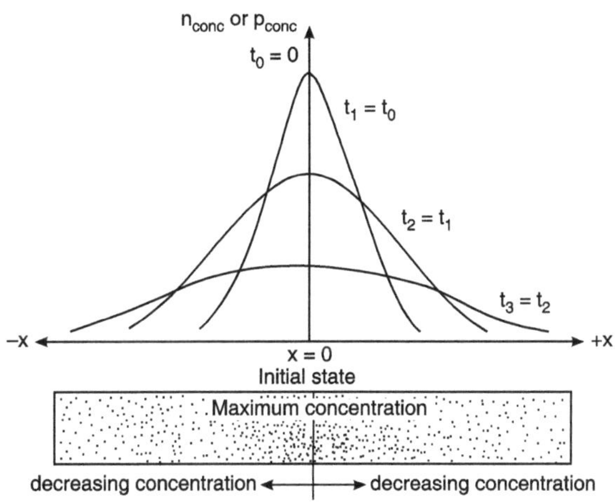

Fig. 4.12 Diffusion process causes spreading of a pulse of electrons

In semiconductors, the carriers diffuse by random thermal motion and scattering from the lattice and impurities. Spreading of a pulse of excess electrons is shown in Fig. 4.12. Initially at t = 0, the excess electrons are concentrated at x = 0, which with lapse of time diffuse to regions of lower concentration.

4.12.1 Determining the Rate of Electron and Hole Diffusion

The rate of electron diffusion in one-dimensional case can be determined by considering small incremental segments as shown in Fig. 4.13. It shows electron concentration n_{conc} as a function of diffusion path x. The hole concentration p_{conc} versus diffusion path can also be represented by this profile. Width of each segment

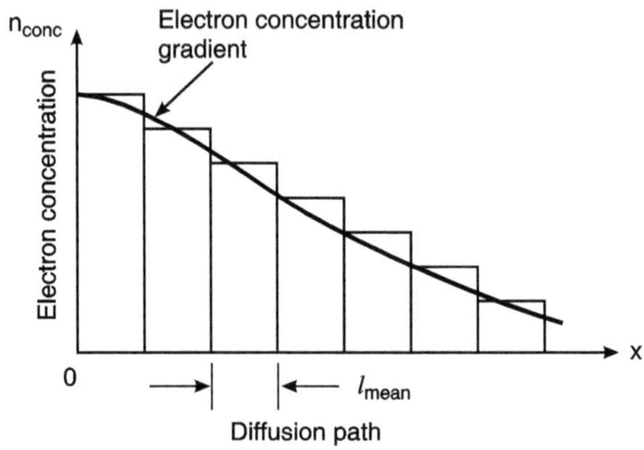

Fig. 4.13 An arbitrary electron concentration gradient along x-direction of semiconductor

is assumed l_{mean} which is equal to mean free length between two successive collisions. The rate of electron flow in positive direction of x i.e. from higher to lower concentration per unit area A may be expressed as follows.

$$\frac{dn}{dt} = -D_{nx} \cdot A \cdot \frac{dn_{conc}}{dx} \tag{4.19a}$$

and

$$\frac{dp}{dt} = -D_{px} \cdot A \cdot \frac{dp_{conc}}{dx} \tag{4.19b}$$

where $\frac{dn}{dt}$ and $\frac{dp}{dt}$ are the number of electrons and number of holes diffusing per unit time into a cross-sectional are A normal to the direction of diffusion path x.

$\frac{dn_{conc}}{dx}$ and $\frac{dp_{conc}}{dx}$ are concentration gradient of electrons and holes, respectively.

D_{nx} and D_{px} are diffusion coefficients of electron and hole respectively. These terms are also known as electron diffusivity and hole diffusivity. Their unit is mm^2/s or cm^2/s. Diffusion coefficients for some semiconductors are given in Table 4.1.

A negative sign in above equations indicates that the net motion of electrons and holes due to diffusion is in the direction of their decreasing concentration.

Equations 4.19a–4.19b can also be written as

$$J_n = \frac{1}{A} \cdot \frac{dn}{dt} = -D_{nx} \frac{dp_{conc}}{dx} \tag{4.20a}$$

$$J_P = \frac{1}{A} \cdot \frac{dp}{dt} = -D_{px} \frac{dp_{conc}}{dx} \tag{4.20b}$$

where J_n and J_p are electron and hole diffusion flux per unit cross-sectional area per unit time (i.e. current density). Its unit is (electron or hole)/(cm^2/s). In general, a change in the value of flux J_n and J_p occurs as a function of position x and time t.

Table 4.1 Electron and hole diffusivity of some semiconductors

Diffusivity	Si	Ge	GaAs
D_{nx} (cm^2/s)	35	100	220
D_{px} (cm^2/s)	12.5	50	10

Therefore, it is written as $J_n(x,t)$ and $J_p(x,t)$ under general situation of a problem. However in steady state condition, it is written as $J_n(x)$ and $J_p(x)$ only because the concentration does not change with time.

4.12.2 *Analysis of Drift and Diffusion Carriers*

We have seen earlier that the conduction mechanism in semiconductors is associated with 'drift of electrons' and 'diffusion of carriers'. Consequently, following two kinds of currents flow in a semiconductor.

1 Drift current due to electric field, and
2. Diffusion current due to gradient of carrier concentration.

Flow of these currents apply to both: to the electrons as well as to the holes. Since the conductivity associated with conduction electrons is $\sigma_e = n_e\, e\mu_e$ (see Chap. 3) and with conduction holes is $\sigma_h = n_h\, e\mu_h$, therefore the current density associated with drift of electrons and holes, due to the applied field E may be written as

$$J_e = n_e e \mu_e E \tag{4.21a}$$

and

$$J_h = n_h e \mu_h E \tag{4.21b}$$

Diffusivity. According to diffusion theory given by Fick's 1st law for steady state condition, the net flow of carriers along x-axis in one-dimensional case is given by

$$\frac{d\eta}{dt} = -D_x A \frac{d\eta}{dx} \tag{4.22}$$

where η is density of carriers, $d\eta/dt$ is the number of carriers diffusing per unit time into a cross- sectional area A normal to the direction of diffusion x, $d\eta/dx$ is concentration gradient, and D_x is diffusion coefficient along x-direction. The term D_x is popularly known as diffusivity. It is a constant and represents the system characteristic. The minus sign in above equation indicates that when concentration increases, the flow of particles due to diffusion occurs in negative x-direction. Equation 4.22 may also be written as

$$J_x = \frac{1}{A}\frac{d\eta}{dt} = -D_x\frac{d\eta}{dx} \tag{4.23}$$

where J_x is diffusion flux per unit cross-sectional area per unit time.

Thus the hole current may be written as the sum of contributions of current obtained from the electric field and from diffusion process, as follows.

$$J_h = n_h e\mu_h E - eD_{xh}\frac{d\eta_h}{dx} \tag{4.24a}$$

where D_{xh} is diffusion coefficient of the holes. The term e for charge appears in diffusion contribution because of the fact that the electric current is equal to the particle current times the charge e per particle. A similar expression may be written for electron also, and is given as

$$J_e = n_h e\mu_e E + eD_{xe}\frac{d\eta_e}{dx} \tag{4.24b}$$

In this case the diffusion contribution term is positive because the charge for electron is −e (not +e as in the case of holes).

The sum of Eqs. 4.24a and 4.24b yields the total (or resultant) current. It is

$$\begin{aligned} J_h + J_e &= (n_h e\mu_h E + n_e e\mu_e E) - \left(eD_{xh}\frac{d\eta_h}{dx} - eD_{xe}\frac{d\eta_e}{dx}\right) \\ &= \text{Drift current}-\text{Diffusion current} \end{aligned} \tag{4.25}$$

The diffusion current plays an important role in rectifier and transistor actions.

4.13 Einstein Relation

If a semiconductor is subjected to an electric field E and a concentration gradient $d\eta/dx$ such that the resultant current given by Eq. 4.25 is zero; an important relation can be established between the diffusion coefficient D_x and mobility μ of the carriers. This relationship is known as Einstein relation. Under the condition of no current, the system remains in thermal equilibrium, and the gradient of hole density is given by

$$\frac{d\eta_h}{dx} = -\frac{e}{KT}\eta_h\frac{dV}{dx} \tag{4.26}$$

where K is Boltzmann constant and $dV/dx = -E(x)$. Here V is potential and E is electric field as a function of direction x.

Since the hole current vanishes in thermal equilibrium, hence using Eqs. 4.24a and 4.26, we obtain

$$D_{xh} = \left(\frac{KT}{e}\right)\mu_h \tag{4.27a}$$

Similarly for electrons, the relationship exists as

$$D_{xe} = \left(\frac{KT}{e}\right)\mu_e \tag{4.27b}$$

Equations 4.27a–4.27b are Einstein relations and can be used to calculate diffusivity at different temperatures, provided the mobilities are known. These values for germanium at room temperature are

- $D_e = 0.0093\ m^2 s^{-2}$, and
- $D_h = 0.0044\ m^2 s^{-2}$.

4.14 Continuity Equation (i.e. Diffusion and Recombination)

Continuity equation describes the behaviour of minority carriers in rectifying junctions and transistors. The minority carriers (electrons) in p-type semiconductor and holes in n-type semiconductor play an important role in understanding the theory of above devices and their design. In p-type semiconductor, the basis of derivation for continuity equation is

$$\begin{bmatrix} \textit{Time rate of} \\ \textit{increase of} \\ \textit{electrons} \end{bmatrix} = \begin{bmatrix} \textit{Rate of thermal} \\ \textit{generation of} \\ \textit{electrons} \end{bmatrix} + \begin{bmatrix} \textit{Rate of increase} \\ \textit{due to recombination} \end{bmatrix}$$

if there is no electron current or hole current. And if there is an electron current flowing through it, the formulation of basic equation will involve

$$\begin{bmatrix} \textit{Total rate of change} \\ \textit{of electron density } n_e \end{bmatrix} = \begin{bmatrix} \textit{Rate of change} \\ \textit{of } n_e \textit{ for } J_e = 0 \end{bmatrix} + \begin{bmatrix} \textit{Rate of chnge} \\ n_e \textit{ due to } J_e \end{bmatrix}$$

Here J_e is current density. A rigorous analysis ultimately leads to following continuity equations which must be satisfied by the minority carriers in their operations.

(i) In p-type semiconductor where electrons are minority carriers,

$$\frac{d\eta_e}{dt} = -\frac{\eta_{eo} - \eta_e}{\tau_e} + \frac{1}{e} \cdot \frac{dJ_e}{dx} \tag{4.28a}$$

(ii) In n-type semiconductor where holes are minority carriers,

$$\frac{d\eta_h}{dt} = -\frac{\eta_{ho} - \eta_h}{\tau_h} - \frac{1}{e} \cdot \frac{dJ_h}{dx} \tag{4.28b}$$

Here J_e and J_h are the current densities as given by Eqs. 4.21a and 4.21b; η_{eo} and η_{ho} are thermal equilibrium densities, and τ_e and τ_h are the lifetime (time constant) of electrons and holes respectively.

4.14.1 Diffusion Equation and Diffusion Length

When the current in a semiconductor is carried by diffusion only (i.e. the drift is negligible), the following expressions of diffusion current can be substituted for the currents in continuity equations.

$$J_e = eD_{xe}\frac{\partial \eta_e}{\partial x} \tag{4.29a}$$

$$J_h = eD_{xh}\frac{\partial \eta_h}{\partial x} \tag{4.29b}$$

Hence, we obtain the following diffusion equations.

$$\text{For electrons :} \quad \frac{\partial \eta_e}{\partial t} = D_{xe}\frac{\partial^2 \eta_e}{\partial x^2} - \frac{\eta_e}{\tau_e} \tag{4.30a}$$

$$\text{For holes :} \quad \frac{\partial \eta_h}{\partial t} = D_{xh}\frac{\partial^2 \eta_h}{\partial x^2} - \frac{\eta_h}{\tau_h} \tag{4.30b}$$

Steady state carrier injection. The above equations can be used to solve transient problems involving diffusion with recombination. However, there are many other problems in which a steady state distribution of excess carriers is maintained. In them the time derivatives are zero, and the diffusion equations given by Eq. 4.30a–4.30b modify to following form.

$$\frac{d^2\eta_e}{dx^2} = \frac{\eta_e}{D_{xe}\tau_e} = \frac{\eta_e}{L_{hdiff}^2} \tag{4.31a}$$

$$\frac{d^2\eta_h}{dx^2} = \frac{\eta_h}{D_{xh}\tau_h} = \frac{\eta_h}{L_{hdiff}^2} \tag{4.31b}$$

Here the term $L_{ediff} = \sqrt{D_{xe}\tau_e}$ is called the electron diffusion length and $L_{hdiff} = \sqrt{D_{xh}\tau_h}$ is known as the hole diffusion length.

4.15 Transport of Charges and Impurity Distribution Profile During Diffusion

The transport of charges in a semiconductor may be accounted for by a phenomenon called 'diffusion'. Diffusion of impurities into silicon chip is the most important process in fabrication of ICs. The essential features of diffusion are briefly discussed below.

Diffusion laws. Diffusion is a flow process that governs the movement of atoms and molecules in solids. It can be microscopic or macroscopic in nature. In doping process, the diffusion occurs at microscopic level. Diffusion processes are governed by two laws, namely (i) Fick's first law, and (ii) Fick's second law. The second law deals with the practical situations of diffusion in materials. It considers the non-steady state flow. The doping process is also governed by 'Fick's second law'.

The governing differential equation of Fick's second law is given by

$$\frac{\partial M}{\partial t} = D_x \frac{d^2M}{dx^2} \tag{4.32}$$

where M is concentration at a distance x and time t, and D_x is diffusivity along x-direction. It is expressed in m^2/s.

4.15.1 Solution by Error Function Method

Fick's second law stated by Eq. 4.32 is simple to solve the problems. We shall take up the solution of Eq. 4.32 by error function method. Its general solution for diffusion across a common interface is obtained as

$$M(x,t) = C_1 - C_2 \text{erf}\left(\frac{x}{2}\sqrt{D_x t}\right) \tag{4.33}$$

where C_1 and C_2 are constants, and are evaluated from initial and boundary conditions of a problem. In Eq. 4.33, 'erf' stands for error function. This is a mathematical function defined as follows :

$$\mathrm{erf}(p) = \frac{2}{\sqrt{x}} \int_0^p e^{(-\lambda)} d\lambda \tag{4.34a}$$

From Eq. 4.33 and 4.34a, it can be seen that the quantity p is expressed as

$$\mathrm{p} = \frac{x}{2\sqrt{D_x t}} \tag{4.34b}$$

where λ is an integration variable. Figure 4.14 shows the plot between $e^{(-\lambda^2)}$and λ. The hatched area under the curve from λ = 0 to p gives the value of 'erf' in Eq. 4.34a. The area under the curve for λ = 0 to λ = ± ∞ comes out to be $\pm\sqrt{\pi/2}$.

Following values of error function should be kept in mind.

$$\mathrm{erf}(0) = 0, \quad \mathrm{erf}(-\infty) = -1$$
$$\mathrm{erf}(+\infty) = +1, \quad \mathrm{erf}(-p) = -\mathrm{erf}(p)$$

Error function of quantity p may be seen in Error function Table as given in Ref. [2].

4.15.2 Complementary Error Function and Gaussian Distribution

The problems of diffusion during doping of semiconductors are of two kinds. These are discussed below as case I and case II.

Case I: Constant surface concentration per unit volume. Dopants are the inpurities that are deliberately diffused in silicon and germanium to make extrinsic (compound) semiconductors. We consider the case of an intrinsic silicon wafer

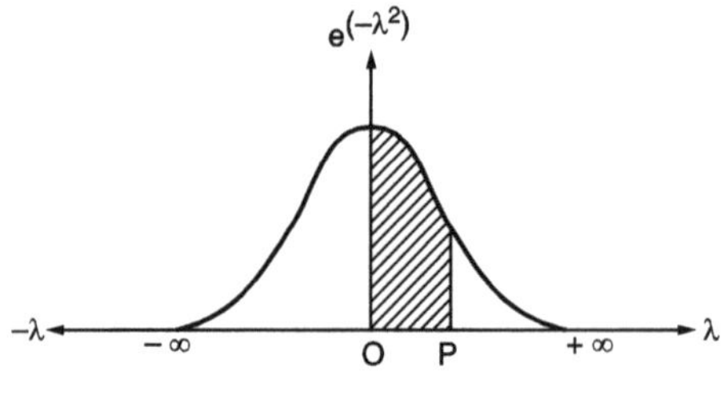

Fig. 4.14 Curve of error function shows that the hatched area is equal to the value of integral in Eq. 4.34a

(chip) which is exposed to gaseous surrounding of phosphorus, a n-type dopant, having a uniform concentration M_0 atoms per unit volume, Fig. 4.15a. Let the diffusion continues for extremely long time so that the silicon is uniformly doped with M_0 phosphorus atoms per unit volume. It is assumed that the surface concentration has

$$M(\mathrm{x}) = 0 \text{ at } \mathrm{t} = 0 \quad \text{for} \quad \mathrm{x} > 0 \tag{4.35a}$$

and

$$M(\mathrm{x}, \mathrm{t}) = \mathrm{M}_0 \quad \text{for all values of} \quad \mathrm{t} \tag{4.35b}$$

Substitution of above boundary conditions in Eq. 4.33 yields

$$\begin{aligned} \mathrm{M}(x, t) &= M_0\left(1 - erf\frac{x}{2\sqrt{D_x t}}\right) \\ &= M_0 erfc\frac{x}{2\sqrt{D_x t}} \end{aligned} \tag{4.36}$$

where 'erfc' means **complementary error function**, and erfc(p) = 1 – erf(p).

Case II: Constant surface concentration per unit area. Diffusional concentration as a function of distance and time may be obtained by Gaussian distribution when total number of atoms M_0 per unit area A is held constant on the surface x = 0 of the silicon wafer. Diffusion of phosphorus atoms into silicon takes place between ends x = 0 and x = ∞ when the material is heated. Following boundary conditions are considered.

$$M(\mathrm{x}) = 0 \text{ at } \mathrm{t} = 0 \quad \text{for} \quad \mathrm{x} > 0 \tag{4.37a}$$

and

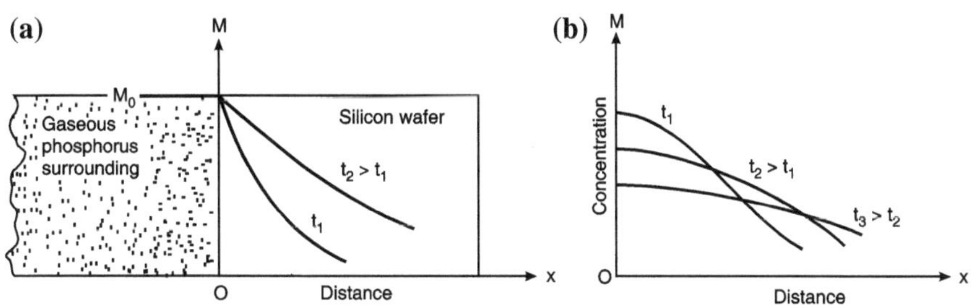

Fig. 4.15 Concentration M as a function of distance x in a silicon wafer for two different diffusion times when surface concentration is held constant **a** at M_0 per unit volume, and **b** M_0 per unit area

$$\int_0^{\infty} M(x,t)dx = \frac{M_0}{A} \tag{4.37b}$$

When above boundary conditions are applied to Eq. 4.33, we find

$$M(x,t) = \frac{M_0}{A\sqrt{\pi D_x t}} e^{-\frac{x^2}{4D_x t}} \tag{4.38}$$

Equation 4.38 is known as **Gaussian distribution**. It is shown in Fig. 4.15b. The area under each curve indicates total amount of phosphorus diffused. The two areas are equal and each is equal to M_0. It also indicates that the surface concentration decreases as the time increases.

Example 4.4 The ratio of rate of diffusion in silicon at 1350 °C to that at 1100 °C was found to be 8 in a doping process. Taking gas constant 8.314 J/mol/K, calculate the activation energy for silver diffusion.

Solution. The ratio of rate of diffusion is expressed as

$$\frac{\frac{dM}{dt} at 1350\,^{\circ}\mathrm{C}(1623\,\mathrm{K})}{\frac{dM}{dt} at 1100\,^{\circ}\mathrm{C}(1373\,\mathrm{K})} = 8 \tag{i}$$

From the Fick's first law, we know that at constant concentration gradient, the rate of diffusion is directly proportional to diffusivity. Hence Eq. (i) may be written as

$$\frac{D_x \ at \ 1623\,\mathrm{K}}{D_x \ at \ 1373\,\mathrm{K}} = 8 \tag{ii}$$

Now we use $\mathrm{D_x = D_0 e^{(-Q/RT)}}$. It gives

$$\mathrm{D_x(at\ 1623\,K)} = \mathrm{D_0.e^{-Q/13496}} \tag{iii}$$

$$\mathrm{D_x(at\ 1373\,K)} = \mathrm{D_0.e^{-Q/11415}} \tag{iv}$$

where RT = 13495 and 11415 respectively.

On dividing Eq. (iii) by (iv), we have

$$\frac{D_x(\mathrm{at 1623\,K})}{D_x(\mathrm{at 1373\,K})} = e^{-\left[\left(\frac{Q}{13495}\right)-\left(\frac{Q}{11415}\right)\right]} \tag{v}$$

Using Eq. (ii), the Eq. (v) may be rewritten as

$$8 = e^{-1.35\times10^{-5}\times Q}$$

Taking log to base e on both sides, we get

$$2.08 = 1.35 \times 10^{-5} \log_e e = 1.35 \times 10^{-5} \times Q \times 1$$

$$\therefore \quad Q = \frac{2.08}{1.35 \times 10^{-5}} = \mathbf{154} \frac{\mathbf{kJ}}{\mathbf{mol}}$$

4.16 Long Diode and Short Diode

Long diodes are those devices in which the diffusion lengths are small as compared to the lengths of p and n regions. But in short diodes, the diffusion lengths are long as compared to the lengths of p and n regions. A comparison between them is summarized below.

Description	Long diode	Short diode
• Diffusion lengths as compared to lengths of p and n regions	Small	Long
• Nature of decay of injected minority carriers on either side of depletion region	Exponentially up to their equilibrium values	Almost linearly to zero value
• Situation of ohmic contact	Far away from the equilibrium value as shown in Fig. 4.16a	At the zero equilibrium value as shown in Fig. 4.16b

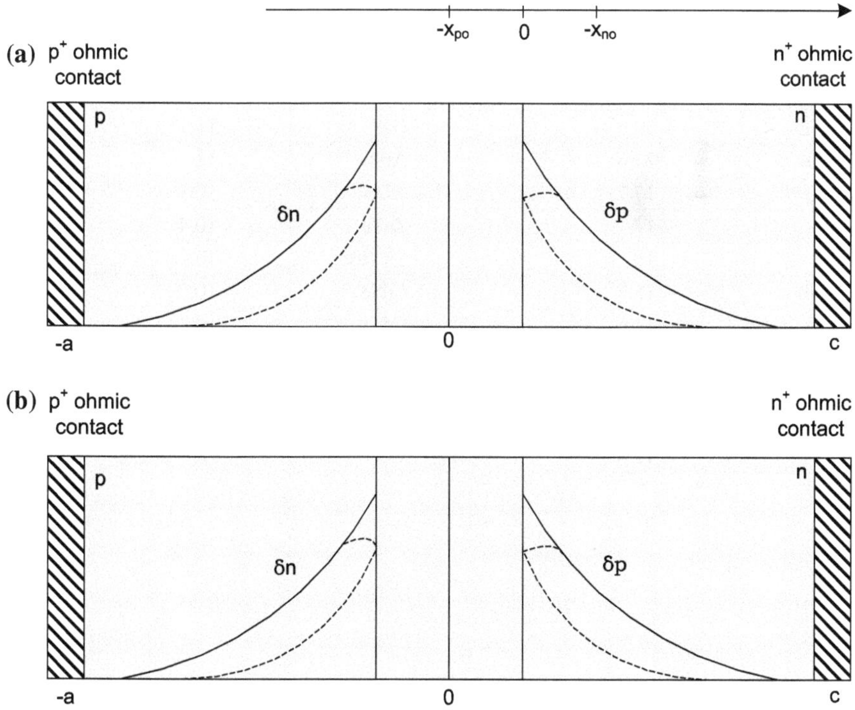

Fig. 4.16 **a** Long diode **b** Short diode

4.16.1 *Voltage-Variable Capacitance*

The junction capacitance C_{junc} is termed as voltage-variable capacitance. It is named so because it varies with voltage, which is given as

$$C_{junc} \propto \frac{1}{\sqrt{V_0 \times V}} \tag{4.39}$$

Some important applications of it are as given below.

- in tuned circuits
- as varactor

4.16.2 *Effect of Dielectric Constant on Width of the Transition Region and Voltage Capacitance*

The effects of dielectric constant on width of transition region and voltage capacitance are as given by Eq. 4.40. These indicate that

$$w \propto \sqrt{\varepsilon_r} \, and \, C_{junc} \propto \sqrt{\varepsilon_r} \tag{4.40}$$

The values of ε_r for various semiconductors vary from about 9–30. A list in this regard is given below for a ready reference.

Semiconductor	Dielectric constant ε_r	Semiconductor	Dielectric constant ε_r
PbTe	30	GaP	11.1
PbSe	23.6	AlSb	11
InSb	17.7	AlAs	10.9
PbS	17	ZnTe	10.4
Ge	16	SiC(α)	10.2
GaSb	15.7	CdSe	10.2
InAs	14.6	CdTc	10.2
GaAs	13.2	AlP	9.8
InP	12.4	ZnSe	9.2
GaN	12.2	ZnS	8.9
Si	11.8		

4.17 Solved Examples

Example 4.5 Energy band gap in a Ge crystal is 0.75 eV. Determine the wavelength at which Ge starts to absorb light.

Solution. Given are $E_g = 0.75\ \text{eV} = 0.75 \times 1.6 \times 10^{-19}$ J

$$\because \quad E_g = hf = \frac{hc}{\lambda}$$

$$\therefore \quad \lambda = \frac{hc}{E_g}$$

$$= \frac{(6.62 \times 10^{-34}) \times (3 \times 10^8)}{0.75 \times 1.6 \times 10^{-19}} = \mathbf{16560\,Å}$$

Example 4.6 What is quantum efficiency?

Solution. Quantum efficiency is defined as the ratio of 'electrons generated' to the 'incident photons absorbed by the material'. It is expressed as

$$\eta_{quantum} = \frac{number\ of\ generated\ electrons\ per\ second}{number\ of\ incident\ photons\ per\ second} = \frac{\eta_{gen}}{\eta_{inc}}$$

Quantum efficiency mainly depends on absorption coefficient. It is always <1 because all the incident photons do not create electron-hole pairs.

Example 4.7 Find the cutoff wavelength for InP having bandgap energy 1.35 eV. Take appropriate values suitably. Energy band gap for indium phosphide is 1.35.

Solution. Given is: $Eg = 1.35\ \text{eV} = 1.35 \times 1.6 \times 10^{-19}$ eV

$$\therefore \quad \lambda = \frac{hc}{E_g} = \frac{(6.63 \times 10^{-34}) \times (3 \times 10^8)}{0.35 \times 1.6 \times 10^{-19}} = \mathbf{0.902\,\mu m}$$

Example 4.8 When 3×10^{12} photons of wavelength 0.85 μm are incident on a photodiode, an average of 1.2×10^{12} electrons are collected at the terminals of the device. Find the (a) quantum efficiency, and (b) responsivity at 0.85 μm.

Solution. Given are: $\lambda = 0.85\ \mu m = 0.85 \times 10^{-6}$ m. We take appropriate values of electronic charge e, Planck's constant h, and velocity of light c.

(a)

$$As\ the\ quantum\ efficiency = \frac{no.\ of\ collected\ electrons}{no.\ of\ incident\ photons}$$

$$\therefore \quad \eta_{quantum} = \frac{1.2 \times 10^{12}}{3 \times 10^{12}} = 0.4\ or\ \mathbf{40}\ \%$$

(b) The responsivity is determined from

$$R = \frac{\eta_{quantum} e\lambda}{hc}$$
$$= \frac{(0.4) \times (1.6 \times 10^{-19}) \times (0.85 \times 10^{-6})}{(6.63 \times 10^{-31}) \times (3 \times 10^{8})} = \mathbf{0.272\,As^{-1}}$$

Example 4.9 Cite an example of photoluminescence with a brief explanation.

Solution. The 'fluorescent lamp' is a common example of photoluminescence. It comprises of a glass tube filled with a gas, which is generally the mixture of mercury vapours and argon. The inside of the tube has a fluorescent coating. When an electric discharge is induced between electrodes of the tube, the atoms of the gas are excited and emit photons. This emission is largely in the visible and U/V regions of the spectrum. The light thus produced is absorbed by the luminescent coating and hence the visible photons are emitted.

Example 4.10 What is the utility of phosphor in the fabrication of a colour T.V. screen?

Solution. The colour of the light emitted by phosphor depends mainly on the impurities present in it. It is so because many radiative transitions involve impurity levels within the band gap of certain materials such as ZnS. Hence, selection of proper phosphors is useful in fabrication of colour T.V. screen.

Example 4.11 A photodiode has quantum efficiency 60 % when photons of energy 1.5×10^{-19} are incident on it. Determine (a) the wavelength at which the photodiode is operating. Also find (b) the incident optical power required to obtain a photon current of 3 μA for the said photodiode.

Solution. (a) $\because \quad \lambda = \frac{hc}{E}$

$$\therefore \quad \lambda = \frac{(6.626 \times 10^{-34}) \times (3 \times 10^{8})}{1.5 \times 10^{-19}} = \mathbf{1.32\,\mu m}$$

(b) $\because \quad R = \frac{\eta_{quantum} e}{hf} = \frac{0.6 \times (1.6 \times 10^{-19})}{1.5 \times 10^{-19}} = 0.64\ \mathrm{AW^{-1}}$

and also $R = \frac{IP}{P_0}$

$$\therefore \quad P_0 = \frac{IP}{R} = \frac{3 \times 10^{-6}}{0.64} = \mathbf{4.67\,\mu W}$$

Example 4.12 Find the maximum energy of the emitted photoelectrons when the light of frequency 1.5×10^{9} MHz falls on the surface of a material whose threshold frequency is 1.2×10^{9} MHz.

Solution. As $E = hf$

$$\therefore \quad E_{max} = h(f - f_0)\,\text{J}$$
$$= \left(6.62 \times 10^{-34}\right) \times \left[\left(1.5 \times 10^{15}\right) - \left(1.2 \times 10^{15}\right)\right]$$
$$= 6.62 \times 10^{-34}(1.5 - 1.2) \times 10^{15}\,\text{J}$$
$$= \frac{6.62 \times 10^{-34} \times 0.3 \times 10^{15}}{1.602 \times 10^{-19}}\,\text{eV}$$
$$= \mathbf{1.24\,eV}$$

Example 4.13 What is meant by trapping centre?

Solution. An impurity or a defect centre in semiconductor is referred to as trapping centre (or trap).

Example 4.14 What is phosphorescence?

Solution. It is a phenomenon of delayed luminescence in which the material emits visible light after some delay (1 s or more) on absorption of optical energy.

Example 4.15 What is lattice absorption edge?

Solution. In pure semiconductors, the absorption coefficient drops rapidly at a wavelength near to or in the intermediate infrared region, and the material becomes transparent at larger wavelengths. This marked drop in absorption coefficient is called lattice absorption edge or only 'absorption edge'. The absorption edges for some semiconductors are given as follows.

- InSi, at $\lambda = 1.1\ \mu\text{m}$
- InGe, at $\lambda = 1.65\ \mu\text{m}$

Example 4.16 What are different laws of diffusion? What are their importance to electronic materials and devices?

Solution. Classical laws of diffusion were propagated by Fick. Hence they are known as Fick's laws. There are two laws viz. Fick's 1st law and Fick's 2nd law. Fick's 1st law describes the flow of electrons and holes etc. under steady state conditions, while the 2nd law states non-steady state flow.

These laws are of great importance to electronic materials and devices because the mechanism of semiconduction by electrons and holes, and the doping process etc. are governed by them.

Example 4.17 What is the significance of diffusion equation?

Solution. Diffusion equations are useful in solving transient problems involving diffusion with recombination. An example of such a problem may be like. "A pulse of electrons in a semiconductor spreading out by diffusion and disappearing by recombination."

Example 4.18 What is Haynes-Shockley experiment? What is its importance?

Solution. The Haynes-Shockley experiment demonstrates the drift and diffusion of minority carriers in semiconductors. With the help of this experiment, the measurement of mobility μ and diffusivity D of minority carriers can be accomplished.

Example 4.19 What are the characteristic properties of photons?

Solution. Photons are the basic constituents of any radiation. They possess the following characteristic properties.

(i) Energy content
(ii) Mass
(iii) Momentum

The energy of photon is independent of intensity, and has non-electrical nature. Various properties are expressed as follows.

- $$E = nhf \quad \text{(i)}$$
  where $n = 1, 2, 3, \ldots$ are principalquantum number.
- $$E = mc^2 \quad \text{where c is speed ofthe light.} \quad \text{(ii)}$$
  For $n = 1, E = hf = mc^2$

$$\therefore \quad f = \frac{c}{\lambda}, \text{ where } \lambda \text{ is wavelength of photon} \quad \text{(iii)}$$

Kinetic mass of photon is given by

$$m_{photon} = \frac{h}{c\lambda} \quad \text{(iv)}$$

and momentum of photon is given by

$$p = mc = \frac{hf}{c} = \frac{h}{\lambda} \quad \text{(v)}$$

Example 4.20 What is meant by dark resistance of a photoconductor?

Solution.It is the resistance of a photoconducting cell when it is un-illuminated. Dark resistance is always higher than the bright resistance (i.e. the resistance of the cell when illuminated).

Example 4.21 What is diffusivity? What is its application in semiconductor materials? How is diffusivity influenced by temperature?

Solution. Diffusivity is a flow process that governs the movement of atoms and molecules in solids. Doping of intrinsic semiconductor (pure elements like Si, Ge) by dopants (e.g. P, As, Ga etc.) is accomplished by diffusion process. Effect of temperature is to enhance the rate of diffusion in materials.

Example 4.22 How can the carrier lifetime be determined?

Solution. The carrier lifetime can be determined by following means.

1. By measuring the storage delay time t_{sd}, experimentally; and then to calculate the lifetime τ_p from Eq. 6.8 (see Chap. 6).
2. By measuring the photoconductive decay.

Example 4.23 What is quasi-steady state? What is its importance in the analysis of transients problems?

Solution. This is a condition intermediatory between the steady and unsteady states. The quasi-steady state solutions of various transient problems are based on gross approximations and give crude results. However, the quasi-steady state approach to a problem provides a comparatively simpler solution. The time involved in the analysis is also less.

Example 4.24 Obtain the expression for junction voltage using quasi-steady state approach.

Solution. The expression for junction voltage V_{junc} can be obtained by assuming an exponential distribution for δp at every instant, during the decay of stored charge. The quasi-steady state does not consider the effects of diffusion, and also neglects the distortion due to slope at $x_n = 0$, during transient conditions.

Hence if we take

$$\delta p(x_n, t) = \Delta p_n e^{-\frac{x_n}{L_p}} \tag{4.41}$$

the value of stored charge at any instant will be

$$Q_P(t) = eA \int_0^\infty \Delta p_n e^{-\frac{x_n}{L_P}} dx_n = eAL_P \Delta p_n \tag{4.42}$$

$$\therefore \quad \Delta p_n = \frac{Q_p(t)}{eAL_P} \tag{4.43}$$

Now on relating Eq. 4.43 with Eq. 6.7, we get

$$\begin{aligned} p_n \cdot \left(e^{\frac{eV}{KT}} - 1\right) &= \frac{Q_p(t)}{eAL_P} \\ \therefore \quad V_{junc} &= \frac{kT}{e} ln\left(\frac{I\tau_p}{eAL_P p_n} e^{-\frac{t}{\tau_P}+1}\right) \end{aligned} \tag{4.44}$$

Review Questions

1. Define the following.
   (a) Photon (b) Optical absorption (c) Luminescence
   (d) Photoconductivity (e) Diffusion (f) Drift
2. What do you mean by optical absorption? State the conditions of photon absorption and discuss its mechanism.
3. What is absorption coefficient? Derive the equation to determine the intensity of transmitted light through a semiconductor sample.
4. Discuss the factors that influence the absorption coefficient of a semiconductor. Enlist semiconductors lying in different regions of spectrum of light.
5. Explain the phenomenon of luminescence. What are its different types? How does fluorescence differ from phosphorescence?
6. What is photo-luminescence? What are its different types? Explain the mechanism of excitation and recombination in respect of it?
7. What is meant by carrier lifetime? How does direct recombination lifetime differ from indirect recombination lifetime?
8. Derive an expression for carrier lifetime in direct recombination mechanism. Write the assumptions and simplifications made therein.
9. What is photoconductivity? What are its applications in semiconductor devices? List various photoconductive materials with their specialities.
10. Discuss the factors that govern the selection of an appropriate photoconducting material for a specific application.
11. How do the following factors influence the selection of photoconductive materials for use in semiconductor devices : (a) wavelength range, (b) time response, and (c) optical sensitivity
12. Sketch and explain the construction and working of a photoconductive cell. How does it work to detect a submarine in the ocean?
13. Explain the working of a fire alarm with the help of a circuit meant for it, that employs a photoconductive cell.
14. How will you determine the rate of electron and hole diffusion. Develop and write expressions in this regard.
15. The current flowing through a junction formed by two dissimilar semiconductors is the sum of diffusion and drift currents. Explain how and why these two different effects happen to develop?

16. What is Einstein relation? Develop expressions to establish relations between diffusion coefficient and mobility of carriers.
17. Discuss and derive the continuity equations. What are their importance for rectifying junctions and transistors?
18. Discuss the governing differential equation to induce transport of charges and impurity distribution profile during diffusion in semiconductors. How is the 'error function method' used to solve such problems?
19. Explain the complementary error function and Gaussian distribution in respect of solution to Fick's diffusion problems.
20. Distinguish between Long and short diodes
21. Distinguish between the following.
    (a) Cathode-luminescence and Chemed-luminescence
    (b) Direct recombination and indirect recombination
    (c) Dift and diffusion of carriers
22. Write notes on the following.
    (a) Electro-luminescence
    (b) Decay of carriers
    (c) Diffusion length
    (d) Photo-multiplier tube
    (e) Quasi-Fermi level

Numerical Problems

1. Calculate the energy content of ultraviolet light photons of wave length 3000 Å in (a) joule, and (b) electron-volt.
   [**Ans** : (a) 6.62×10^{-19} J (b) 4.13 eV]
2. Determine the energy of photons of a spectral emission line whose wavelength is 210 mm. [**Ans** : 5.9×10^{-6} eV]
3. Find the frequency and amount of energy associated with a quantum of electromagnetic radiation at each of the following wavelengths (a) 1 m, (b) 1 mm, (c) 5000 Å, and 1 Å.
   [**Ans** : (a) 3×10^{8} Hz, 1.25×10^{-6} eV (b) 3×10^{11} Hz, 1.25×10^{-3} eV, 6×10^{14} Hz, 1.25 eV (d) 3×10^{18} Hz, 1.25×10^{4} eV]
4. A compound semiconductor sample, 0.46 μm thick is illuminated by a monochromatic light of photon energy = 2 eV. The power incident on the sample is 10 mW. If the absorption coefficient is 5×10^{4} cm^{-1}, calculate (a) the total energy absorbed by the same per second, (b) the rate of excess thermal energy given up by the electrons to the lattice before recombination, and (c) the number of photons given off per second from recombination. The quantum efficiency is perfect. [**Ans** : (a) 9 mW (b) 2.57×10^{-3} J/s (c) 2.81×10^{16} photons/s]

5. The wavelength of light from a GaP laser is 5490Å. Compute the energy gap of it. [**Ans** : 2.26 eV]

6. The holes and electron mobilities in a Si single crystal at 300 K are 0.025 m^2/Vs and 0.17 m^2/Vs respectively. Determine the diffusion coefficient of holes and electrons at this temperature.
[**Ans** : $D_h = 0.00091\ m^2/s$, $D_e = 0.0044\ m^2/s$]

Objective Questions

1. Consider the following statements.
   A. Excess charge carriers can be created by electron bombardment also, instead of optical excitation.
   B. CdS lies in infrared region of spectrum of light.
   C. Mean lifetime of electron-hole pair recombination is of the order of 10^{-8} s.
   D. Drift current flows due to electric field.
   Of these, the correct answers are
   (a) A and C (b) B and D
   (c) A, B and C (d) A, C and D

2. Which of the following is incorrect?
   (a) An application of photoconductivity is found in automatic door opener.
   (b) ZnS coated with Ag is an example of phosphorescence material.
   (c) CdSe lies in red visibility of spectrum of light.
   (d) Photons are absorbed by a semiconductor, if the energy of photon is less than the band gap energy of semiconductor.

3. Match List I with List II and select the correct answer using the codes given below the lists.

List I	List II
A. Diffusion	1. Conduction band
B. Drift	2. Valence band
C. Free electrons	3. Fick's law
D. Holes	4. Electric field

Codes :

	A	B	C	D		A	B	C	D
(a)	3	4	2	1	(b)	4	3	2	1
(c)	3	4	1	2	(d)	4	3	1	2

4. Which of the following pairs is not correctly matched?
   A. Diffusion current flows due to gradient of carrier concentration
   B. Einstein relation Dxh = (KT/e). μh
   C. Continuity equation behaviour of minority carriers in rectifying junctions
   D. Fick's 2nd law steady state condition

5.

List 1 (Semiconductor Parameters)	List 2 (Physical Process)
A. Impurity concentration	1. Recombination
B. Carrier mobility	2. Band to band transition
C. Carrier life time	3. Scattering
D. Intrinsic carrier	4. Ion implantation

Code:

A	B	C	D
3	4	2	1
4	3	2	1
3	4	1	2
4	3	1	2

Answers

1. (d) 2. (d) 3. (a) 4. (d) 5. (c)

References

1. Gupta, K.M., Gupta, N.: Advanced Electrical And Electronics Materials. Scrivener Wiley Publishing, USA (2015). (Some of the matter of this chapter has been reproduced from this book)
2. Gupta, K.M.: Engineering Materials Research, Applications and Advances. CRC Press, Taylor and Francis Group, USA (2014)

Part II
Junction and Interfaces

Chapter 5
P-N Junctions and Their Breakdown Mechanisms

Abstract P-N junction is an important control element for the performance of semiconductor devices. Devices such as amplifiers, rectifiers, linear and digital ICs etc. employ one or more P-N junctions for their working. Therefore, for a good understanding of all these, the P-N junction diode: their applications and working at equilibrium conditions are presented. In this chapter, relations have been established between contact potential and doping concentrations, for Fermi level at equilibrium, for the width of depletion region and penetration depth. Biased junctions: their working when not connected to a battery, forward-biased and reverse biased p-n junctions and their voltage-current characteristics are described. Poisson's equation, junction breakdown mechanisms, Zener and avalanche breakdown, and junction capacitance are discussed. Rectifying diodes, half-wave rectifiers, Zener diode for electric meter protection and as peak clipper are given. The breakdown diode; specifications of p-n junction diode, their physical structures and terminal identifications are also elaborated. Minute insight into the various topics are given through solved numerical and theoretical examples. Review questions, numerical problems and objective type questions are also given with their answers.

Keywords Junction diode · Biased junction · Diode equation · Poission's equation · Junction breakdown mechanisms · Rectifying diodes · Half-wave rectifier · Full-wave rectifier · Zener diode · The breakdown diode

5.1 Junction Diode

Majority of semiconductor devices are made such that there is at least one junction between n-type and p-type extrinsic materials. This is called p-n junction. Although there are many other semiconductor devices having 2,3 or more junctions and known as p-n-p, n-p-n, p-n-p-n junctions etc., but the p-n junctions is most basic among them. It formulates the fundamental performance of all devices. A p-n junction basically performs the following functions in electronic circuits.

K.M. Gupta and N. Gupta, *Advanced Semiconducting Materials and Devices*, Engineering Materials, DOI 10.1007/978-3-319-19758-6_5

1. Switching
2. Rectification
3. Amplification

In practical applications, the semiconductor devices are subjected to varying states of flow of electrons and holes across the junctions. These states may be

(i) equilibrium state, and
(ii) transient state etc.

In this chapter, we shall study their behaviour under equilibrium state. The behaviour under transient state will be studied in later chapters.

5.1.1 P-N Diode (or P-N Junction Diode)

P-N diode is a two-terminal electronic device consisting of a p-n junction, formed by Si or Ge crystals. The p-type and n-type regions are referred to as anode and cathode respectively. A p-n junction diode is a one-way device as it conducts current in one direction only. In other (reverse) direction, it offers a very high resistance.

By themselves, p-type and n-type semiconductors are of very limited use separately, but when they are joined together the crystal structure maintains continuity at the boundary and offers appreciable properties. Special fabrication techniques are used to form a p-n junction. p-n diodes are commercially available in a variety of physical structures with different kinds of terminals and having identification marks. Practically all semiconductor devices contain at least one p-n junction.

5.1.2 Applications of P-N Diode

The p-n junction diodes are used in following main applications.

1. As rectifier for converting a.c. into d.c.
2. In d.c power supplies such as for transistor radio.
3. As zener diode for operating in breakdown region (2–200 V) without damage, such as in

   (i) voltage regulator
   (ii) meter protection
   (iii) peak clipper
   (iv) reshaping a waveform

4. As tunnel diode for conducting large current even when the forward bias is very low (0.05 V or so), such as
   (i) ultrahigh speed switches
   (ii) logic storage memory devices
   (iii) microwave oscillators
   (iv) relaxation oscillators circuits
5. As light-emitting diode (LED) to emit light when in forward-bias, such as in
   (i) burglar alarm circuit
   (ii) video displays
   (iii) optical communication
   (iv) numeric displays in calculators

5.2 Equilibrium Conditions

By equilibrium condition we mean a state in which the p-n junction has neither any external excitation, nor any net flow of current. However, there is a potential difference between n- and p type materials on account of difference in doping on each side of the junction. For equilibrium, the following components of current flowing across the junctions have zero net value.

$$I_{drift\ of\ electron} + I_{diffusion\ of\ electron} + I_{drift\ of\ hole} + I_{diffusion\ of\ hole} = 0 \tag{5.1}$$

The properties of a p-n junction under equilibrium condition are shown in Fig. 5.1a, b. It shows an isolated n and p type materials and their energy bands in Fig. 5.1a. The space charge in depletion (i.e. transition) region marked w alongside the electric Field ζ, electrostatic (or contact) potential V_0, and reshuffled energy bands are shown in Fig. 5.1b.

5.2.1 Electrostatic (or Contact) Potential

When p and n regions are joined, there starts the diffusion of holes from p to n side and of electrons from n to p side. This results in building-up of diffusion current. But the current cannot continue to attain an infinite value. It is because of an opposite electric field created at the junction. On account of diffusion of holes and electrons, a region of positive space charge develops near n-side of the junction and of negative charge near-p side. Therefore the resulting electric field is directed from positive to negative charge. Thus electric field ζ acts in a direction that is opposite to that of diffusion current for n and p carriers. Hence, the field creates a drift

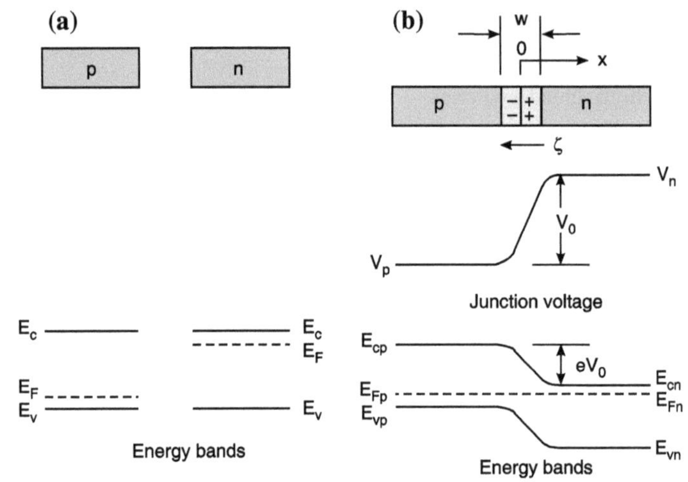

Fig. 5.1 Properties of a p-n junction under equilibrium showing **a** isolated n and p type semiconductors along with their energy bands, and **b** space charge, electric field, contact potential and the changed energy bands

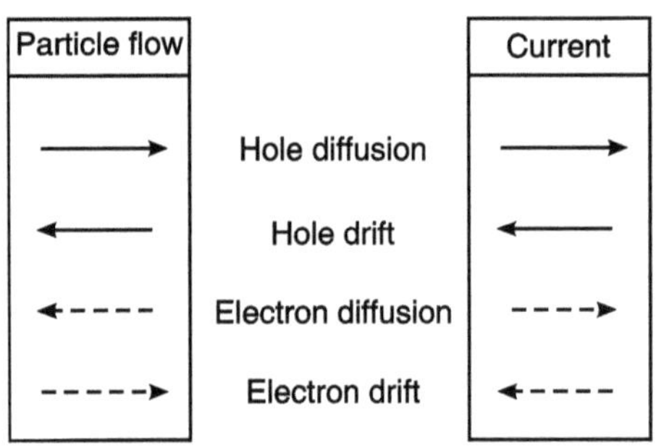

Fig. 5.2 Directions of flow of holes and electrons within depletion region under diffusion and drift, and the directions of resulting currents

component of current from n to p, whose nature is to oppose the diffusion current. This is shown in Fig. 5.2.

Since at equilibrium, there cannot be a net flow of current across the junction, and also there cannot be any net build-up of electrons or holes as a function of time, therefore the drift and diffusion currents cancel-out for both, the electron and the hole carriers. Hence

$$J_{P\ drift} + J_{P\ diffusion} = 0 \tag{5.2a}$$

and

$$J_{n\,drift} + J_{n\,diffusion} = 0 \tag{5.2b}$$

However, the electric field appears in a region of width w in a very close vicinity of the junction. This causes an equilibrium potential difference V_0 across the depletion width w. The potential difference V_0 is known as contact potential or electrostatic potential. It is equal to

$$V_0 = V_n - V_P \tag{5.3}$$

where V_n and V_p are electrostatic potential in neutral n and neutral p respectively.

5.2.2 Establishing the Relation Between Contact Potential and Doping Concentrations

The magnitude of contact potential depends on doping concentrations. In order to establish a relation between them, we consider the requirements of equilibrium in drift and diffusion currents in one-dimensional case. For that we take up the case of hole current along x-direction only. Knowing that at equilibrium, the drift and diffusion currents must cancel, therefore

$$J_p(x) = 0 \quad i.e. \quad e\left(\mu_p p\zeta - D_P\frac{dp}{dx}\right) = 0 \tag{5.4a}$$

where μ_p is mobility of holes, p is concentration of holes in valence band, ζ is electric field, D_p is diffusion coefficient for holes. On rearranging this equation, we get

$$\frac{\mu_p}{D_p}\zeta = \frac{1}{p}\cdot\frac{dp}{dx} \tag{5.4b}$$

Here, x denotes the direction which is taken from p-type material towards n-type. Since the electric field and potential gradient are related as $\zeta = -\,dV/dx$, and Einstein relation is $\frac{\mu_p}{D_p} = e/kT$, therefore on substituting it in Eq. 5.4b, we get

$$-\frac{e}{kT}\cdot\frac{dV}{dx} = \frac{1}{p}\cdot\frac{dp}{dx} \tag{5.5}$$

Solution of Eq. 5.5. Equation 5.5 is solved by integrating it over the limits of electrostatic potential V_p and V_n for hole and electron, and hole concentration p_p and p_n on both sides of the junction near the edge of depletion region. Hence we write

$$-\frac{e}{kT}\int_{V_P}^{V_n} dV = \int_{V_P}^{V_n} \frac{1}{p} dp$$

$$-\frac{e}{kT}\left(-V_n - V_p\right) = lnp_n - lnp_p = ln\left(\frac{p_n}{p_p}\right) \tag{5.6}$$

As the contact potential $V_0 = V_n - V_p$, therefore Eq. 5.6 may be written as

$$V_0 = \frac{kT}{e} \cdot ln\left(\frac{p_n}{p_n}\right) \tag{5.7}$$

This is a relation between contact potential and hole concentration at equilibrium.

Other useful forms of Eq. 5.7. Equation 5.7 can also be expressed in terms of other useful quantities and forms as given below.

(i) If the junction is made of a material having acceptor concentration Na on p-side and donor concentration Nd on n-side, then

$$V_0 = \frac{kT}{e} \cdot ln\frac{N_a}{\left(\frac{n_i^2}{N_d}\right)} = \frac{kT}{e} ln\left(\frac{N_d N_d}{n_i^2}\right) \tag{5.8}$$

(ii) Equation 5.7 may also be expressed as

$$\frac{P_p}{P_n} = e^{\left(\frac{eV_0}{kT}\right)} \tag{5.9}$$

(iii) Equation 5.9 may be written in terms of electron concentrations on both sides of the junction. Hence on considering the equilibrium condition given by $p_p n_p = n_i^2 = p_n n_n$, we may write

$$\frac{P_p}{P_n} = \frac{n_n}{n_p} = e^{\left(\frac{eV_0}{kT}\right)} \tag{5.10}$$

These relations are very useful in determining the V-I characteristics of p-n junction.

Example 5.1 A p-n junction diode has electrons concentration of 10^{15} cm^{-3} on p-side and holes concentration of 10^{13}cm^{-3} on n-side. If intrinsic concentration of electrons is 1.5×10^{10} cm^{-3}, determine the contact potential at 300 K.

Solution. Given are: $N_a = 10^{15}/\text{cm}^3$, $N_d = 10^{13}/\text{cm}^3$, $n_i = 1.5 \times 10^{10}/\text{cm}^3$, and T = 300 K. Using Eq. 5.8, the contact potential is obtained as

$$V_0 = \frac{kT}{e}\ln\left(\frac{N_a N_a}{n_i^2}\right)$$

$$= 0.0259 \times \ln\left[\frac{10^{15} \times 10^{13}}{(1.5 \times 10^{10})^2}\right]$$

$$\text{as } \frac{kT}{e} = 0.0259 \text{ at } 300\,\text{K}$$

$$= 0.0259 \times \ln\left[\frac{10^{28}}{2.25 \times 10^{20}}\right]$$

$$= 0.0259 \times 17.609 = \mathbf{0.456}\,\text{eV}$$

5.3 Fermi Level at Equilibrium

At equilibrium, the Fermi level remains constant throughout the semiconductor device. This fact matches with the observations of the previous article, and so the Eq. 5.9 may be related as

$$\frac{p_p}{p_n} = e^{\frac{eV_0}{kT}} = \frac{N_v \cdot e^{-\frac{E_{F_p} - E_{vp}}{kT}}}{N_v e^{-\frac{E_{F_n} - E_{vn}}{kT}}} \tag{5.11a}$$

where N_v is effective density of state at the edge of valence band, E_v is valence band edge (in Joule or eV), E_{Fp} and E_{Fn} are Fermi energy on p-side and n-side of the junction.

The above equation can be simplified as

$$\begin{aligned} e^{\frac{eV_0}{kT}} &= e^{-\frac{(E_{F_n} - E_{vn})}{kT}} \cdot e^{-\frac{(E_{F_p} - E_{vp})}{kT}} \\ &= e^{-\frac{(E_{F_n} - E_{F_p})}{kT}} \cdot e^{-\frac{(E_{vp} - E_{vn})}{kT}} \end{aligned} \tag{5.11b}$$

As the Fermi levels on both sides of the junction are equal at equilibrium i.e. $E_{Fn} - E_{Fp} = 0$, therefore we can write

$$E_{v_p} - E_{vn} = eV_0 \tag{5.12}$$

The potential barrier may increase or decrease in comparison to the value of contact potential, when a bias is applied on the junction.

Example 5.2 A silicon p-n junction has acceptor concentration of 10^{17} cm^{-3} on p-side and donor concentration of 10^{16} cm^{-3} on n-side. At 27° C, calculate (a) the

Fermi levels, (b) potential barrier, and (c) draw the band diagram at equilibrium condition. Assume anything else, if required.

Solution. Given are: $N_a = 10^{17}/cm^3$, $N_d = 10^{16}/cm^3$, T = 27 °C + 273 = 300 K
We take $n_i = 1.5 \times 10^{10}/cm^3$, Boltzmann constant $k = 1.38 \times 10^{-23}$ J/K
$= 8.62 \times 10^{-5} eV/K$

(a) Fermi level on p-side of p-n junction is calculated as

$$E_{ip} - E_F = kTln\frac{p_p}{n_i}$$
$$= 0.0259 \times ln\left(\frac{10^{17}}{1.5 \times 10^{10}}\right) \qquad (kT = 0.0259 \text{ at } 300\text{ K})$$
$$= \mathbf{0.407} \text{ eV}$$

Similarly on n-side, it is found as

$$E_F - E_{in} = kTln\frac{n_n}{n_i}$$
$$= 0.0259 \times ln\left(\frac{10^{16}}{1.5 \times 10^{10}}\right) = \mathbf{0.347} \text{ eV}$$

(b) The potential barrier is obtained from

$$V_0 = (E_{ip} - E_F) + (E_F - E_{in})$$
$$= 0.407 + 0.347 = \mathbf{0.754} \text{ eV}$$

(c) The equilibrium band diagram will be as shown below.

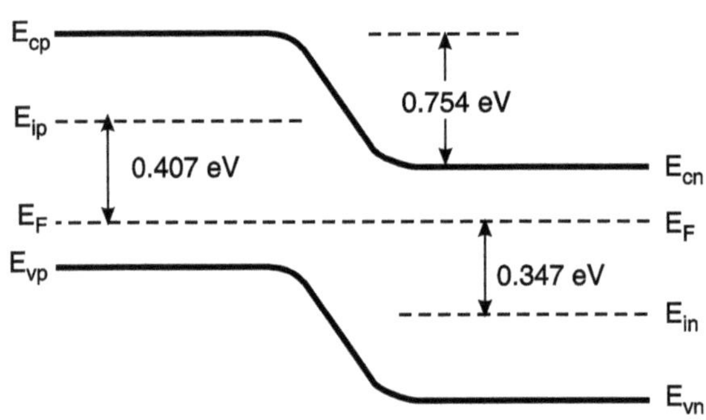

5.3.1 Space Charge at Junction

In equilibrium condition, the electrons and holes move from one side of junction to other, within the depletion (or transition) region. While some electrons diffuse from n to p, others are carried away from p to n by electric field. The space charge within the transition region is due to uncompensated acceptor and donor ions. Therefore on neglecting the carriers within space charge region, we find

- the charge density on n-side = eN_d, and
- negative charge density on p-side = $-eN_a$.

In case the p-side is lightly doped than n-side (i.e. $N_a < N_d$), the space charge region extends into p-side than into n-side of the junction. This is shown in Fig. 5.3a. But if the p-side is heavily doped than n-side (i.e. $N_a > N_d$), the space charge region extends into n-side than p-side of the junction. This is shown in Fig. 5.3b.

Thus for a sample of cross-sectional area A, the total uncompensated charge Q on each side of the junction will be

$$Q_{positive} = eAx_{p\ pen}N_a \tag{5.13a}$$

and

$$Q_{negative} - Q_{positive} = eAx_{n\ pen}N_d \tag{5.13b}$$

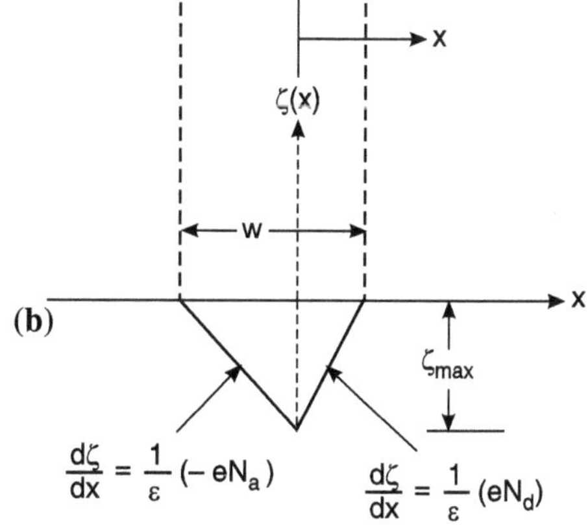

Fig. 5.3 Distribution of space charge and electric field within depletion region **a** space charge extended more towards p-side, and **b** electric field distribution

Here $x_{p\ pen}$ is penetration of space charge into p-side and x_n pen is penetration into n-side. And also the width of depletion region is equal to the sum of these penetrations. Hence

$$w = x_{p\ pen} + x_{n\ pen} \tag{5.14}$$

5.3.2 Determining the Maximum Value of Electric Field

To determine the distribution of electric field ζ within the depletion region, we consider the region of constant space charge given as below.

$$\frac{d\zeta}{dx} = \frac{e}{\varepsilon} N_d \quad for \quad 0 < x < x_{n\ pen} \tag{5.15a}$$

and

$$\frac{d\zeta}{dx} = \frac{e}{\varepsilon} N_d \quad for \quad - x_{p\ pen} < x < 0 \tag{5.15b}$$

Here ε is permittivity of semiconductor. In formulating the above equations, the contribution of carriers towards space charge has been neglected. The variation of ζ as a function of x within the depletion region is shown in Fig. 5.3b. It has a maximum value ζ_0 at $x = 0$.

On integrating above equations within the appropriate limits, we get

$$\int_{\zeta 0}^{0} d\zeta = \frac{e}{\varepsilon} N_d \int_{0}^{x_n\ pen} dx \quad for\, 0 \quad < x < x_{n\ pen} \tag{5.16a}$$

and

$$\int_{\zeta 0}^{0} d\zeta = \frac{e}{\varepsilon} N_a \int_{-x_n\ pen}^{0} dx \quad for \quad - x_{p\ pen} < x < 0 \tag{5.16b}$$

Further simplifying the above equations, we get the maximum value of electric field as

$$\zeta_0 = -\frac{e}{\varepsilon} N_d x_{n\ pen} = -\frac{e}{\varepsilon} N_a x_{p\ pen} \tag{5.17}$$

5.4 Determining the Width of Depletion Region and Penetration Depth

The width w of depletion region can be determined in terms of contact potential V_0, doping concentrations N_a and N_D, permittivity ε, and electronic charge as follows.

We know that the electric field at a distance x is related to potential gradient by

$$\zeta(x) = -\frac{dV}{dx}, \quad or = V_0 - \int_{-x_{p\ pen}}^{-x_{n\ pen}} \zeta(x)dx \tag{5.18}$$

From Fig. 5.3, it is clear that the above equation refers to the area formed under a triangle. So the relation between contact potential and width w of depletion region may be established as

$$V_0 = -\frac{1}{2}\zeta_0 w \tag{5.19a}$$

On substituting Eq. 5.17 for ζ_0 in above equation, we get

$$V_0 = \frac{1}{2}\cdot\frac{e}{\varepsilon}N_d x_{n\ pen} w \tag{5.19b}$$

Further from the consideration of balance of charge and considering Eqs. 5.13a, 5.13b, we may write $x_{n\ pen} = \frac{wN_a}{(N_a+N_d)}$ in Eqs. 5.19a, 5.19b. It yields

$$V_0 = \frac{1}{2}\cdot\frac{e}{\varepsilon}\frac{N_a N_d}{N_a + N_d}w^2 \tag{5.20}$$

On simplifying it, we get

$$w^2 = \frac{2\varepsilon V_0}{e}\cdot\frac{(N_a + N_d)}{N_a N_d}$$

or

$$w = \sqrt{\frac{2\varepsilon V_0}{e}\left(\frac{N_a + N_d}{N_a N_d}\right)} = \sqrt{\frac{2\varepsilon V_0}{e}\left(\frac{1}{N_a} + \frac{1}{N_d}\right)} \tag{5.21}$$

On using Eq. 5.8, the above equation may also be written as

$$w = \sqrt{\frac{2\varepsilon kT}{e^2}\cdot ln\left(\frac{N_a N_d}{n_i^2}\right)\left(\frac{1}{N_a} + \frac{1}{N_d}\right)} \tag{5.22}$$

In above equations $\varepsilon = \varepsilon_0\varepsilon_r$. The values of ε_r for important semiconductors lie between 12–16.

5.4.1 Determining the Penetration Depth in Depletion Region

The depth of penetration of depletion region into n and p semiconductors may now be obtained by substituting Eq. 5.21 in Eq. 5.19a, 5.19b. Hence we obtain

$$x_{p\ pen} = \frac{wN_d}{N_a + N_d} = \sqrt{\frac{2\varepsilon V_0}{e}\left[\frac{N_d}{N_a(N_a + N_d)}\right]} \tag{5.23a}$$

and

$$x_{n\ pen} = \frac{wN_a}{N_a + N_d} = \sqrt{\frac{2\varepsilon V_0}{e}\left[\frac{N_a}{N_d(N_a + N_d)}\right]} \tag{5.23b}$$

Example 5.3 Boron is implanted into a n type Si sample having donor concentration of $10^{16}/cm^3$, to form an abrupt junction. If the acceptor concentration in p-type region is $4 \times 10^{18}/cm^3$, determine the (a) width of depletion region, and (b) depth of penetration on n-side and p-side materials, at equilibrium. Take room temperature as 27 °C, $n_i = 15 \times 10^{10}/cm^3$, and relative permittivity of boron as 11.8.

Solution. Given are: $N_d = 10^{16}/cm^3$, $N_a = 4 \times 10^{18}/cm^3$, $n^i = 1.5 \times 10^{10}/cm^3$, and T = 300 K

We take $e = 1.6 \times 10^{-19}C$, $\varepsilon_0 = 8.85 \times 10^{-12}$ F/m $= 8.85 \times 10^{-14}$ F/cm

(a) The width of depletion region is found from Eq. 5.21 as

$$w = \sqrt{\frac{2\varepsilon V_0}{e}\left[\frac{N_a + N_d}{N_aN_d}\right]} \quad \text{where } \varepsilon = \varepsilon_0\varepsilon_r \tag{i}$$

As

$$V_0 = \frac{kT}{e} ln\frac{N_aN_d}{n_i^2} = 0.0259 \ln\left(\frac{4 \times 10^{18} \times 10^{16}}{(1.5 \times 10^{10})^2}\right)$$
$$= 0.0259 \ln\left(1.78 \times 10^{14}\right) = \mathbf{0.85}\ \mathrm{V}$$

therefore on substituting this value in Eq. (i), we find

$$
\begin{aligned}
w &= \sqrt{\frac{2 \times (11.8 \times 8.85 \times 10^{-14}) \times 0.85}{1.6 \times 10^{-19}} \cdot \left(\frac{(4 \times 10^{18}) + (10^{16})}{(4 \times 10^{18}) + (10^{16})}\right)} \\
&= \sqrt{(110.956 \times 10^{5}) \times (100.25 \times 10^{-18})} \\
&= \sqrt{(11123.339 \times 10^{-13})} \\
&= 3.34 \times 10^{-5} \mathrm{cm} \; = \mathbf{0.334}\, \mu\mathrm{m}
\end{aligned}
$$

(b) The depth of penetration is found from Eqs. 5.23a, 5.23b as

$$
\begin{aligned}
x_{n\,pen} &= \frac{wN_a}{N_a + N_d} = \frac{\left(3.34 \times 10^{-5}\right) \times (4 \times 10^{18})}{(4 \times 10^{18}) + (10^{16})} \\
&= \frac{13.36 \times 10^{13}}{401 \times 10^{16}} = 0.0333 \times 10^{-3} \mathrm{cm} = \mathbf{0.333}\, \mu\mathrm{m}
\end{aligned}
$$

$$
\begin{aligned}
x_{n\,pen} &= \frac{wN_a}{N_a + N_d} = \frac{\left(3.34 \times 10^{-5}\right) \times \left(10^{16}\right)}{(4 \times 10^{18}) + (10^{16})} \\
&= \frac{3.34 \times 10^{11}}{401 \times 10^{16}} = 8.3 \times 10^{-8} \mathrm{cm} = \mathbf{8.3}\, \text{Å}
\end{aligned}
$$

Example 5.4 Refer Example 5.2, in which for the boron implant into an n-type Si-sample, the abrupt junction formed is of square cross-section having 0.2 mm^2 area. Taking all other data as the same, calculate (a) the total positive charge, and (b) maximum electric field. (c) Also determine the value of total negative charge?

Solution. Since the given data are

$$
A = 0.2\, \mathrm{mm}^2 = 0.2 \times 10^{-2}\, \mathrm{cm}^2 = 2 \times 10^{-3}\, \mathrm{cm}^2,
$$

$$
N_d = 10^{16}/\mathrm{cm}^3, \quad \text{and} \quad \varepsilon_r = 11.8
$$

The assumed values are

$$
e = 1.6 \times 10^{-19}\, \mathrm{C}, \quad \textit{and} \quad \varepsilon_0 = 8.85 \times 10^{-14} \frac{\mathrm{F}}{\mathrm{cm}}
$$

and calculated values from Example 5.2 are

$$
x_{n\,pen} = 0.33\, \mu = 3.34 \times 10^{-5}\, \mathrm{cm}
$$

(a) Hence on substituting these values in Eqs. 5.13a, 5.13b, we find

$$\begin{aligned} Q_{positive} &= e.Ax_{n\,pen} \cdot N_d \\ &= \left(1.6 \times 10^{-19}\right) \times \left(2 \times 10^{-3}\right) \times \left(3.33 \times 10^{-5}\right)\left(10^{16}\right) \\ &= \mathbf{1.07 \times 10^{-10}}\,\mathrm{C} \end{aligned}$$

(b) From Eq. 5.17, we get the maximum electric field as

$$\begin{aligned} \zeta &= \frac{-e.N_d \cdot x_{n\,pen}}{\varepsilon} \\ &= \frac{-\left(1.6 \times 10^{-19}\right) \times \left(10^{16}\right) \times \left(3.3 \times 10^{-5}\right)}{\left(8.85 \times 10^{-14}\right)(11.8)} \\ &= \mathbf{-5.1 \times 10^4}\,\mathrm{V/cm} \end{aligned}$$

(c) The total negative charge at equilibrium will be the same as total positive charge. Hence

$$-Q_{negative} = Q_{positive} = \mathbf{-1.07 \times 10^{-10}}\,\mathrm{C}$$

5.5 Biased Junctions

Behaviour of a p-n junction is different in its inherent form, Fig. 5.4a (i.e. when not connected to a battery) from its behaviour under ‘biasing’ (i.e. when connected to a battery). It can be connected to a battery in following two ways.

1. Under forward biasing, and
2. Under reverse (or backward) biasing.

Under forward biasing, the positive terminal of battery is connected to p-side and negative terminal to n-side, Fig. 5.4b. But if the connection is just opposite of it i.e. negative terminal connected to p-side and positive terminal connected to n-side, it is known to be in reverse biasing as shown in Fig. 5.4c.

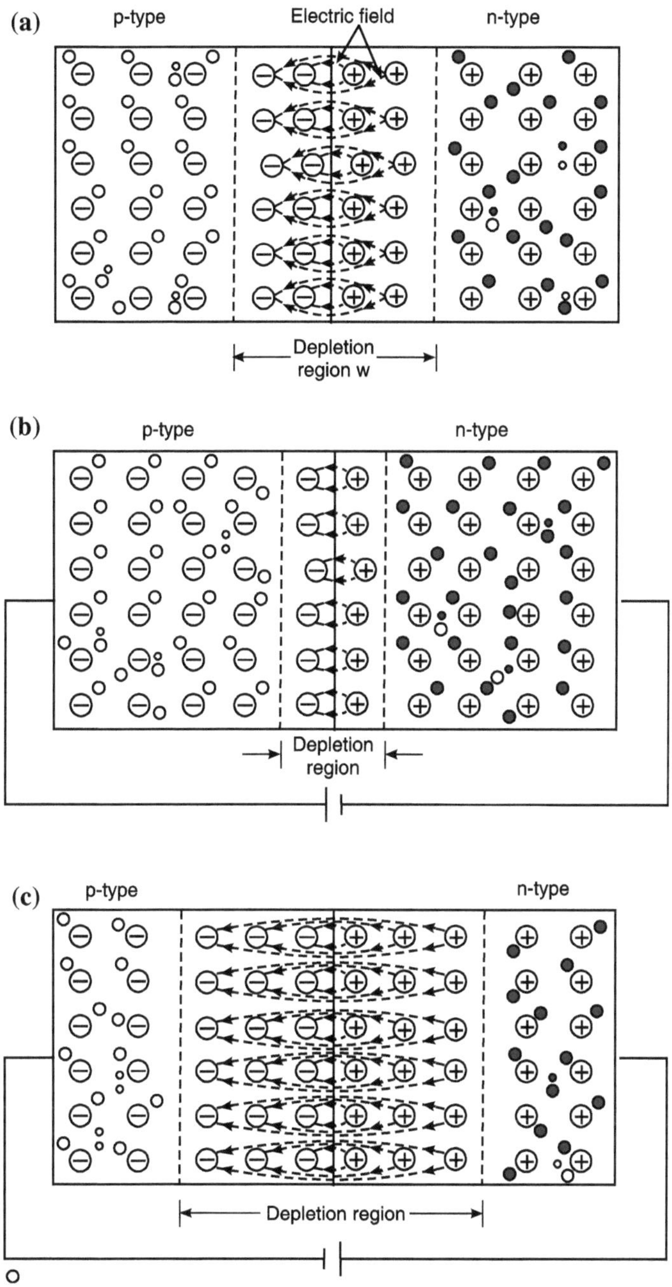

Fig. 5.4 A p-n diode **a** when not connected to a battery, **b** under forward-bias, and **c** under reverse-bias. Reproduced from [1]

5.6 Working of p-n Diode When not Connected to a Battery

Figure 5.4a shows a single crystal of p-n junction without any external voltage. In it the holes and electrons are mobile charges, but the ions are immobile. The crystal is electrically neutral, and the p-region and n-region are separate. As soon as the p-n junction is formed, the following processes are initiated.

5.6.1 *Diffusion of Holes and Electrons*

Holes from p-region diffuse into n-region and combine with free electrons available there. Similarly, the free electrons from n-region diffuse into p-region and combine with holes. Diffusion occurs for a short duration due to thermal energy and also due to difference in concentration of holes and electrons in their regions.

5.6.2 *Set-Up of Barrier*

After a few recombination of holes and electrons in close vicinity of the junction, a restraining force called barrier is set up automatically. This barrier stops further diffusion of holes and electrons from one side to other.

5.6.3 *Formation of Depletion (or Space Charge) Region*

Recombination of holes and electrons eliminates a hole and a free electron from their respective regions. Consequently, the negative acceptor ions in p-region and positive donor ions in n-region are left uncompensated. Additional holes trying to diffuse into n-region and additional electrons trying to diffuse into p-region are repelled by uncompensated positive and negative charges, respectively. As a result, further recombination of holes and electrons cannot occur.

The region containing the uncompensated acceptor and donor ions is called depletion region. It is named so because there is a depletion of mobile charges (holes and electrons) in this region. The electric field between the acceptor and donor ions is called a barrier. The physical distance between two sides of the barrier is called width of the barrier and the difference in potential between them is referred to as height of the barrier.

The width of the barrier is of the order of fraction of millimetre and the barrier height is of the order of tenths of a volt. For silicon p-n junction, the barrier potential is 0.7 V and for germanium it is about 0.3 V.

5.6.4 *Flow of Drift and Diffusion Current*

The barrier discourages the diffusion of majority carriers across the junction but helps the minority carriers (few free electrons in p-region and few holes in n-region) to drift across the junction. The minority carriers are constantly generated due to thermal energy. The drift of minority carriers is counterbalanced by diffusion of the same number of majority carriers across the junction. Since the majority carriers have sufficiently high kinetic energy, therefore they overcome the barrier and cross the junction.

5.7 Forward Biased p-n Junction

In the case shown in Fig. 5.5, the junction is said to be biased in forward direction. The current flow may be explained in the following two ways:

(i) As soon as the battery connection is made, the holes are repelled by positive battery terminal and electrons by negative battery terminal. It results in movement of electrons and holes towards junction where they recombine. This movement of electrons to the left and of holes to the right of the junction constitutes the flow of a large current through the semiconductor. Obviously, the crystal offers low resistance in forward direction. It may be noted that though there is a movement of both electrons and holes inside the crystal, only free electrons move in external circuit i.e. in the battery-connected wires.

(ii) Due to the applied external voltage, the barrier potential is reduced which allows more current to flow across the junction. The forward bias reduces the thickness of depletion layer as shown in Fig. 5.6.

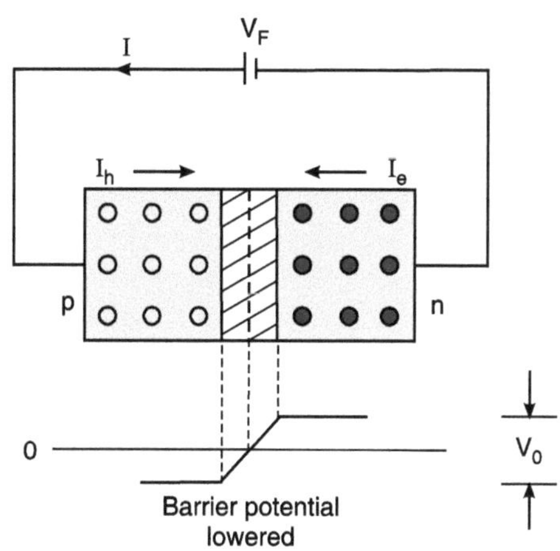

Fig. 5.5 A reduced barrier potential in forward biasing

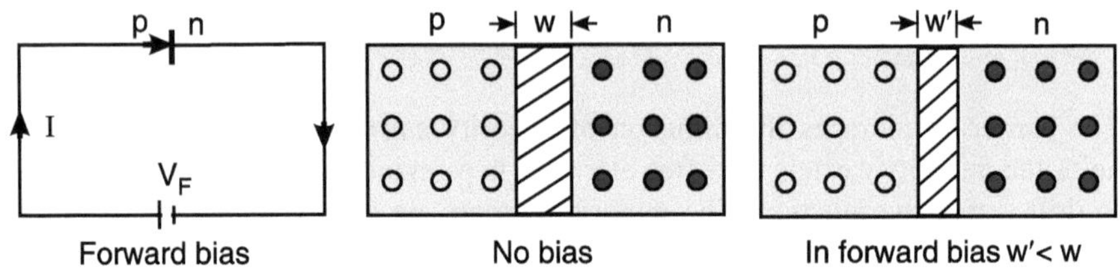

Fig. 5.6 The thickness of depletion layer w decrease to w′ in forward bias

5.7.1 Voltage-Current Characteristics

A typical V-I characteristic for a forward-biased p-n junction is shown in Fig. 5.7. It is shows that the forward current rises exponentially with the applied forward voltage. However at room temperature, a potential difference of about 0.3 V is required before a reasonable amount of forward current starts flowing in Ge junction. This voltage is known as threshold (or cut-in) voltage V_{thresh}. Its value for Si junction is about 0.7 volt. For $V < V_{thresh}$, the current flow is negligible. But as the applied voltage increases beyond threshold value, the forward current increases sharply. And if the forward voltage is increased beyond a certain safe value, it produces an extremely large current that may destroy the junction due to overheating. Ge devices can stand junction temperature around 100 °C, whereas Si devices can function up to 175 °C.

5.7.2 Voltage-Ampere Equation and Its Temperature Dependence

For a p-n junction under forward biasing, the current is related to the applied voltage by following equation.

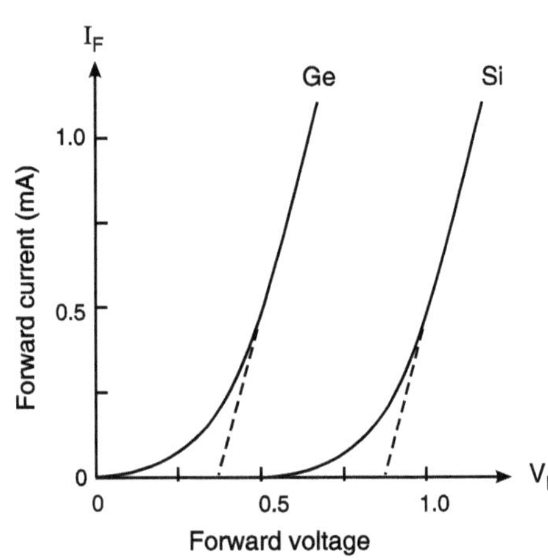

Fig. 5.7 Volt-current characteristic in forward biasing

$$J_e = J_{eo}\left(e^{\frac{eV}{KT}} - 1\right) \tag{5.24a}$$

and in reverse biasing by

$$J_h = J_{ho}\left(e^{\frac{eV}{KT}} - 1\right) \tag{5.24b}$$

where J_{eo} represents the magnitude of two compensating electron currents: one associated with electrons moving up the hill and the other with electrons moving down hill. The suffix ho has a similar meaning for holes. Thus the total current through the junction is the sum of Eqs. 5.24a and 5.24b, and in forward biasing it is equal to

$$J = (J_{eo} + J_{ho})\left(1 - e^{\frac{eV}{KT}} - 1\right) \tag{5.25a}$$

This relation in case of reverse biasing is

$$J = (J_{eo} + J_{ho})\left(1 - e^{\frac{eV}{KT}}\right) \tag{5.25b}$$

Conclusion. Since (J_{eo} + J_{ho}) is a constant, the current density increases exponentially with the voltage in forward-bias. The reverse current approaches a saturation value with increase in the reverse bias. In both the cases, the system has a non-linear V-I characteristics. The magnitudes of J_{eo} and J_{ho} in terms of material properties may be determined from the continuity equation, described later in this chapter. The temperature dependence is self-explanatory from the above equations.

5.8 Reverse Biased P-N Junction

When the junction is connected to the battery as shown in Fig. 5.8, it is said to be in reverse bias. In this case, the holes are attracted by negative battery terminal and electrons by positive terminal, so that the holes and electrons move away from the junction and away from each other. Since there is no electron-hole combination, the current does not flow and the junction offers a high resistance.

In this case, the applied voltage increases the barrier potential, thereby blocking the flow of majority carriers as shown in Fig. 5.8. Under reverse bias conditions, the width of depletion region increases because of increased barrier potential, Fig. 5.9.

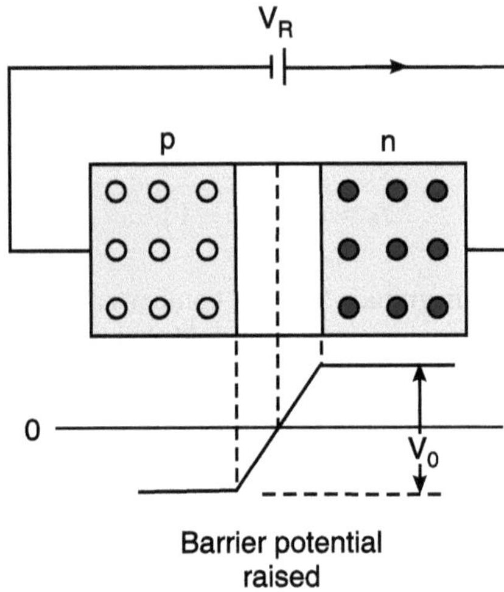

Fig. 5.8 The barrier potential is increased in reverse biasing

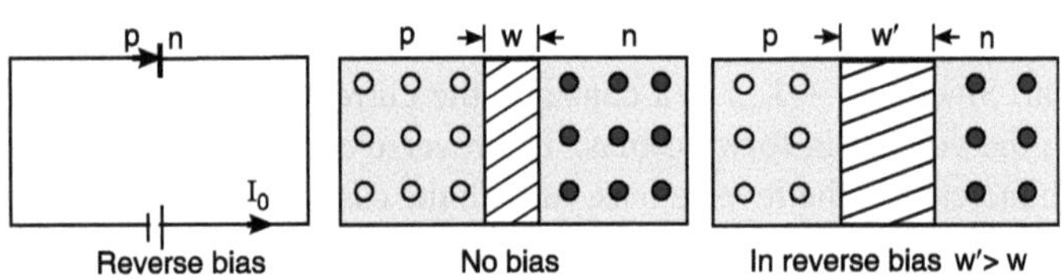

Fig. 5.9 The thickness of depletion layer w increases to *w′* in reverse biasing

5.8.1 Reverse-Bias Characteristics of a p-n Diode

To obtain the reverse-bias characteristics, we use a circuit as shown in Fig. 5.10. It consists of a voltmeter V, a microammeter along with the diode. The voltage can be varied by a pointer at P. On conducting the experiment, the i_R is observed for different values of V. Hence the V_R versus I_R plot comes in the form as shown in Fig. 5.14a (see Fig. 5.11).

Fig. 5.10 A circuit diagram to plot reverse-bias characteristics of a diode

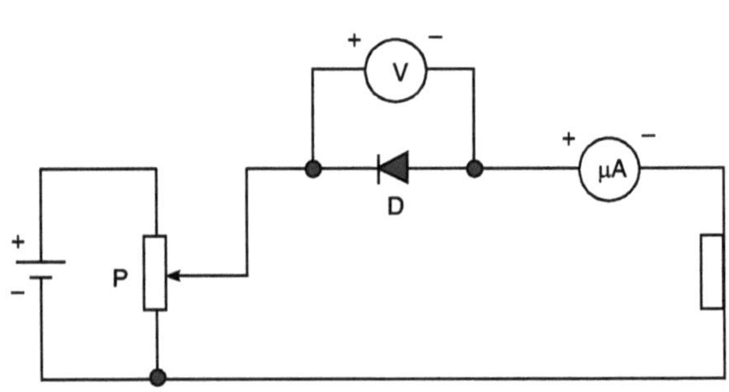

5.8.2 *Reverse Saturation Current*

Although practically there is no current due to majority carriers, yet there is a small amount of current of a few μA only due to the flow of minority carriers across the junction. Due to thermal energy, some holes are always generated in n-type region and some electrons in p-type region. The applied voltage acts as forward bias for these minority carriers though it is a reverse bias for the majority carriers.

The battery drives these minority carriers across the junction thereby producing a small current called reverse current (or reverse saturation current) I_{rs}. Since minority carriers are thermally-generated, I_{rs} is extremely temperature-dependent. For the same reason, the forward current is also temperature-dependent but to a lesser degree. It is because the minority current forms a very small percentage of majority current. The temperature dependence of I_{rs} for Si and Ge is as follows.

- With Si, it doubles for every 6 °C rise in temperature.
- With Ge, it doubles for every 10 °C rise in temperature.

The value of I_{rs} is of the order of 10^{-9}A for Si, and 10^{-6} A for Ge.

5.8.2.1 Voltage-Current Characteristics

The reverse saturation current is also referred to as leakage current of p-n junction. Figure 5.11 shows V-I characteristics of a reverse-biased p-n junction. It can be seen that as the reverse voltage is increased from zero, the reverse current quickly rises to its maximum (or saturation) value. Keeping the temperature constant as the reverse voltage is increased, I_{rs} is also found to increase slightly. This slight increase is due to the impurities on the surface of the semiconductor, which behaves as an effective resistance and hence obeys Ohm's law. This gives rise to very small current called surface leakage current. Unlike the main leakage (or saturation) current, the surface leakage current is independent of temperature, but depends on the magnitude of reverse voltage.

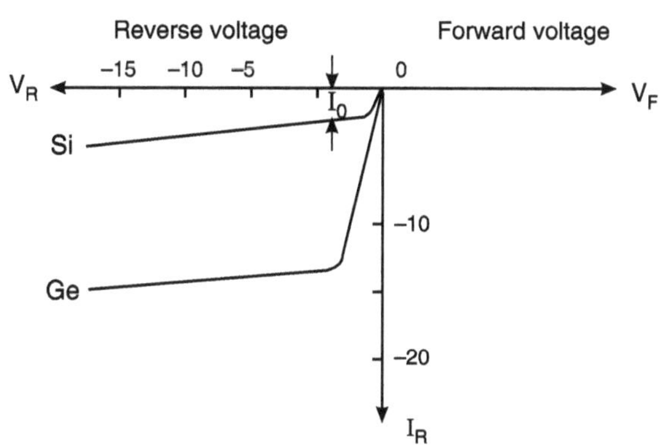

Fig. 5.11 Voltage-current characteristics in reverse biasing

5.9 Comparison of the Effects of No-bias, Forward-Bias and Reverse-Bias at a P-N Junction

The above effects in respect of the (a) width of transition region, (b) electric field, (c) potential barrier, (d) energy band diagrams, and (e) particle flow and current flow directions are shown in Fig. 5.12 whose details are self-explanatory.

5.9.1 Diode Equation

The diode equation is given below.

$$I = eA\left(\frac{D_{xh}}{L_h}p_n + \frac{D_{xe}}{L_e}n_p\right)\left(e^{\frac{eV}{KT}} - 1\right) = I_0\left(e^{\frac{eV}{KT}} - 1\right) \tag{5.26a}$$

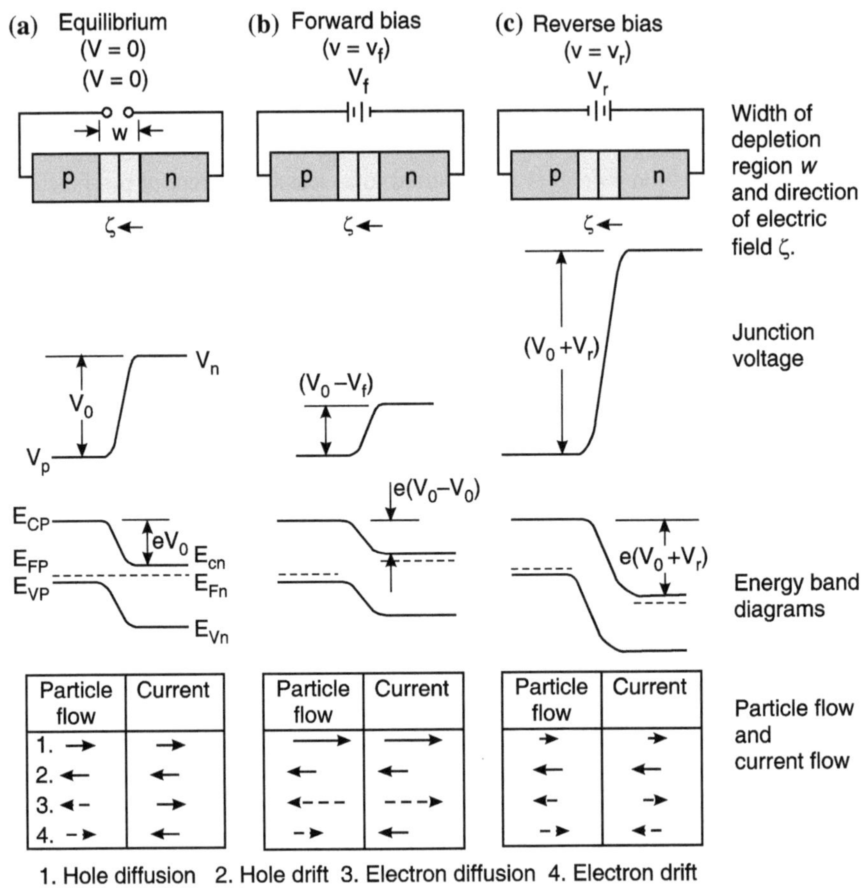

Fig. 5.12 Effects of no-bias (**a**), forward-bias (**b**), and reverse-bias (**c**) at a P-N junction

where I is total current in diode, A is cross-sectional area of the junction,

D_{xh} and D_{xe} are diffusion coefficients of holes and electrons,
L_h and L_e are diffusion lengths for holes and electrons,
p_n is equilibrium concentration of holes in n-type material,
n_p is equilibrium concentration of electrons in p-type material, and
I_0 is reverse-saturation current.

The diode equation describes the total current passing through the diode for either forward-bias or reverse-bias. Equation 5.26a given above is for forward-bias. To write it for reverse-bias, the voltage V has to be replaced by $-V_r$. Hence Eq. 5.26a becomes

$$I = eA\left(\frac{D_{xh}}{L_h}p_n + \frac{D_{xe}}{L_e}n_p\right)\left(e^{-\frac{eV}{KT}} - 1\right) = I_0\left(e^{-\frac{eV}{KT}} - 1\right) \tag{5.26b}$$

5.9.2 *Poisson's Equation*

The Poisson's equation is used to calculate the distribution of electric field within transition region of a p-n junction. It relates the gradient of electric field with local space charge at any location x. This is expressed as

$$= \frac{d\zeta(x)}{dx} = \left(\frac{e}{\varepsilon_0\varepsilon_r}\right)\left(N_d^+ - N_a^- + p - n\right) \tag{5.27}$$

5.10 Volt-Ampere Characteristics

The voltage-current (V-I) characteristics of a p-n junction illustrates its response when connected to an electrical circuit. This response is between the applied voltage across its terminal and the current that flows through it. The responses are different in forward and reverse biasing, and are shown in Fig. 5.13.

5.10.1 *In Forward Biasing*

Figure 5.13 shows that in forward biasing the diode current in very small for the first few tenths of a volt. It is because the diode does not conduct well until the external voltage overcomes the barrier potential. As we approach the cut-in or knee voltage V_0, a large number of free electrons and holes start crossing the junction and hence the current starts increasing rapidly. The cut-in voltage

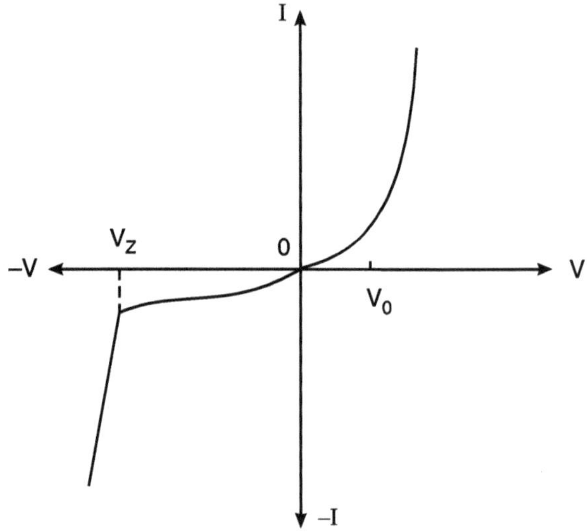

Fig. 5.13 Voltage-ampere characteristics of a p-n diode in forward and reverse biasing

- for Si diode is $V_0 = 0.7$ V, and
- for Ge diode is $V_0 = 0.3$ V

Even a small increase in the above cut-in voltage produces a sharp increase in current. However if this current it too large, it can destroy the diode due to excessive heat. Hence, the diodes are specified to carry a certain maximum current rating. This current may be of the order of 1 A.

5.10.2 In Reverse Biasing

In this case, the diode current is very small. It is of the order of 10^{-6} A in germanium diode and of 10^{-9} A in silicon diode. It remains small and almost constant for all voltages up to breakdown voltage Vz. At breakdown, the current increases rapidly for a small increase in voltage. At this voltage, the crystal structure breaks down. Therefore in normal applications, this condition is avoided. The junction breakdown may occur due to either of the following reasons.

1. Zener breakdown, or
2. Avalanche breakdown.

Zener breakdown is caused due to high electric field which generates a large number of carriers. Carriers induce a flow of large current that ultimately results in zener breakdown.

In case of avalanche breakdown, the increased electric field causes increase in the velocity of minority carriers. These high energy carriers break the covalent bond of Si and Ge, and generate more carriers. Again, these generated carriers are accelerated by the electric field. They break more covalent bonds during their travel. Thus a chain reaction is established which gives rise to a high reverse current. This mechanism of breakdown is called 'avalanche breakdown'.

5.11 Junction Breakdown Mechanisms

If the reverse bias applied to a p-n junction is increased, a point is reached when the junction breaks down and reverse current rises sharply. The rise in the value of this current is limited by the resistance connected in series with the junction, Fig. 5.14a. This critical value of the voltage is known as breakdown voltage (V_{BR}). It is found that once the breakdown has occurred, a very little increase in voltage is required to increase the current up to a relatively high value. The junction offers almost zero resistance at this point. The breakdown voltage depends on the width of the depletion region which, in turn, depends on the doping level.

The following two mechanisms are responsible for breakdown under increasing reverse voltage.

5.11.1 Zener Breakdown

This form of breakdown occurs in junctions which, being heavily doped, have narrow depletion layers. The breakdown voltage sets up a very strong electric field (about 10^8 V/m) across the narrow layer. This field is strong enough to break or rupture the covalent bonds, thereby generating the electron-hole pairs. Even a small increase in reverse voltage is capable of producing large number of current carriers. That is why the junction has a very low resistance in the breakdown region.

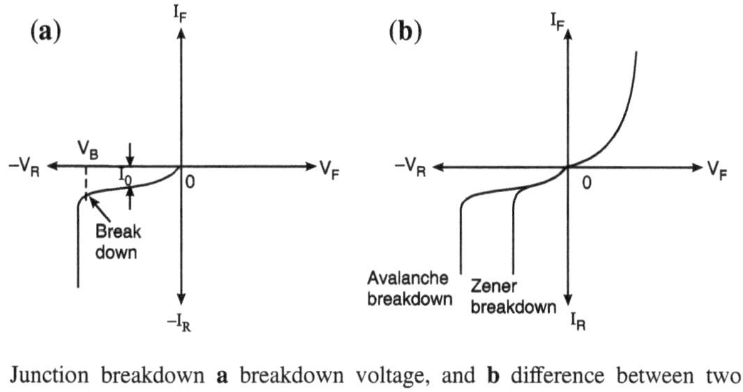

Fig. 5.14 Junction breakdown **a** breakdown voltage, and **b** difference between two kinds of breakdown

5.11.2 Avalanche Breakdown

This form of breakdown occurs in junctions which being lightly-doped, have wide depletion regions where the electric field is not strong enough to produce Zener breakdown. Instead, the minority carriers (accelerated by this field) collide with the semiconductor atoms in the depletion region. Upon collision, the covalent bonds are broken and electron-hole pairs are generated. These newly-generated charge carriers are also accelerated by the electric field, resulting in more collision and hence further producing of charge carriers. This leads to avalanche of charge carriers and to a very low reverse resistance. The two breakdown phenomena are shown in Fig. 5.14b.

Example 5.5 A silicon diode has a forward voltage drop of 1.2 V for a forward dc current of 100 mA. It has a reverse current of 1 μA for a reverse voltage of 10 V. Calculate the (a) bulk and reverse resistance of diode, and (b) ac resistance at forward dc current of 2.5 and 25 mA, and forward dc voltage of 25 mV.

Solution. Given are:

$$\mathrm{V_F} = 1.2\,\mathrm{V},\ \mathrm{I_F} = 100\,\mathrm{mA} = 100 \times 10^{-3}\,\mathrm{A},$$

$$\mathrm{I_R} = 1\,\mathrm{mA} = 1 \times 10^{-6}\,\mathrm{A},\ \mathrm{V_R} = 10\,\mathrm{V}$$

We take junction voltage for Si as $\mathrm{V_j} = 0.7$ V. To find: bulk resistance $\mathrm{R_B}$, reverse resistance $\mathrm{P_R}$ and ac resistance $\mathrm{R_{ac}}$.

(*a*)

$$R_B = \frac{V_F - V_j}{I_F} = \frac{1.2 - 0.7}{100 \times 10^{-3}} = \mathbf{5\,\Omega}$$

$$R_R = \frac{V_R}{I_R} = \frac{10}{1 \times 10^{-6}} = \mathbf{10\,\Omega}$$

(b) At 2.5 mA,

$$R_j = \frac{V_{dcf}}{I_{dcf}} = \frac{25\,\mathrm{mV}}{2.5\,\mathrm{mA}} = \mathbf{10\,\Omega}$$

$$R_{ac} = R_B + R_J = 5 + 10 = \mathbf{15\,\Omega}$$

At 25 mA,

$$R_J = \frac{25 \text{ mV}}{25 \text{ mA}} = \mathbf{1\,\Omega}$$

$$R_{ac} = 5 + 1 = \mathbf{6\,\Omega}$$

5.12 Junction Capacitance

By junction capacitance we mean the capacitive effects that are exhibited by p-n junctions, when they are forward and reverse biased. Junction capacitance may be of two kinds.

1. Transition capacitance or space-charge capacitance
2. Diffusion or storage capacitance

Transition capacitance. When a p-n junction is reverse-biased, the depletion region acts like an insulator or as a dielectric material, which is essential for making a capacitor. The p-and n-type regions on either side have low resistance and act as the plates. We, therefore, have all the components necessary for making a parallel-plate capacitor. This junction capacitance is called transition or space charge capacitance C_T. It is voltage-dependent.

Diffusion capacitance. This capacitive effect is present when the junction is forward-biased. It accounts for the time delay in moving the charge across the junction by diffusion process. Due to this fact, this capacitance cannot be identified in terms of a dielectric and plates. It varies directly with the magnitude of forward current.

5.13 Rectifying Diodes

Conversion of alternating current (or voltage) into direct current (or voltage) is called rectification. A diode is well-suited for this job because it conducts in one direction only i.e. only when forward-biased. Several kinds of rectifiers are in use. These are

- half-wave rectifier which uses one diode
- full-wave rectifier which uses two diodes
- bridge rectifier which uses four diodes.

In India, the electrical energy is available at an alternating voltage of 220 V (r.m.s. value) at 50 Hz. However, for the operation of most of the electronic equipments, dc voltage is essentially needed. Hence, practically every electronic equipment includes a circuit, which provides it with a stabilized dc output voltage from the input ac voltage. In fact, this circuit forms the power supply of the equipment and consists of the following.

1. Transformer: for either stepping-up or stepping-down the ac supply voltage.
2. Rectifier: for converting the stepped-up (or down) ac voltage into dc voltage.
3. Smoothing filter: for removing any variation or ripples in dc output voltage.
4. Stabilizer: for keeping output dc voltage constant even when the input voltage or load current changes.

5.13.1 *Half-Wave Rectifier*

A simple circuit of a half-wave rectifier is shown in Fig. 5.14 along with the input and output voltage wave-forms. An ac voltage is applied to a single diode connected in series with a load resistor R_L.

Working. During the positive half-cycle of input ac voltage i.e. when point A is positive, the diode *D* is forward-biased (i.e. in ON state) and conducts. While conducting, the diode acts as a short-circuit so that the circuit current flows and hence, positive half-cycle of the input ac voltage is dropped across R_L. It constitutes the output voltage V_{out} as shown in Fig. 5.15b. The waveform of the diode current, which equals load current, will be similar to the voltage waveform.

During negative input half-cycle i.e. when point A becomes negative, the diode is reverse-biased (i.e. in OFF state) and so, does not conduct i.e. there is no current flow. Hence, there is no voltage drop across R_L i.e. $i_L = 0$ and $V_L = V_{out} = 0$. Obviously the negative input half-cycle is suppressed i.e. it is not utilized for delivering power to the load. As seen, the output is not a steady dc but only a pulsating dc wave having a ripple frequency equal to that of the input voltage frequency. This wave can be observed by an oscilloscope connected across R_L. When measured by a dc meter, it will show some average positive value both for

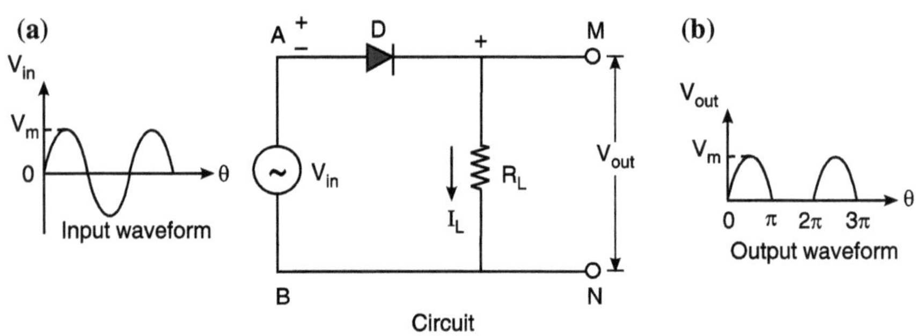

Fig. 5.15 A half wave rectifier showing **a** diode in the circuit, and **b** rectification

voltage and current. Since only one half-cycle of the input wave is used, it is called a 'half-wave rectifier'. It should be noted that the forward voltage drop across the diode has been neglected in above discussion. We have, in fact, assumed an ideal diode.

5.13.2 Full-Wave Rectifier

In this case, both half-cycles of the input are utilized with the help of two diodes working alternately. For full-wave rectification, the use of a centre-tap transformer is essential (though it is optional for the half wave rectification). A full-wave rectifier circuit using two diodes and centre-tapped transformer is shown in Fig. 5.16a. The centre-tap is usually taken as the ground (or zero) voltage reference point *G*.

Working. When input ac supply is switched on, the ends A and B of the secondary transformer become positive and negative alternately. During the positive half-cycle of the ac input, the terminal A is positive, point G is at zero potential and terminal B is at negative potential. Hence being forward-biased, the diode D_1 conducts (but not D_2 which is reverse-biased) and current flows along AD_1CMNG (Fig. 5.16a). As a result, the positive half-cycle of the voltage appears across RL.

During the negative half-cycle, when terminal B becomes positive, then the diode D_2 conducts (but not D_1) and current flows along BD_2CMNG. Thus, we find that the current keeps on flowing through R_L in the same direction (i.e. form M to N) in both half-cycles of the input. It means that both half-cycles of the input a-c supply are utilized as shown in Fig. 5.16b. Also, the frequency of the rectified output voltage is twice the supply frequency.

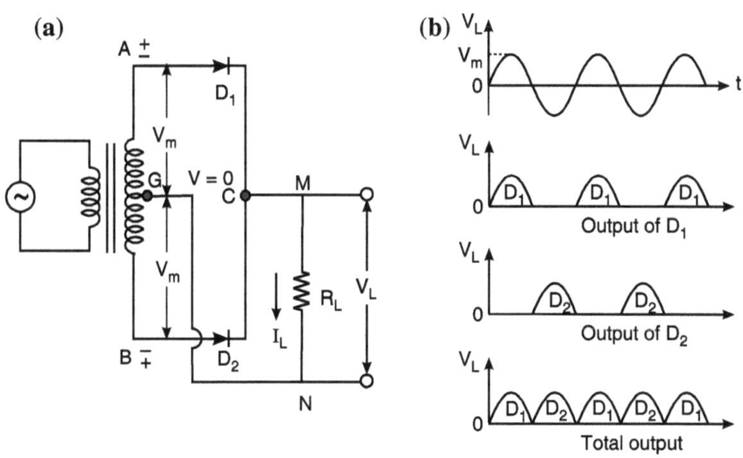

Fig. 5.16 A full wave rectifier showing **a** diodes in the circuit, and **b** stepwise process of full-rectification

5.14 Zener Diode

Zener diodes are designed to operate in the breakdown region without damage. By varying the doping level, it is possible to produce zener diodes with breakdown voltages ranging from about 2–200 V. As discussed earlier, the large current at breakdown is brought by two factors, known as the zener and avalanche effects. When a diode is heavily doped, the depletion layer is very narrow. When the voltage across the diode is increased (in reverse bias) the electric field across the depletion layer becomes very intense. When this field is about 3×10^7 V/m, electrons are pulled from the covalent bonds. A large number of electron-hole pairs is thus produced and the reverse current sharply increases. This is known as the zener effect.

Avalanche effect occurs because of a cumulative action. The external applied voltage accelerates the minority carriers in depletion region. They attain sufficient kinetic energy to ionize atoms by collision. This creates new electrons, which are again accelerated to high-enough velocities to ionize more atoms. In this way, an avalanche of free electrons is obtained, and the reverse current sharply increases.

Whereas the zener effect is predominant for breakdown voltages less than about 4 V, the avalanche breakdown is predominant for voltages greater than 6 V. Between 4 and 6 V, both effects are present. It is the zener effect that was first discovered, and the term 'zener diode' is in wide use for a breakdown diode whether it uses zener effect or avalanche effect, or both. If the applied reverse voltage exceeds the breakdown voltage, a zener diode acts like a constant voltage source. For this reason, a zener diode is also called voltage reference diode.

The circuit symbol of a zener diode is shown in Fig. 5.17. Zener diode is specified by its breakdown voltage and the maximum power dissipation. The most common application of a zener diode is in voltage stabilizing or regulator circuits. The other uses are as

- voltage regulator
- for meter protection
- as peak clipper
- in dc power supplies

5.14.1 Zener Diode for Meter Protection

Zener diodes are frequently used in volt-ohm-milliammeters (VOM) for protecting the meter against burn-out from accidental overloads. If VOM is set to its 2–5 V

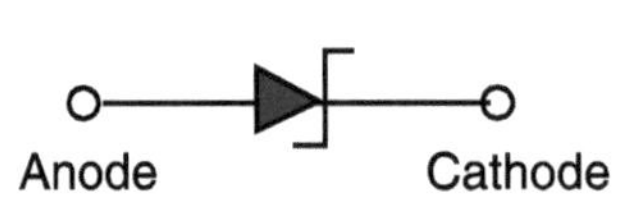

Fig. 5.17 Circuit symbol of a zener diode

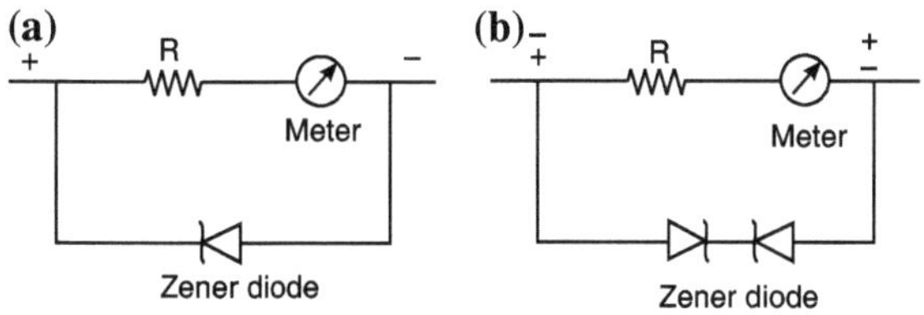

Fig. 5.18 Use of Zener diode for meter protection

range and the test leads are accidentally connected to a 25 V circuit, an unprotected meter will burn out or will get severely damaged.

This hazard can be avoided by connecting a Zener diode in parallel with the meter as shown in Fig. 5.18a. In the event of an accidental overload, most of the current will pass through the diode. Two Zener diodes connected as shown in Fig. 5.18b can provide overload protection regardless of the applied polarity.

5.14.2 Zener Diode as Peak Clipper

Use of Zener diodes in wave-shaping circuits is illustrated in Fig. 5.19. Here, a semi-square wave is produced by clipping a sine wave. Two Zener diodes are shunted across a sine wave source. The shunt resistance remains very high until the positive and negative voltage across the diodes exceed the Zener value. Thereafter, the diodes become low-resistance shunt circuits and the increasing voltage values are dropped across the series resistance R. Consequently the output wave is clipped on both peaks.

Example 5.6 Breakdown voltage of a Zener diode is 9.1 volt. It can dissipate a maximum power of 364 mW. Determine the maximum current the diode can handle.

Solution. Given are: $V_Z = 9.1$ V, $P = 364$ mW $= 364 \times 10^{-3}$ W

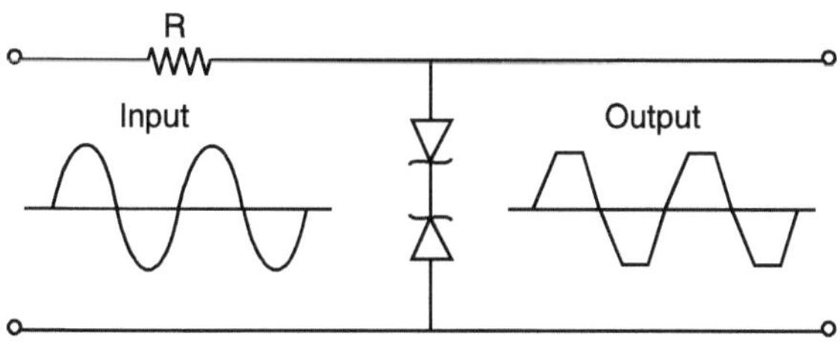

Fig. 5.19 Use of Zener diodes in wave-shaping circuits

Therefore, the maximum safe current will be

$$I_{Zmax} = \frac{P}{V_z} = \frac{364 \times 10^{-3}}{9.1}$$

$$= 40 \times 10^{-3}\,\text{A} = \mathbf{40}\,mA$$

5.15 The Breakdown Diode

As we discussed earlier in this section, the reverse-bias breakdown voltage of a junction can be varied by choice of junction doping concentrations. The breakdown mechanism is the Zener effect (tunnelling) for abrupt junctions with extremely heavy doping. However, the more common breakdown is avalanche (impact ionization), typical of more lightly doped or graded junctions. By varying the doping we can fabricate the diodes with specific breakdown voltages ranging from less than one volt to several hundred volts. If the junction is well designed, the breakdown will be sharp and the current after breakdown will be essentially independent of voltage (Fig. 5.20a). When a diode is designed for a specific breakdown voltage, is called a breakdown diode. Such diodes are also called Zener diodes, despite the fact that actual breakdown mechanism is usually the avalanche effect.

Breakdown diodes can he used as voltage regulators in circuits with varying inputs. The 15 V breakdown diode of Fig. 5.20 holds the circuit output voltage v_O constant at 15 V, while the input varies at voltages greater than 15 V. For example, if v_s is a rectified and filtered signal composed of a 17 V dc component and a 1 V ripple variation above and below 17 V, the output v_o will remain constant at 15 V. More complicated voltage regulator circuits can be designed using breakdown diodes, depending on the type of signal being regulated and the nature of the output load. In a similar application, such a device can be used as a reference diode. Since the breakdown voltage of a particular diode is known, the voltage across it during breakdown can be used as a reference in circuits that require a known value of voltage.

5.15.1 *Use of Diodes in DC Power Supplies*

Electric energy is available in homes and industries in the form of alternating voltage. For operation of most of the electronic devices, a dc voltage is needed. For instance, a transistor radio requires a dc supply for its operation, usually, this supply is provided by dry-cells. But sometimes we use battery eliminator in place of dry

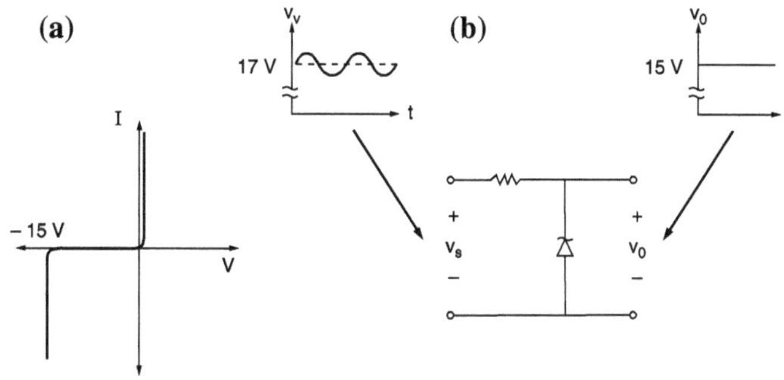

Fig. 5.20 A breakdown diode: **a** I-V characteristic, and **b** application as a voltage regulator

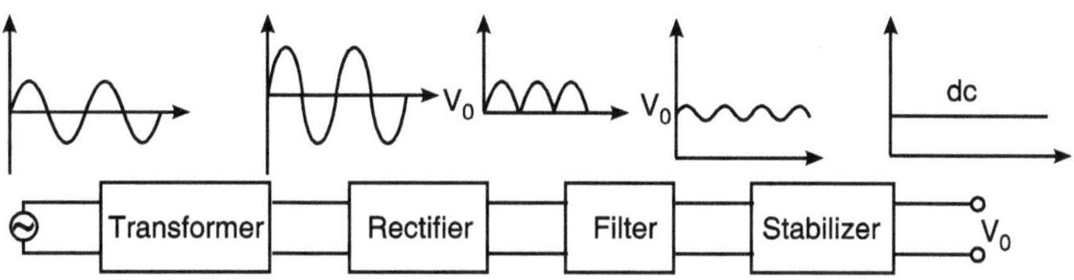

Fig. 5.21 Block diagram of a power supply

cell. The battery eliminator converts the ac voltage of main supply into dc voltage and thus eliminates the need for dry cells. Now-a-days, almost all-electronic equipments include a circuit which converts ac voltage of main supply into dc voltage. This part of the equipment is called power supply. In general, at the input of the power supply, there is a power transformer. It so followed by a diode circuit called rectifier. The output of the rectifier goes to a smoothing filter, and then to a voltage regulator circuit. A block diagram of such a power supply is shown in Fig. 5.21. The rectifier circuit is the heart of power supply.

5.16 Solved Examples

Example 5.7 Write the two major factors which govern the operation of semiconductor devices.

Solution. The operation of semiconductor devices mainly depends upon the following two factors.

1. Injection of minority carriers, and
2. Extraction of minority carriers.

Example 5.8 Explain briefly the following terms.

(i) Depletion region
(ii) Space-charge region
(iii) Barrier
(iv) Width and height of barriers.

Solution. The terms are defined as follows.

(i) In a p-n junction, the region containing the uncompensated acceptor and donor ions is called depletion region. There is a depletion of mobile charges (holes and free electrons) in this region.
(ii) Since this region has immobile (fixed) ions which are electrically charged, it is also the called space-charge region.
(iii) The electric filed between the acceptor and donor ions is called a barrier.
(iv) The physical distance from one side of the barrier to the other side is referred to as width of the barrier. The difference of potential from one side of the barrier to the other side is referred to as height of the barrier.

With no external batteries connected, the barrier height is of the order of tenths of a volt. For a silicon p-n junction, the barrier potential is about 0.7 V, whereas for a germanium p-n junction it is approximately 0.3 V.

Example 5.9 Explain depletion capacitance of a p-n diode.

Solution. The reverse-biased pn-junction has a region of high resistivity (space-charge or depletion region) sandwiched in between two regions (p and n region away from the junction) of relatively low resistivity. The p and n regions act as the plates of a capacitor, and the space-charge region acts as the dielectric. Thus, the pn-junction in reverse-bias has an effective capacitance, called transition or depletion capacitance.

Example 5.10 In performing diffusion process during fabricating a p-n junction, the Si wafers are placed inside a diffusion furnace, and the impurity atoms into the gas. Which impurity source materials are used to introduce B, P etc.?

Solution. Common source materials for introducing B and P during diffusion in Si are the following.

- For boron B: B_2O_3, BBr_3, BCl_3
- For phosphorus P: PH_3, P_2O_5, $POCl_3$

Example 5.11 A silicon diode dissipates 3 W for a forward dc current of 2 A. Calculate the forward voltage drop across the diode and its bulk resistance.

Solution. Given are: $P = 3$ W, $I_F = 2$ A, and junction voltage for Si is $V_j = 0.7$ V

Forward drop

$$V_F = \frac{P}{I_F} = \frac{3}{2} = 1.5\,\text{V}$$

Since

$$V_F = V_j + r_B I_F$$

$$\therefore \quad 1.5 = 0.7 + (r_B \times 2)$$

Or

$$r_B = \mathbf{0.4\,\Omega}$$

Example 5.12 What is Zener effect?

Solution. Zener effect is referred to as a state where the reverse breakdown of a p-n diode occurs, when the electric field in junction region reaches a critical value. The Zener effect is operative at low voltages (for a reverse bias up to a few volts).

Example 5.13 What is avalanche effect?

Solution. Avalanche effect means reverse breakdown of a p-n diode, when electric field in junction region crosses the critical value of voltage. The avalanche effect is operative at higher voltages, which may range from a few volts to thousands volts.

Example 5.14 What is the effect of doping on breakdown of a p-n junction diode?

Solution. The effect of doping on breakdown of a p-n junction diode is as follows.

- It suffers Zener breakdown when heavily doped.
- It suffers avalanche breakdown when lightly doped.

Example 5.15 Under which conditions does a p-n junction possess the diffusion capacitance and transition capacitance?

Solution. a p-n junction possesses above types of capacitance under following conditions.

- Diffusion capacitance when forward-biased.
- Transition capacitance when reverse-biased.

Example 5.16 Write the specifications of a p-n junction diode.

Solution. A p-n junction diode is specified by the following quantities/parameters.

- Construction material e.g. Si-diode, Ge-diode etc.
- Power rating i.e. the maximum power that can be dissipated safely, e.g. 5 mA diode, 10 mA diode etc.

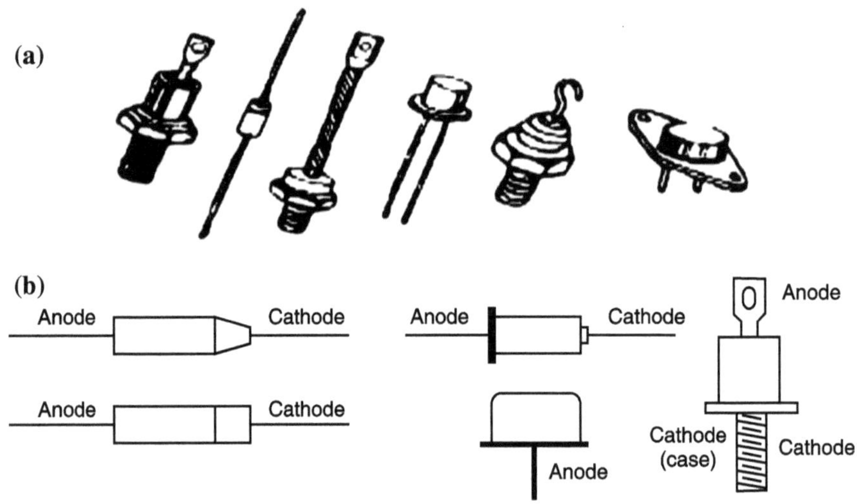

Fig. 5.22 **a** Physical structures of diodes. **b** their terminal identifications

- Maximum forward current I_{Fmax} that the diode can carry, e.g. 1 mA, 2 mA etc.
- Maximum reverse current I_{rmax} e.g. 1 μA, 2 μA etc. It should be a minimum possible value.
- Peak inverse voltage that the diode can withstand under reverse-biasing.
- Junction capacitance e.g. 1 pF (10^{-12} F), 2 pF etc.
- Packaging style

Example 5.17 Display the various physical structures of diodes and their terminal identifications.

Solution. Figure 5.22a shows different physical structures and Fig. 5.22b shows various terminals identifications for diodes.

Example 5.18 Some p-n diodes are designed with a very narrow n-region—why?

Solution. By making a diode with narrow n-region, the stored charge problems can be reduced. If the n-region is much narrower than the hole diffusion length L_{Dh}, then a very little charge will be stored in the junction. Consequently, the time required to switch the diode 'off and on' will be much less.

Example 5.19 How can the switching of a p-n diode be made faster by choosing an appropriate material? Explain its reason also.

Solution. The switching process of a p-n diode can be made faster by adding gold (Au) atoms in Si. It is because the gold acts as recombination centre, thereby increases the recombination rate and decreases the carrier lifetime.

Review Questions

1. Define the following.

(a) Equilibrium conditions	(b) Depletion region
(c) Rectification	(d) Steady state condition
(e) Contact potential	(f) Biasing

2. What is the necessity of forming a p-n junction? What are its salient features? Enlist various important applications of a p-n diode.
3. Explain the requirements of equilibrium conditions. What is the state of drift and diffusion of electrons and holes at this condition?
4. Establish a relation between contact potential and doping concentrations, stating the assumptions and approximations made therein. Express this relation in different alternative forms.
5. Show that the maximum value of electric field in a p-n junction is given as proportional to penetration depth in p-type material and inversely proportional to relative permittivity of the semiconductor.
6. Derive an expression to obtain the width of depletion region and depth of penetration on either side of the p-n junction.
7. Explain the formation of depletion region in an open circuited p-n junction.
8. What cause majority carriers to flow at the moment when p-region and n-region are brought together? Why does this flow not continue until all the carriers have recombined?
9. What do you understand by barrier potential across a p-n junction? Also explain its significance.
10. The barrier potential developed across an open-circuited p-n junction aids the flow of majority carriers. Explain as to how the flow of charge carriers is counterbalanced.
11. Describes the construction and working of a p-n junction diode in its inherent state, stating as to how (a) the diffusion of holes and electrons occur in it, (b) barrier is set-up, and (c) depletion region is formed.
12. Explain the meaning of forward biasing of a p-n junction. What significant features are noticed after biasing?
13. Explain the meaning of reverse biasing of a p-n junction; discuss the significant features of this biasing.
14. Write notes on the following.
    (a) Fermi level in p-n junction at equilibrium.
    (b) Space change in p-n junction at equilibrium.
    (c) Reverse saturation current in a reverse biased p-n junction.
15. Show the temperature of V-I relation for a p-n junction diode.
16. Explain the voltage-current characteristics of a p-n diode under forward and reverse biasing.

17. Sketch on the same axes, the typical static characteristics for germanium and silicon diodes. Label clearly the values of forward voltage drop and reverse saturation current.
18. What limits the numbers of reverse current carriers?
19. Why is the reverse current in a silicon diode much smaller than that in a comparable germanium diode?
20. What do you mean by breakdown of a p-n junction? What are different types of breakdown mechanisms?
21. Discuss the mechanism of Zener and avalanche breakdowns.
22. Name the two types of reverse breakdowns that can occur in a p-n junction diode. Which occurs at a lower voltage?
23. Explain the process of avalanche breakdown that occurs in a p-n junction diode. How is it different from zener breakdown?
24. What do you know about rectifying diodes? Name different kinds of rectifying diodes.
25. With the help of suitable diagram and circuit, explain the operational mechanism of diode in a half-wave rectifier.
26. How do the twin diodes govern the operation of a full-wave rectifier?
27. Describe the construction and working of a Zener diode. Write its applications also.
28. With the help of suitable line or block diagram or circuit, explain as how the zener diode (a) helps in meter protection, and (b) functions as a peak clipper.
29. Which carriers conduct forward current in a diode? Draw the symbol of a p-n junction diode showing the direction of forward current.
30. Draw the block diagram of a regulated power supply, and explain in brief the function of each block.
31. Draw the circuit diagram of a voltage regulator using a zener diode. Explain its working. Is there any limitation on the value of the series resistor used in this circuit?
32. Sketch and explain the construction and working of a breakdown diode.
33. Explain why is it necessary to use a voltage regulator circuit in a power supply.
34. Explain as to why a p-n junction possesses capacitance.
35. For distinguishing the response of a p-n junction under no-bias and reverse bias conditions, draw suitable diagrams at one place showing the width of depletion region, potential barrier, electric field, energy band diagrams, and particle flow and current flow.
36. Write notes on the following.
    (a) Heterojunction device
    (b) Negative resistivity
    (c) Dielectric constants for semiconductors
    (d) Quasi-steady state

Numerical Problems

1. Boron is implanted into a n-type Si sample (Nd = $10^{16}/cm^3$). If the acceptor concentration in p- type region is Na = 4×10^{18} per cm^3, calculate the contact potential for this junction at equilibrium (300 K). Take ni = $1.5 \times 10^{10}/cm^3$.

   [**Ans** : 0.85 V]
2. A centre-tap full-wave rectifier supplies a load of 1 kW. The ac voltage across the secondary is 200-0-200 V (r.m.s.). If diode resistance is neglected, find the dc voltage across the load and dc load current.

   [**Ans** : 180 V, 0.18 A]
3. In a centre-tap full-wave rectifier, load resistance is 2 kW. The ac supply across the primary winding is 220 Sin 314t. Taking transformer turn ratio N1/N2 = 1/2 and neglecting diode and secondary winding resistance, compute the dc load voltage and current.

   [**Ans** : 140 V, 70 mA]

Objective Questions

1. Consider the following statements.
   A. Barrier potential is 0.3 V in silicon p-n junction.
   B. A Zener diode may be used for meter protection.
   C. A p-n diode conducts current in one direction only.
   D. Drift current flows due to electric field.

   Of these, the correct statements are

   (a) A and B (b) C and D
   (c) B,C and D (d) A, B and D
2. In an unbiased p-n-junction, the junction current at equilibrium is
   (a) due to diffusion of minority carriers only
   (b) due to diffusion of majority carriers only
   (c) zero, because no charges are crossing the junction
   (d) zero, because equal but opposite carriers are crossing the junction

3. In a p-n junction diode, if the junction current is zero, it means that
   (a) the potential barriers has disappeared
   (b) there are no carriers crossing the junction
   (c) the number of holes diffusing from the p-region equals the number of electrons diffusing from the n-region.
   (d) the number of majority carriers crossing the junction equals the number of minority carriers crossing the junction.
4. When a p-n junction is reverse-biased,
   (a) its barrier potential decreases (b) its depletion layer becomes narrow
   (c) it offers high resistance (d) it breaks down.
5. The number of minority carriers crossing the junction of a diode depends primarily on the
   (a) rate of thermal generation of electron-hole pairs
   (b) magnitude of the potential barrier
   (c) magnitude of the forward-bias voltage
   (d) concentration of doping impurities.
6. Avalanche breakdown in a semiconductor diode occurs when
   (a) forward current exceeds a certain value
   (b) forward bias exceeds a certain value
   (c) reverse bias exceeds a certain value
   (d) the potential barrier is reduced to zero
7. The forward bias applied to a p-n junction diode is increased from zero to higher values. Rapid increase in the current flow for a relatively small increase in voltage occurs
   (a) immediately
   (b) when the flow of minority carriers is sufficient to cause an avalanche breakdown
   (c) only after the forward bias exceeds the potential barrier
   (d) when the depletion area becomes larger than the space- charge area
8. A zener diode
   (a) has a sharp breakdown at low reverse voltage
   (b) has a high forward-voltage rating
   (c) is useful as an amplifier
   (d) has a negative resistance

9. Match list I with list II, and choose the correct answer from the codes given below the lists.

List I	List II
A. Breakdown range for zener diode	1. $e^{(-eV_0/kT)}$
B. p_p/p_p	2. (0.2 to 2 V)
C. Width of depletion region decreases	3. reverse bias
D. Avalanche breakdown	4. (2 to 200 V)
	5. $= e^{(-eV_0/kT)}$
	6. forward bias

Codes :

	A	B	C	D		A	B	C	D
(a)	4	5	6	3	(b)	2	1	3	6
(c)	1	3	6	2	(d)	5	6	3	2

10. The thickness of the depletion layer in an PN junction is of the order of
    (a) 10^{-6} m.
    (b) 10^{-10} m.
    (c) 10^{-4} m.
    (d) None of these.

11. The most important factor which can influence the operation of a PN junction is
    (a) impurity atom of a PN concentration.
    (b) temperature.
    (c) mobility of change carriers.
    (d) All of these.

12. The potential existing across a PN junction
    (a) facilitates recombination of electrons and holes.
    (b) prevents flow of minority carriers.
    (c) prevents total recombination of holes and electrons.
    (d) prevents neutralization of acceptor and donor ions.

13. The rms value of a half-wave rectified symmetrical square wave current of 2 A is
    (a) A
    (b) 1 A
    (c) 1/A
    (d) A

14. Which of the following are the effects of increasing the reverse bias voltage across a P-N junction?
    1. Decrease in junction capacitance of the diode.
    2. Increase in carrier generation current.
    3. Increase in carrier recombination in the depletion layer.

Select the correct answer using the codes given below:
Codes:

(a) 1 alone (b) 2 and 3
(c) 1 and 2 (d) 1, 2 and 3

Answers

1. (c) 2. (d) 3. (d) 4. (c) 5. (a) 6. (b) 7. (c) 8. (a)
9. (a) 10. (a) 11. (b) 12. (c) 13. (a) 14. (a)

Reference

1. Gupta, K.M., Gupta, N.: Advanced Electrical and Electronics Materials. Scrivener Wiley Publishing, USA (2015) (Some of the matter of this chapter has been reproduced from this book)

Chapter 6
Different Types of Diodes, Ideal and Real Diodes, Switching Diodes, Abrupt and Graded Junctions

Abstract In this chapter, different types of specific purpose diodes viz. crystal diode, thermal diode, constant current diode, PIN diode, Schottky diode, gold-doped diode, super barrier diode, varicap diode, Esaki diode, laser diode, TVS diode and snap-off diode etc. are introduced besides many others. Their symbolic representation, diode model, ideal diode and real diode are also presented. Different conditions of working of p-n junctions and transient conditions are included. Time dependent variation in space charge, and solution for stored charge is mathematically elaborated. Switching diodes, narrow base diode, reverse recovery transient, and method to improve the switching speed are discussed. P-N junction capacitance, derivation of its expression; Graded junctions, linearly graded, abrupt and hyperabrupt junctions and their doping profile are given. Minute insight into the various topics are given through solved numerical and theoretical examples. Review questions, numerical problems and objective type questions are also given with their answers.

Keywords Different types of diodes · Symbolic representation of different diodes · Diode model · Real diode · Switching diode · Graded junctions · Abrupt junction

6.1 Examples of Diodes

There are several types of *p-n* junction diodes. They either emphasize different physical aspects of diode such as geometric scaling, doping level, or just an application of diode in a special circuit, or different devices like the Gunn diode, laser diode and the MOSFET etc.

Normal (*p-n*) diodes are usually made of doped silicon or rarely of germanium. Before the development of modern silicon power rectifier diodes, earlier the cuprous oxide and later selenium were used. They had low efficiency, much higher forward voltage drop, and required a large heat sink. They were much larger than a silicon diode of the same current ratings. The vast majority of all diodes are the *p-n* diodes found in CMOS integrated circuits, which include two diodes per pin and many other internal diodes. These are briefly discussed below.

K.M. Gupta and N. Gupta, *Advanced Semiconducting Materials and Devices*, Engineering Materials, DOI 10.1007/978-3-319-19758-6_6

6.1.1 Zener Diodes

These are the diodes that can be made to conduct backward. This effect, called Zener breakdown, occurs at a precisely defined voltage, allowing the diode to be used as a precision voltage reference. In practical voltage reference circuits, Zener and switching diodes are connected in series and opposite directions to balance the temperature coefficient to near-zero. Some devices labelled as high-voltage Zener diodes are actually avalanche diodes. Two (equivalent) Zener in series and in reverse order, in the same package, constitute a transient absorber (or Transorb, a registered trademark). The Zenger diode is named for Dr. Clarence Melvin Zener of Carnegie Mellon University, inventor of the device. More details have already been described in Chap. 5.

Other uses for semiconductor diodes include sensing temperature, and computing analog logarithms.

6.1.2 Avalanche Diodes

These are the diodes that conduct in the reverse direction when the reverse bias voltage exceeds the breakdown voltage. These are electrically very similar to Zener diodes, but break down by a different mechanism, the *avalanche effect*. This occurs when the reverse electric field across the *p-n* junction causes a wave of ionization, reminiscent of an avalanche, leading to a large current. Avalanche diodes are designed to break down at a well-defined reverse voltage without being destroyed. The difference between the avalanche diode (which has a reverse breakdown above about 6.2 V) and the Zener is that the channel length of the former exceeds the mean free path of the electrons, so there are collisions between them on the way out. The only practical difference is that the two types have temperature coefficients of opposite polarities. More details are already discussed in Chap. 5.

6.1.3 Cat's Whisker (or Crystal) Diodes

These are a type of point-contact diode. The cat's whisker diode consists of a thin or sharpened metal wire pressed against a semiconducting crystal, typically galena or a piece of coal. The wire forms the anode and the crystal forms the cathode. Cat's whisker diodes were also called crystal diodes and found application in crystal radio receivers. These diodes are generally obsolete.

6.1.4 Thermal Diodes

This term is used both for conventional *p-n* diodes used to monitor temperature due to their varying forward voltage with temperature, and for Peltier heat pumps for thermoelectric heating and cooling. Peltier heat pumps may be made from semiconductor, though they do not have any rectifying junctions. They use the differing behaviour of charge carriers in *N* and *P* type semiconductor to move heat.

6.1.5 Constant Current Diodes

These are actually a JFET with the gate shorted to the source. These function like a two- terminal current-limiter analog to the Zener diode, which is limiting voltage. They allow a current through them to rise to a certain value, and then level off at a specific value. These are also called as CLDs, *constant-current diodes, diode-connected transistors, or current-regulating diodes*.

6.1.6 Photodiodes

All semiconductors are subject to optical charge carrier generation. This is typically an undesired effect. So most semiconductors are packaged in light blocking material. Photodiodes are intended to sense light (photodetector), so they are packaged in materials that allow light to pass, and are usually PIN (the kind of diode most sensitive to light). A photodiode can be used in solar cells, in photometry, or in optical communications. Multiple photodiodes may be packaged in a single device, either as a linear array or as a two-dimensional array. These arrays should not be confused with charge-coupled devices.

6.1.7 PIN Diodes

A PIN diode has a central un-doped (or *intrinsic layer*) forming a *p*-type/intrinsic/*n*-type structure. They are used as radio frequency switches and attenuators. They are also used as large volume ionizing radiation detectors and as photodetectors. PIN diodes are also used in power electronics, as their central layer can withstand high voltages. Furthermore, the PIN structure can be found in many power semiconductor devices such as IGBTs, power MOSFETs, and thyristors.

6.1.8 Schottky Diodes

Schottky diodes are constructed from a metal to semiconductor contact. They have a lower forward voltage drop than *p-n* junction diodes. Their forward voltage drop at forward currents of about 1 mA is in the range 0.15–0.45 V, which makes them useful in voltage clamping applications and prevention of transistor saturation. They can also be used as low loss rectifiers, although their reverse leakage current is in general higher than that of other diodes. Schottky diodes are majority carrier devices and so do not suffer from minority carrier storage problems that slow down many other diodes. So they have a faster reverse recovery than *p-n* junction diodes. They also tend to have much lower junction capacitance than *p-n* diodes, which provides for high switching speeds. They are also used in high-speed circuitry and RF devices such as switched-mode power supply, mixers, and detectors.

6.1.9 Gold-Doped Diodes

As a dopant, gold (or platinum) acts as recombination centers which help a fast recombination of minority carriers. This allows the diode to operate at signal frequencies, at the expense of a higher forward voltage drop. Gold-doped diodes are faster than other p-n diodes (but not as fast as Schottky diodes). They also have less reverse-current leakage than Schottky diodes (but not as good as other p-n diodes).

6.1.10 Super Barrier Diodes

Super barrier diodes are rectifier diodes that incorporate the low forward voltage drop of the Schottky diode with the surge-handling capability and low reverse leakage current of a normal *p-n* junction diode.

6.1.11 Varicap (or Varactor) Diodes

These are used as voltage-controlled capacitors. These are important in PLL (phase-locked loop) and FLL (frequency-locked loop) circuits, allowing tuning circuits such as those in television receivers, to lock quickly, replacing older designs that took a long time to warm up and lock. A PLL is faster than an FLL, but prone to integer harmonic locking (if one attempts to lock to a broadband signal). They also enabled tuneable oscillators in early discrete tuning of radios where a cheap and stable but fixed-frequency, crystal oscillator provided the reference frequency for a voltage-controlled oscillator.

6.1.12 Gunn Diodes

These are similar to tunnel diodes in that they are made of materials such as GaAs or InP that exhibit a region of negative differential resistance. With appropriate biasing, the dipole domains are formed and travel across the diode, allowing high frequency microwave oscillators to be built.

6.1.13 Esaki (or Tunnel) Diodes

These have a region of operation showing negative resistance caused by quantum tunnelling, allowing amplification of signals and very simple bistable circuits. Due to the high carrier concentration, tunnel diodes are very fast. They may be used at low (mK) temperatures, high magnetic fields, and in high radiation environments. Because of these properties, they are often used in spacecraft.

6.1.14 Light-Emitting Diodes (LEDs)

In a diode formed from a direct band-gap semiconductor such as gallium arsenide, carriers that cross the junction emit photons when they recombine with the majority carrier on the other side. Depending on the material, wavelengths (or colours) from the infrared to the near ultraviolet may be produced. The forward potential of these diodes depends on the wavelength of the emitted photons e.g. 2.1 V corresponds to red and 4.0 V to violet. The first LEDs were red and yellow, and higher-frequency diodes have been developed over time. All LEDs produce incoherent, narrow-spectrum light white LEDs are actually combinations of three LEDs of a different colour, or a blue LED with a yellow scintillator coating. LEDs can also be used as low-efficiency photodiodes in signal applications. An LED may be paired with a photodiode or phototransistor in the same package, to form an opto-isolator (Fig. 6.1).

6.1.15 Laser Diodes

When an LED-like structure is contained in a resonant cavity formed by polishing the parallel end faces, a laser can be formed. Laser diodes are commonly used in optical storage devices and for high speed optical communication.

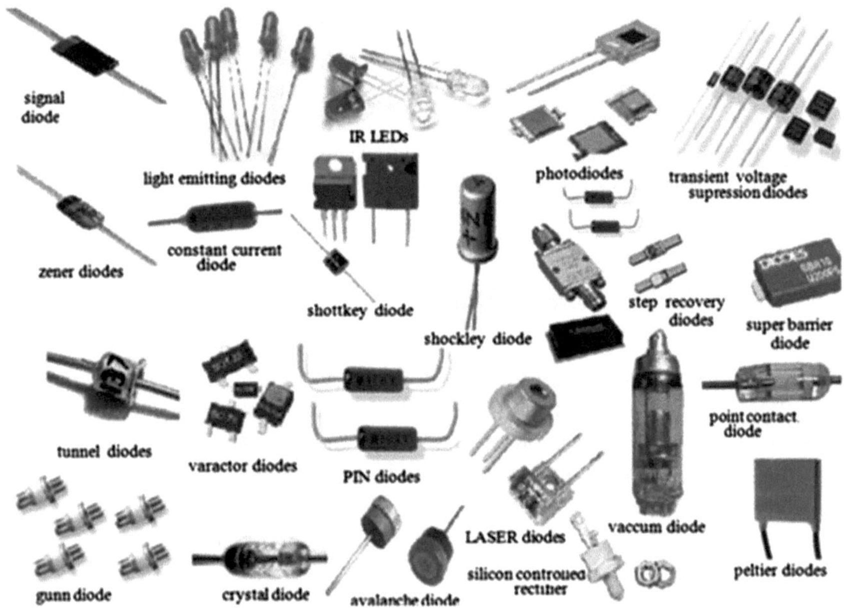

Fig. 6.1 Several types of diodes [1]

6.1.16 *Transient Voltage Suppression (TVS) Diode*

These are avalanche diodes designed specifically to protect other semiconductor devices from high-voltage transients. Their *p-n* junctions have a much larger cross-sectional area than those of a normal diode, allowing them to conduct large currents to ground without sustaining damage.

6.1.17 *Snap-Off (or Step Recovery) Diodes (SRD)*

The term *step recovery* relates to the form of the reverse recovery characteristic of these devices. After a forward current has been passing in an SRD and the current is interrupted or reversed, the reverse conduction will cease very abruptly (as in a step waveform). SRDs can, therefore, provide very fast voltage transitions by the very sudden disappearance of the charge carriers.

6.2 Symbolic Representation of Different Diodes

These are given in Fig. 6.2.

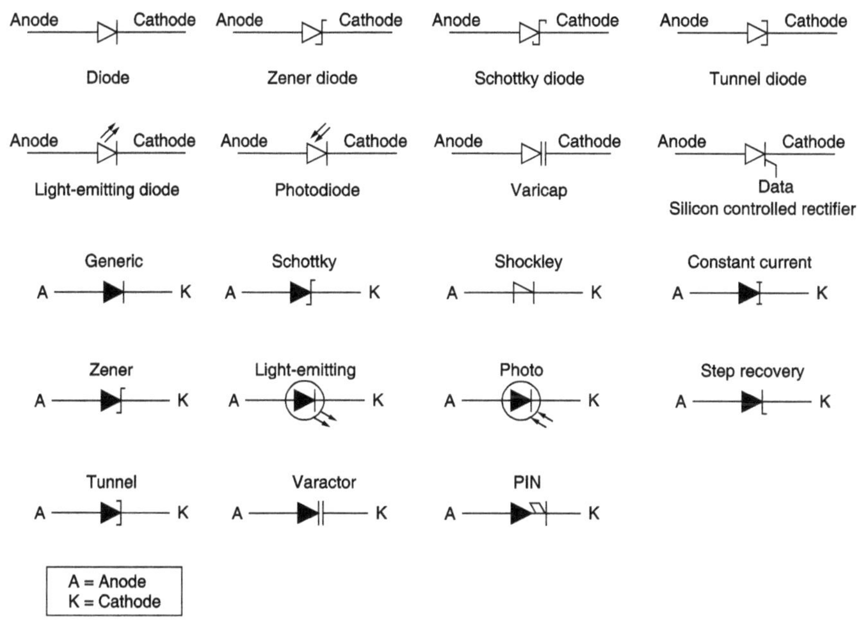

Fig. 6.2 Some diode symbols

6.3 The Diode Model

In the study of electronic and semiconductor devices, the diode modelling refers to the mathematical models that are used to approximate the actual behaviour of real diodes, to enable calculations and circuit analysis. As a diode's current–voltage (I-V) curve is non-linear, it complicates the calculations in circuits involving diodes. Therefore, the simpler (or ideal) models are often required.

Ideal diodes differ from real diodes in the way that they are assumed to have negligible generation-recombination inside the depletion region. The *p-n* junction diode that we come across, does not exactly fit under the ideal diode model but their I-V characteristics can be approximated by the ideal diode in series with other circuit elements to form an equivalent circuit.

The ideal model of a diode can be thought of as a simple switch. When the diode is forward biased, it acts like a closed switch (short circuit) and when reverse biased, it acts like an open switch (open circuit). For example, most diodes when forward-biased exhibit an offset voltage say E_0 which can be approximated in a circuit model by a battery in series with the ideal diode, Fig. 6.3a. The series battery in the model keeps the ideal diode turned off for all applied voltages less than E_0. The I-V characteristics of an ideal diode is shown in Fig. 6.3b.

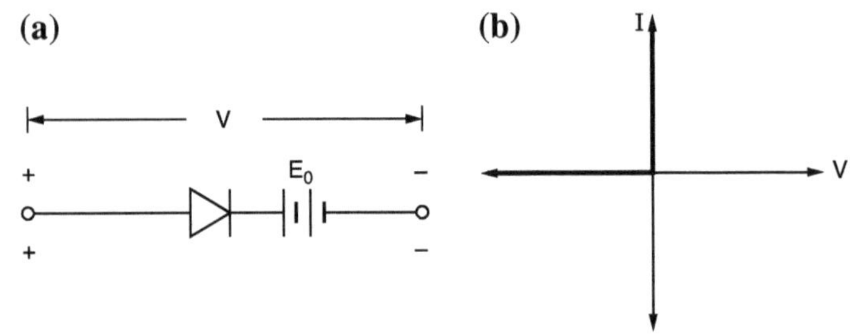

Fig. 6.3 a Circuit model of an ideal diode. b I-V characteristic

6.4 Real Diodes

Real diode is a non-linear device. Opposite to an approximated diode (ideal diode), a real diode has non-linear voltage drop across it when current flows in the forward direction. It allows only a small flow of current in reverse direction and breaks down to allow large reverse current if its rated reverse voltage is exceeded. On the other hand, an ideal diode allows current to flow in the forward direction without any resistance to the flow of current and completely prevents any current flow in the reverse direction. The real diode is modelled as 3 components in series viz. an ideal diode, a voltage source and a resistor. The Fig. 6.4a shows a real diode I-V curve being approximated by a two-segment piecewise linear model.

Real diodes do not follow the ideal diode equation because of physical limitations of device fabrication or design techniques. For ideal diodes, there would be no reverse breakdown, and no reverse leakage current I_s. They would also have a right-angle break at the forward voltage V_f, going from no conduction to full

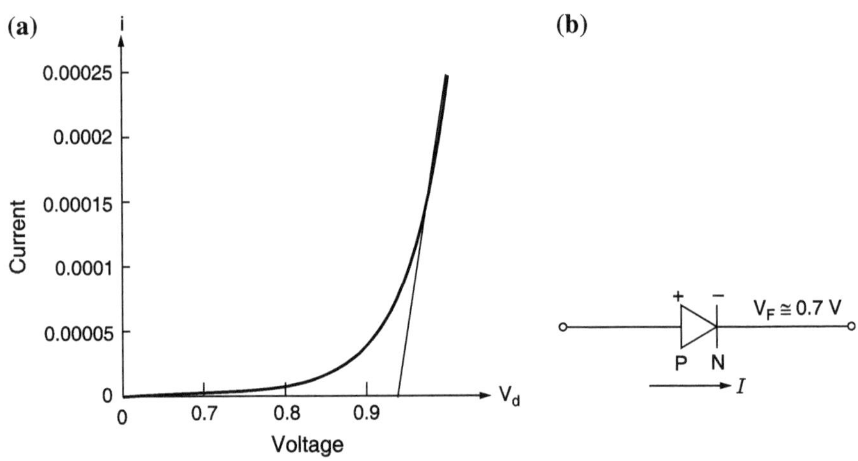

Fig. 6.4 A real diode depicting its a current–voltage curve, and b symbolic representation

conduction (zero resistance), whereas a forward voltage drop VF appears across a real diode when current flows through it. One can neither calculate the voltage nor the current across/through a real diode, instead, must use the diode's characteristic curve given on datasheet (datasheet not shown here), and linearize it. The symbol of a real diode is shown in Fig. 6.4b.

In a real diode:

- R_f is of the order of a few ohms.
- $V_f = 0.6$ for silicon and 0.3 for germanium.
- Reverse bias resistance R_r is of the order of a few kilo ohms.

6.4.1 *Ideal Diode Versus Real Diode*

Difference between an ideal diode and a real diode are given below.

S. No.	Description	Ideal diode	Real diode
1.	Assumption	Negligible generation-recombination inside the depletion region	Can be modelled as ideal diode, a voltage source and a resistor in series
2.	Behaviour under forward-biasing	Acts like a closed switch (short circuit)	Forward voltage drop appears when current flows through it
3.	Behaviour under reverse-biasing	Acts like an open switch (open circuit)	Reverse-bias resistance is of the order of a few kΩ

More differences can be self-judged by the details given in Fig. 6.5.

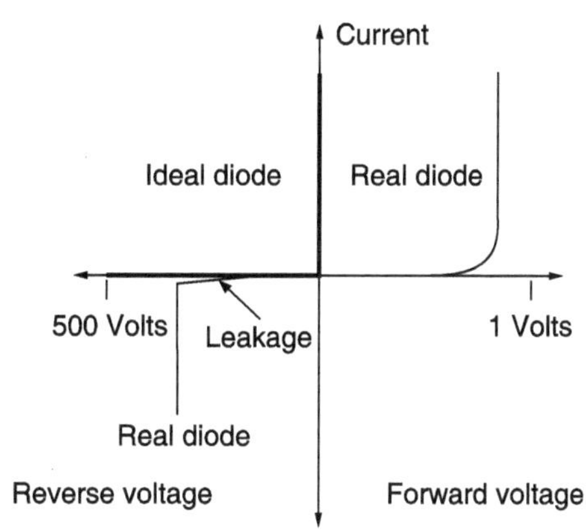

Fig. 6.5 Depiction of difference between ideal and real diodes by current–voltage representation

6.5 Different Conditions of the Working of P-N Junctions

The *p-n* junctions work under vivid conditions in real life applications. They are also employed for a variety of applications of widely varying needs. Main conditions under which the *p-n* junctions have to work are the following.

1. Under equilibrium conditions
2. Under steady state conditions
3. Under quasi-steady state conditions
4. Under transient conditions
5. Under alternating current (a-c) conditions

The *p-n* junctions have to work under both; the time independent and time dependant states. A variety of junction diodes are used to fulfil these functions. Most of them are made of semiconductors alone, but many others are made of metal-semiconductor combination. The semiconductors used in making those diodes are also of different characteristics. For example, some of their combinations have different band gaps; some have different work functions and electron affinities; while some others have almost equal values.

Junctions Under Transient Conditions. The working of *p-n* junctions under equilibrium and steady-state conditions has already been described in Chap 5. Now, we shall describe their response under transient conditions. The requirements of transient conditions are met by different kinds of diodes such as given below.

- Switching diodes
- Varactor diode
- Metal-semiconductor junctions
- Schottky diode
- Heterojunction diode

Now we shall study their behaviour under transient conditions.

6.6 Transient Conditions

Although most of the basic concepts of *p-n* junction devices can be understood from their behaviour under equilibrium conditions and steady state as described in Chap 5, but this is not sufficient to know all about them. The study of their behaviour under transition conditions change in behaviour as a function of time) is also very essential. It is more so important because most solid state devices that are used for switching or processing the a-c signals, undergo the time-dependent processes. The problems of *p-n* junction devices under transient conditions generally require solution of various current flow equations involving space and time variables. Some such time dependent applications of junction devices are under mentioned.

1. Switching of diode from forward conduction state to reverse-biased state, and vice versa.
2. Processing of small a-c signals to determine equivalent capacitance of a p-n junction.
3. Influence of excess carriers such as given below.

- Time variation of stored charge
- Decay of stored charge in n-region
- Excess hole distribution in n-region
- On switching signal, etc.

6.7 Time Dependent Variation in Space Charge

Transient problems of *p-n* junctions can be solved by using the time-dependent continuity equations, given below.

$$-\frac{\partial J_P(x,t)}{\delta x} = e.\frac{\delta p(x,t)}{\tau_p} + e.\frac{\partial p(x,t)}{\partial t} \tag{6.1}$$

Here J_p is current density of holes with respect to position x and time t; δp is hole concentration, p is concentration of holes in valence band and δp is recombination lifetime for holes with respect to x and t.

In order to obtain the instantaneous current density, Eq. 6.1 is integrated with respect to x at time t. It then yields

$$J_P(0) - J_P(x) = e\int_0^x \left\{\frac{\delta p(x,t)}{\tau_p} + \frac{\partial p(x,t)}{\partial t}\right\}dx \tag{6.2}$$

When we consider the injection of current from p to n region, the following conditions can be assumed to exist.

(i) At $x = 0$ in n region, all current is hole current, and
(ii) At $x = \infty$ in n region, $J_p = 0$.

On denoting $x = 0$, t in n-region by $x_n = 0$, t; the total injected current can be written as

$$i(t) = i_P(atx_n = 0, t) = \frac{eA}{\tau_p}\int_0^x \delta p(x_n, t)dx_n + eA\frac{\partial}{\partial t}\int_0^x \delta p(x_n, t)dx_n \tag{6.3}$$

Since we know that

$$eA \int_0^x \delta_P(x_n, t) dx_n = Q_P(t)$$

and

$$eA\partial \int_0^\infty \delta_P(x_n, t) dx_n = dQ_P(t)$$

in which Q_p is charge stored in hole distribution, the Eq. 6.3 may be expressed as follows.

$$i(t) = \frac{Q_P(t)}{\tau_p} \frac{dQ_P(t)}{dt} \tag{6.4}$$

This is *turn-on time equation* for *p-n* junction under transient condition. Important results of this equation are as follows.

- Total diode current i.e. the hole current injected across the *p-n* junction comprises of two charge storage effects as given below.
  (i) Recombination term, and Q_p/τ_p, and
  (ii) Charge build-up or depletion term dQ_p/dt.

Transformation of Eq. 6.4 **into steady state**. For steady state, the term $dQ_p/dt = 0$. Therefore Eq. 6.4 reduces to

$$i(t) = \frac{Q_P(t)}{\tau_p} \tag{6.5}$$

6.7.1 Determining the Solution for Stored Charge

The solution for stored charge may be found as a function of time t for a given transient in current. A step turn-off transient is shown in Fig. 6.6a in which the current is suddenly removed at $t = 0$. It leaves the diode with stored charge. Its solution can be determined with the help of Laplace transforms. Hence using

$$\mathrm{i} = 0 \text{ for } \mathrm{t} > 0 \quad \text{and} \quad \mathrm{Q_p(0)} = \mathrm{I}\tau_\mathrm{p}$$

we obtain

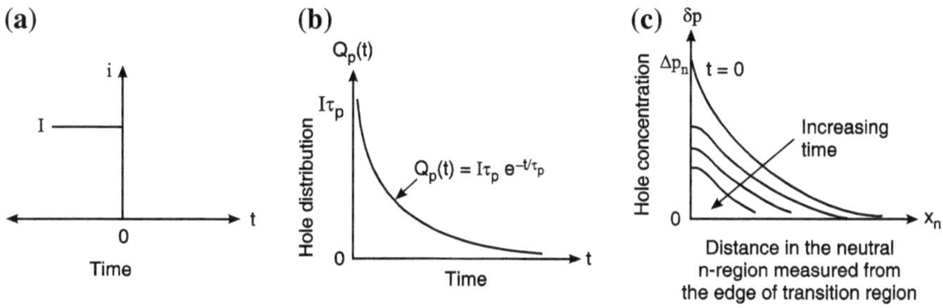

Fig. 6.6 Effects of a step turn-off transient on a *p-n* junction showing **a** the current-time relation, **b** die-out of stored charge-time behaviour, and **c** excess hole distribution-time response

$$Q_P(t) = I\tau_p e^{\left(-\frac{t}{\tau_p}\right)} \tag{6.6}$$

It depicts an exponential decay of $I\tau_p$ with time as shown in Fig. 6.6b. The distribution of excess holes in *n*-region of a *p-n* junction during transient is also shown in Fig. 6.6c. This represents a non-exponential distribution. The slopes of distribution curves are zero at $x_n = 0$. The behaviour with time shows a decrease in δp. It is because of recombination of excess electrons and holes. The change in δp as a function of time can be expressed as

$$\Delta p_n(t) = p_n \cdot \left(e^{\frac{eV(t)}{KT}} - 1\right) \tag{6.7}$$

in which *e* is electronic charge, *V(t)* is applied voltage.

6.8 Reverse Recovery Transient

When a *p-n* diode is switched-off from forward conduction to reverse-biased state or vice versa a reverse current flows into the junction. This current may be much larger than the normal reverse saturation current. The flow of reverse current occurs during the period that is required for readjustment of stored charge. The reason for unusual large reverse current is that the stored charge cannot be changed instantaneously at the moment, when diode voltage is reversed.

Later when the stored charge is depleted, the junction exhibits a negative voltage. As time proceeds, the magnitude of reverse current becomes smaller. Ultimately it becomes zero. The time required for the stored charge to become zero is termed as 'storage delay time' t_{sd}. This is also the time when junction voltage becomes zero. It is given by

$$t_{sd} = t_p \left[\frac{1}{erf\left(\frac{I_f}{I_f + I_r}\right)} \right] \tag{6.8}$$

where *erf* is error function, I_f is steady state current in forward bias, and I_r is reverse current. A smaller t_{sd} as compared to switching time is desired for an efficient operation.

6.9 Switching Diodes

Switching diodes are the p-n junctions which are used to serve the time-dependent processes under transient conditions. They are generally used to switch off rapidly from conducting state to non-conducting state, and vice versa. Such a rapid switching is normally required in following applications.

- Voltage regulator
- Operational amplifier (Op-amp)

Since the time response is an important factor in many applications, some special considerations are given to their charge control properties. Main among these considerations are the following.

1. The power losses should be minimum under forward bias.
2. The current should be minimum under reverse bias.
3. Very little charge should be stored in neutral region for steady forward currents.
4. They should have a very short carrier lifetime.
5. They should have a high switching speed.
6. They should have a lower switching time.

6.9.1 Improving the Switching Speed

In order to develop an efficient switching diode, it is desired that its switching speed should be high so that the switching time is less. The improvement in them can be accomplished by following approaches.

1. By adding efficient recombination centres to the bulk material such as Au in Si-diode.
2. By making the lightly-doped neutral region shorter than the minority carrier diffusion length. It implies that a 'narrow base diode' is prepared.

The usefulness of Au doping in Si diode can be understand by the following explanation.

In a *p-n* Si diode, the recombination lifetime of holes τ_p and reverse recovery time τ_{sd} before and after doping of Au are as follows.

(i) Before doping:	$\tau_p = 1\ \mu s$	$\tau_{sd} = 0.1\ \mu s$
(ii) After doping of 10^{14} Au atoms/cm^3:	$\tau_p = 0.1\ \mu s$	$\tau_{sd} = 0.01\ \mu s$
(iii) After doping of 10^{15} Au atoms/cm^3:	$\tau_p = 0.01\ \mu s$	$\tau_{sd} = 1$ ns

Since the carrier lifetime varies with the reciprocal of concentration of recombination centre, the increased doping of Au as stated above is fruitful. However, there is a limit of everything. This process of doping cannot be continued infinitely. Otherwise due to large concentration of Au centres, the following ill-effects can crop-in.

- The reverse current becomes appreciably large.
- Also the equilibrium carrier concentration near the junction gets affected.

6.9.2 *Narrow Base Diode*

By making a comparatively shorter and lightly-doped narrow region, the stored charge for forward conduction becomes very small. It is due to diffusion of injected carriers through lightly- doped region to end contact. Consequently on switching to reverse conduction, the time taken to eliminate the stored charge in narrow neutral region becomes very small. Hence the required purpose is served.

6.10 Capacitance of P-N Junctions

The capacitance is an important quantity in designing of *p-n* junction devices operating under the time-varying signals. Basically, there are two types of capacitance associated with a junction. These are

1. Junction capacitance which arises due to dipole in transition region. It is dominant under reverse-bias conditions.
2. Charge storage capacitance which arises due to charge storage effects and lagging of voltage with change in the current. It is dominant under forward-bias conditions.

In many applications, above capacitances are the deciding factors for the usefulness of *p-n* junction devices.

6.10.1 Derivation of the Expression for Junction Capacitance

To determine the expression for capacitance, it is quite common to use $C = Q/V$. This is true for the case of parallel-plate capacitors in which the charge is a linear function of voltage. But this relation does not hold good for *p-n* junction, as the charge Q on both sides of depletion region varies non-linearly with the applied voltage. It can be verified by recalling the relation between width of the transition region w and equilibrium voltage V_O as given in Chap. 5, and reproduced as

$$w = \sqrt{\frac{2\varepsilon_0\varepsilon_r V_0}{e}\left(\frac{N_a + N_d}{N_a N_d}\right)}\text{(under equilibrium)} \tag{6.9}$$

Since we are dealing with non-equilibrium state of the junction subjected to applied voltage V (which is different from V_O), therefore Eq. 6.9 modifies to

$$w = \sqrt{\frac{2\varepsilon_0\varepsilon_r (V_0 - V)}{e}\left(\frac{N_a + N_d}{N_a N_d}\right)}\text{(under biasing)} \tag{6.10}$$

In this expression, V can be either +ve or −ve depending upon whether the junction is in forward- bias or reverse-bias. It is worth recalling that the value of w will increase for reverse-bias and will decrease for forward-bias. This causes variation in uncompensated charge Q on either side of the junction. Consequently, we can write

$$|Q| = eAx_{no}N_d = eAx_{po}N_a \tag{6.11}$$

Since

$$x_{no} = w\frac{N_a}{N_a + N_d} \tag{6.12a}$$

and

$$x_{po} = w\frac{N_d}{N_a + N_d}\text{(see in Chap.5)} \tag{6.12b}$$

therefore, Eq. 6.11 with the help of Eqs. 6.12a, 6.12b, modifies to

$$|Q| = eAw\frac{N_a N_d}{N_a + N_d} = A\sqrt{\left[2e\varepsilon_0\varepsilon_r (V_0 - V)\frac{N_a N_d}{N_a + N_d}\right]} \tag{6.13}$$

For the charge to be a nonlinear function of voltage, the capacitance may be defined as

$$C = \left|\frac{dQ}{dV}\right| \tag{6.14}$$

Hence using (V_0–V) for dV, the junction capacitance may be expressed as

$$C_{\text{junc}} = \left|\frac{dQ}{d(V_0 - V)}\right| = \frac{A}{2}\sqrt{\frac{2e\varepsilon_0\varepsilon_r}{(V_0 - V)} \cdot \frac{N_a N_d}{N_a + N_d}} \tag{6.15}$$

Reduction of Eq. 6.15 **to other cases**. This is given as follows.

- For the form of a parallel-plate capacitor, it is

$$C_{\text{junc}} = \varepsilon_0\varepsilon_r A\sqrt{\frac{2}{e\varepsilon_0\varepsilon_r(V_0 - V)} \cdot \frac{N_a N_d}{N_a + N_d}} = \frac{\varepsilon_0\varepsilon_r A}{w} \tag{6.16}$$

- For a *p-n* junction having $N_a >> N_d$, and $x_{no} \approx w$

$$C_{\text{junc}} = \frac{A}{2}\sqrt{\frac{2e\varepsilon_0\varepsilon_r}{(V_0 - V)}}.N_d \tag{6.17}$$

6.11 Linearly Graded, Abrupt and Hyperabrupt Junctions

The variation of capacitance is different for different types of junctions. Whereas for an abrupt *p-n* junction, the junction capacitance C_{junc} varies as square root of reverse bias voltage V_r; in a graded junction, it is expressed as

$$C_{\text{junc}} \propto V_r^{-n} \, for \, V_r \gg V_0 \tag{6.18}$$

A higher value of the exponent n in above equation is desired for high sensitivity. Its value varies from 1/3 to 2. For different types of junctions, the value of n is as follows.

• For linearly graded junction	n = 1/3
• For abrupt junction	n ≈ 1/2, and
• For hyperabrupt junction	n > 1/2

Thus the hyperabrupt junction has the greatest voltage sensitivity. When they are used with an inductance L in a resonant circuit, the resonant frequency is obtained as

$$w_{\text{resonant}} = \frac{1}{\sqrt{LC}} \tag{6.19a}$$

It varies linearly with the voltage applied to the varactor, and is

$$w_{\text{resonant}} \propto V_r \text{ for } n = 2 \tag{6.19b}$$

The ***hyperabrupt junctions*** are the linearly graded junctions having greater voltage sensitivity. They are generally fabricated by epitaxial growth technique. This technique is preferable because the epitaxial layer and substrate doping profile may be chosen such that the junction having exponent $n > 1/2$ (in Eq. 6.18) can be produced.

6.11.1 Doping Profile

Doping profiles for linearly graded, abrupt, and hyperabrupt *p-n* junctions are shown in Fig. 6.7. The donor distribution on *n*-side is expressed by

$$N_{\text{d}} = Cx^{\lambda} \tag{6.20}$$

where C is a constant and exponent λ has the following values.

$\lambda = 1$ for linearly graded junction.
$\lambda = 0$ for abrupt junction.
$\lambda = -3/2$ for hyperabrupt junction.

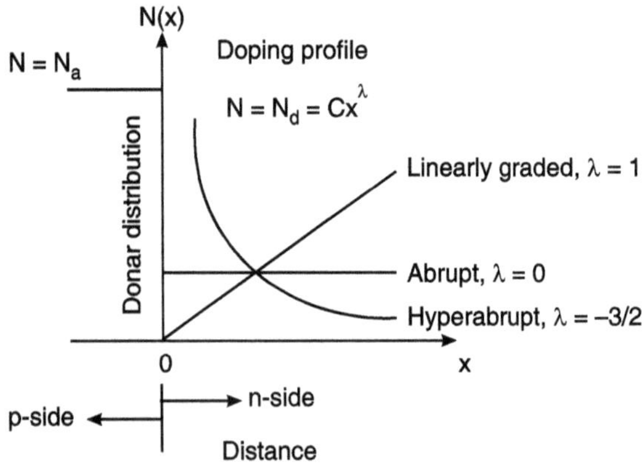

Fig. 6.7 The effect of doping on linearly graded, abrupt, and hyperabrupt junctions

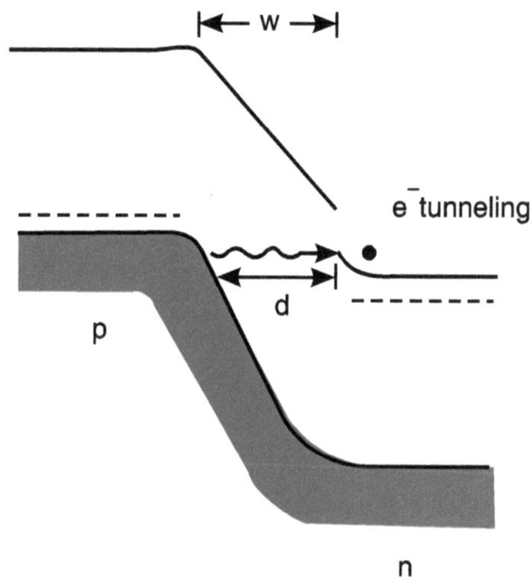

Fig. 6.8 The heavily doped junction at equilibrium, in reverse bias with electron tunnelling from p to n

The exponent n in Eq. 6.18 is related to exponent λ of Eq. 6.20 by $n = 1/(\lambda + 2)$. From this relation we find the value of n for hyperabrupt junctions as $n = 2$. Since by choosing different doping profiles, a wide variety of C_{junc} versus V_r dependencies can be produced, therefore the varactor diodes can be fabricated for vivid specific applications.

Example 6.1 What do you mean by abrupt junction? Under which conditions the transition region will be too wide for tunnelling?

Solution. The basic requirements for tunnelling current are a large number of electrons separated from a large number of empty states by a narrow barrier of finite height. Since the tunnelling probability depends on the width of the barrier W_B as shown in Fig. 6.8, it is important that the metallurgical junction be sharp and the doping high. It causes the transition region w to extend only a very short distance from each side of the junction. The transition region w will be too wide for tunnelling, if

(i) the junction is not abrupt, or
(ii) either side of the junction is lightly doped.

6.12 Graded Junctions

Although the abrupt junction approximation accurately describes the properties of many epitaxially grown junctions, but it does not do so for fused or implanted junction devices. For shallow diffusions in which the diffused impurity profile is very steep (Fig. 6.9a), the abrupt approximation is usually acceptable but if the impurity profile is spread out into the sample, a graded junction results in

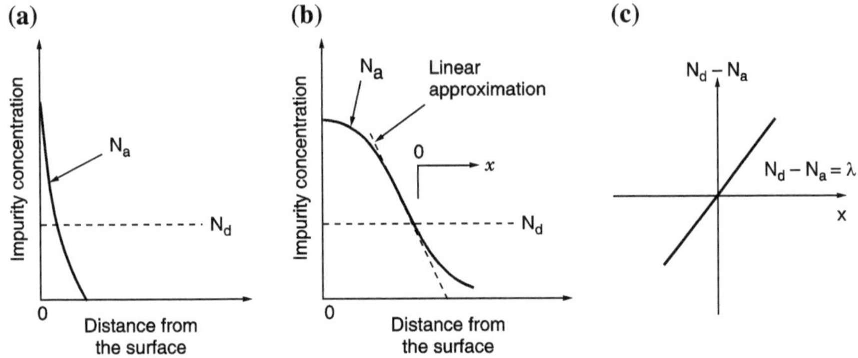

Fig. 6.9 Approximations to diffused junctions: **a** shallow diffusion (abrupt); **b** deep drive-in diffusion with source removed (graded); **c** linear approximation to the graded junction

(Fig. 6.9b). The graded junction problem can be solved analytically if a linear approximation of the net impurity distribution is made near the junction (Fig. 6.9c). Hence, the graded region can be expressed approximately by

$$N_d - N_a = \lambda x \tag{6.21}$$

where λ is a grade constant giving the slope of the net impurity distribution.

In Poisson's equation described earlier, the linear approximation becomes

$$\frac{d\xi}{dx} = \frac{q}{\varepsilon}\left(p - n + N_d^+ - N_a^-\right) \simeq \frac{q}{\varepsilon}\lambda x \tag{6.22}$$

within the transition region. In this equation, we assume complete ionization of the impurities and neglect the carrier concentrations in the transition region. The net space charge varies linearly over W, and the electric field distribution is parabolic. The expressions for contact potential and junction capacitance are different from the abrupt junction case (Fig. 6.10), since the electric field is no longer linear on each side of the junction.

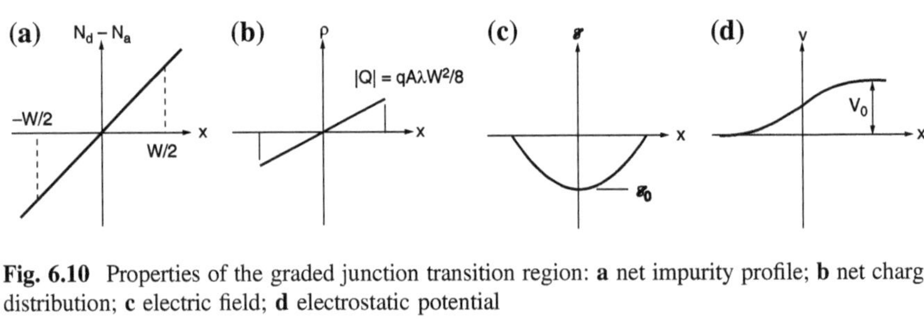

Fig. 6.10 Properties of the graded junction transition region: **a** net impurity profile; **b** net charge distribution; **c** electric field; **d** electrostatic potential

In a graded junction the usual depletion approximation is often inaccurate. If the grade constant λ is small, the carrier concentrations (p-n) can be important in Eq. 6.22. Similarly, the assumption of negligible space charge outside the transition region is questionable for small λ. It would be more accurate to refer to the regions just outside the transition region as quasi-neutral rather than neutral. Thus the edges of the transition region are not sharp as shown in Fig. 6.10, but are spread out along x-axis. These effects complicate the calculations of junction properties, and for that a computer will be needed to solve the problem accurately.

6.13 Solved Examples

Example 6.2 Consider an abrupt P-N junction with donor density $N_d = 10^{17}$ atoms/cm^3 and acceptor density $N_a = 0.5 \times 10^{16}$ atoms/cm^3. If $V_B = 0.7$ V, $\varepsilon_r = 10$ for the material and $\varepsilon_0 = 8.85 \times 10^{-12}$ F/m, calculate the junction width.

Solution. Given that,

$$N_d = 10^{17} \times 10^6 = 10^{23}\ \text{atoms/m}^3$$
$$N_a = 0.5 \times 10^{16}\ \text{atoms/cm}^3 = 0.5 \times 10^{22}\ \text{atoms/m}^3\ \text{and}$$
$$\varepsilon = \varepsilon_r\varepsilon_0 = 10 \times 8.85\ \text{x}10^{-12}\ \text{F/m} = 8.85 \times 10^{-11}\ \text{F/m}$$

Now

$$W = \left[\frac{2\varepsilon V_B}{e}\left(\frac{1}{N_a} + \frac{1}{N_d}\right)\right]^{1/2}$$
$$= \left[\frac{2 \times (8.85 \times 10^{-11}) \times (0.7)}{(1.6 \times 10^{-19})}\left(\frac{1}{0.5 \times 10^{22}} + \frac{1}{10^{23}}\right)\right]^{1/2}$$
$$= 0.393 \times 10^{-6}\ \text{m} = 0.393\ \mu\text{m}$$

Example 6.3 Consider a Si abrupt P-N junction at 300 K with $N_a = 10^{18}$ cm^{-3} and $N_d = 10^{15}$ cm^{-3}. Taking $n_i = 1.5 \times 10^{10}$ cm^{-3}, calculate the value of V_B. For Si, $C = 1.05 \times 10^{-12}$ F/cm. Also calculate the zero-bias depletion width.

Solution. We know that,

$$V_B = \frac{kT}{e} \text{loge}\left(\frac{N_a N_d}{n_i^2}\right)$$

$$\therefore \quad V_B = 0.0259 \text{ loge}\left[\frac{10^{18} \times 10^{15}}{(1.5 \times 10^{10})^2}\right]$$

$$= \mathbf{0.752\,V}$$

We know that,

$$W = \left[\frac{2\varepsilon V_B}{e}\left(\frac{1}{N_a} + \frac{1}{N_d}\right)\right]^{1/2}$$

On substituting the given values, we get

$$= \left[\frac{2 \times (1.05 \times 10^{-12}) \times 0.752}{(1.6 \times 10^{-19})}(10^{-18} + 10^{-15})\right]^{1/2}$$

$$= \mathbf{9.93 \times 10^{-5}\ cm}$$

In terms of depletion width W,

$$|x_n| = \left(\frac{N_a}{N_d + N_d}\right)W = \left(\frac{10^{18}}{10^{15} + 10^{18}}\right) \times 9.93 \times 10^{-5}$$

$$= \mathbf{9.92 \times 10^{-5}\ cm}$$

and

$$|x_p| = \left(\frac{N_d}{N_d + N_a}\right)W = \left(\frac{10^{15}}{10^{15} + 10^{18}}\right) \times 9.93 \times 10^{-5}$$

$$= \mathbf{9.92 \times 10^{-8}\ cm}$$

Review Questions

1. Define the following.
   (a) Transient conditions
   (b) Storage delay time
   (c) Narrow base diode
   (d) Hyperabrupt junction
2. Enumerate different kinds of diodes and briefly discuss about them.
3. What do you mean by crystal diode, super barrier diode, constant current and gold-doped diodes? Briefly, explain them.
4. Explain the step-recovery diode, TVS diode, laser diode and thermal diode. Sketch their symbolic representation also.
5. Explain the ideal diode model and real diodes. Also differentiate between them.
6. Discuss the various conditions in which a p-n junction works in real applications. Name the junctions that are used under transient conditions.
7. What is meant by time-dependent response of junctions? Enlist some p-n junction devices which exhibit time-dependent response. Briefly write their functions.
8. Explain the physical significance of time variation of stored charge, and derive expression for it. Discuss the implications of this equation.
9. Deduce an expression to determine the 'turn-on time' equation for a p-n junction under transient condition. Transform this equation for the case of steady state condition.
10. What is stored charge? Obtain its solution under a turn-off transient condition and discuss the results on suitable diagrams.
11. Explain the reverse recovery transient and storage delay time, and discuss their physical implications.
12. Describe the functions and requirements of a switching diode. Suggest methods to improve the switching speed.
13. Why is gold preferred as dopant in silicon switching diode? What are its effects on recombination lifetime and reverse recovery time of holes?
14. What is the utility of narrow base diode? How does it work?
15. What do you mean by capacitance of a p-n junction? What is its importance in design of semiconductor devices?

16. Derive an expression to obtain capacitance of the junction. Write assumptions and approximations made therein.
17. Explain the effect of doping profile on linearly graded, abrupt and hyperabrupt p-n junctions. Support your answer with the help of suitable diagram, data and illustrations.
18. Discuss the diffused graded junction stating the approximations of shallow diffusion and deep diffusion. Also explain the properties of the transition region of the graded junction.
19. Distinguish between the following.
    (a) Abrupt and hyperabrupt p-n junctions
    (b) Switching and Varactor diodes
    (c) Long and short diodes.

Objective Questions

1. Match list I with list II, and choose the correct answer from the codes given below the lists.

List I	List II
A. Hyperabrupt junction	1. Low voltage sensitivity
B. Recombination term	2. Q_p/τ_p
C. Rectifying type metal-semi-conductor Junction	3. dQ_p/dt
D. Heterojunctions	4. Ballistic transport
	5. High voltage sensitivity
	6. Schottky diode

Codes:

	A	B	C	D
(a)	1	3	4	6
(b)	5	2	6	4
(c)	1	2	6	4
(d)	5	3	4	6

2. Given below are some statements.
   A. Influence of excess carriers is to cause decay of stored charge in n-region of the junction devices.
   B. Switching diode should have a very short carrier lifetime.
   C. A varactor diode helps in reducing the size of the circuits.
   D. A heterojunction is formed of a semiconductor, a metal and an insulator

   The correct statements are

   (a) A and D (b) C only
   (c) A, B and C (d) A, C and D

3. Which of the following is incorrect?
   (a) $\frac{dQ_p}{dt} = 0$ for steady state.
   (b) A homojunction is formed within a single semiconductor
   (c) In a switching diode, the power losses should be minimum under forward bias.
   (d) Doping of Au in Si switching diode lowers the recombination. lifetime of holes, but increases the reverse recovery time. increases the reverse recovery time.

4. A power diode is in the forward conduction mode and the forward current is now decreased. The reverse recovery time of the diode is t_r and the rate of fall of the diode current is di/dt. What is the stored charge?

   (a) $\left(\frac{di}{dt}\right).t_r$ (b) $\left(\frac{di}{dt}\right).t_r{}^2$
   (c) $\left(\frac{di}{dt}\right).t_r{}^2$ (d) $\left(\frac{di}{dt}\right).t_r$

Answers

1. (b) 2. (c) 3. (d) 4. (b)

Reference

1. http://www.instructables.com/id/Types-of-Diodes/

Part III
Majority Carrier Diodes, Microwave Diodes, and Optoelectronic Devices

Chapter 7
Majority Carrier Diodes (Tunnel Diode, Backward Diode, Schottky Barrier Diode, Ohmic Contacts, and Heterojunctions)

Abstract In the recent past, the development of microwave devices having low noise, high frequency, greater bandwidth, lesser switching time etc. has yielded devices with improved performance. These devices have been used in drying machines for textile, food and papers industries etc., for biomedical applications, for electronic warfare etc. In this regard, several microwave devices have been developed. Therefore in this chapter, the majority carrier diodes for use at microwave frequencies are included. Tunnel diode: its response under zero-bias, reverse-bias, forward-bias and increased forward-bias conditions, its response beyond negative resistance region and characteristics are presented. Transitive devices, transit time effects, and requirements of a good transit time device are discussed. The backward diode, its characteristics and applications are given. Metal-semiconductor junctions, Schottky diodes: its operating mechanism, characteristics, limitations and applications are explained. Ohmic contacts, heterojunctions: their unique behaviour, band discontinuities and band bending, and potential well in it are described. Minute insight into the various topics are given through solved numerical and theoretical examples. Review questions, numerical problems and objective type questions are also given with their answers.

Keywords Negative conductance devices · Tunnel diode · Transit time device · The backward diode · Metal-semiconductor junctions · Schottky diode · Ohmic contacts · Heterojunctions

7.1 Introduction to Microwave Devices

Microwave devices are used to provide amplification, oscillation and other functions at microwave frequencies. They operate in the frequency range of about 10^8–10^{11} Hz. The varactor diodes and high-frequency transistors are such devices which are useful in microwave circuits. The high-frequency transistors are very common and important among these devices. In case we need them to be used at higher

K.M. Gupta and N. Gupta, *Advanced Semiconducting Materials and Devices*, Engineering Materials, DOI 10.1007/978-3-319-19758-6_7

frequencies (above 10^{11} Hz), they are not useful. It is because of transit time and other effects (see art. 7.6.1). Therefore other devices are used to perform the high frequency switching and conversion of d-c power into microwave power.

There are several kinds of instabilities that occur in semiconductors. The phenomenon of negative conductance is involved with one such instability. Some microwave devices for high frequency applications make favourable use of this instability. They are therefore known as 'negative conductance devices'. Main among these devices are the following.

1. *Tunnel (or Esaki) diodes*: whose working depends upon the quantum-mechanical tunnelling effect.
2. *Transit time diodes (such as IMPATT diodes)*: whose working depends upon the combined phenomenon of transit time effect and carrier injection.
3. *Gunn diodes*: whose working depends upon the phenomenon of transfer of electrons from a high-mobility state to a low-mobility state.

Each of above diodes is a two-terminal device and can be operated in a negative conductance mode.

7.2 Tunnel Diodes

A tunnel diode is a *p-n* junction device that operates by quantum-mechanical tunnelling of electrons through the potential barrier of the junction. This operation occurs only in those regions of I-V characteristic of the junction, which exhibits negative resistance. The tunnelling process for reverse current is the same as Zener effect, but a negative reverse bias is required to initiate the tunnelling in these diodes.

A tunnel diode is used in following applications at microwave frequencies.

- as logic storage memory device
- as ultra-high speed switch in high-speed switching operations
- as a microwave oscillator
- in relaxation oscillator circuits
- to provide amplification.

A tunnel diode is symbolized as given in Fig. 7.1.

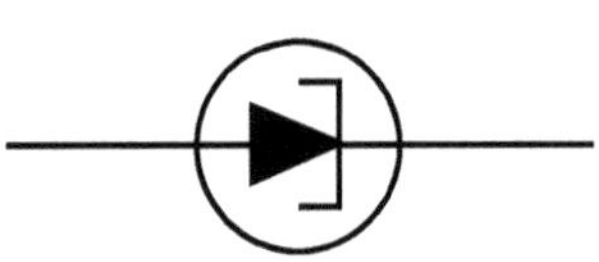

Fig. 7.1 Symbol of a tunnel diode

7.3 Tunnel Diode Operation

Operation of a tunnel diode may be described with the help of Figs. 7.2–7.4. These depict the band diagrams and I-V characteristics under different bias conditions viz. zero-bias, small reverse and small forward bias, increased forward bias, and beyond the tunnel current region. The step-wise explanation under each condition is given as follows.

7.3.1 Equilibrium or Zero-Bias Condition (Fig. 7.2a)

It shows a *p-n* junction between two degenerate semiconductors. The Fermi energy levels E_{Fp} and E_{Fn} in *p*-type and *n*-type respectively are constant throughout the junction. The E_{Fp} is below the energy level of valence band E_v on *p*-side, and E_{Fn} is above the energy level of conduction band E_c on *n*-side. The energy bands are thus overlapping.

As the doping concentration is high, the depletion region w is very narrow and the electric field ζ at the junction is quite large. This causes a favourable condition for electron tunnelling to take place i.e. empty states separated by a narrow potential barrier V_B of finite height. At equilibrium, there is equal tunnelling from *n* to *p* and from *p* to *n*. Hence effectively there is no tunnelling and the net current is zero. In I-V diagram, a dot indicates the zero current and zero voltage, since it is an equilibrium state.

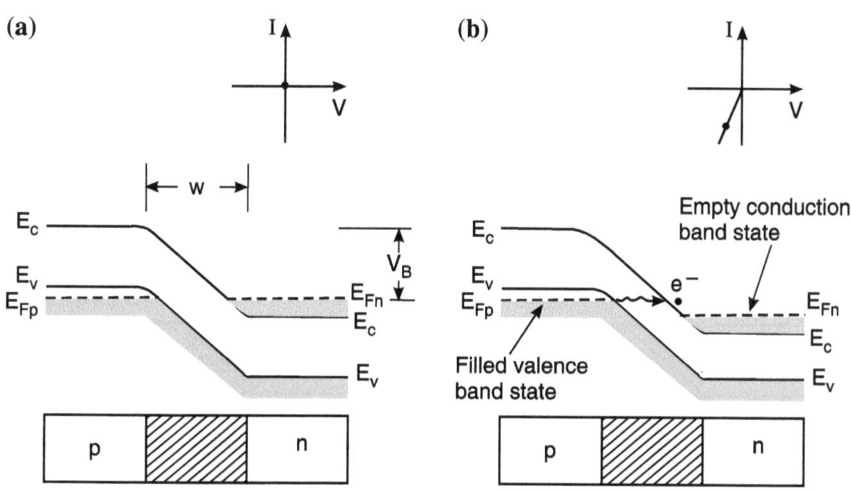

Fig. 7.2 Band diagrams and I-V characteristics of tunnel diode under **a** equilibrium (or zero bias) condition, and **b** small reverse bias condition

7.3.2 Small Reverse Bias Condition (Fig. 7.2b)

On putting the degenerate *p-n* junction under a small reverse bias, the electron tunnelling occurs from the filled valence band states below E_{Fp} to the empty conduction band states above E_{Fn}. When the reverse bias is increased, E_{Fn} starts moving down with respect to E_{Fp} (see in Fig. 7.2b that the energy line of E_{Fn} is lower to that of the E_{Fp}). It places more filled states on *p*-side, opposite to empty states on *n*-side. This results in an increase in the tunnelling of electrons from *p*-side to *n*-side, on increasing the reverse bias.

The resulting conventional current is opposite to the flow of electron i.e. from *n* to *p*, and is shown in I-V diagram.

7.3.3 Small Forward Bias Condition (Fig. 7.3a)

On applying a small forward bias, the energy level E_{Fn} moves up with respect to energy level E_{Fp}. This can be seen in Fig. 7.3a. The amount of rise in energy is equal to $e \times V$. Now the electron tunnelling takes place from *n*-side to *p*-side, and electrons below E_{Fn} are placed opposite to empty states above E_{Fp}.

As the tunnelling occurs from n to p, the conventional current flows from p to n. It is shown in I-V diagram.

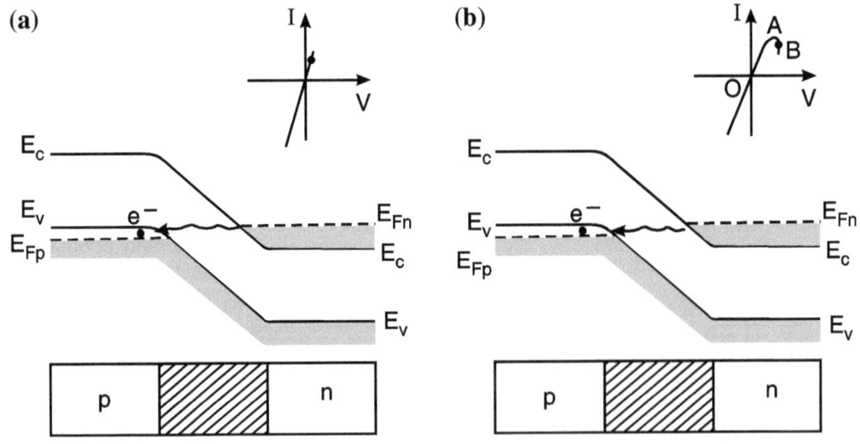

Fig. 7.3 Band diagrams and I-V characteristics of tunnel diode under **a** small forward bias condition, and **b** increased bias condition

7.3.4 *Increased Forward Bias Condition (Fig. 7.3b)*

With increased forward biasing, more filled states are placed opposite the empty states. However, the moving up of E_{Fn} as compared to E_{Fp} continues until a point is reached where the bands begin to pass by each other. Consequently, the number of filled states opposite the empty states decreases.

The forward tunnelling current continues to increase with increased bias as shown by *OA* in Fig. 7.3b; but the resulting tunnelling current decreases, as shown by *AB* in the same figure. The region *AB* depicts a region of negative slope i.e. *dV/dI* is negative. This negative resistance region is an important characteristic of tunnel diode and is useful in a number of applications.

7.4 Response of a Tunnel Diode Beyond the Negative Resistance Region

Response of tunnel diode beyond the negative resistance region is shown in Figs. 7.4a, b. It depicts that when the forward bias is increased further, the tunnelling current begins to increase and the negative slope *AB* (see Fig. 7.4b) continues to extend up to *C* as shown in Fig. 7.4b. Once the bands have passed each other, the forward current is dominated by the diffusion current. It is because

- the electrons surmount the potential barrier from n to p, (Fig. 7.4a), and
- the holes surmount their potential barrier from p to n.

In fig. 7.4b, the tunnelling current is shown by firm line and the diffusion current by dotted line. Although the diffusion current is present in forward tunnelling region, but its value is negligible as compared to tunnelling current.

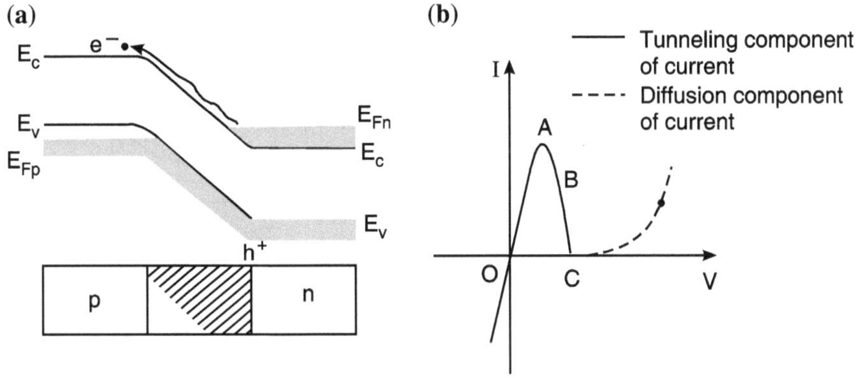

Fig. 7.4 Behaviour of a tunnel diode beyond negative resistance region showing **a** band diagram, and **b** *I-V* characteristic

7.5 Total Tunnel Diode Characteristic

The total tunnel diode characteristic is shown in Fig. 7.5 on I-V curve. The curve resembles the English letter '*N*', therefore this characteristic is often referred to as exhibiting a *type 'N' negative resistance*. Salient features of this curve are the following.

(i) In *OA* region, the peak tunnelling current I_{pt} is proportional to peak tunnelling voltage V_{pt}. This is a positive resistance region.
(ii) The peak voltage is taken at *A*, which is the point of maximum forward tunnelling.
(iii) In *AB* region, the current decreases with increases in voltage. This is negative resistance region. It is also called as a *voltage-controlled negative resistance.*
(iv) Point *B* is valley of the curve.
The value of negative resistance slope is determined by I_{pt} and valley current I_v.
(v) From *B* to *C*, there is a very little rise in current for a substantial increase in voltage, but there is almost a sharp rise in current beyond *C* up to *D*.
(vi) Point *D* is indicative of peak tunnelling current I_{pt}.

The current and voltage ratios mentioned below are often used as figure of merit for the tunnel diode.

- I_{pt}/I_v = Ratio of peak tunnelling current to valley current.
- V_{pt}/V_f = Measure of voltage spread between the two positive resistance regions.

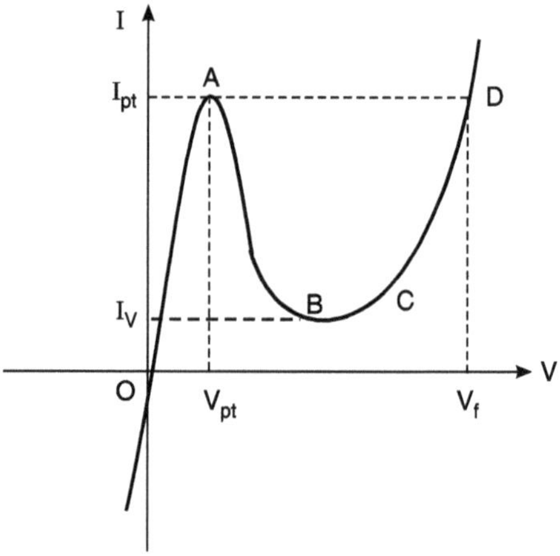

Fig. 7.5 Depiction of total tunnel diode characteristics

7.6 Transit Time Device

'Transit' means movement and 'transit time' means the time taken in this movement. A transit time device is a high-frequency device that operates at or above the microwave frequencies. They are very useful for following purposes.

- For converting d–c to microwave a–c signals with high efficiency
- In generation of microwave power for various applications.

The meaning of transit time may be explained with the help of a *p-n-p* transistor which is connected in a *CE* circuit. Since the n-type base region is electrostatically neutral in it, the excess holes are present for transit from emitter to collector. It calls for compensating excess electrons from the base contact. But the times which the holes and electrons spend in the base, are different. The average excess hole spends a time τ_t which is termed as *transit time* from emitter to collector.

7.6.1 Transit Time Effects

The transit time of carriers across the base region is a big limitation in the operation of high- frequency transistors. In a *p-n-p* transistor, the transition time τ_t required for holes to diffuse from emitter to collector can determine the maximum frequency of operation of the device. For a transistor with normal biasing and emitter injection efficiency $\gamma = 1$, the transit time is given by

$$\tau_t = \frac{w_b^2}{2D_p} \tag{7.1}$$

where w_b is base width and D_p is diffusion coefficient for holes.

On taking the reasonable values of $w_b \approx 0.1$ μm and of D_p for Si as 10 cm^2/s, the transit time for a typical device may be obtained to be about 0.5×10^{-11} s. Therefore by approximating the upper frequency limit of $1/2\pi\tau_t$, the transistor can be used to operate at a frequency of about 30 GHz.

7.6.2 Requirements of a Good Transit Time Device

A good transit time device should possess the following characteristics.

1. The transit time should be as less as possible. It can be reduced
   - by making the use of field-driven currents in the base.
   - by making a narrow base width.
   - by increasing the doping gradient in the base.

2. The maximum operational frequency of the transistor should be high.
3. Their physical size should be small.
4. The junction capacitance should be less. This can be accomplished by making the emitter and collector of smaller area.
5. The emitter, base and collector resistances should be kept to a minimum value. It is desired so because these resistances affect the *RC* charging times.
6. The breakdown voltage for collector junction should be high. It can be achieved by
   - heavy doping of substrate and low doping of collector material.
7. The collector depletion region should be kept as small as possible to reduce the transit time. This can be accomplished by
   - making the collector region narrow, and
   - doping the collector region lightly.

Generally a heterojunction transistor or *n-p-n* transistors in Si are preferred to make transit time devices. Their packaging also requires careful considerations to avoid any parasitic effects of resistance, inductance or capacitance at high frequencies.

7.7 The Backward Diode

In semiconductor devices, the backward diode is a variation on a tunnel diode or Zener diode having a better conduction for small reverse biases (for example −0.1 to −0.6 V) than for forward bias voltage. It is similar to a tunnel diode except that it has no forward tunnel current. The reverse current in such a diode is by tunnelling, which is also known as the **Zener effect**.

Alternately, the backward diode can also be thought of as an abrupt junction in which the doping on *p*-side is degenerate and the Fermi level on *n*-side is aligned with the bottom of conduction band.

7.7.1 I-V Characteristics of Backward Diode

The forward I-V characteristic is the same as that of an ordinary P-N diode. The breakdown starts when the reverse voltage is applied to it. In the case of Zener breakdown, it starts at a particular voltage. In this diode the voltage remains relatively constant (independent of current) when it is connected in reverse bias. In backward diode, the tunnelling phenomenon is imperfectly formed and the negative resistance region virtually disappears. Then the forward current is very small and becomes equivalent to the reverse current of a conventional diode.

The schematic symbol for the backward diode is shown in Fig. 7.6.

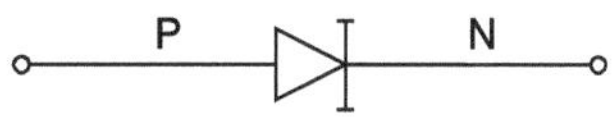

Fig. 7.6 Symbolic representation of a backward diode

7.7.2 Applications of Backward Diode

Backward diodes find use in following main applications.

1. *As Detector*: Since it has low capacitance and low charge storage effect, it can be used as a detector up to 40 GHz.
2. *As Rectifier*: It can be used to rectify the weak signals with peak amplitudes of 0.1–0.7 V.
3. *As Switch*: It can be used in high speed switching applications.

7.8 Metal-Semiconductor Junctions

Metal-semiconductor junctions are those junctions which are fabricated by forming a compatible contact between a metal and a semiconductor. They are mainly used for high-speed rectification purposes. The metal-semiconductor junctions may of following types.

1. Rectifying contact type, and
2. Non-rectifying (or ohmic) contact type

A metal-semiconductor contact is made by depositing a metal film on to a semiconductor surface, followed by making the contact pattern photolithographically.

Examples of *rectifying contact type* metal-semiconductor junction are the Schottky barrier devices. Good Schottky barriers are formed of various metals such as Au or Pt, for Si. These are suitable for use in densely packed ICs. *Ohmic contacts* are formed by heavy doping of the semiconductors in contact region. For example,

- Au containing a small percentage of Sb may be used for alloying with n-type Si, to form an excellent ohmic contact.
- Ohmic contact for n-type In As can be made by depositing, almost any metal on the surface.

7.9 Schottky Diodes

A Schottky diode (or Schottky barrier diode) works on the principle of *Schottky effect*. The Schottky effect may be stated as below.

> When the negative charges are brought near a metal surface, the positive charges are induced in the metal. It causes introduction of 'charge forces'. When the metal is subjected to an

electric field, the charge forces combine with it and reduce the effective work function eW_f to some extent. Such reduction in work function (or barrier) is known as Schottky effect.

The reduction in barrier is the basis of origination of potential barrier arising in metal- semiconductor contacts. Its mechanism is explained below for two specific cases viz.

1. when semiconductor is of *n*-type
2. when semiconductor is of *p*-type

7.9.1 Case I: Mechanism of Schottky Diode When the Semiconductor Is of n-Type

We consider a metal having a work function eW_{fm} and a *n*-type semiconductor whose work function is eW_{fs}. Further, we assume that the work function of metal is greater than the work function of semiconductor i.e. $W_{fm} > W_{fs}$ as shown in Fig. 7.7a. When the metal and semiconductor are brought in contact with each other

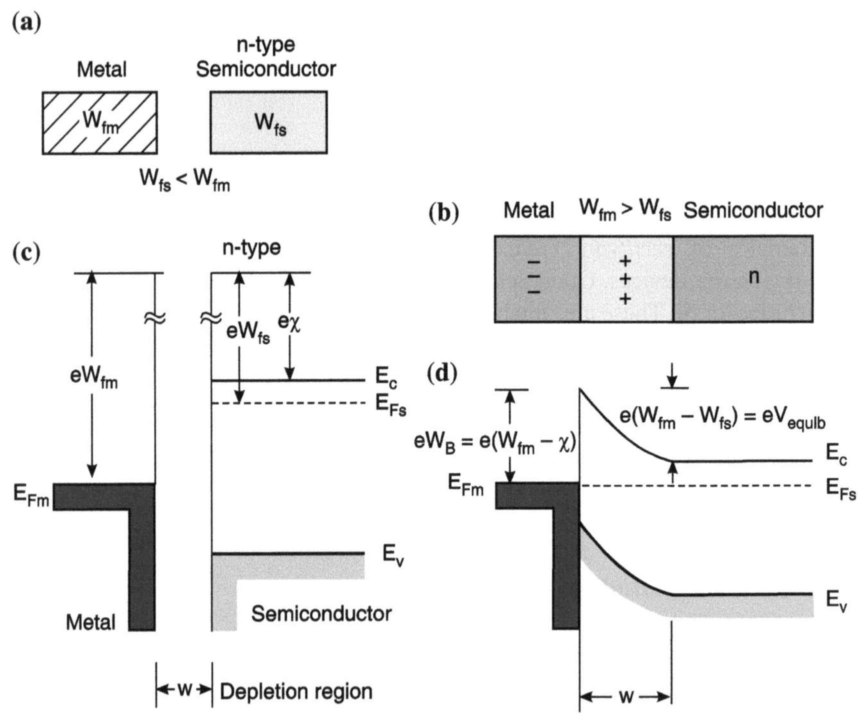

Fig. 7.7 A Schottky barrier is formed by **a** a metal and a n-type semiconductor, **b** the metal has a greater work function than that of the n-type semiconductor i.e. $W_{fm} > W_{fs}$, **c** band diagrams before making the contact, and **d** band diagram for the junction after joining

(Fig. 7.7b), transfer of charge takes place between them. Since for $W_{fm} > W_{fs}$, the Fermi level of semiconductor E_{Fs} is higher than the Fermi level of metal EFm, therefore for Fermi levels in both of them to be aligned at equilibrium (Fig. 7.7c), the electrostatic potential of semiconductor has to be raised. It implies that the electron energies must be lowered with respect to that of the metal.

Now when the contact between metal and semiconductor is established, equilibrium contact potential V_{equlb} prevents diffusion from the conduction band of semiconductor into the metal. The equilibrium contact potential is equal to the difference in work function potentials of the metal and semiconductor, and is given by

$$V_{equlb} = W_{fm} - W_{fs} \tag{7.2}$$

Also the potential barrier height WB (Fig. 7.6d) for injection of electron from the metal into conduction band of semiconductor is given by

$$W_B = W_{fm} - \chi \tag{7.3}$$

where χ is electron affinity. The value of equilibrium potential difference V_{equlb} may be altered as per the requirements of the application, by applying a forward-bias or reverse-bias voltage.

7.9.2 Case II: Mechanism of Schottky Diode When the Semiconductor Is of *p*-Type

A Schottky barrier for *p*-type semiconductor is shown is Fig. 7.8. Unlike the Schottky barrier with *n*-type semiconductor, it has the work function of metal less than the work function of semiconductor i.e. $W_{fm} < W_{fs}$. In this case, the aligning of Fermi levels at equilibrium requires a positive change on metal side and negative change on semiconductor side of the junction. The other details are analogous to case I, and are self-explanatory.

7.9.3 Limitations of Schottky Barrier Junctions

The description of Schottky barrier given in previous articles is mainly based on ideal metal- semiconductor contacts. But the ideal condition is rarely met in actual applications, because of the following surface and interfacial reasons.

1. Discontinuity exists between metal and semiconductor crystals.
2. A thin interfacial oxide layer of about 10–20 Å remains coated over the contact, even after etching.

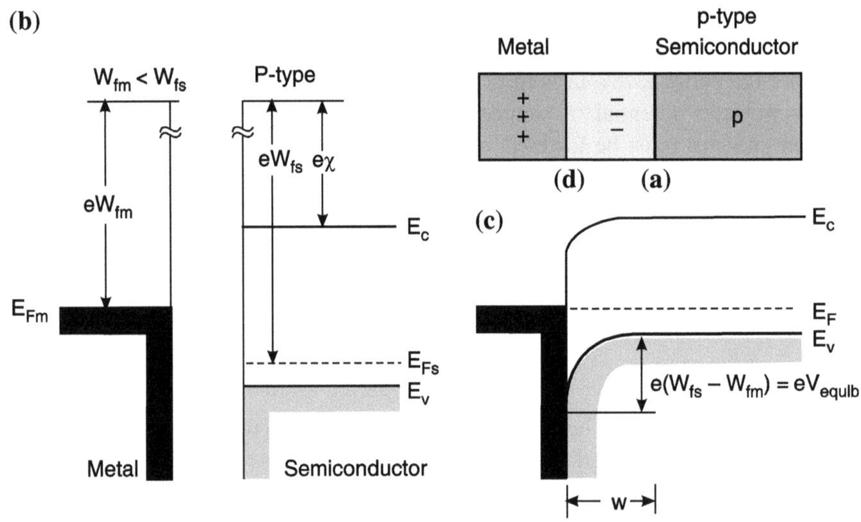

Fig. 7.8 A Schottky barrier **a** between a metal and a *p*-type semi-conductor, **b** band diagrams before contacts are made, and **c** band diagrams of the junction at equilibrium

3. Metal deposition on Si surface may leave a glassy layer at the junction.
4. Several other surface and interfacial inhomogeneities make it difficult to fabricate a junction having barriers near to the ideal values.

Therefore, metals like Au and Pt are used to form Si based Schottky barriers. Heat treatment is also imparted to Pt so as to make a reliable Schottky diode.

For a practical Schottky barrier diode, the forward current equation is of the following form.

$$I = ABT^2 e^{\left(\frac{-eW_B}{KT}\right)} \cdot e^{\left(\frac{eV}{nKT}\right)} \tag{7.4}$$

where A is area, B is a constant related to junction properties, n is a number lying between 1 and 2, W_B is Schottky barrier whose value is approximately ≈0.85 V on n-type Si.

7.9.4 Characteristics of Schottky Diode

These devices have negligible storage charge. A metal, say platinum, acts as an acceptor material for electrons when bonded to n-type silicon. Thus, when the metal is connected to n-type silicon,

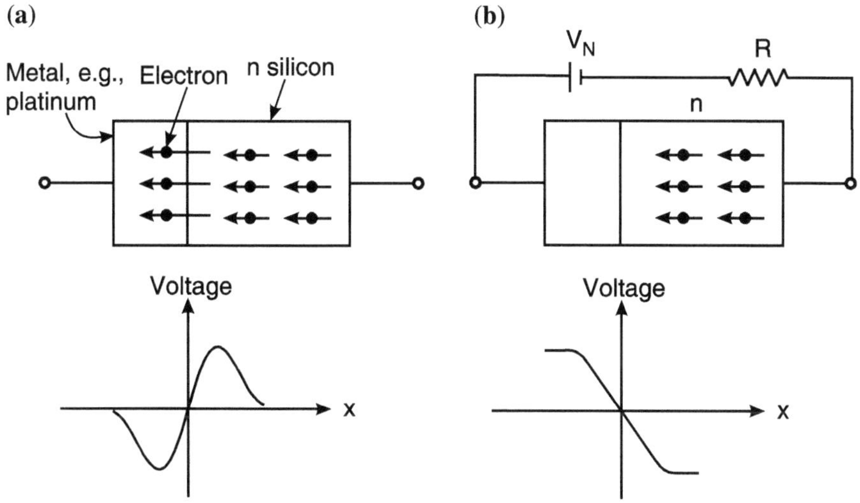

Fig. 7.9 Schottky diodes potential distribution after **a** initial diffusion, and **b** application of a positive voltage

(i) the electrons initially diffuse from the silicon into the metal. As shown in Fig. 7.9a, this diffusion results in depletion of electrons. The positive voltage inhibits further diffusion of electrons. On the other hand,

(ii) when a sufficiently large positive voltage is applied externally across the diode as shown in Fig. 7.9b, the electrons in *n* region find a positive potential towards metal side of the junction and start flowing.
In a rectifying contact, negligible current flows until V_N exceeds a certain minimum voltage V_λ. V_λ is the voltage needed to flatten the voltage curve shown in Fig. 7.9a. In a *p-n* silicon diode, the voltage V_λ is approximately 0.65 V. A small increase in voltage V_N, above V_λ, produces a very large change in current.

(iii) When the voltage applied to the diode is reversed so that the *n* material is made positive with respect to platinum (or *p* material), the voltage on *n* side of the junction increases (Fig. 7.10a) beyond the level indicated in Fig. 7.9a and there is no current flow. When an ohmic contact is made, there is no initial diffusion of electrons across the junction, since the two materials are such that the density and energy of electrons on both sides of the junction are same.

(iv) The circuit symbol for a Schottky diode is shown in Fig. 7.10b. The diode has a *V-I* characteristic similar to that of an ordinary *p-n* silicon diode except that the forward break voltage of the diode is 0.3 V.

It is interesting to note that there is a metal and *n*-silicon junction formed even in a *p-n* diode, since a metal is attached to the *n* material to make the connection with the external circuit. To prevent this connection from behaving like a diode, the

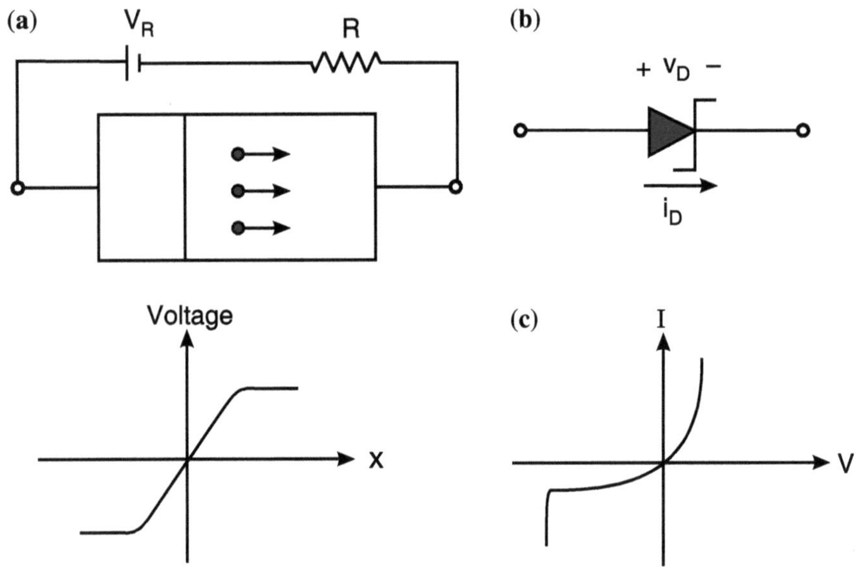

Fig. 7.10 Schottky diodes **a** potential distribution after application of a negative voltage, **b** circuit symbol, and **c** V-I characteristic

n-type silicon is doped so that it has surplus electrons at the end to be bonded to the metal. This excess electron region is called *n*+ region. After the initial diffusion of electrons to the acceptor metal, the *n* and *n*+ region take on the same characteristics, as the metal and contact become ohmic. When the Schottky diode is operated in forward mode, the current due to electrons starts moving from *n* type silicon across the junction, through the metal. Since electrons move relatively unimpeded through the metal, the recombination time τ is very small. Typically, it is of the order of 10 picosecond (ps).

7.9.5 Applications

Schottky diodes are used in following main applications.

1. Low noise mixer. Schottky barrier diodes are being increasingly used as a low noise mixer replacing the tunnel diodes and varactors, due to low cost, simplicity and reliability A noise level of 4–5 dB is common with them.
2. Balanced mixer in a CW Radar.
3. Microwave detectors.

7.10 Ohmic Contacts

In many cases, it is desired to have an *ohmic* metal-semiconductor contact, having a linear *I-V* characteristic in both biasing directions. For example, the surface of a typical integrated circuit must be contacted and interconnected. For that, such contacts the ohmic, with minimal resistance and no tendency to rectify signals.

Ideal metal-semiconductor contacts are ohmic when the charge induced in the semiconductor in aligning the Fermi levels is provided by majority carriers, Fig. 7.11. For example, in the $\Phi_m < \Phi_s$ (*n*-type) case of Fig. 7.11a, the Fermi levels are aligned at equilibrium by transferring electrons from the metal to the semiconductor. This raises the semiconductor electron energies (lowers the electrostatic potential) relative to the metal at equilibrium, Fig. 7.11b. In this case the barrier to electron flow between the metal and the semiconductor is small and easily overcome by a small voltage. Similarly, the case $\Phi_m > \Phi_s$ (*p*-type) results in easy hole flow across the junction, Fig. 7.11c. Unlike the rectifying contacts, no depletion region occurs in the semiconductor in these cases since the electrostatic potential difference required to align the Fermi levels at equilibrium desires accumulation of majority carriers in the semiconductor.

A practical method for forming ohmic contacts is by doping the semiconductor heavily in the contact region. Thus if a barrier exists at the interface, the depletion width is small enough to allow carriers to tunnel through the barrier. For example, Au containing a small percentage of Sb can be alloyed to *n*-type Si, forming an *n*+ layer at the semiconductor surface and an excellent ohmic contact. Similarly, *p*-type material requires a *p*+ surface layer in contact with the metal. In the case of Al on *p*-type Si, the metal contact also provides the acceptor dopant. Thus the required *p*+ surface layer is formed during a brief heat treatment of the contact after the Al is deposited.

7.11 Heterojunctions

The name heterojunction refers to such a *p-n* junction which is formed between two lattice-matched semiconductors having different energy band gaps. The *heterojunctions* are quite different from *homojunctions* which are formed within a single semiconductor. A heterojunction is different from a metal-semiconductor *p-n* junction also. Since it is made of lattice-matched semiconductors, the interface between them is virtually free of defects. It is also possible to form a single or multiple heterojunctions from continuous crystals.

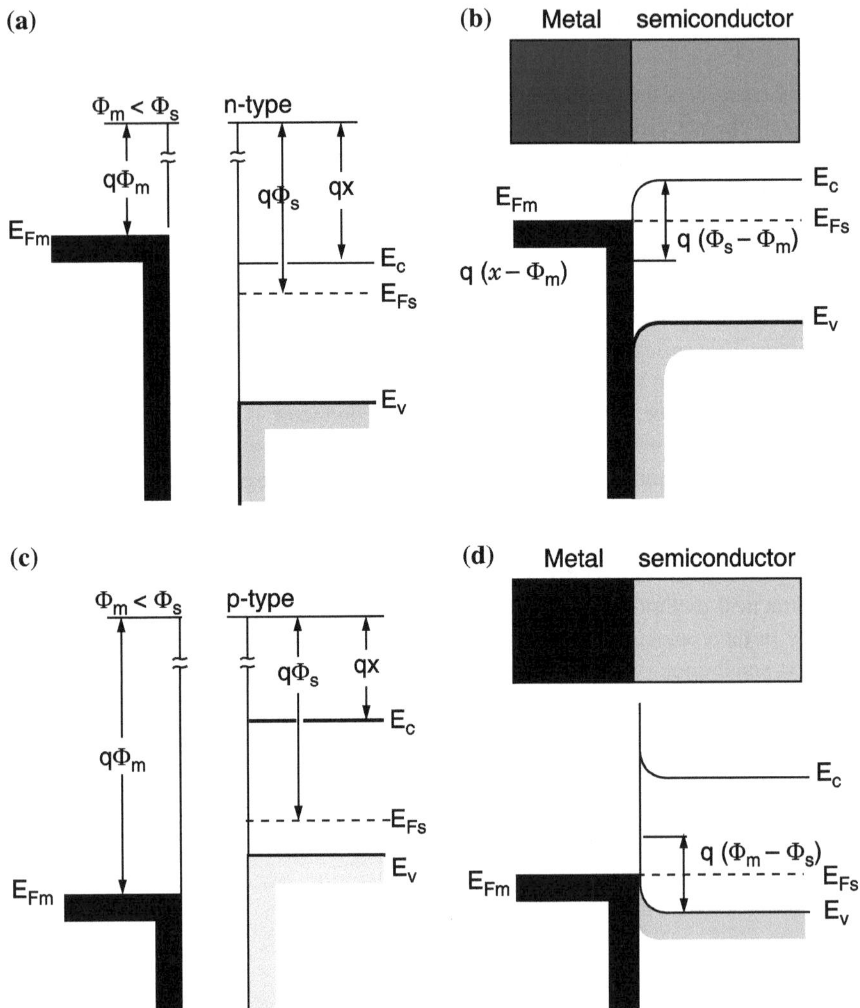

Fig. 7.11 Ohmic metal-semiconductor contacts **a** $\Phi_m < \Phi_s$ for an *n*-type semiconductor, **b** the equilibrium band diagram for the junction, **c** $\Phi_m > \Phi_s$ for a *p*-type semiconductor, and **d** the junction at equilibrium

7.11.1 Unique Behaviour of Heterojunctions

A heterojunction device is a microelectronic device. It is made up of multiple, alternate layers of gallium arsenide (GaAs) and gallium aluminium arsenide (GaAlAs). Each layer is only a couple of atoms thick. Its unique behaviours are as follows.

(i) **negative resistivity**. This remarkable phenomenon begins to emerge with a small structure of heterojunctions. In this phenomenon, the current decreases instead of increasing when the voltage is increased.
(ii) **ballistic transport**. This phenomenon allows an electron to pass from one side of a barrier to another without striking any atoms in between.

The availability of heterojunctions and multilayer structures in compound semiconductors has opened a wide range of possibilities for the development of electronic devices. These mainly include the following.

- Heterojunction bipolar transistors
- Field-effect transistors
- Semiconductor lasers, etc.

Band diagrams. The band diagram for any semiconductor heterojunction (and also for homojunctions) can be drawn if the following material parameters are known.

- *Energy band gap* that depends on material property but not on doping.
- *Electron affinity* that also depends on material property but no on doping.
- *Work function* that depends on material property as well as on doping.

7.11.2 Band Discontinuities and Band Bending

In making the heterojunctions, different types of semiconductors are used. They generally have different energy band gaps, different work functions, and different electron affinities. Therefore, when they are brought together to form a junction, the discontinuities are developed in energy bands. It is because of the lining-up of Fermi levels at equilibrium. The band diagrams shown in Fig. 7.12a are related to a *p*-type *wide* band gap semiconductor, with a *n*-type *narrower* band gap semiconductor, before their joining. Figure 7.12b shows the band discontinuities.

The discontinuities in conduction band ΔE_c and in valence band ΔE_v accommodate the difference in band gap ΔE_g between the two semiconductors.

Band bending. A vacuum level E_{vac} is taken as the reference for work function and electron affinity. The vacuum level is a potential energy reference for an electron which is being taken out of the semiconductor to infinity. Since there acts no force, hence the vacuum level remains constant as shown in Fig. 7.12b.

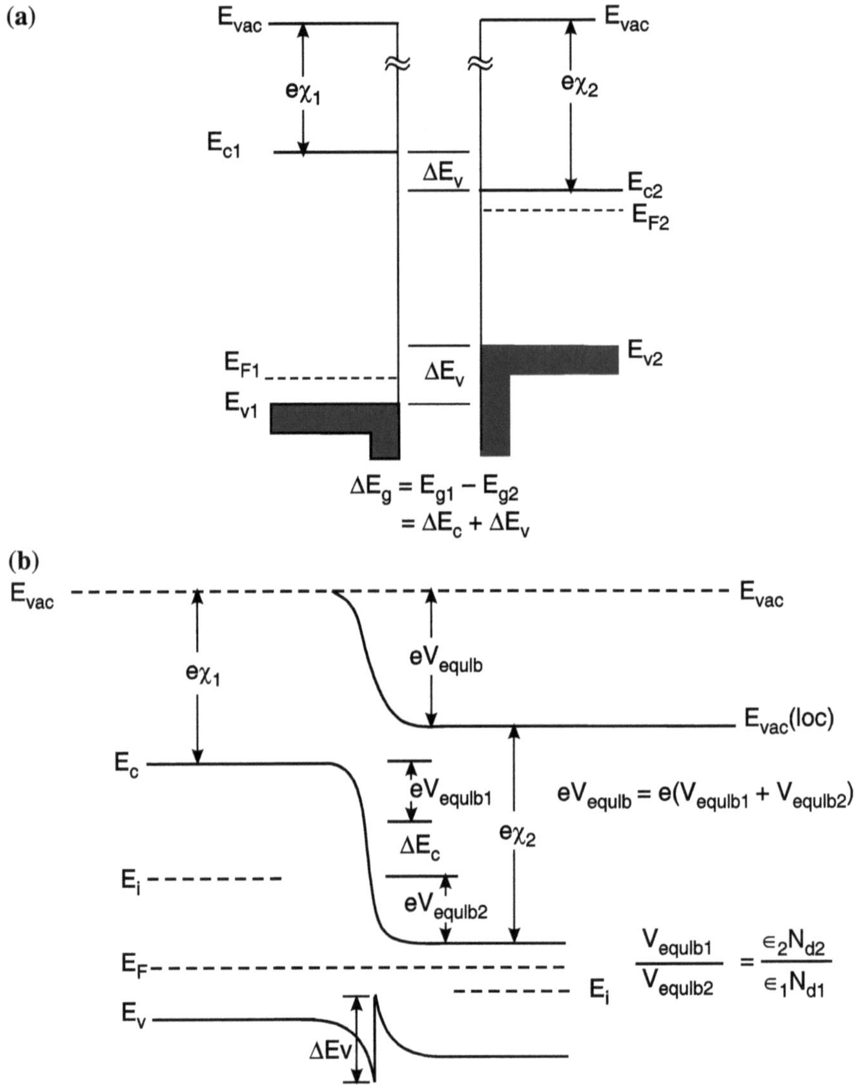

Fig. 7.12 A heterojunction between a *p*-type (wide band gap) and *n*-type (narrow band gap) semiconductors showing **a** band diagrams before contacts are made, and **b** band discontinuities and band bending at equilibrium

7.12 Potential Well in Heterojunction

A particularly important example of a heterojunction is shown in Fig. 7.13 in which the heavily *n*-type AlGaAs is grown on lightly doped GaAs. In this example the discontinuity in the conduction band allows electrons to spill over from the N^+-AlGaAs into the GaAs, where they become trapped in the potential well. As a result, the electrons collect on the GaAs side of heterojunction and move the Fermi level above the conduction band in the GaAs near the interface. These electrons are confined in a narrow potential well in the GaAs conduction band.

If we construct a device in which the conduction occurs parallel to the interface, the electrons in such a potential well form a two-dimensional electron gap with very interesting device properties. Electron conduction in such a potential well can result in very high mobility electrons. This high mobility is due to the fact that the electrons in this well come from the AlGaAs and not from doping the GaAs. As a result, there is negligible impurity scattering in the GaAs well, and the mobility is controlled almost entirely by lattice scattering (phonons). At low temperatures, where phonon scattering is low, the mobility in this region can be very high. If the band-bending in GaAs conduction band is strong enough, the potential well may be extremely narrow, so that discrete states such as E_1 and E_2 in Fig. 7.13 are formed.

Feature. Another feature of Fig. 7.13 is that the concept of a contact potential barrier eV_0 for both electrons and holes in a homojunction is no longer valid for the heterojunction. In Fig. 7.13, the barrier for electrons eV_n is smaller than the barrier for holes eV_p. This property of a heterojunction can be used to alter the relative injection of electrons and holes.

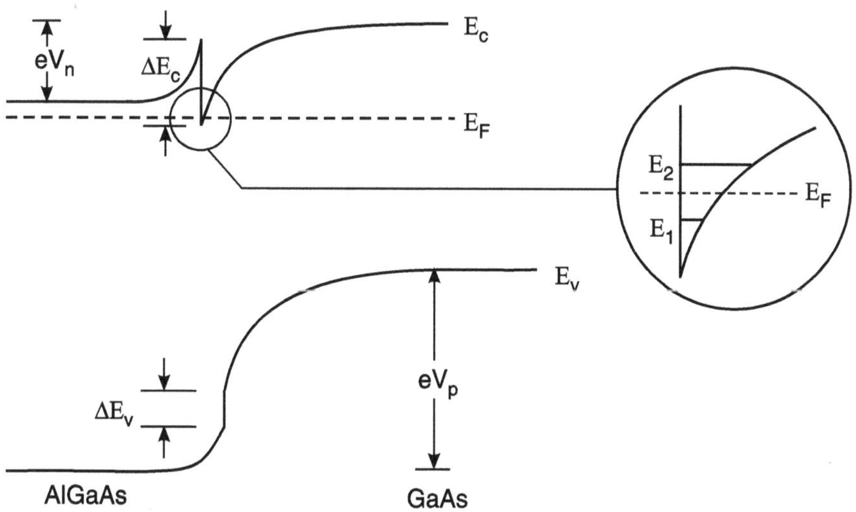

Fig. 7.13 A heterojunction between N^+-AlGaAs and lightly doped GaAs, illustrating the potential well for electrons formed in the GaAs conduction band

Review Questions

1. Define the following terms.
   (a) Quantum-mechanical tunnelling. (b) Negative resistance.
   (c) Bulk negative differential conductivity. (d) Unijunction transistor
2. What is meant by 'negative conductance microwave device'? What is its significance? Name different types of such devices and the principle employed in their working.
3. What is a tunnel diode? What are its applications? Briefly explain its working.
4. Discuss the effects of following conditions on the working of a tunnel diode. Show the behaviour on I-V diagrams also.
   (a) Zero bias condition (b) Small reverse bias condition
   (c) Small forward bias condition (d) Increased forward bias condition
5. Show the band diagram and I-V characteristic of a tunnel diode to explain its response beyond the negative resistance region.
6. Describe the total tunnel diode characteristic stating the salient features.
7. What are transit time devices? Why are they named so? Name some such devices and write their applications.
8. Enlist the requirements of a good transit time device. How can these requirements be met in construction of such devices?
9. What for is a backward diode used? What are its I-V characteristics? Write the applications of backward diode.
10. Write notes on the following.
    (a) Heavy doping of a semiconductor (b) type 'N' negative resistance
    (c) Transit time effects (d) Satellite valleys in band diagram of GaAs.
11. Define the following.
    (a) Transient conditions (b) Schottky effect
    (c) Ballistic transport (d) Storage delay time
12. Explain the metal-semiconductor junctions. What are their different types?

13. Describe the functions and working mechanism of a Schottky diode. What changes are noticed in a Schottky barrier when its semiconductor is of n-type or p-type?
14. What are the limitations of a practical Schottky barrier junction? How to overcome them?
15. Discuss the characteristics of a Schottky diode under positive and negative voltages. Also explain the V-I characteristic.
16. Sketch and explain the construction and working of ohmic contacts. How are such ohmic contacts formed?
17. Explain heterojunctions. What are the unique phenomena associated with them? Enlist their applications.
18. Explain the formation of potential well in a heterojunction. What is its physical significance?

Chapter 8
Microwave Diodes (Varactor Diode, p-i-n Diode, IMPATT Diode, TRAPATT Diode, BARITT Diode, etc.)

Abstract Microwave diodes are used to provide amplification, oscillation and other functions at microwave frequencies. They have a large bandwidth, therefore more information can be transmitted in microwave frequency range. It is the current trend to use microwaves more and more in long distance communication such as telephony, TV network, space communication, radars, microwave ovens etc. Therefore in this chapter, the Varactor diode; its construction, characteristics, performance and applications are presented, different types of semiconductor photodiodes are classified, and the IMPATT diode, TRAPATT diode and BARITT diode are elaborated along with their operation, performance and applications. The construction, operational characteristics and applications of p-n photodiodes, p-i-n photodiode, avalanche photodiode, and mid-infrared photodiodes are described. The transferred electron mechanism, valleys in conduction band and dependence of electron velocity on electric field are discussed. The Gunn diode: its fabricational materials, dielectric relaxation time etc. are given. The Dovett diode is also discussed. Minute insight into the various topics are given through solved numerical and theoretical examples. Review questions, numerical problems and objective type questions are also given with their answers.

Keywords Varactor diode · Photodiodes · IMPATT diode · TRAPATT diode · BARITT diode · Gunn diode · Dovett diode

8.1 Varactor Diode

The term varactor is derived from variable reactor (i.e. **var** + **actor** = varactor). It refers to voltage-variable capacitance of a reverse-biased p-n junction. A varactor diode is used to exploit the voltage-variable properties of junction capacitance, such as in following applications.

- Tuning stage of a radio receiver
- Harmonic generation

K.M. Gupta and N. Gupta, *Advanced Semiconducting Materials and Devices*,
Engineering Materials, DOI 10.1007/978-3-319-19758-6_8

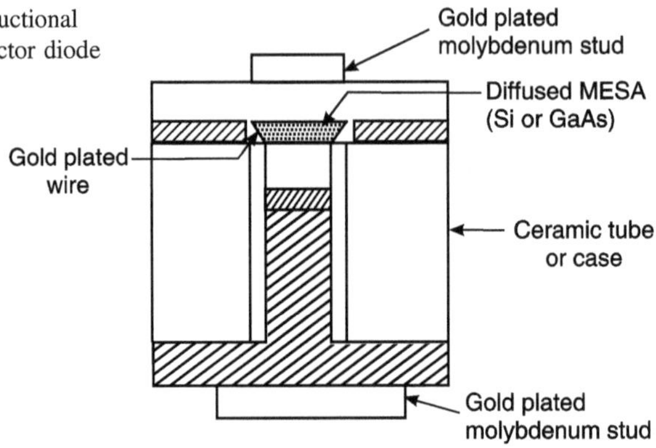

Fig. 8.1 Constructional details of a varactor diode

- Microwave frequency multiplication
- Active filters
- Tuned circuit, etc.

A varactor diode may be designed to have the following favourable features.

(i) It can replace the variable plate capacitor which is bulky.
(ii) The size of circuits can be greatly reduced.
(iii) The dependability of circuit is much improved.

Varactor diodes are generally fabricated by epitaxial growth technique, or by ion implantation.

Construction. The constructional details of a varactor diode are shown in Fig. 8.1. The diode encapsulation contains electrical leads attached to the semiconductor wafer and a lead to the ceramic tube or case. Diffused junction Si diodes are widely used at microwave frequencies. They are capable of handling larger powers, large reverse breakdown voltages, and have low noise. Frequency limit of Si varactor diodes is up to about 25 GHz. Varactors made of GaAs have high operating frequency (over 90 GHz) and better functioning at the lowest temperatures. However, the manufacturing techniques are easier for Si.

8.1.1 V–I Characteristics of Varactor Diode

As already said, varactor diode is a semiconductor device in which the junction capacitance can be varied as a function of reverse voltage (bias) of the diode (Fig. 8.2). Losses in this non-linear element are almost negligible. The junction capacitance depends on the applied voltage and junction design. In some cases a junction with fixed reverse bias may be used as a capacitance of a fixed value. The V–I characteristics of a typical varactor diode is shown in Fig. 8.3. Commonly used schematic symbols are shown in Fig. 8.4.

Fig. 8.2 Variation in junction capacitance as a function of voltage

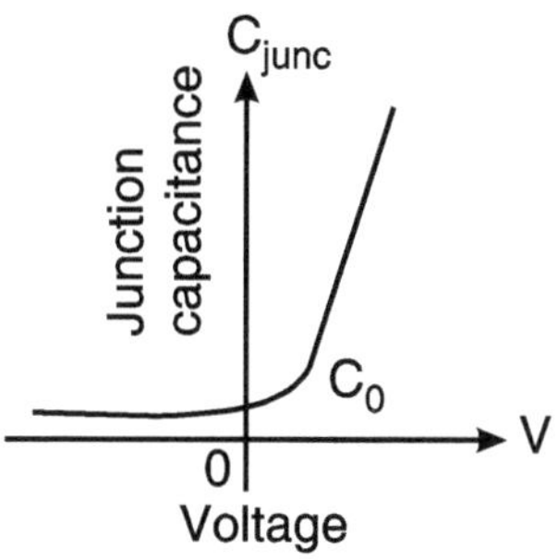

Fig. 8.3 V–I characteristics of varactor diode

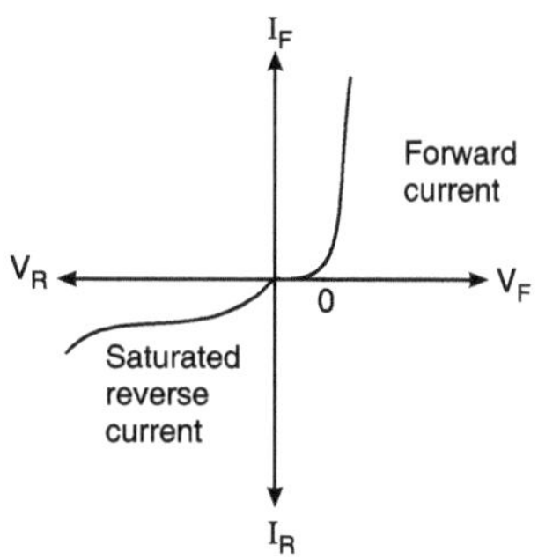

Fig. 8.4 Commonly used symbols for varactor diodes

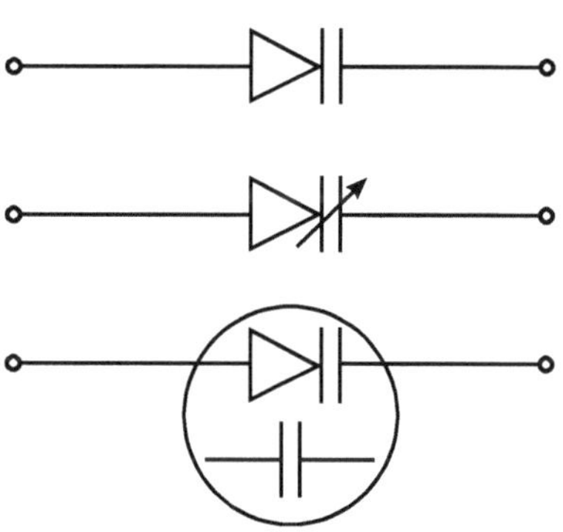

This clearly shows that C_j versus V_r curve can be effectively used for choosing specific doping profiles and varactor diodes can be designed for specific applications.

8.1.2 Performance Characteristics of Varactor Diode

The figure of merit of any device is the quality from which vivid aspects of performance may be predicted. These are

(i) ***Static cut-off frequency***. At a specific reverse bias V_r, it is given by

$$f_{crb} = \frac{1}{2\pi R_s C_{junc}} \tag{8.1a}$$

where C_{junc} is junction capacitance at voltage V_r, R_s is series resistance of the diode. The values of f_{crb} are

- for Si = 250 GHz, and
- for GaAs = 900 GHz

At zero bias,

$$f_{czb} = \frac{1}{2\pi R_s C_{junc}} \tag{8.1b}$$

It values are

- for Si f_{czb} = 25 GHz, and
- for GaAs f_{czb} = 90 GHz

(ii) ***Quality factor***. At a specific bias voltage V and frequency f, it is defined as

$$Q_f = \frac{f_{cv}}{f} \tag{8.2}$$

where Q_f is quality factor at a bias voltage V, f_{cv} is cut-off frequency at a bias voltage V, and f is any frequency of interest at which Q_f is measured.

(iii) ***Dynamic cut-off frequency***. It is the cut-off frequency at which the device is operated between bias extremes. It is given by

$$f_c = \left(\frac{1}{C_{jmin}} - \frac{1}{C_{j0}}\right)\frac{1}{2\pi R_s} \tag{8.3}$$

where C_{jmin} = capacitance of device near the reverse breakdown voltage, and C_{jo} = junction capacitance corresponding to zero bias.

(iv) ***Dynamic quality factor Q***. It is given by

$$Q = \frac{S_1}{\omega R_s} \tag{8.4a}$$

where S_1 is first Fourier component of the time-dependent elastance (i.e. reciprocal of capacitance), $\omega = 2\pi f$.

Also

$$S_1 = \frac{1}{C_1} = \frac{\Gamma}{C_{jv}} \tag{8.4b}$$

where 0.17 < Γ < 0.25 for most varactor junctions. It is 0.17 for graded junction, and 0.25 for step junction.

8.1.3 Applications of Varactor Diodes

They have several applications such as in following.

1. Harmonic generation
2. Microwave frequency multiplication (up-conversion)
3. Low noise amplification (i.e. parametric amplification)
4. Pulse generation and pulse shaping.
5. Tuning stage of a radio receiver (replacing the bulky variable plate capacitor)
6. Active filters.
7. Switching circuits and modulation of a microwave signal

8.2 Photodiodes

Photodiodes are two-terminal semiconductor devices which respond to photon absorption when radiations fall on them. Accordingly due to such response, they serve important functions as electronic devices. Their ability to serve important functions is made possible by changing the conductivity of bulk semiconductor sample in proportion to optical generation rate. Photodiodes are used to improve speed of responses and sensitivity of optical and high-energy radiation detectors. Some of their applications are in the following fields.

- High speed reading of punched cards
- High speed reading of film sound track.
- Optical switches etc.

A pocket size optical switch can handle millions of telephone conversations at any instant, simultaneously.

Types. There are several kinds of photodiodes. Some of them are 'without internal gain' (i.e. those generating a single electron-hole pair per absorbed photon) and others are 'with internal gain'. A detailed classification is illustrated in Fig. 8.5.

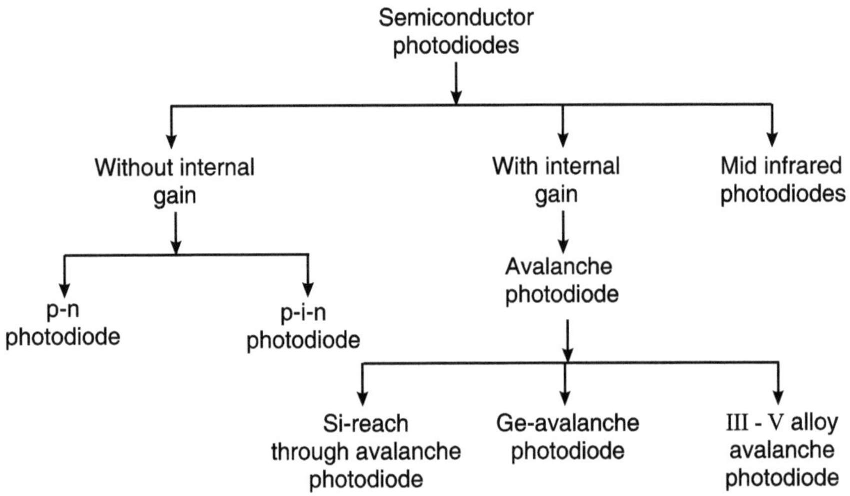

Fig. 8.5 Classification of photodiodes

8.2.1 Basic Construction of a Photodiode

A semiconductor photodiode is essentially a reverse-biased junction diode. The light is permitted to fall on one surface of the device across the junction, keeping the remaining sides unilluminated. In this condition, the diode current varies almost linearly with the light flux. Schematic construction of a photodiode is shown in Fig. 8.6. It is a small sized construction.

The p-n junction is embedded in a plastic package. All sides of the plastic capsule except the illuminating sides are painted black or enclosed in a metallic case. When the light falls on p-n junction, the electron-hole pairs are generated

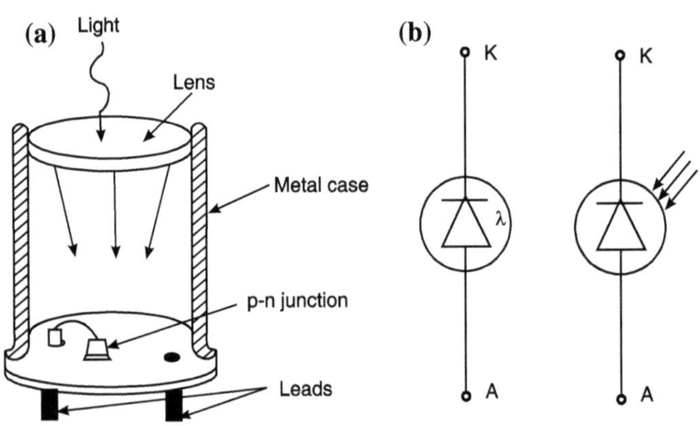

Fig. 8.6 Construction of a **a** semiconductor photodiode, and **b** its symbol

which increase the minority carrier concentration. The concentration of generated minority carriers varies according to the quanta of incident light. Hence, the reverse current increases.

8.2.2 p-n Photodiode

Figure 8.7 shows a reverse biased p-n photodiode with both: the depletion region and diffusion region. The depletion region is formed by immobile negatively ionized acceptor atoms in p-type material and positively ionized donor atoms in n-type region. The mobile carriers in depletion layer are swept away to their majority sides under the influence of electric field. The width of depletion layer depends upon the doping concentrations for a given applied reverse bias i.e. *the less the doping, the wider the depletion layer.*

Photons may be absorbed in both the depletion and diffusion regions as shown in Fig. 8.7. The absorption region width depends upon the energy of incident photons and on the material from which the photodiode is fabricated. In case of a weak absorption, the absorption region may extend throughout the device. The e–h pairs are therefore generated in both: the depletion layer and in diffusion layer. In depletion region the carrier pairs separate and drift under the influence of the electric field. Outside this region, the hole diffuses towards depletion region to be collected. The diffusion process is quite slow as compared to drift and thus limits the response of photodiode.

Output characteristics. It is therefore important that the photons are absorbed in depletion layer. Thus, it is made possible by decreasing the doping in n-type material. The depletion layer width in a photodiode is normally 1–3 μm and is optimized for the efficient detection of light at a given wavelength. For silicon devices, it is in the visible spectrum (0.4–0.7 μm); while for Ge devices, it is in the near infra-red region (0.7–0.9 μm). Typical output characteristics for a reverse biased p-n photodiode are shown in Fig. 8.8. It shows different operating conditions for no light input to a high light level.

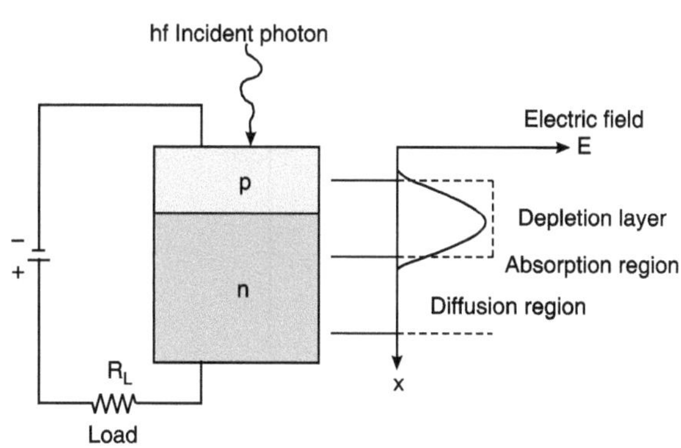

Fig. 8.7 A p-n photodiode showing depletion and diffusion regions

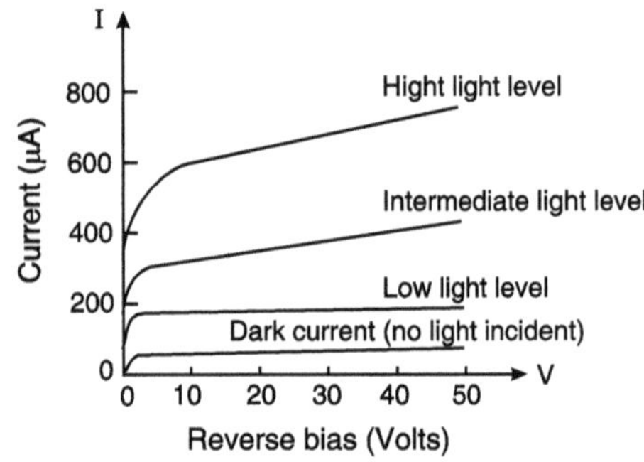

Fig. 8.8 Output current of a typical p-n photodiode

8.2.3 p-i-n Photodiode

In order to allow operation at longer wavelengths, where the light penetrates more deeply into semiconductor materials, a wider depletion layer is required. To achieve this target, the *n*-type material is doped so lightly that it can be considered intrinsic. And also to make a low resistance contact, a heavily doped *n*-type (n^+) layer is added. This completes a *p–i–n* structure. Figure 8.9 shows that all the absorption occurs in depletion layer.

Structures of photodiodes. Figures 8.10a–b show the structures of two different types of Si p-i–n photodiodes for operation in shorter wavelength band, below 1.09 μm. These are

1. Front illuminated photodiode.
2. Side illuminated photodiode.

The front illuminated photodiode (Fig. 8.10a) when operating in 0.8–0.9 μm band, requires a depletion layer of 20 to 50 μm in order to attain high quantum efficiency (about 85 %) together with fast response (<1 ns) and low dark current (≤1 nA). Dark current arises from the surface leakage currents as well as generation recombination results in depletion region in the absence of illumination. The side

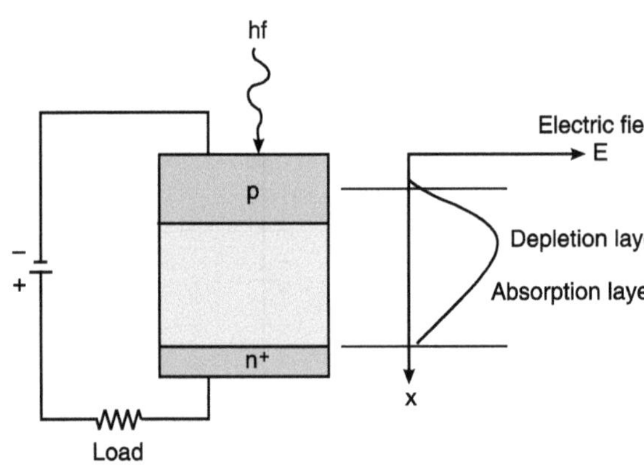

Fig. 8.9 A p-i-n photodiode showing combined absorption and depletion layer

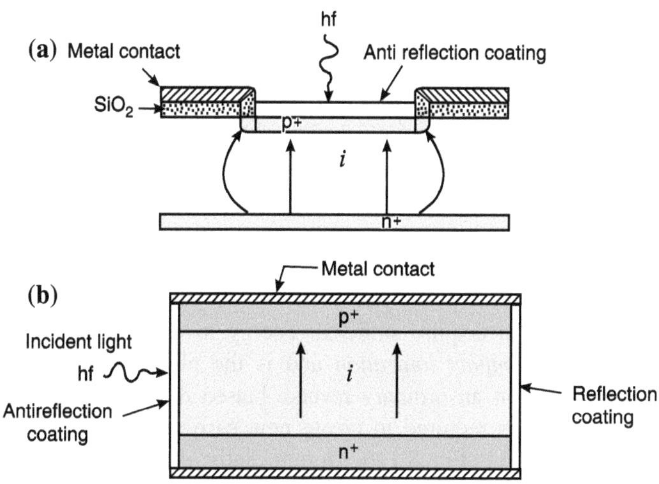

Fig. 8.10 Structure of a **a** front-illuminated p-i-n photodiode, and **b** side illuminated (parallel to junction) p-i-n diode

illuminated structure of Fig. 8.10b exhibits a large absorption width of about 500 μm. Hence, it is sensitive to those wavelengths which are close to the band gap limit which are (1.09 μm), where the absorption coefficient is relatively small.

8.2.4 Avalanche Photodiode

A avalanche photodiode (APD) has more sophisticated structure than a p-i-n photodiode. It is used to create extremely high field region ($\approx 3 \times 10^5$ V/m). A typical avalanche photodiode is shown in Fig. 8.11a. In the high field region, the

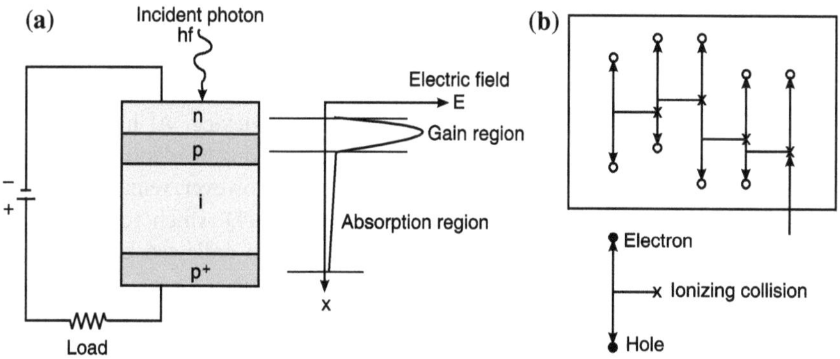

Fig. 8.11 Avalanche photodiode showing **a** high-electric field (gain region), and **b** e-h pair multiplication region

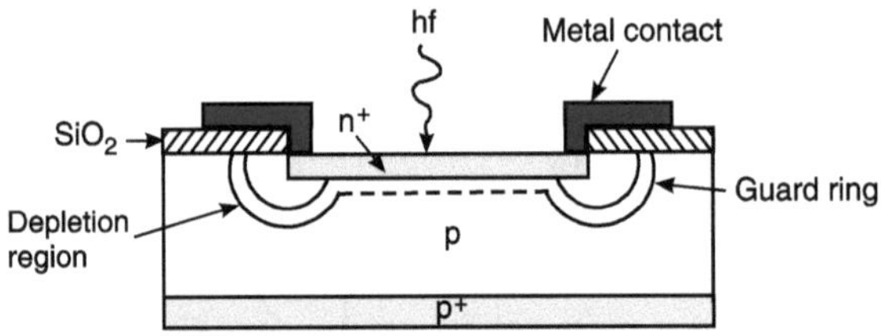

Fig. 8.12 Structure of a Si avalanche photodiode with guard ring

electrons and holes can acquire sufficient energy to excite new e–h pairs. This process is known as *impact ionization* and is the phenomenon which leads to avalanche breakdown in an ordinary reverse biased diode. Often a high reverse voltage of 50–400 V is required to create new carriers by impact ionization by themselves as illustrated in Fig. 8.11b. More recently, devices operating at voltages 15–25 V have also been developed.

Carrier multiplication factor as large as 104 may be obtained using defect-free material to ensure uniformity of carrier multiplication over the entire photo-sensitive area. There are other factors also which affect the gain. *Micro plasmas*, which are small areas with lower breakdown voltages than remainder of the junction, should be reduced by selection of defect-free materials. A large carrier multiplication factor may also be achieved with careful device processing and fabrication. In addition, excessive leakage at the junction edges can be eliminated by the use of a guard ring structure as illustrated in Fig. 8.12. Both Si and Ge APDs are commercially available.

Operation at high speed. Operation of these devices at high speed requires full use of the absorption region. When carriers are generated outside the depletion region, they are collected slowly by diffusion process. This has the effect of producing a long diffusion tail on a short optical pulse. When the APD is fully depleted by applying electric field $>10^4$ V/m, all the carriers drift at saturation limited velocities. The response time of the device is limited by (a) the transit time of carriers across the absorption region (depletion width), (b) the time taken by carriers to perform the avalanche multiplication process, and (c) the RC time constant incurred by the junction capacitance of the diode and its load.

At low gain, the transit time and RC effects dominate giving a definitive response time and hence a constant bandwidth for the device. At high gain, the avalanche build-up time dominates and the device bandwidth decreases proportionately with increasing gain. Gain bandwidth product, however, remains constant.

An asymmetric pulse shape is obtained from the APD which results from a relatively fast rise time. It is because the electrons are collected and fall time dictated by the transit time of the holes travelling with slower speed. The rise time lies between 150 to 200 ps (1 ps = 10^{-12} s) and fall time ≥1 ns.

8.2.5 *Mid-Infrared Photodiodes*

The development of photodiodes for mid-infrared transmission systems is at its infancy, though several potential devices have been demonstrated in recent years. Obtaining suitable lattice matching for III-V alloy materials is a problem when operating at wavelength >2 μm. A lattice matched InGaAsSb/GaSb material system has been used in a p-i-n high speed photodiode operation at wavelengths up to 2.8 μm.

An alternative approach uses Indium (In) alloys. High indium contents are mismatched with respect to InP substrate above 2 μm which cause dislocation-induced junction leakage and low quantum efficiency. These problems have been reduced by introducing a compositionally graded buffer layer to accommodate the lattice mismatch. One technique has replaced p-i-n homojunctions with InGaAs/AlInAs heterojunctions. The wider bandgap p-type AlInAs layer acts as a transparent window at long wavelengths to ensure that the optical absorption occurs in a lightly doped n-type region of the device. This device exhibits a useful response at a wavelength of 2.4 μm with quantum efficiency 95 % and dark current as low as 35 nA. Mesa structure photodiodes which operate at 2.55 μm are produced using $InGa_xAs_{1-x}/InAs_yP_{1-y}$. A compositionally graded region of InGaAs or InAsP accommodates the lattice mismatch between the ternary layers and InP substrate. The devices having $In_{0.85}Ga_{0.15}As$ absorbing layers and lattice matched $InAs_{0.68}P_{0.32}$ capping layers have exhibited quantum efficiency of 52 % at a wavelength of 2.55 μm and dark currents 10–20 μA.

8.3 The IMPATT Diode

The term IMPATT refers to Impact Avalanche Transit Time. Accordingly, the IMPATT diode means an 'impact avalanche transit time diode'. An IMPATT diode is a type of microwave negative conductance device that operates by combined action of transit time effects and carrier injection into drift region. To form an IMPATT diode, a simple p-n junction diode or diodes with variation in their structures are biased to obtain tunnelling or avalanche breakdown. This is achieved by an a-c voltage, superimposed upon d-c bias.

Construction. The IMPATT operation may be obtained in diodes having simple structures. But its best illustration is a 'Read diode'. This is a n^+-p-i-p^+ structure diode employing transit time effects. It is named 'Read' in the honour of its inventor W.T. Read. A Read diode working on IMPATT effect is shown in Fig. 8.13a. Its construction involves the following regions.

(i) The n^+-p region where avalanche multiplication takes place.
(ii) The intrinsic i-region through which the drifting of generated holes occur in order to move to p^+ contact.

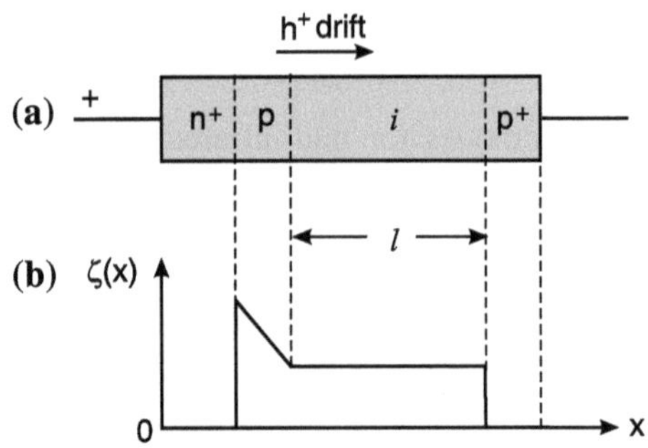

Fig. 8.13 A 'Read diode' working on IMPATT principle showing **a** its basic configuration, and **b** distribution of electric field under reverse bias

An IMPATT diode can be constructed in another configuration also, such as p^+-n-i-n^+. In this configuration, instead of holes the electrons are generated from avalanche multiplication and drift through i-region. The p^+-n-i-n^+ configuration is advantageous over n^+-p-i-p^+ configuration due to higher mobility of electrons than holes.

8.3.1 *Operational Mechanism*

An IMPATT diode operates in negative conductance mode. This causes the current to lag behind the voltage in time. The negative conductance occurs because of following processes.

1. A delay due to avalanche process, and
2. a further delay due to transit time of carriers across the drift region.

The above operations desire that the a-c component of current is negative over a portion of cycle during which the a-c voltage is positive. It is also true for a vice versa situation.

Figure 8.13b shows a d-c reverse bias on the Read diode. In this condition, the holes drift in the direction of the field (from left to right). Further, if an a-c voltage is superimposed such that the applied electric field ζ decreases during the negative half cycle, then the drift of holes also decreases. But in IMPATT operation, the drift of holes through i-region increases and the a-c field decreases. This is due to avalanche and drift effects.

8.4 Trapatt Diode

The TRAPATT diode (Trapped Plasma Avalanche Transit-Time diode) is a high frequency component of micro-electrons. It gives high output and high efficiency. It is a p-n junction diode whose operation is different from that of an IMPATT diode.

It is characterized by the formation of a trapped space-charge plasma within the junction region. It is vastly used in the generation and amplification of microwave power. The key phenomena taking place in TRAPATT diode are the following.

(i) Carrier generation by impact ionization producing a current pulse of phase delay of 90°.
(ii) An additional phase shift introduced by the drift of carriers.
(iii) Operate at frequencies ranging from 400 MHz to about 12 GHz.

The structure of a TRAPATT diode is as shown in Fig. 8.14a.

8.4.1 *Plasma Formation in TRAPATT Diode*

During the operation of TRAPATT diode, a high field avalanche zone propagates through the depletion region and fills the layer with dense plasma of electrons and holes which get trapped in the low field region behind the zone.

The diode is placed in a circuit containing high-Q resonators and is reversed biased into avalanche. As the diode breaks down, a highly conducting electron-hole plasma quickly fills the n region and the voltage across the diode drops to a low value. The plasma trapped in the low E-field is then extracted, causing a large current flow even though the voltage is low. Once the extraction of the plasma is completed, the current becomes very small and the voltage rises to a high value.

8.4.2 *Operation*

The operation of TRAPATT diode is self-explanatory through Fig. 8.14b–c.

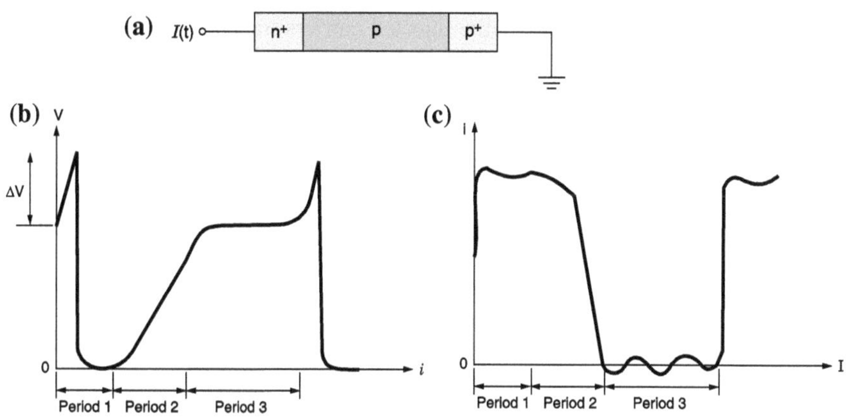

Fig. 8.14 **a** Structure of a TRAPATT diode. **b** ΔV-pulse overvoltage triggering the operation. **c** current vs time variation during TRAPATT operation

8.4.3 Advantages

The TRAPATT diodes possess the following advantages:

(i) The reactive energy stored in the resonant circuits raises the voltage above breakdown and the cycle repeats.
(ii) Most are $p^+ nn^+$ or the opposite that are fabricated from diffused epitaxial silicon. Well suited for generation of high powers up to several hundreds of watts are L-band to several tens of watts at X and Ku-band frequencies.
(iii) Best efficiencies from about 60 % at L to 30 % at X-band.

8.4.4 Applications of TRAPATT Devices

(a) Phased-array Radar systems.
(b) Intermediate frequency transmitters.
(c) Proximity fuse sources.
(d) Radio altimeters.
(e) Microwave levelling systems.
(f) Marine radars.
(g) Used in airborne.

8.5 BARITT Diode

A BARITT (Barrier Injection Transit Time) diode uses the injection and transit time characteristics of minority charge carriers to generate a negative resistance at microwave frequencies. The basic mechanism behind the formation of these diodes is the thermionic injection and diffusion of minority carriers across a forward-biased barrier and a transit time delay of the injected carriers traversing the drift region.

Since there is no a valance, the BARITT diodes deliver less power and efficiency, but show better noise properties and stability. An Optimum BARITT has low-hi-low structure in the depleted region, i.e. $p^+ nn^+ np^+$ or (reversed for electron flow).

8.5.1 Structure of BARITT Diode

When no bias voltage is applied, the electric field profile of this structure is determined by the built-in field region of the p + n and np + junction. This is shown in Fig. 8.15a. When bias voltage is applied, the conditions change drastically. One of the junctions becomes forward biased and the other reverse biased. It is shown in Fig. 8.15b.

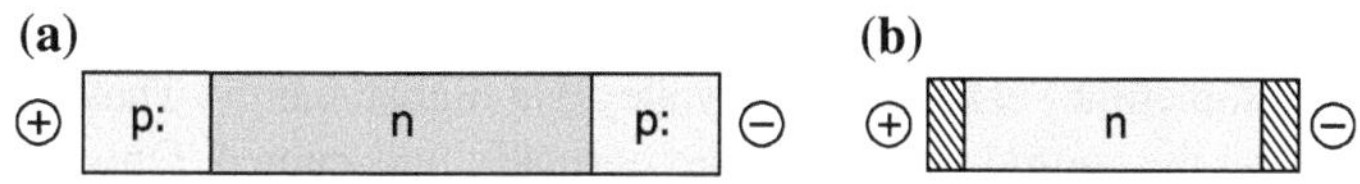

Fig. 8.15 **a** Structure of BARITT. **b** condition on bias voltage application

8.5.2 Performance of BARITT Diode

1. The possibilities offered by the BARITT diode are restricted to around 8 GHz as compared to 100 GHz offered by the IMPATT diode.
2. As compared to LED, the performance of BARITT diode is better since it has high power efficiency.
3. Its simpler fabrication technique using a less sophisticated material is a major advantage.
4. BARITT diodes have very low noise level.

8.6 Transferred Electron Mechanism

In Chap. 3, we have discussed that the Ohm's law is valid in carrier drift processes. It implies that the drift current is proportional to the electric field. Although this assumption is valid over a wide range of electric field, but is not true for large electric fields ($\zeta > 10^3$ V/cm). It is because the high field effects exhibit a sub linear dependence of current and non-linearity of mobility. The carriers reach a scattering limited velocity. This behaviour is typical of Si, Ge and other semiconductors. However in some materials, a decrease in electron velocity can be achieved at high electric fields, which results in negative conductivity and current instability in the sample. In them the energy of electrons can be raised by applied field, if they transfer from lower region of conduction band to higher region. Since the negative resistivity in some band structures can result from this electron transfer, the process is referred to as 'transferred electron mechanism'. GaAs is one such material.

Thus, the transferred electron mechanism refers to a process in which 'the conduction of electrons of some semiconductors are shifted from a state of high mobility to a state of low mobility under the influence of a high electric field'. Consequently the negative conductance operation can be achieved in a diode. Such diodes are two-terminal device having no p-n junction.

8.6.1 Valleys in Conduction Band

A simplified band diagram for GaAs is shown in Fig. 8.16. In n-type GaAs the valence band is filled, and the lower valley (or minimum) of the conduction band at (k = 0) contains the conduction electrons. There is a set of subsidiary minima at upper valley (sometimes called satellite valleys) at higher energy. These minima are many kT above the central valley and are normally unoccupied. Only one satellite valley has been shown for convenience. There are other equivalent valleys for different directions in k-space. The effective mass ratio of 0.55 refers to the combined satellite valleys. The direct band gap at lower valley and the energy bands centered at k = 0 are generally used to describe the conduction processes in GaAs. The presence of the satellite valleys is crucial to the Gunn effect. If the material is subjected to an electric field above some critical value (about 3 kV/cm), the electrons in the central lower valley of Fig. 8.16 gain more energy than the 0.3 eV separating the valleys. Therefore, there is considerable scattering of electrons into the higher-energy satellite valley at upper valley.

8.6.2 Negative Differential Conductivity

Once the electrons have gained enough energy from the field to be transferred into the higher-energy valley, they remain there as long as the field is greater than the critical value. It is because of the fact that the combined effective density of states for the upper valleys is much greater than for the central valley. It seems reasonable that the probability of electron scattering between valleys should depend on the density of states available in each case. As a result, once the field increases above the critical value, most conduction electron in GaAs reside in the satellite valleys and exhibit properties typical of that region of the conduction band.

The effective mass for electrons in the higher upper valleys is almost eight times (0.55/0.067 = 8.2) as great as in the central valley, and the electron mobility is much lower. This is an important result for the negative conductivity mechanism. As the electric field is increased, the electron velocity increases until a critical field is reached; then the electrons slow down with further increase in the field. The electron transfer process allows electrons to gain energy at the expense of velocity over a range of values of the electric field. Taking current density as $J = ev_d n$, it is clear that the current also drops in this range of increasing field, giving rise to a negative differential conductivity $dJ/d\zeta$.

8.6.3 Dependence of Electron Velocity on Electric Field

A possible dependence of electron velocity versus electric field for a material capable of electron transfer is shown in Fig. 8.17. For low values of field, the electrons reside in the lower valley of the conduction band, and the mobility ($\mu_r = v_d/\zeta$) is high and constant with field. For high values of field, the electrons transferred to the satellite valleys where their velocity is smaller and their mobility is lower. Between these two states is a region of negative slope on the v_d versus ζ plot, indicating a negative differential mobility $dv_d/d\zeta = -\mu^*$.

8.7 The Gunn Diode

The Gunn diodes are the microwave devices which operate by transferred electron mechanism. This is named so in the honour of its originator J.B. Gunn. The Gunn diodes are two-terminal devices which do not involve a p-n junction. A p-n junction is not required because they utilize the bulk instabilities of semiconductor materials. The Gunn diodes are simple and homogeneous samples having ohmic contacts on each end. They find use in following main applications.

- Microwave circuits
- Microwave oscillation

Operational requirements. A Gunn diode is required to have the following abilities.

1. High mobility
2. Few lattice defects

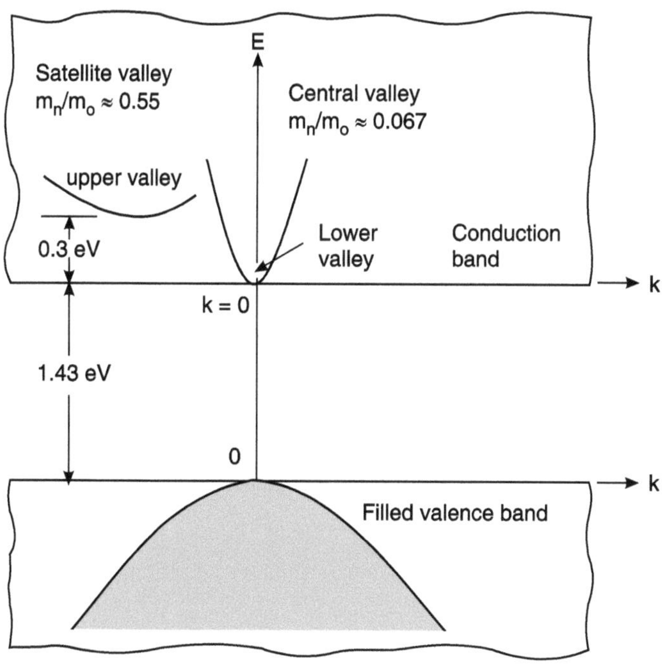

Fig. 8.16 Simplified band diagram for GaAs, illustrating the lower and upper valleys in conduction band

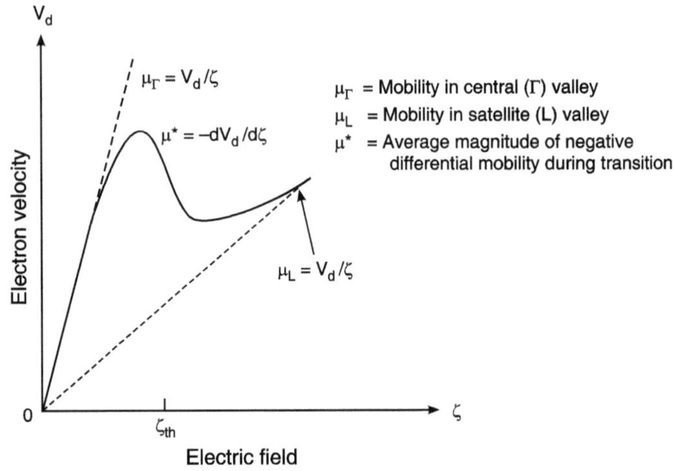

Fig. 8.17 Illustration of a possible characteristic of electron drift velocity versus electric field for a semiconductor exhibiting the transferred electron mechanism

3. Homogenous doping
4. Carrier concentration in the range of about 10^{13}–10^{16}/cm^3.
5. Exact sample lengths. In a typical configuration, the n-type epitaxial layer may be about 10 μm thick and n^+ substrate wafer of 100 μm thickness.
6. High rate of heat removal. The power dissipation may be of the order of 10^7 W/cm^3 or more.

With rise in temperature, the carrier concentration and mobility also vary. Therefore they affect the efficiency of the diode.

8.7.1 Materials and Fabrication

The Gunn diodes are made of a number of such materials which have an appropriate band structures for electron transfer. They are also made from such materials whose band structure can be altered to exhibit properties which are favourable for electron transfer. Various materials suitable for making the Gunn diodes are the following.

- GaAs
- InP
- GaAsP
- CdTe
- ZnSe
- InAs

In order to put in favourable properties, the energy bands of InAs are distorted by applying high pressure on its crystal. As a result, a set of satellite valleys become available for electron transfer. If the pressure applied is not sufficiently high, the upper valleys will be far above the lower valley and hence the electron transfer may not take place.

Although a Gunn diode can be made of bulk samples, but the use of ingot material as a substrate for epitaxial layer is desired. It is because the material properties of epitaxial layers are superior to those of bulk samples. External contracts of Gunn diodes are made by evapourating a thin layer of following materials on each surface.

- Au–Sn
- Au–Ge

8.7.2 *Dielectric Relaxation Time in Respect of the Gunn Diode*

When a sample of GaAs is biased such that the applied electric field falls in negative conductivity region, it results in instabilities in the space charge. As result, the Gunn diode cannot be maintained in d-c stable condition. The formations of instabilities are due to exponential die-out of space charge with time. It is given as follows.

$$Q_{sc}(t) = Q_0 e^{-t/\tau_{dr}} \tag{8.5a}$$

Here $Q_{sc}(t)$ is instantaneous space charge at an arbitrary time t, Q_0 is initial space charge, t is arbitrary time, and τ_{dr} is dielectric relaxation time. The τ_{dr} depends on permittivity ε and conductivity σ of the semiconductor. It is expressed as

$$\tau_{dr} = \frac{\varepsilon}{\sigma} \tag{8.5b}$$

The value of τ_{dr} for GaAs sample of 1 Ω is about 10^{-12} s.

8.7.3 *Dependence of Electron Drift Velocity on Electric Field for a Gunn Diode*

GaAs and InP are commonly used semiconductors in Gunn diode. Dependence of drift velocity vd in them as a function of applied electric field ζ is shown in

Fig. 8.18. It depicts that the negative resistance due to electron transfer occurs at a higher field for InP, and a higher peak value is achieved by electrons before their transfer from central to satellite valley.

8.8 Dovett Diode

The double velocity transit time (DOVETT) diode is similar to the BARITT diode, except that the velocity of carriers near the injection contact is much less than that near the collection contact. A heterostructure DOVETT diode is symbolically shown in Fig. 8.19. The device is biased, so that P-N anisotype heterojunction is completely punched through, and large current can be injected over the forward biased N^+-P junction. Because of the transit-time delay in the low-velocity injection region, the negative resistance of this diode is higher than that of BARITT diode. This diode can be operated at higher dc current densities than BARITT diodes. These diodes have efficiency above 25 %.

8.9 Solved Examples

Example 8.1 What is a tunnel diode? What are its specialities?

Solution. A tunnel diode is a p-n diode having negative resistance region. In this region, an increase in terminal voltage results in reduction in diode current. The main speciality of a tunnel diode is its very thin depletion layer (of the order of about 10^{-6} cm) through which the carriers flow in such a way as if, they are tunnelling through a very narrow passage.

Fig. 8.18 Dependence of drift velocity on electric field in a Gunn diode

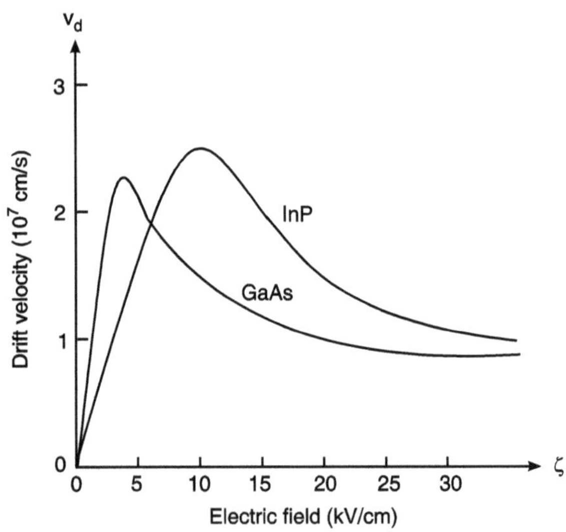

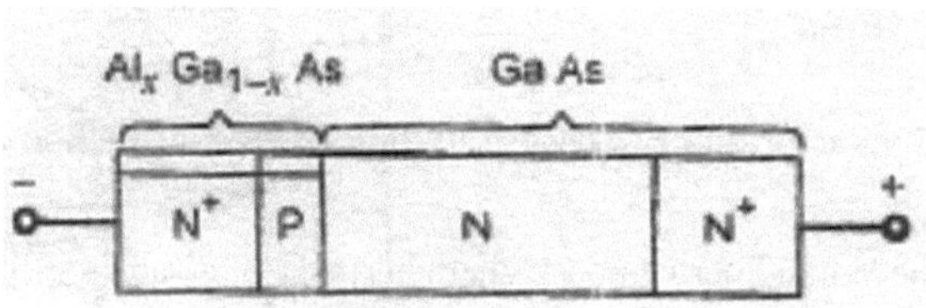

Fig. 8.19 DOVETT diode

Example 8.2 Why is a tunnel diode preferred for high-speed circuit applications?

Solution. The negative resistance of a tunnel diode can be used in a number of ways. These are as in switching, amplification, oscillation etc. The advantageous feature of a tunnel diode is the tunnelling process in it, which occurs without any time delay in flow of carriers during drift and diffusion. This provides almost an instantaneous operation that makes the tunnel diode a suitable device for high speed circuit applications.

Example 8.3 State the condition under which the electron tunnelling takes place.

Solution. Electron tunnelling takes place when there exist filled and empty states in the material separated by a narrow potential barrier of finite height.

Example 8.4 What purpose is served by a transit time device?

Solution. A transit time device can convert d-c into microwave a-c signals with a high efficiency. It can also generate microwave power for many applications.

Example 8.5 Can we obtain a negative conductance in simple p-n junctions or in p-i-n devices?

Solution. Yes, a negative conductance can be obtained in simple p-n junctions and also in p-i-n devices. In p-i-n devices, most of the applied voltage occurs across the i-region. It serves as a uniform avalanche region as well as a drift region. Thus the process of delay is distributed within the i-region of p-i-n device. It implies that the electrons and holes both play a role in avalanche and drift processes.

Example 8.6 A varactor diode with a linearly graded doping profile has a capacitance of 50 pF when no bias is applied to the diode. Determine the junction capacitance for the silicon diode when the reverse bias applied to the diode is of 8 V.

Solution. Capacitance of diode with zero bias, $C_T(0)$ = 50 pF

Reverse bias applied to the diode, V_R = 8 V

and n = 1/3

Barrier potential for Si is, V_k = 0.7 V

So, the junction capacitance is given as

$$C_T = \frac{C_T(0)}{\left(1+\frac{V_R}{V_k}\right)^n} = \frac{50}{\left(1+\frac{8}{0.7}\right)^{1/3}} = 21.59\,\text{pF}$$

Review Questions

1. What is a varactor diode? Discuss its functions, construction and working mechanism.
2. Explain the V-I and performance characteristics of a varactor diode.
3. What is photodiode? What are its different types? Describe the basic construction of a photodiode.
4. Discuss the requirements of a good photodiode material and enlist suitable materials for its construction.
5. Describe the construction and working of a p-n photodiode. Discuss its characteristics and applications.
6. Describe the construction and working of an avalanche photodiode. Discuss its characteristics and applications.
7. Describe briefly the mid-infrared photodiode and phototransistor.
8. Describe the construction and working of a p-i-n photodetector. Write its characteristics and applications also.
9. What is meant by IMPATT? Explain the construction and working of an IMPATT diode.
10. Elaborate the transferred electron mechanism. Explain it on a band diagram of GaAs stating the sense of lower and upper valleys in conduction band.
11. State the characteristics of a semiconductor exhibiting the transferred electron mechanism.
12. Elaborate it on electron velocity versus electric field diagram.
13. Explain the TRAPATT diode. How is plasma formed in it? Discuss its operation, advantages and applications.
14. Describe a BARRITT diode explaining its structure and performance.
15. What is a Gunn diode? What are its operational requirements? Suggest suitable materials for its construction.
16. Sketch and explain the construction and working of a Dovett diode. How does it different from the BARRIT diode ?

Objective Questions

1. Consider the following statements.
   A. Negative conductance is involved with instability of semiconductors.
   B. The Gunn diodes employ the combination of carrier injection and transit time.
   C. Atomic density of semiconductor is influenced due to heavy doping.
   D. At equilibrium, there is equal tunnelling from n to p and from p to n, in a tun nel diode.

   Of these, the correct statements are

   (a) A, B and C (b) A, C and D
   (c) A, B and D (d) B, C and D

2. Match list I with list II, and choose the correct answer from the codes given below the lists.

List I	List II
A. Region between E_v and E_F in degenerate p-type semiconductor	1. dV/dI is negative
B. On applying small forward bias to tunnel diode	2. GaAs
C. The Gunn diode	3. almost completely filled with holes
D. IMPATT diode	4. p^+-n-i-n^+
	5. EFn moves up with respect to EFp
	6. for only a part, filled with holes

Codes:

	A	B	C	D		A	B	C	D
(a)	1	6	4	2	(b)	1	5	2	4
(c)	3	5	2	4	(d)	3	6	4	2

3. Match list I with list II, and choose the correct answer from the codes given below the lists.

List I	List II
A. Hyperabrupt junction	1. Low voltage sensitivity
B. Recombination term	2. Q_p/T_p
C. Rectifying type metal-semi-conductor Junction	3. dQ_p/dt
D. Heterojunctions	4. Ballistic transport
	5. High voltage sensitivity
	6. Schottky diode

Codes:

	A	B	C	D		A	B	C	D
(a)	1	3	4	6	(b)	5	2	6	4
(c)	1	2	6	4	(d)	5	3	4	6

4. Given below are some statements.
   A. Influence of excess carriers is to cause decay of stored charge in n-region of the junction devices.
   B. Switching diode should have a very short carrier lifetime.
   C. A varactor diode helps in reducing the size of the circuits.
   D. A heterojunction is formed of a semiconductor, a metal and an insulator

   The correct statements are
   (a) A and D (b) C only
   (c) A, B and C (d) A, C and D

5. Which of the following is incorrect?
   (a) $\frac{dQ_p}{dt} = 0$ for steady state.
   (b) A homojunction is formed within a single semiconductor
   (c) In a switching diode, the power losses should be minimum under forward bias.
   (d) Doping of Au in Si switching diode lowers the recombination lifetime of holes, but increases the reverse recovery time.

6. Photovoltaic emf of semiconductor photodiode is that voltage at which the resultant current flowing through the device becomes
   (a) minimum. (b) maximum.
   (c) zero. (d) twice the value of the normal current.

7. A surface having a work-function of 4.3 eV isradiated with mercury line of 25730 A. The maximum velocity of the emitted photoelectrons will be
   (a) 1.55 105 m/sec. (b) 2.45 105 m/sec.
   (c) 4.56 105 m/sec. (d) 8.35 105 m/sec.

8. Find the correct match between Group 1 and Group 2:

Group 1	Group 2
A. Varactor diode.	1. Voltage reference.
B. PIN diode.	2. High-frequency switch.
C. Zener diode.	3. Tuned circuits.
D. Schottky diode.	4. Current controlled attenuator.

(a) A-4, B-2, C-1, D-3 (b) A-2, B-4, C-2, D-3

(c) A-3, B-4, C-1, D-2 (d) A-1, B-3, C-2, D-4

Answers

1. (b) 2. (c) 3. (b) 4. (c) 5. (d)

6. (c) 7. (c) 8. (c)

Chapter 9
Optoelectronic Devices

Abstract Optoelectronics refers to the studies of optics and electronics. Optoelectronic devices involve the interaction of photons of semiconductors. These devices find important applications such as sensors, couplers, isolaters and emitters. In this chapter, the Optoelectronic semiconductor devices are enumerated. Optical properties and characteristics of illuminated junction, are elaborated. Solar cell: its construction, working, factors affecting its efficiency, fabrication and materials, advantages and limitations, and applications are narrated. Further, different types of photodetectors are classified, requirements of a good photodetector and methods to achieve fast responsing speed are discussed. Working, construction and choice of materials are given for p-i-n photodetector, silicon heteroface photodetector and photoconductive detector. Light emitting diode: its construction and working, advantages, applications, specifications, and suitable materials for them are described. Semiconductor lasers: their working, merits, characteristics and properties, applications, and suitable materials for them are elaborated. The lasers diodes, light dependent resistor, differences between homojunction laser and heterojunction laser, and spontaneous emission and stimulated emission are explained. Importance of optoelectronic pair and population inversion are briefly presented. Minute insight into the various topics are given through solved numerical and theoretical examples. Review questions, numerical problems and objective type questions are also given with their answers.

Keywords Optical properties · Solar cells · Photodetectors · Different types of photodetectors · Light emitting diodes · GaN LED · Semiconductor lasers · Laser diodes · Light dependent resistors · Overlight detector

9.1 Introduction to Optoelectronic Devices

The devices, based on the technology which combines optics and electronics, are known as optoelectronic devices. The optoelectronic device convert optical energy into electrical energy and vice versa. Therefore, these devices can be classified into following two main categories:

K.M. Gupta and N. Gupta, *Advanced Semiconducting Materials and Devices*,
Engineering Materials, DOI 10.1007/978-3-319-19758-6_9

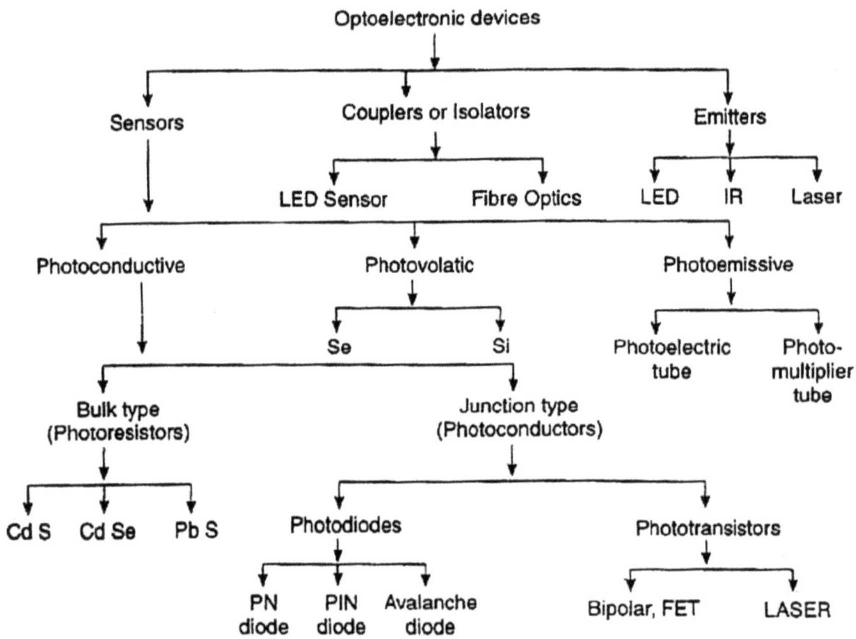

Fig. 9.1 Showing the family of optoelectronic devices

1. **Sensors**: The devices that convert optical radiation into electrical energy are known as sensors. The mechanism of converting optical radiation into electrical energy is known as photovoltaic effect. The devices are also known as light-sensitive devices. Two important devices based on photovoltaic effect are solar cell and photodiode. Light sensitive devices also include light sensitive resistors which have optosensitive resistivity.
2. **Emitters**: The devices that convert electricity into light are known as emitters. The phenomenon is called as electroluminescence (emission of optical radiation by converting electrical energy into light). The devices are also called as light generating optoelectronic devices. Light emitting diode (LED) and laser are examples of such devices.

A detailed classification of optoelectronic devices is displayed in Fig. 9.1.

9.1.1 Salient Applications

Optoelectronics is the study of electronic devices that source, detect and control the light. It is a sub-field of photonics. In this context, the light often includes invisible forms of radiation such as gamma rays, X-rays, ultraviolet and infrared, in addition to visible light. Optoelectronic devices are electrical-to-optical or optical-to-electrical transducers, or instruments that use such devices in their operation. Optoelectronics is based on the quantum mechanical effects of light on electronic materials; especially semiconductors, sometimes in the presence of electric fields. Broadly, their applications are found in following areas.

Photoelectric or photovoltaic effect based, used in:

- photodiodes (including solar cells)
- phototransistors
- photomultipliers
- optoisolators
- integrated optical circuit (IOC) elements

Photoconductivity based, used in:

- Photoresistors
- photoconductive camera tubes
- charge-coupled imaging devices

Stimulated emission based, used in:

- injection laser diodes
- quantum cascade lasers

Lossev effect (or radiative recombination) based, used in:

- light-emitting diodes or LED
- OLEDs

Photoemissivity based, used in

- photoemissive camera tube

Other Important applications of optoelectronics include:

- optocoupler
- optical fibre communications

9.1.2 Optoelectronic Semiconductor Devices

Optoelectronic devices are those devices whose functions involve the interaction of photons with semiconductors. These devices are different from the electronic devices. They provide optical sources and optical detectors for different devices. In their working, some of the optoelectronic devices detect photons, while some others emit photons. The optoelectronic devices may broadly be grouped into following two categories.

1. Devices that convert optical energy into electrical energy, such as
   - Photodiodes
   - solar cells
2. Devices that emit photons, such as
   - incoherent sources like light-emitting diodes (LEDs)
   - coherent sources such as lasers.

Details of these devices are discussed in subsequent sections.

Compound semiconductors are produced by combining equal atomic fractions of elements of III and V columns, or II and VI columns. Their structures are of diamond cubic (DC) type. They are widely used in a number of optoelectronic devices such as given below.

1. Photodiodes
2. Photon detectors
3. Photocells
4. Solar cells
5. Light-emitting diodes
6. Lasers
7. Optical fibres
8. Integrated circuits (ICs)

Photodiodes. A photodiode is a two terminal device, which responds to photon absorption.

Photoconductors (or photo detectors) are used to detect and measure the quanta of light such as in automatic door opener, in switching the street lights, and as burglar alarm. CdS, CdSe and CdTe are commonly used photoconducting semiconductors.

Photocells convert the light energy into electrical energy. They are fabricated from CdS, Se and $PbSO_4$. Photocells are used in cinematography, fire-alarms and television cameras etc.

Solar cells made from semiconducting materials are of immense utility in satellites and space-going vehicles. They are also used in calculators, solar power generation, and solar auto vehicles.

Light-emitting diode is an incoherent light source, which is used as a light source in fibre optic systems and other devices.

Laser is the source of a highly directional, monochromatic coherent light which is used as a light source for various optical and electronic devices.

Optical fibre is a means of transmitting optical signals from a source to a detector.

Integrated circuits are manufactured on single chip that contains diodes, transistors, resistors and capacitors etc. They are generally monolithically constructed using either unipolar or bipolar techniques.

9.2 Optical Properties

By optical property we mean the response of a material against exposure to electromagnetic radiations, particularly to visible light. When light falls on a material, several processes such as reflection, refraction, absorption, scattering etc. take place due to interaction between the light and material. The e.m.f. of the light radiation interacts with the changes contained in a material to exhibit its different optical properties. The nature and distribution of charges in a material play an important role in determining the optical properties.

When light of a suitable wavelength falls on materials, some of them emit electrons. This phenomenon of emission of electrons is called photoelectric effect. Under some appropriate conditions, the matter is capable of amplifying instead of absorbing an incident ray. This property is the base of modern laser. Double refraction (or birefringence) occurs when the index of refraction is a function of crystal orientation. Spectrum of electromagnetic radiations is displayed in Fig. 9.2 which is helpful in designing of optoelectronic devices.

9.2.1 *Current and Voltage Characteristics of an Illuminated Junction*

The well-known diode equation describing the flow of current in it due to thermal and optical generations may be expressed as

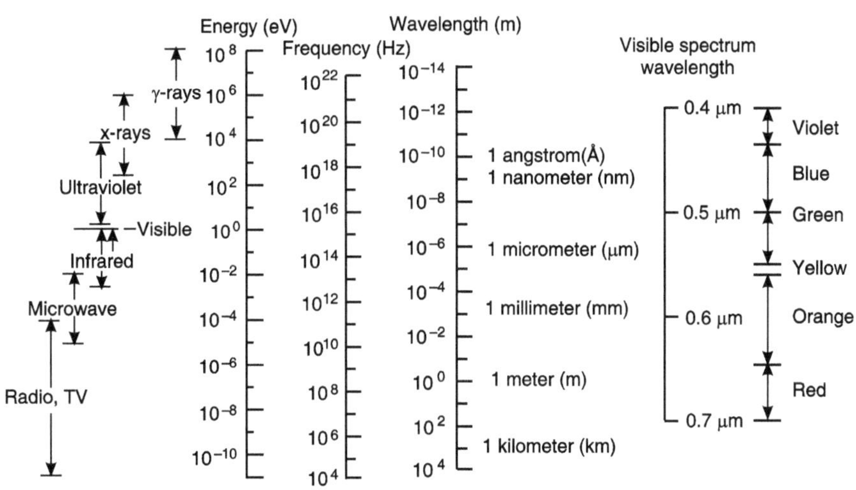

Fig. 9.2 The spectrum of electromagnetic radiations, wavelength ranges for various colours in visible spectrum. Reproduced from [1]

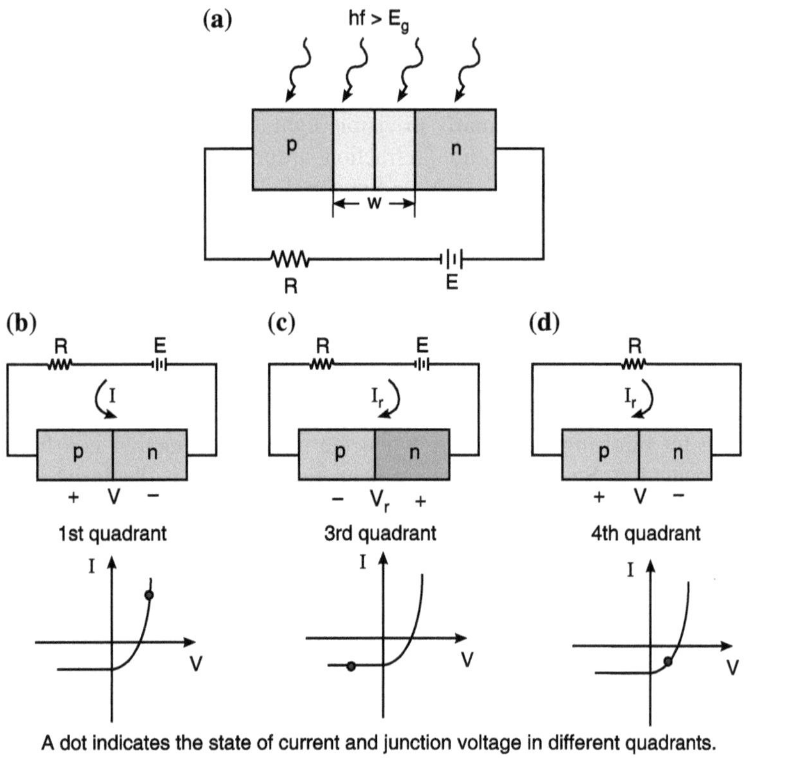

Fig. 9.3 **a** Absorption of light by a p-n junction, **b–d** operation of an illuminated junction showing I–V characteristics in different quadrants

$$I = eA\left(\frac{L_p P_n}{\tau_p} + \frac{L_n n_p}{\tau_n}\right) \cdot e^{\left(\frac{eV}{KT}-1\right)} - eAg_{\text{opr}}\left(L_p + L_n + \omega\right) \tag{9.1}$$

where A is cross-sectional area of the junction, L_p and L_n are hole diffusion length and electron diffusion length respectively, τ_p and τ_n are recombination lifetime for holes and electrons, g_{opr} is optical generation rate, and w is width of depletion region.

The optical generation of carriers in a p-n junction, when light is absorbed by the device, is shown in Fig. 9.3a. The I–V characteristics of the illuminated junction are shown in Figs. 9.3b–d. This p-n junction (or the photodiode) can be used in different ways. Depending upon the desired applications, the junction shown in Fig. 9.3a can be operated in such a way that the I–V characteristics displayed in Figs. 9.3b–d may be effectively utilized in different devices. These characteristics indicate that

(i) the power will be delivered 'to the device' from the external circuit, when the current and junction voltage are both positive or both negative, as shown in Fig. 9.3b, c. However,
(ii) the power will be delivered 'from the junction' to the external circuit when the junction voltage is positive, but the current is negative as shown in Fig. 9.3d.

The I–V characteristic of Fig. 9.3c is being utilized for the working of photodetector as the junction is reverse biased for its operation. Similarly the characteristics of Fig. 9.3b, d are utilized for other optoelectronic devices.

9.3 Solar Cells

A solar cell is used to convert solar energy into electrical energy. It is a direct energy conversion system, based on photovoltaic action. A solar cell is a self-generating arrangement which employs semiconductor contacts against metals. The current generated depends on the illuminated area. As an illustration, the current generated is about 10–100 mA from a junction of about 1 cm^2 area. Usually the amount of current generated from a single solar cell is too less for certain applications, therefore many solar cells are used together to form a 'solar battery'. The power produced by a solar battery is significant. The arrays of p-n junction solar cells are widely used to supply electrical power to space satellites, as shown in Fig. 9.4a.

9.3.1 *Working Principle*

Working of a solar cell is based on photovoltaic action. As the power can be delivered to an external circuit by illuminated junction, therefore the solar energy can also be converted into electrical energy. Accordingly, when light is incident upon the semiconductor-metal combination (see Fig. 9.4b), an internal voltage is generated that causes current to flow through the internal circuit even though not a single battery is connected externally. The e.m.f. thus generated depends on the radiant energy received by the solar cell.

9.3.2 *Construction and Working*

The most commonly used photovoltaic cells are of the barrier-layer type like iron-selenium cells or Cu-CuO_2 cells. In the iron-selenium cell as shown in Fig. 9.4b, the selenium layer is placed on an iron disc and then an extremely thin

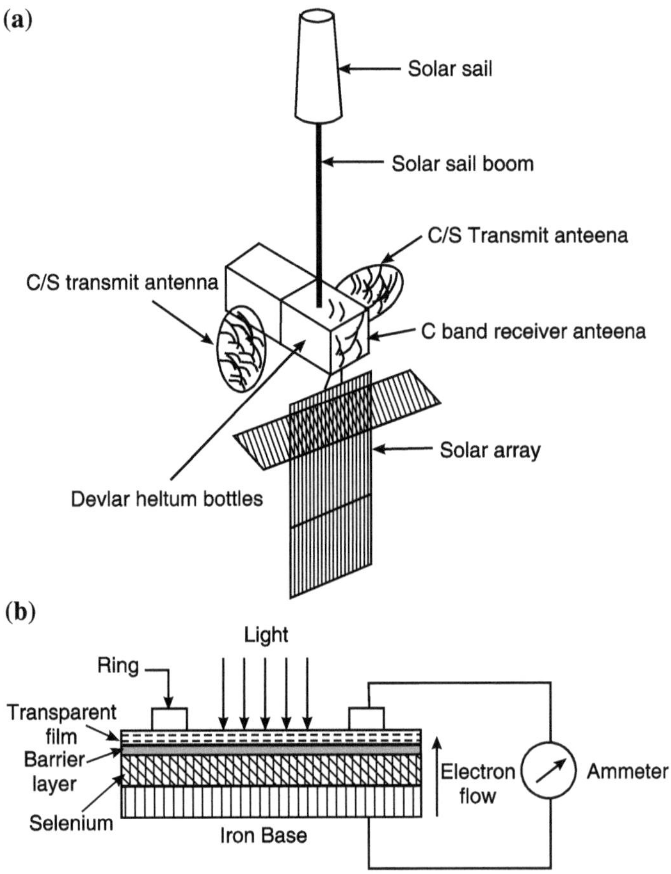

Fig. 9.4 A solar cell showing **a** the arrays of p-n junctions in a satellite, and **b** a typical construction. Reproduced with permission from [1]

transparent layer of gold or silver is formed on the Selenium to act as a front electrode. The barrier layer is formed by cathode-sputtering the semi-transparent film on the selenium. A contact ring on the silver layer acts as one electrode and the iron base as the other.

When luminous flux or radiant flux falls on the semiconductor i.e. selenium, it ejects electrons which travel from selenium to the front silver electrode through the barrier layer as shown in Fig. 9.4b. The flow in the opposite direction is not permitted by the (boundary) barrier layer because it acts as a rectifier. The e.m.f. generated is almost directly proportional to the incident flux.

9.4 Factors Affecting the Efficiency of Solar Cells

There are three main factors which influence the efficiency of solar cells. These are

1. the energy gap E_g of the emitter/absorber material.
2. the absorption coefficient α of the absorber, and
3. the diffusion length of minority carrier.

9.4.1 *Effect of Energy Gaps*

E_g determines the threshold frequency f_0 and the maximum current which can be generated. The energy gap also determines the maximum emf generated by the solar cell. It thus follows that the energy gap would largely control the maximum efficiency that can be obtained using a given absorbing material. The maximum conversion efficiency is found to occur around the energy gap of 1.4 eV for GaAs material.

9.4.2 *Effect of Absorber*

The degree of absorption of solar flux by the absorber depends on the thickness of specimen and the absorption coefficient α of the material. The minimum thickness required to absorb almost all the useful photon flux having energy greater than E_g depends on the nature of the material. The absorption is higher for direct band materials than for the indirect band materials because of reduced transition. Silicon being an indirect band semiconductor requires a minimum thickness of 300 μm from the consideration of optical absorption. It has been observed that the absorbing material for high efficiency must have high values of α over most of the region of high solar spectral density.

It is also observed that the direct band GaAs has higher values of α as compared to Si. Amorphous Si has a higher value of α as compared to the crystalline state. This occurs on account of the presence of large density of states in the energy gap in amorphous state which is of the order of 10^{24}–10^{26} m^{-3} eV^{-1}. Recently, hydrogenated amorphous silicon has been found to possess high absorption coefficient ($\approx 5 \times 10^6$ m^{-1}) with low density of states in the gap ($\approx 10^{21}$ m^{-3} eV^{-1}). This material can also be doped and made p- or n-type amorphous semiconductor.

Anti-reflection coatings are also used to reduce reflection loss at the top surface of the cell.

9.4.3 Effect of Diffusion Length

Third factor which controls the efficiency of solar cells, the diffusion length of minority carriers. After the electron-hole pairs are produced by absorption of light quanta from the solar flux, the carriers have to be diffused to the junction to get separated. Therefore the diffusion length has to be greater than the thickness of absorption layer. Since this thickness is already a fixed value from the consideration of maximum absorption, the diffusion length has to optimise by controlling the physical parameters of the material. The diffusion length L_{diff} is proportional to $\sqrt{\mu\tau}$ where μ is mobility and τ is lifetime of the charge carrier. The value of τ depends on the cross-section for recombination and the density of defects, μ depends on temperature, density and types of defects. All these factors have to be taken into account while choosing a suitable material for the solar cell.

9.5 Solar Cell Fabrication and Materials

We consider a p-n junction solar cell. This has three main parts viz. the emitter/absorber, the junction and the collector. These are shown in Fig. 9.5. The emitter/absorber region is that in which the light is absorbed and electron-hole pairs are created. In a p-type semiconductor, the minority carriers are electrons as shown in Fig. 9.5.

The fabrication of a solar cell necessarily produces surface states at the top layer of the cell. These surface states reduce the lifetime of the electron-hole pairs. Consequently in this region, the recombination takes place even before collection can occur. This is called a dead layer. In a material like GaAs, the presence of a dead layer of 1 μm thickness can reduce the available power by more than 90 %. From this consideration, Si is the best choice.

Some of the best solar cell materials with efficiencies lying between 7 and 10 % are the following.

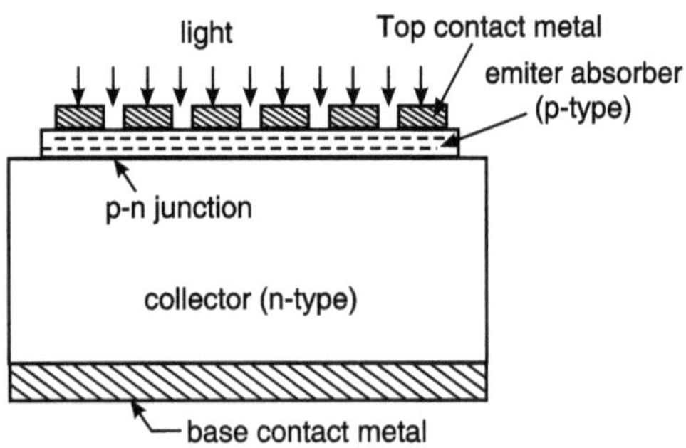

Fig. 9.5 Basic construction of a solar cell. Reproduced with permission from [1]

Absorber material	***Collector material***
Si	Si
Cu_2S	CdS
CdTe	CdS
Cu_2S	ZnCdS
$CuInSe_2$	CdS
GaAs	AlGaAs

Whereas the Si cells lose efficiency at high temperatures, the GaAs compound can be used at 100 °C or more. The GaAs-AlGaAs heterojunction cell can operate at elevated temperatures, which is a common requirement of solar concentration systems.

9.5.1 Advantages and Limitations of Solar Cells

A solar cell possesses several advantages which are given below.

1. It is a self-generating device; therefore it does not require any external power source such as a battery or electric motor etc.
2. The internal e.m.f. and the current generated by it are large enough to be measured on a galvanometer.
3. These cells can supply power to a space satellite for a long period of time. This is a distinct advantage over the acid or alkaline batteries.
4. It is pollution-free energy conversion system.

However, the solar cells have certain limitations also. These are

(i) It does not convert all solar radiation into electric energy.
(ii) Efficiency of a well-made Si cells is about 10 % only under full illumination.
(iii) The efficiency of Si cells lowers down at high temperatures.
(iv) Its use in terrestrial applications is less effective than outer space applications because a considerable amount of solar radiation is absorbed by the atmosphere.

9.5.2 Applications of Solar Cells

Solar cells are used in a wide range of outer space and terrestrial applications. Besides them, they are used in a variety of applications. The main among them are the following.

- In direct-reading illumination meters.
- In portable exposure meters.

- In low-resistance relays for 'ON-OFF' operations.
- In various industrial monitoring operations.
- In space satellites and deep space vehicles.
- For supplying electrical power to various electronic equipment aboard a satellite.

9.5.3 Importance of Fill Factor in Design of Solar Cell

Figure 9.6 shows the I–V characteristics of an illuminated solar cell. The open-circuit voltage Voc and short-circuit current I_{sc} for a given light level is plotted therein. Theoretically a solar cell will deliver maximum power to a load when the product VI_r is a maximum. Practically, the maximum delivered power by a solar cell for a given light level is equal to the hatched area as shown in Fig. 9.6. This area is less than the product of $V_{oc}I_{sc}$. Thus we define the fill factor as

$$fill\,factor = \frac{V_m I_m}{V_{oc} I_{sc}} \tag{9.2}$$

Its value is less than 1. This is an important parameter for solar cell design, as it is a 'figure of merit' for the cell.

9.6 Photodetectors

Photodetectors are those optoelectronic semiconductor devices which are used to detect optical signals, measuring illumination levels, and converting time-varying optical signals into electrical signals.

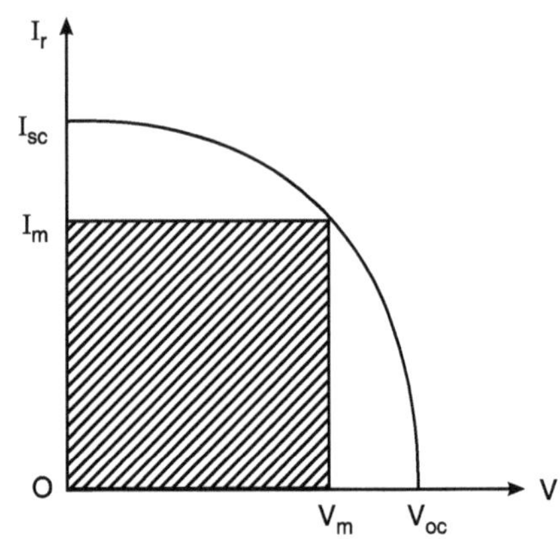

Fig. 9.6 I–V characteristics of an illuminated solar cell

9.6.1 Working Principle

The principle of detection in an intrinsic absorber is illustrated in Fig. 9.7. The p-n junction is reverse biased and the electric field developed across the junction sweeps the mobile carriers (holes and electrons) to their respective majority sides. A depletion layer is created on either side of the junction. This barrier stops the majority carriers to cross the junction in opposite direction of the field. The field accelerates minority carriers from both sides to opposite side of the junction, forming reverse leakage current of the diode. Thus intrinsic conditions are created in the depletion layer.

A photon incident in or near the depletion region of this device, which has an energy $\geq E_g$ of the fabricating material (i.e. hf $\geq E_g$), excites an electron-hole (carrier) pair as illustrated in Fig. 9.7a. Pairs so generated near the junction are separated and swept (drift) under the influence of electric field. It produces a displacement of current in the external circuit in excess of the reverse leakage current as shown in Fig. 9.7b. Figure 9.7c depicts generation and separation of e-h pair in the depletion layer of this reverse biased p-n junction.

The depletion layer should be sufficiently thick to allow a large fraction of incident light to be absorbed so that the maximum number of e-h pairs may be generated. Since the long carrier drift time in depletion layer restricts the speed of operation of the photodiode, it is necessary to limit the width. Hence, a trade off exists between the number of photons absorbed and the speed of response.

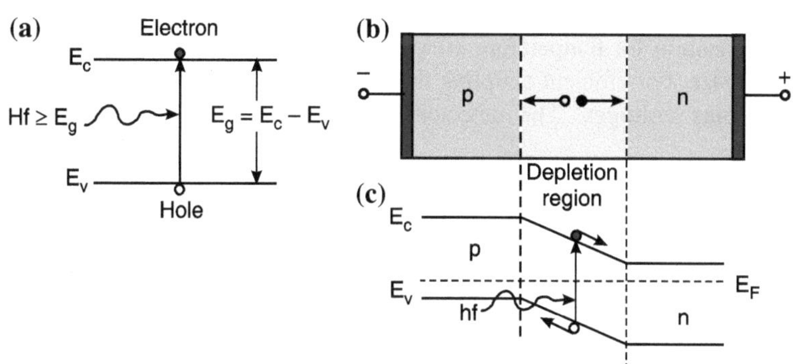

Fig. 9.7 Operation of a p-n photodiode showing **a** generation of an e-h pair, **b** illustrating e-h drift in reverse biased p-n junction, and **c** energy band diagram of reverse biased p-n junction showing separation and drift of e-h pair

9.6.2 Requirements of a Good Photodetector

A good and efficient photodetector should possess the following performance and compatibility requirements.

1. **High sensitivity at the operating wavelengths**. The first generation systems have wavelengths between 0.8 and 0.9 μm Considerable advantage is gained at the detector from second generation sources with operating wavelength above 1.1 μm, as the fibre attenuation and dispersion are reduced. Research activities are going on especially at wavelength around 1.3 μm where material dispersion and attenuation can be minimized.
2. **High fidelity**. To reproduce the received signal waveform with fidelity for analog transmission, the response of detector must be linear with optical signal over a wide range.
3. **Large electrical response to the received optical signal**. The optical detector should produce a maximum electrical signal for a given amount of optical power i.e. its quantum efficiency should be large.
4. **Short response time to obtain suitable bandwidth**. The systems should have a large operating bandwidth. Present systems operate at hundreds of MHz. Future system will operate at several GHz using signal mode fibre.
5. **Minimum noise introduced by the detector**. Dark current, leakage current and shunt conductance must be low. Moreover, the gain mechanism associated with the detector should have low noise.
6. **Stability of performance characteristics**. Ideally the performance characteristics should be independent of changes in ambient conditions. However, the present detectors have characteristics that vary with temperature; therefore the compensation for temperature effects is often necessary.
7. **Small size**. For efficient coupling the physical size of detector should be small.
8. **Low bias voltages**. The detector should operate on low bias voltages or currents.
9. **High reliability**. The detector should be capable of continuous stable operation at room temperature for many years.
10. **Low cost**. Economic conditions are often of prime importance in any large scale communication system application.

9.6.3 Method to Achieve a Fast Responding Speed

The most important requirement of a photodetector is its speed of response. It means that the photogenerated minority carriers must diffuse to the junction, and be swept across to other side in a time much less than the time required to respond to a series of light pulses. In a typical case the speed of response may be 1 nano second (1 ns) or less than that. Since the carrier diffusion process is time consuming, the above objective is fulfilled by providing a large width of depletion region w. Consequently, most of the photons are absorbed within w instead of p and n regions. Now when an electron-hole pair is created in depletion region, the electric field sweeps the hole to p-side and electron to n-side. Its effect is to cause carrier drift in a very short time; therefore the response of photodetector is also very fast.

9.7 Different Types of Photodetectors

Based on the width of depletion layer and method to control it, the photodetectors can be classified into different kinds as given below.

1. Depletion layer p-n photodetector
2. p-i-n photodetector
3. Avalanche photodetector
4. Heterojunction photo-transistors photodetector
5. Photoconductive photodetector

These photodetectors are intrinsic as well as extrinsic in nature.

In a **depletion layer p-n photodetector**, the carriers are generated within the depletion layer w. The width of w has to be of an appropriate value, which is neither too wide nor too narrow. The **p-i-n photodetector** offers a convenient method of controlling the width of depletion region. When put in reverse-bias, most of photogenerated carriers in this device are collected by p and n regions. The **avalanche photodetector** is used to detect low-level optical signals. Hence it is operated in avalanche region of photodiode's characteristic. The **intrinsic photodetectors** are those which are sensitive to photons whose energies are close to the energies of band gap. The **extrinsic photodetectors** are those which are sensitive to longer wavelengths. In them the photons can excite electrons into or out of the impurity levels.

9.7.1 Construction of a p-i-n Photodetector

Schematic arrangement of a p-i-n photodetector is shown in Fig. 9.8. Besides p and n regions, it has an i-region (intrinsic region) also in between. The i-region has to be made intrinsic if the resistivity is low, but need not be truly intrinsic if the resistivity is high. In a typical arrangement, it can be grown epitaxially on a n-type substrate, whereas p-region may be formed by implantation.

When this photodetector is subjected to reverse biasing, the applied voltage appears across the i-region. If in comparison to drift time the carrier lifetime within i-region is long, then most of photogenerated carriers are collected by n and p-regions.

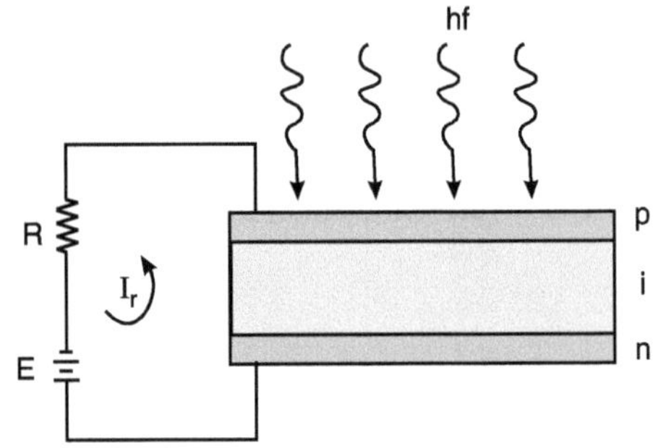

Fig. 9.8 Schematic representation of a p-i-n photodiode

9.7.2 Construction of a Silicon Heterointerface Photodetector (SHIP)

In optical communication systems, the response time and sensitivity of a photodetector are of critical importance. But optimization of these properties is difficult because of the following reasons.

(i) the gain depends on the ratio of carrier lifetime to transit time, but
(ii) the frequency and bandwidth varies inversely with carrier lifetime.

Therefore a compromise has to be made between the above two desirable characteristics. Hence, the detectors are designed with following considerations of merits.

1. Gain-bandwidth product, and
2. Signal-to-noise ratio.

Various photodetectors briefly described above have the disadvantages of one kind or the other. For example, a p-i-n photodetector suffers with the limitations of not having the gain mechanism, although the noise is considerably lower. Although the avalanche photodetector are advantageous in providing gain, but they have the disadvantage of increased noise as compared to p-i-n photodetectors. These deficiencies can be minimized by incorporating alternate materials and improved fabrication techniques etc. One such design, which overcomes this problem, is by making a Si heterointerface photodetector (SHIP). Schematic cross-section of a typical SHIP is shown in Fig. 9.9.

9.7.3 Photoconductive Detector

It is based on the principle that resistivity of semiconductor materials like selenium (Se), cadmium sulphide (CdS), lead sulphide (PbS) and thalmium-sulphide (TlS) etc. is decreased when illuminated. It means that such materials have 'dark' high resistance and illuminated low resistance.

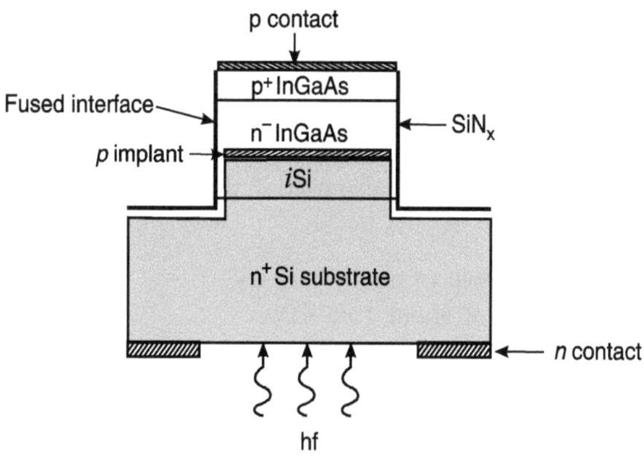

Fig. 9.9 Cross-section of a silicon heterointerface photodetector (SHIP). The light passes through Si, but is absorbed by InGaAs

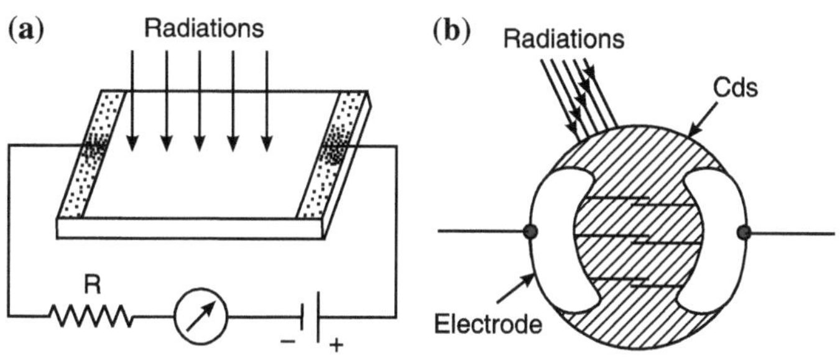

Fig. 9.10 **a**, **b** Schematic arrangement of a photoconductive detector

Construction. The simplest form of such a cell (detector) using selenium is shown in Fig. 9.10a, b. It consists of the semiconductor material with two electrodes attached to it. The resistance of the cell when unilluminated (dark) is always high so that the current through the circuit shown in Fig. 9.10a is low. When cell is illuminated, its resistance is decreased and circuit current become large. The shape of the semiconductor material is so made as to obtain a large ratio of 'dark to light' resistance.

A commonly used CdS cell is shown in Fig. 9.9b. It has a very high 'dark to light ratio', and gives maximum response at 5000 A.U. As shown, the two electrodes are extended in an inter-digital pattern in order to increase the contact area with the sensitive material.

Uses. Photoconductive cells employing TlS and PbS have been used for detecting the ships and aircrafts by means of the radiations given out by their exhausts or funnels. They have also been used for telephony by modulated infra-red light.

9.7.4 *Choice of Materials for Photodetectors*

Since basic element in formation of a photodetector is a photodiode, most of the materials used in making of photodiodes are also used in making the photodetectors. In addition, other materials are also used in their formation. In this regard, a list of materials is given below.

- InGaAs having a band gap of about 0.75 eV.
- InP having a band gap of about 1.35 eV.
- InAlAs.

9.8 Light Emitting Diodes

Light-emitting diodes are the devices that act as the generator of light. They are the light source for fibre optic systems. Their working is based on the effect called as 'injection electroluminescence'. According to this effect, the semiconductors characterized by direct recombination give-off light from their junctions under forward-bias. It is in contrast with the semiconductors having indirect band gaps (e.g. Si or Ge), in which only heat is released on recombination and not the light.

9.8.1 *Construction and Working of LED*

It is a p-n junction which when forward biased, emits light. Charge carrier recombinations occur at the junction as the electrons cross from n-side and recombine with holes on p-side. When this recombination takes places, charge carriers give up energy in the form of light and heat. If the semiconductor material used is translucent, the light is emitted and the junction acts as source of light.

Cross-sectional view of a typical LED is shown in Fig. 9.11. In it, the recombination takes place in p type material. Thus p-region becomes surface of the device. For maximum light emission, a metal film anode is deposited around the

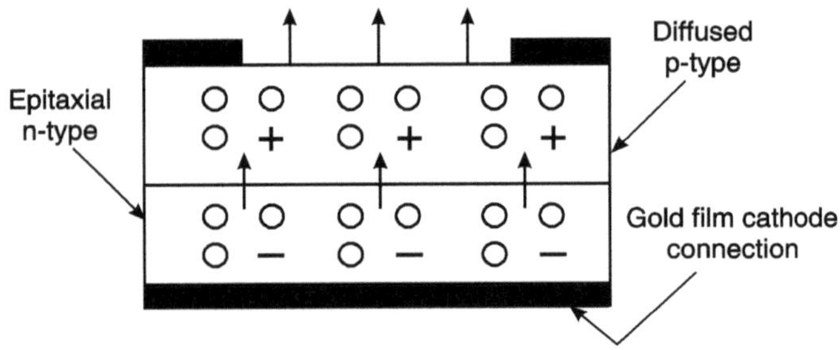

Fig. 9.11 Cross-section of a typical LED. Reproduced with permission from [1]

edge of p type material. The cathode connection is usually a gold film at the bottom of n-type region. This helps in reflecting the light on surface. Most common p-type semiconductor materials used for manufacture of LED are gallium arsenide and phosphide, which emit red or yellow light, and gallium arsenide which emits green or red light emission.

9.8.2 Construction of a LED for Fibre-Optic System

A light-emitting diode is a light source for fibre optic system and others. It is a multilayer construction of GaAs and AlGaAs as shown in Fig. 9.12. As shown, the nGaAs layer is on the top and p^{+}GaAs on the bottom of its cross-section. In it the optical fibre is held in an etched well by means of epoxy resin. It is on the backside of diode.

Since in a LED, the photons are emitted from the junction randomly in all directions, which are not in-phase with each other, therefore LED generated signals can be transmitted in several modes. Generally the fibres in single-mode are of about 5 μm diameter and of about 25 μm diameter in multi-mode.

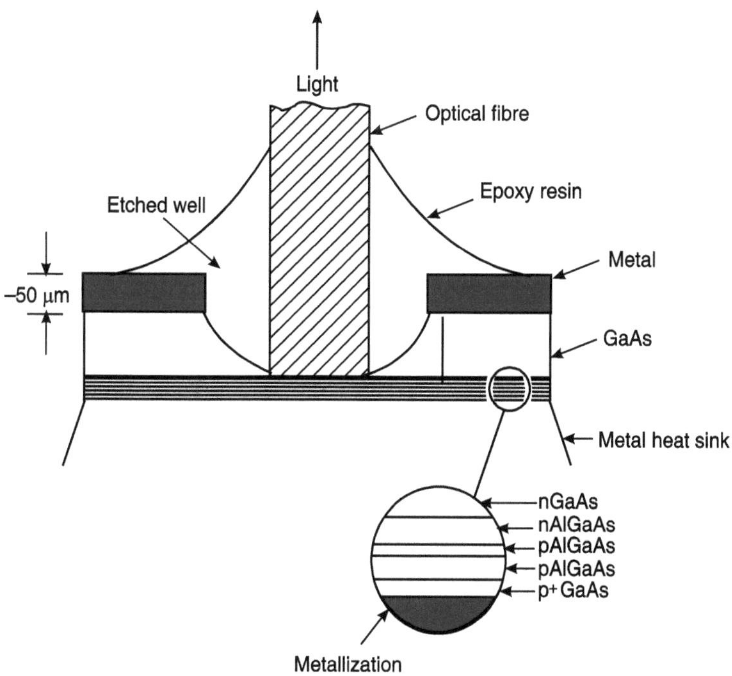

Fig. 9.12 Cross-section of a GaAs-AlGaAs LED for fibre-optic applications

The optical power emitted by an LED varies linearly with the input current over a wide range. The digital information transmitted by them can be easily modulated by analog signals. When the information is to be transmitted over a long distance, losses occur in the fibres. Therefore, many LEDs and photodetectors are required along the path. These are housed into various repeater stations at periodic intervals along the distance.

9.8.3 Advantages, Applications and Specifications of LEDs

The merits of LEDs over conventional incandescent and other types of lamps are the following.

1. Small size and weight.
2. No warm-up time.
3. Low working current and voltage
4. Less power consumption
5. Very fast switching
6. Immune to mechanical vibration
7. Long life
8. Less fragile

Applications of LEDs. LEDs are used in following main applications.

- As display devices
- In optical communication systems
- Several LEDs are added in series to form segmental and dot matrix displays.

Specifications of LED. The main specifications of a LED are as follows.

(i) Colour
(ii) Wavelength at peak emission
(iii) Forward voltage
(iv) Forward current
(v) Power dissipation
(vi) Luminous intensity
(vii) Maximum reverse current

9.8.4 Light Emitting Materials

A large variety of light emitting semiconductors are known to exist. They include n-type, p-type, III–V and II–VI group materials, ternary and quaternary compounds and alloys. They are available in a wide range of photon energy emission extending from ultraviolet to infrared and visible range of spectrum. A detailed list of them is given below.

n-type binary compounds

- ZnSe
- ZnS
- CdSe
- CdS

p-type binary compounds

- ZnSe
- ZnTe

Infrared emitter compounds and alloys

- Ga As
- InP
- GaN (Eg = 3.4 eV) for blue wavelength
- InSb (Eg = 0.18 eV)
- Mixed alloys of GaAs and InP

Ternary compounds

- AlGaAs
- AlGaP system for yellow and green wavelengths
- Gallium arsenide-phosphide ($GaAs_{1-x}P_x$) alloy

Quaternary compounds

- InGaAsP system
- AlGaAsSb system

Multilayer heterojunction systems

- (Zn, Cd) system
- (S, Se)

For a good light-emitting material its band gap energy should be high. This band gap in compound/ alloy semiconductors can be increased sufficiently by using ternary and quaternary compounds. For example, the $In_xGa_{1-x}As_yP_{1-y}$ quaternary alloy when grown on InP substrate, will result in band gaps ranging from 0.75 to 1.35 eV.

9.8.5 A New Generation LED: Gallium Nitride Based Light Emitting Diodes [2]

Light-emitting diodes (LEDs) fabricated from gallium nitride (GaN) have led to the realization of high-efficiency white solid-state lighting. Currently, the GaN white LEDs exhibit luminous efficacy greater than 150 lm W^{-1}, and external quantum efficiencies higher than 60 %. This has enabled LEDs to compete with traditional lighting technologies, such as incandescent and compact fluorescent (CFL) lighting. It is now possible to actually generate the white light using GaN LEDs. The three most popular and well-established approaches to using LEDs to generate white light are shown in Fig. 9.13.

They are blue LED with yellow phosphors; a UV LED with blue and yellow phosphors (or red, green and blue phosphors); and a device that combines red, green and blue LEDs. Currently, the blue GaN LED plus yellow phosphor dominates the white LED industry. The advantage of using a blue LED and yellow phosphor is the ease of manufacture and its high theoretical efficacy; which is attractive for the creation of a cheap, bright white light source. However, this benefit comes at the expense of a lower colour rendering index value which is typically low, and makes such devices undesirable for indoor use. Typically an additional red or green phosphor is then added to the phosphor mix to get broader spectral emission and hence higher colour rendering index (CRI) values. Violet and UV LEDs with tri-colour phosphor mixtures provide better CRI values and are suitable for indoor applications but at the expense of poorer efficiency.

If one desires to dynamically control the white light, the third approach of a combination of three (or more) LEDs of different wavelengths is attractive. It will enable higher efficacies than the UV phosphor LEDs, but will generally be the most

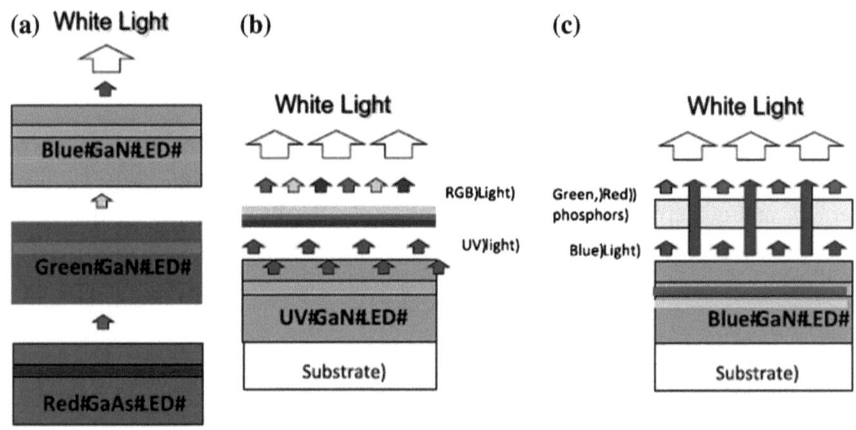

Fig. 9.13 Three ways of generating *white* light from GaN LEDs: **a** combination of *red*, *green* and *blue* LEDs, **b** UV GaN LED plus three phosphors, and **c** *blue* GaN LED plus yellow phosphor. (Reproduced from [2])

expensive option of the three. The first two white LED approaches use phosphors for creating broad-band white light by using a material which absorbs light of one wavelength and emits at longer wavelength.

Phosphors are commonly used for this task and a selected few have received considerable attention, such as rare-earth doped yttrium aluminum garnets (YAG: RE, $Y_3Al_5O_{12}$). For example, cerium-doped YAG can absorb blue and UV light and emit yellow light relatively efficiently. Crucial to this process is the fact that the higher energy light (e.g. UV or blue) is converted to lower energy (e.g. yellow or red). Therefore, LEDs emitting red light, the lowest energy colour in the visible spectrum, are not feasible for white light generation using phosphors; instead, a short-wavelength UV, violet or blue LED is required. In addition, yellow LEDs made from InGaN are still not as efficient as blue InGaN LEDs pumping yellow phosphors.

9.9 Semiconductor Lasers

The word LASER is an abbreviation of "light amplification by stimulated emission of radiation". It is a quantum electronic device that generates intense electromagnetic radiation. It is a source of highly directional, monochromatic, coherent light. This light can be a continuous beam of low or medium power, or a very short burst of intense light. The short burst may deliver a power worth millions of watt. The laser is of great interest in optics and optoelectronics, particularly in fibre-optic communications.

9.9.1 Classification of Lasers

The lasers can be classified in a number of ways, as given below.

I. On the basis of state of the laser medium

- Solid lasers: such as ruby laser, Nd YAG (neodymium doped with yttrium aluminium garnet) laser, semiconductor (e.g. GaAs) laser
- Liquid lasers
- Gaseous lasers: such as CO_2 laser. He-Ne laser

II. On the basis of manner of pumping

- Flash light lasers
- Chemical action lasers
- Electric discharge lasers

III. On the basis of nature of output

- Pulsed wave lasers
- Continuous (or direct) wave lasers

IV. On the basis of spectral region of light

- Ultraviolet lasers
- Visible lasers: such as GaAsP laser
- Infrared lasers: such as GaAs laser

9.9.2 Merits of Semiconductor Lasers

Although the ruby laser (chromium ions in alumina), CO_2 laser, He-Ne laser etc. are still used in many applications; but the semiconductor lasers are more efficient than these lasers. GaP, GaAs and GaSb are most significant compounds for this purpose. The reasons for superiority of semiconductors (or p-n junction) lasers over non-semiconductor lasers are the following.

1. They are remarkably miniature in size (typically of the order of 0.1 mm × 0.1 mm × 0.3 mm).
2. Their efficiency is high.
3. Their output can be easily modulated by controlling the junction current.
4. They operate at low power as compared to ruby or CO_2 lasers.
5. Their power output is comparable with He-Ne lasers.
6. They can provide a portable source of radiation.
7. They provide an easily controllable source of low-power coherent radiation.

The semiconductor lasers are particularly suitable for fibre-optic communication systems.

9.9.3 Characteristics and Working

In semiconductor lasers, the light emission is obtained through a p-n junction as a result of recombination of electrons and holes. An extremely narrow and perfectly straight junction line provides beneficial laser action. Miniature semiconductor lasers employ various methods for excitation such as injection of electrons and holes through the p-n junction, optical pumping, and a beam of fast electrons. A higher energy gap of semiconductor material results in emissions of shorter wavelengths as shown in Fig. 9.14.

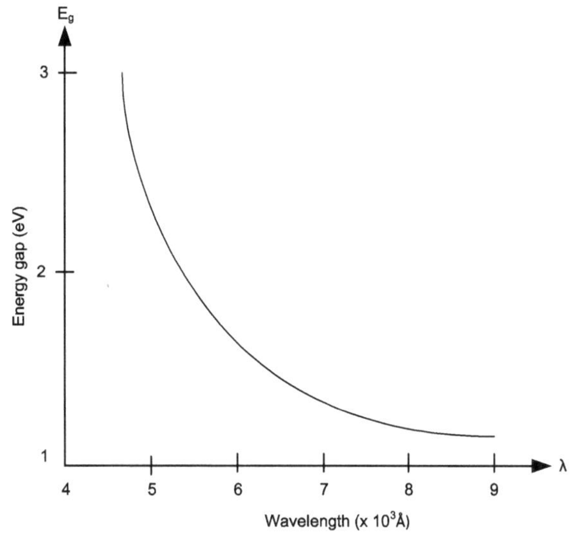

Fig. 9.14 Effect of forbidden gap on the emission of wavelength in a semiconductor laser. Reproduced with permission from [1]

9.9.4 *Properties of Laser Light*

Various properties of a laser light are the following.

1. **Directionality**. The lasers emit radiation in a highly directional, collimated beam with a low angle of divergence. This is important because the energy carried by the laser beam can be collected easily and focussed onto a small area.
2. **Line width**. The laser beam is extremely monochromatic, but the spectral content of its radiation may extend over a wide range of florescent line width. It means that the line width of an individual cavity mode may be extremely small.
3. **Beam coherence**. One of the characteristics of stimulated emission is that the stimulated wave is in-phase with the stimulating wave. It means that the spatial and temporal variation of the electric field of the two waves is the same. Thus for a perfect laser, we would expect the electric field to vary with time in an identical fashion for every point on the beam cross-section.
4. **Brightness**. The primary characteristic of laser radiation is that the lasers have a higher brightness than any other light source.
5. **Focusing**. The minimum spot size to which a laser beam can be focused is determined by diffraction. A single mode beam can be focused onto a spot which has dimensions of the order of the wavelength of light.
6. **Tuning**. Some lasers can be used to emit radiations over a range of wavelengths. Laser tunability leads to applications in photochemistry, high resolution and roman spectroscopy.

9.9.5 Laser Applications

Lasers are being put to a number of uses in different fields. A few examples of these applications are listed below.

1. **Communication**. Modulated laser beams are being used for transmitting messages. Due to high degree of coherence, the loss of transmitted energy is comparatively much less.
2. **Surgery**. Laser beam is being used extensively for bloodless surgery. e.g. lasers can be used
   - to weld the detached retinas
   - for drilling the teeth
   - for removal of tumors
   - for removal of infected cell
   - for preventing the tooth decay by depositing hard material on the surface of the tooth.
3. **Measurements of long distances**. It enables us to determine the distance of the moon from the earth with an error within 0.6 m.
4. **Nuclear fusion**. The laser beams can be used to induce the nuclear fusion. By concentrating the laser beam on to a very narrow spot, the temperature may rise to about 10^8 K and nuclear fusion can occur at this temperature.
5. **Scientific research**. A modified version of the Michelson Morley experiment was conducted to test for the ether drift. Here the beams of two infrared lasers of slightly different frequencies were combined by means of a beam splitter and the beam frequency was determined. No change in beat frequency was observed on rotating the apparatus through 90°. Thus the null result of Michelson Morely experiment could be confirmed.
6. **Industry**. Lasers can be focussed onto a very fine beam, which result in raising the temperature to about 1000 K. It can be used for drilling the holes and fusing or melting of metals.
7. **Weather forecasting**. Pictures of the clouds, wind movements etc. can be obtained with the help of laser beam and the data so obtained can be used in weather forecasting.
8. **National defence**. Lasers can be used for guiding the missiles and satellites. They can also be used for detecting enemy aeroplanes. The laser beams can also be used as weapons to destroy enemy tanks, aeroplanes etc.
9. **Other uses**. The other uses of laser are
   - in holography for the production of three-dimensional images.
   - in cutting of microelectronic circuits.
   - for cutting and sealing the nerves.
   - to raise the temperature of localized area of tissue i.e., physiotherapy.

Laser can also be used for the following purposes.

- Unblocking the heart valves.
- Breaking the kidney stones, etc.

9.9.6 Materials for Semiconductor Lasers

A good material for the purpose of semiconductor lasers must fulfil certain requirements. Main among these are the following.

1. It must be efficient light emitter.
2. It should be amenable to the formation of p-n junctions.
3. It should be amenable to the formation of heterojunction barriers.

Based on various desirable properties, the following materials are suitable for semiconductor lasers applications.

II–VI compounds. These are very efficient for the purpose of light emission, but formation of their junction is difficult. However using nitrogen (N) as acceptor, the junctions can be grown easily in the following materials by crystal growth techniques like MBE (molecular beam epitaxy) and MOVPE (metal-organic vapour-phase epitaxy).

- ZnS
- ZnSe
- ZnTe
- Alloys of these

Lasers made of these materials emit in the region of green and blue-green light spectrum.

Large bandgap semiconductors. Semiconductor compounds having a large band gaps (about 2 to 5 eV) offer good candidature as semiconductor lasers materials. Main among these are the following.

- InN (E_g = 2 eV)
- GaN (E_g = 3.4 eV)
- AlN (E_g = 5 eV)

This covers a wide range of light emission from blue to UV spectrum.

Ternary alloy system. These are well suited for the lasers used in fibre-optic communication systems. Main among these are the following.

- AlGaAs
- PbSnTe
- InGaSb

The PbSnTe is able to provide laser output in the wavelength range from 7 μm to above 30 μm at low temperatures. The InGaSb is suitable for use at intermediate wavelengths.

Quaternary alloy system. These are specifically versatile in fabricating the laser diodes by allowing flexibility in lattice matching. They also provide a choice of wide range of wavelength. Main examples of these materials are the following.

- InGaAsP

Short wavelength emitters. These blue/UV semiconductor lasers are very suitable for storage purposes such as compact discs (CDs), digital versatile discs (DVDs) etc. By reducing the laser wavelength i.e. by using short wavelength emitters, the storage density of these devices can be increased many times. An example of a recent semiconductor laser has used the under mentioned material.

- InGaN

This laser has multi-quantum-well heterostructures.

9.9.7 Homojunction Laser and Hetero-Junction Laser

The differences between these lasers are given below in the chart.

Description	Homojunction laser	Heterojunction laser
• Type of material	Single type	Multiple layers type
• Number of junctions	Only one	More than one
• Operation at room temperature	Less convenient	More convenient
• Population inversion build-up i.e. initiation of laser action	At high current levels	At lower current levels
• Efficiency	Less	More

9.10 Laser Diode

A laser diode (or LD) is an electrically pumped semiconductor laser in which the active medium is formed by a p-n junction of a semiconductor diode similar to that found in a light-emitting diode. Indium arsenide finds its use in manufacturing of diode lasers because of its unique electrical and optical properties. The laser diode is the most common type of laser. This finds wide use in

- Telecommunication as easily modulated and easily coupled light sources for fibre optics communication.
- Various measuring instruments such as range finders.
- In barcode readers.

- Visible lasers, typically red but green also, as laser pointers.
- Both low and high-power diodes in the printing industry as light sources for scanning (input) of images, and or very high-speed and high-resolution printing plate (output) manufacturing.
- Infrared and red laser diodes are common in CD players, CD-ROMs and DVD technology.
- Violet lasers are used in HD DVD technology.
- Diode lasers have also found many applications in laser absorption spectrometry (LAS) for high-speed, low-cost assessment or monitoring of the concentration of various species in gas phase.
- High-power laser diodes are used in industrial applications such as heat treating, cladding, seam welding and for pumping other lasers such as diode-pumped solid-state lasers.

9.11 Light Dependent Resistors (LDRs)

Light dependent resistors (LDRs) are made from cadmium sulphide containing no or very few free electrons, when not illuminated. Its resistance is then quite high. When it absorbs light, the electrons are liberated and the conductivity of material increases. Cadmium sulphide (CdS) is a photoconductor. The approximate relationship between the resistance and illumination is

$$R = A \cdot E - \alpha \tag{9.3}$$

where E is illumination in lux, R is resistance in ohms, A and α are constants. The value of α depends on the cadmium sulphide used and on the manufacturing process. Values around 0.7–0.9 are quite common.

The basic construction and symbol for LDR are shown in Fig. 9.15a, b respectively. The device consists of a pair of metal film contacts separated by a snake-like track of cadmium sulphide film, designed to provide the maximum possible contact area with the two metal films. The structure is housed in a clear plastic or resin case, to provide free access to external light. Practical LDRs are available in a variety of sizes and package styles, the most popular size having a face diameter roughly 10 mm.

LDRs are sensitive, inexpensive, and readily available devices. They have good power and voltage handling capabilities, similar to those of a conventional resistors. Their significant defect is that they are fairly low acting, taking tens or hundreds of milli-seconds to respond to sudden changes in light level. Useful practical applications of LDRs include light and dark-activated switches and alarms, light beam alarms and reflective smoke alarms etc.

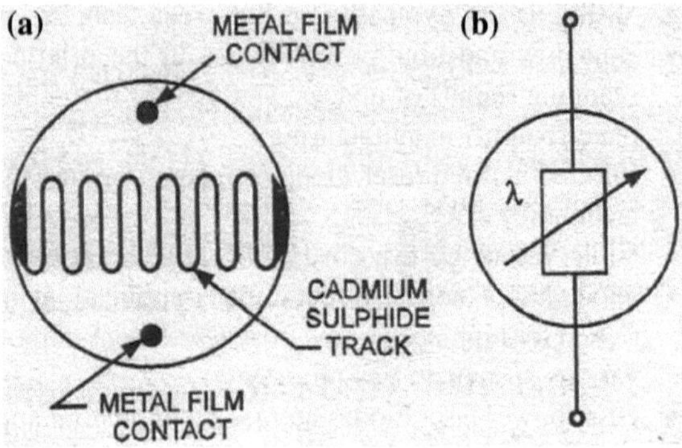

Fig. 9.15 A light dependent resistor **a** Basic structure. **b** Symbol. (Reproduced from [3])

9.12 Overlight Detector

An overlight detector is shown in Fig. 9.16. The LDR is a variable resistor whose resistance decreases with the increase in light intensity. When the light falling on an LDR has normal intensity, its resistance is large enough and the voltage across R1 is insufficient to trigger the SCR. However, when light falling on LDR is of large intensity, the resistance of LDR falls and voltage drop across R1 becomes large enough to trigger the SCR. Consequently, the buzzer sounds the alarm. It may be noted that if the strong light disappears, the buzzer continues to sound the alarm. It is because once the SCR is fired, the gate loses all control.

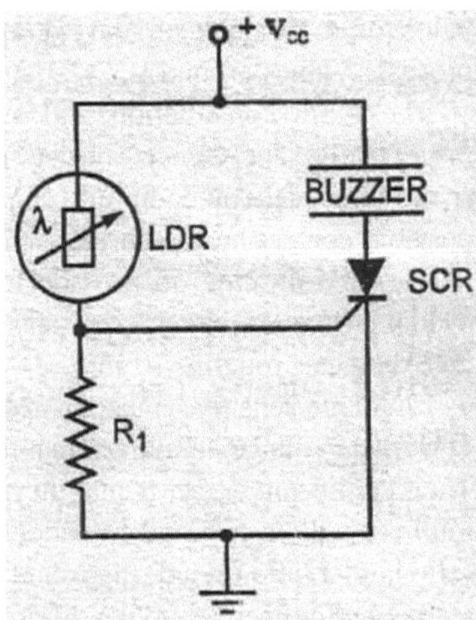

Fig. 9.16 Circuit of a typical overlight detector

9.13 Phototransistor

A phototransistor is similar to an ordinary BJT, except that no base terminal is provided. Instead of a base current, the input to the transistor is in the form of light. The current induced by photoelectric effects is the base current of the transistor.

Figure 9.17 shows the standard symbol of a phototransistor, which can be regarded as a conventional transistor housed in a case that enables its collector-base junction to be exposed to external light. The device is usually packaged in a metal can with a lens on top. Clear plastic encapsulated phototransistors are also available. A phototransistor with the terminal identification and the angular alignment is given in Fig. 9.18a, b, c.

The device is normally used with its base open, in either of the configuration shown in Fig. 9.19a, b. It functions as follows.

In Fig. 9.19a the base-collector junction of the transistor is effectively reverse biased and acts as a photodiode. The photo-generated currents of the base collector junction feed directly into the base of the device, and the nominal current-amplifying transistor action causes the output current to appear (β_{dc} times the base current I_λ), as the collector current and resistor R causes this current to provide an output voltage as illustrated. The resulting collector current, on an approximate basis, is

$$I_C = (I + \beta_{dc})I_\lambda \approx \beta_{dc} \times I_\lambda \qquad (9.4)$$

In practice, the collector and emitter currents of a transistor are virtually identical and, since base is open, the device is not subjected to significant negative feedback. Consequently, the alternative circuit shown in Fig. 9.19b, in which resistor R is connected to the emitter of the transistor, gives a virtually identical performance to that of circuit shown in Fig. 9.19a.

9.13.1 Differences between Phototransistor and Photodiode

The main difference between a phototransistor and a photodiode is the current gain hfe or β_{dc}. The same amount of light striking both devices produces β_{dc} times more

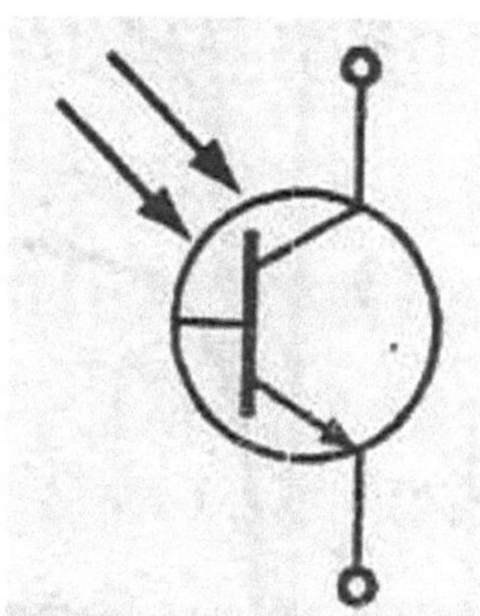

Fig. 9.17 Standard symbol of phototransistor

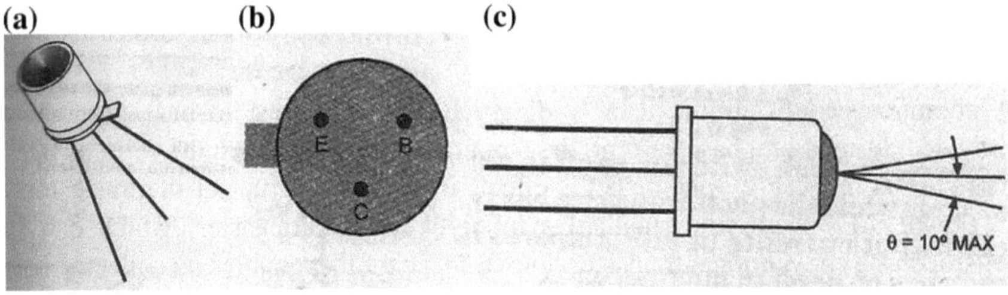

Fig. 9.18 Terminal identification and angular alignment of a typical phototransistor

current in a phototransistor than in a photodiode. The increased sensitivity of a phototransistor is a big advantage over a photodiode. However, photodiodes are the faster of the two. A photodiode has typical output currents in microamperes and can switch ON or OFF in nanoseconds, while the phototransistor has typical output currents in milliamperes but switches ON or OFF in microseconds.

A phototransistor can be converted into a photodiode by using only its base and collector terminals and ignoring the emitter, Fig. 9.19a. Alternatively, the sensitivity (and operating speed) can be controlled with a variable base return resistor, as shown in Fig. 9.19b. With R_B, the open-circuit phototransistor operation is obtained while with R_B short-circuited, photodiode operation is available.

Applications

Some of the areas of application for the phototransistor include

- Punch-card reader
- Computer logic circuitry
- Lighting control
- Level indication
- Relays, and
- Counting system, etc.

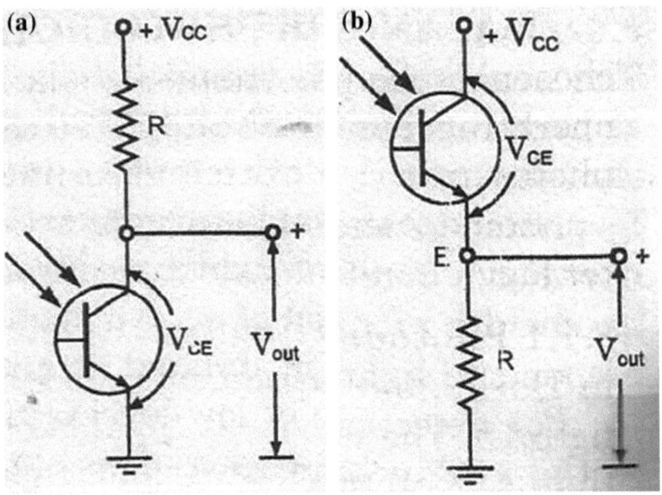

Fig. 9.19 Alternative phototransistor configurations

9.14 Solved Examples

Example 9.1 Describe stepwise the method of fabrication of a p-n junction laser, briefly.

Solution. A p-n junction laser is fabricated as follows.

(i) The junctions are formed with a highly doped, direct semiconductor. The GaAs compound is an appropriate choice for it.
(ii) The junctions are isolated by etching or cutting.
(iii) Individual junction is then cleaved (or cut) into devices to make the front and back faces flat and parallel.
(iv) Each individual junction is now mounted over the header, and lead is drawn out from it. Finally, the adequate heat sinking is provided to the device for large forward current levels.

Example 9.2 Which material is most suitable for producing blue LEDs? Explain why? Write its applications also.

Solution. Blue LEDs are short wavelength emitters. GaN is most suitable material for it, as it gives a very high efficiency. Blue LEDs are used in outdoor displays and TV screens. They are combined with LEDs of other colours also to form very intense white light sources. Luminous efficiency of such light source exceeds over the conventional light bulbs.

Example 9.3 What is the major source of noise in photoconductors?

Solution. The major source of noise in photoconductors is random fluctuations in the dark current. This is termed as noise current, and is also known as Johnson noise. This current increases with increase in temperature and conductance of material, in the dark. Therefore, the noise in a photoconductor at a given temperature can be reduced by increasing the dark resistance. It also results in increase in the 'gain of photoconductor' and decrease in bandwidth.

Example 9.4 What is meant by optoelectronic pair? What is its importance?

Solution. An optoelectronic pair means a pair of a light emitter and a photodiode. It provides electrical isolation between input and output. In an optoelectronic isolator, both these devices are mounted on a ceramic substrate and packed together to form a compact unit. This unit, while maintaining isolation, passes the information.

Example 9.5 A silica optical fibre has refractive index of its core as 1.50 and that of cladding as 1.47. If the core diameter is large enough, determine the critical angle at the core-cladding interface.

Solution. The critical angle at the core-cladding interface is given by μ cladding

$$\sin \phi_c = \frac{\mu_{cladding}}{\mu_{core}}$$

$$\phi_c = \sin^{-1}\left(\frac{1.47}{1.50}\right) = \mathbf{78.5}°$$

Example 9.6 Define responsivity of a photodiode and explain its dependence.

Solution. Responsivity R of a photodiode may be defined as the ratio of its output photocurrent to incident optical power. Thus

$$R = \frac{output\ photocurrent}{incident\ opticalpower} = \frac{I_{opc}}{P_{iop}}\ \text{A/W} \tag{9.5a}$$

It is a performance parameter and illustrates the transfer characteristics.

The dependency of responsivity of photodiode can be expressed as

$$R = \frac{e\lambda\eta_{quantum}}{hc} \tag{9.5b}$$

where e is charge on an electron, λ is incident wavelength on photodiode, η_{quantum} is quantum efficiency, h is Planck constant, and c is velocity of light.

Example 9.7 Photon of energy 1.5×10^{-19} J is incident on a photodiode whose quantum efficiency is 60 %. Determine (a) the wavelength at which the photodiode is operating, (b) responsivity of photodiode, and (c) incident optical power to obtain a photon current of 3 μA.

Solution. Given are η quantum = 60 % = 0.6, I_{opc} = 3 μA = 3×10^{-6} A, E = 1.5×10^{-19} J

(a)

$$\because \quad \lambda = \frac{hc}{E}$$

$$\therefore = \frac{(6.626 \times 10^{-34}) \times (3 \times 10^{8})}{1.5 \times 10^{-19}} = \mathbf{1.32\ \mu m}$$

(b) Using Eq. 9.5b,

$$R = \frac{(1.6 \times 10^{-19}) \times (1.32 \times 10^{-6}) \times (0.6)}{(6.626 \times 10^{-34}) \times (3 \times 10^{8})}$$
$$= \mathbf{0.64\ AW^{-1}}$$

(c) Using Eq. 9.5a,

$$P_{iop} = \frac{I_{opc}}{R}$$
$$= \frac{3 \times 10^{-6}}{0.64} = \mathbf{4.67\ \mu W}$$

Example 9.8 What do you mean by absorption and emission of radiation? How does spontaneous emission differ from stimulated emission?

Solution. The interaction of light with matter takes place in discrete energy packets called quanta or photons. When photon is absorbed, the electron from lower energy state E_1 is excited to higher energy state E_2 as shown in Fig. 9.20a. The energy of photon hf for the process is given by

$$\Delta E = E_2 - E_1 = hf$$

where f is frequency of photon and h is Plank's constant. When the electron makes a transition from E_2 to E_1, emission of light occurs. Emission is of two types.

1. Spontaneous emission in which the atom returns to lower energy state in an entirely random manner, Fig. 9.20b.
2. Stimulated emission when a photon having energy equal to the energy difference between two states ($E_2 - E_1$), interacts with the atom in the higher energy state E_2 causing it to return to the lower energy state E_1 with creation of another photon, Fig. 9.20c. Both the photons happen to be in phase and have the same polarization. Hence the amplitudes are added up i.e. the energy is increased, which is called light amplification.

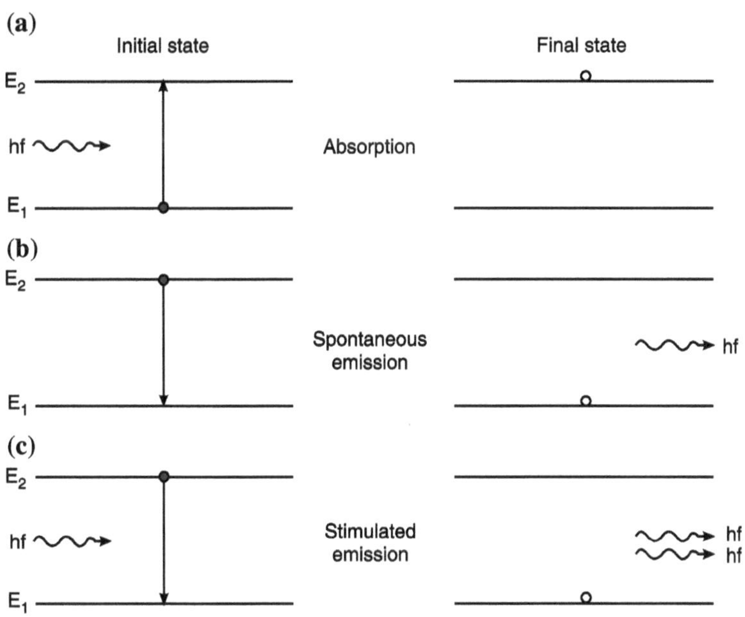

Fig. 9.20 Illustration of **a** absorption, **b** spontaneous emission, and **c** stimulated emission

Example 9.9 What is population inversion? What is its importance in respect of lasers?

Solution. In thermal equilibrium, the lower levels contain more atoms than the higher levels. To achieve optical amplification, it is necessary to create a non-equilibrium distribution of atoms such that the population of higher energy levels is more than that of the lower energy levels (i.e. $N_2 > N_1$). Creation of this condition is termed as population inversion. This can be achieved by exciting the atoms from a lower energy state to a higher energy state using an external energy source. This source is known as pump and the process is called pumping. Use of intense radiation (optical pumping) and electrical excitation are the common means of pumping (excitation).

In respect of lasers, the population inversion may be obtained with three or four energy levels as shown in Fig. 9.21a, b. The central level is a metastable state where the atoms spend unusually long time $\approx 10^{-3}$ s. It is from the metastable level that the stimulated emission takes place. In Fig. 9.21a, E_2 is a normal state and E_1 is metastable state. The pumping forces the atoms to go to excited state E_2. The electrons from E_2 rapidly decay to E_1, or directly to E_0 by non-radioactive process. The metastable state exhibits much longer lifetime. The transition from E_1 to E_0 causes lasing action in 3-level system. The drawback in a 3-level lasing system is that it generally requires high pumping rate and terminal state of laser transition (i.e. ground state). Therefore, more than half the ground state atoms have to be pumped into metastable state to achieve population inversion.

Example 9.10 What is ruby crystal?

Solution. Ruby crystal $Al_2O_3 \cdot Cr_2O_2$ is a crystal of aluminium oxide in which some aluminium atoms are replaced by chromium atoms. Cr^{+2} ions are active material in ruby rod which is an essential part in ruby laser. A ruby crystal containing about 0.05 % chromium is of pink colour.

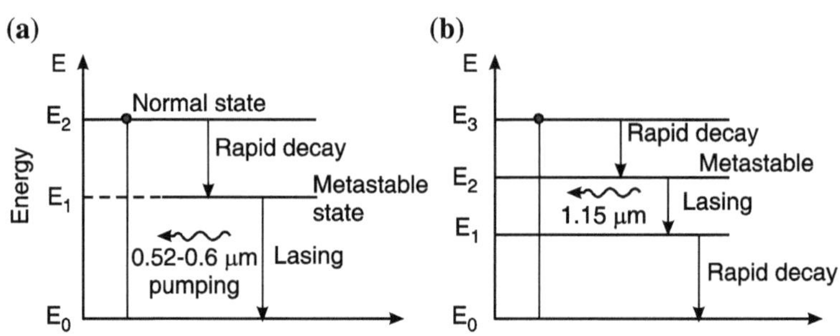

Fig. 9.21 Energy level diagram showing population inversion and lasing action for a **a** 3-level laser, and **b** 4-level laser

Review Questions

1. Define the following.
   (a) Electro-optic effect (b) Photoelectric effect
   (c) Photoemissivity (d) Translucent materials
   (e) Efficiency of solar cell (f) Birefringence
2. Enumerate the various optoelectronic semiconductor devices and give a brief account of each of them.
3. Explain the current and voltage characteristics of an illuminated junction. State the conditions when the power is delivered (a) from the external circuit to the device, and (b) from the junction to the external circuit.
4. What is solar cell? What is its working principle? Describe the construction and working of a solar cell.
5. Discuss the factors that affect the efficiency of a solar cell. Suggest suitable materials for them.
6. Explain the advantages and limitations of a solar cell. Write their various applications.
7. What is meant by a photodetector? What is its working principle? Discuss the requirements of a good photodetector.
8. Enumerate different types of photodetectors. How do they differ from each other?
9. Describe the construction and working of a p-i-n photodetector. Write its characteristics and applications also.
10. Describe the construction and working of a silicon heterointerface photodetector. Write its characteristics and applications also.
11. What is light-emitting diode? What are its applications? Describe the construction and working of any LED.
12. Discuss the advantages and specification of LED. Enlist the various materials used in their construction.
13. What do you mean by laser? What are the merits of semiconductor la sers on conventional lasers?
14. What are different types of semiconductor lasers? Discuss their characteristics and working.
15. Elaborate the properties and applications of lasers.
16. Suggest suitable materials for semiconductor lasers.
17. Distinguish between the following.
    (a) Coherent and incoherent devices
    (b) Fibre-optics and optical fibre
    (c) Solar cell and solar battery
    (d) Homojunction laser and heterojunction laser

18. Write notes on the following.
    (a) Photoconductive detector
    (b) Fill factor in solar cell
    (c) Blue LED
    (d) Chromatic dispersion
19. Explain the phenomena of absorption, spontaneous and stimulated emission.
20. Sketch and explain the construction and working of a phototransistor. How does it differ from photodiode? Write its application also.
21. What is meant by population inversion? What is its significance in respect of lasers?

Objective Questions

1. Match list I with list II, and choose the correct answer from the codes given below the lists.

List I	List II
A. LCDs	1. Numeric display in calculators
B. LEDs	2. Fibre-optic communication systems
C. Semiconductor lasers	3. Optical switches
D. Photodiodes	4. InAlAs
	5. Watches

Codes

	A	B	C	D		A	B	C	D
(a)	1	2	3	5	(b)	1	5	2	3
(c)	2	3	4	1	(d)	3	5	4	2

2. Consider the following statements.
   A. Birefringence occurs when the index of refraction is a function of crystal orientation.
   B. A good photdetector should possess short response time to obtain suitable bandwidth.
   C. Efficiency of Si solar cells increases at high temperatures.
   D. Output of semiconductor lasers can be modulated by controlling the junction current.

   Of these, the correct statements are

   (a) A, B and C
   (b) B, C and D
   (c) A, B and D
   (d) A, B, C and D

3. Which of the following pairs of devices and materials does not match?
   (a) Solar cell : $CuInSe_2$ as absorber and CdS as collector
   (b) LED : GaAs-AlGaAs
   (c) Optical fibre : Vitreous high-silica glass doped with germania
   (d) Semiconductor laser : Nd YAG

4. The losses in optical fibre at a given wavelength is determined by

(a) $I(x) = I_0e^{-\alpha x}$ (b) $I(x) = I_0e^{\alpha x}$
(c) $I(x) = I_0e^{-2\pi x}$ (d) $I(x) = I_0e^{2\pi x}$

5. Match list I with list II, and choose the correct answer from the codes given below the lists.

List I	List II
A. Hyperabrupt junction	1. Low voltage sensitivity
B. Recombination term	2. Q_p/τ_p
C. Rectifying type metal-semiconductor Junction	3. dQ_p/dt
D. Heterojunctions	4. Ballistic transport
	5. High voltage sensitivity
	6. Schottky diode

Codes:

	A	B	C	D		A	B	C	D
(a)	1	3	4	6	(b)	5	2	6	4
(c)	1	2	6	4	(d)	5	3	4	6

6. Given below are some statements.
   A. Influence of excess carriers is to cause decay of stored charge in n-region of the junction devices.
   B. Switching diode should have a very short carrier lifetime.
   C. A varactor diode helps in reducing the size of the circuits.
   D. A heterojunction is formed of a semiconductor, a metal and an insulator The correct statements are

(a) A and D (b) C only
(c) A, B and C (d) A, C and D

7. Which of the following is incorrect?
   (a) $dQ_p/dt = 0$ for steady state.
   (b) A homojunction is formed within a single semiconductor.
   (c) In a switching diode, the power losses should be minimum under forward bias.
   (d) Doping of Au in Si switching diode lowers the recombination lifetime of holes, but increases the reverse recovery time.

8. Consider the following statements.
   A. Negative conductance is involved with instability of semiconductors.
   B. The Gunn diodes employ the combination of carrier injection and transit time.
   C. Atomic density of semiconductor is influenced due to heavy doping.
   D. At equilibrium, there is equal tunnelling from n to p and from p to n, in a tunnel diode.

Of these, the correct statements are

(a) A, B and C (b) A, C and D
(c) A, B and D (d) B, C and D

9. Match list I with list II, and choose the correct answer from the codes given below the lists.

List I	List II
A. Region between E_V and E_F in degenerate p-type semiconductor	1. dV/dI is negative
B. On applying small forward bias to tunnel diode	2. GaAs
C. The Gunn diode	3. almost completely filled with hc
D. IMPATT diode	4. p^+-n-i-n^+
	5. E_{Fn} moves up with respect to E_F
	6. for only a part, filled with holes

Codes:

	A	B	C	D		A	B	C	D
(a)	1	6	4	2	(b)	1	5	2	4
(c)	3	5	2	4	(d)	3	6	4	2

Answers

1. (b) 2. (c) 3. (d) 4. (a) 5. (b) 6. (c)

7. (d) 8. (b) 9. (c)

References

1. Gupta, K.M., Gupta, N.: Advanced Electrical and Electronics Materials. Scrivener Wiley Publishing, USA (2015) (The authors are thankful to the copyright holder 'Scrivener Wiley Publishing', USA (2015) for permitting to use matter of many articles of Ref. [1])
2. DenBaars, P.S., et al.: Development of gallium-nitride-based light-emitting diodes (LEDs) and laser diodes for energy-efficient lighting and displays. Acta Materilia **61**(3), 45–951 (2013)
3. Jojo: Light Dependent Resistors, August 2009. http://www.circuitstoday.com/ldr-light-dependent-resistors

Part IV
BJT and FET Transistors, and Power Devices

Chapter 10
Bipolar Junction Transistors

Abstract In this chapter, the transistors have been grouped into two major categories viz. Bipolar junction transistors (BJTs) and Field effect transistors (FETs). They have been further classified into p-n-p and n-p-n types; MESFET, MISFET, MODFET, MOSFET, n-channel, p-channel types etc. The details of BJTs such as their construction, fundamentals of operation, meaning and symbols of n-p-n and p-n-p are given. Forward and reverse biasing, transistor currents, circuit configurations and their characteristics are described. Comparison between CB, CE and CC configurations is presented. Amplification with different configurations of BJTs, their characteristics and uses, and phase reversal are discussed. Specifications of transistors and their terminal identifications are given. Meanings of cascade amplifier, Debye screening length, and Kirk effect are explained. Minute insight into the various topics are given through solved numerical and theoretical examples. Review questions, numerical problems and objective type questions are also given with their answers.

Keywords Types of transistors · BJT construction · Transistor biasing · Transistor circuit configurations · Amplification with BJTs · BJT fabrication · Specification of transistors

10.1 Introduction

A transistor is a semiconductor device. It is a junction device. Unlike a diode which has only one junction, a transistor has *two* junctions. It is one of the most important micro-electronic devices which has completely revolutionalized the scenario of modern electronic industry. The transistor was invented in 1948 by John Bardeen, Walter Brattain and William Shockley at Bell Laboratory in USA. Since then, there has been a rapidly expanding efforts to utilize and develop many types of semiconductor devices such as FET, MOSFET, UJT, SCR, etc.

Transistors offer several advantages over earlier used gas or vacuum tubes. A few of them are the following.

K.M. Gupta and N. Gupta, *Advanced Semiconducting Materials and Devices*, Engineering Materials, DOI 10.1007/978-3-319-19758-6_10

1. They are much smaller in size, and are light in weight.
2. They consume little power, resulting in greater circuit efficiency.
3. They use very low operating voltages.
4. No heater or filament is required. Hence, there are no heating delays and no heating power needed.
5. They have long life with no ageing effect.
6. They are essentially shock-proof.

10.1.1 Types of Transistors

Different types of transistors are used in vivid applications. They serve different functions in numerous applications such as in amplifiers, oscillators etc. Broadly, they are classified into two main categories, given as below.

1. Field effect transistors (i.e.FET)
2. Bipolar junction transistors (i.e. BJT)

A detailed classification is shown in Fig. 10.1 [1], and their description follows in subsequent sections.

10.2 Bipolar Junction Transistor (BJT)

A bipolar junction transistor is a sandwiched form of construction in which, one type of semiconductor (say *n*-type) is placed between two layers of other type of semiconductor (i.e. *p*-types). Thus, the formation will be a *p-n-p* type construction. Similarly when the *p*-type semiconductor is sandwiched between the two *n*-types, it forms a *n-p-n* type transistor.

10.2.1 Construction of BJT

A transistor is basically formed of a Si or Ge crystal containing three separate regions. These regions are separated by two junctions. These regions and junctions are the following.

1. Emitter E
2. Base B, and
3. Collector C.

The junctions are

1. Emitter-base (or ES) junction, and
2. Collector-base (or CB) junction.

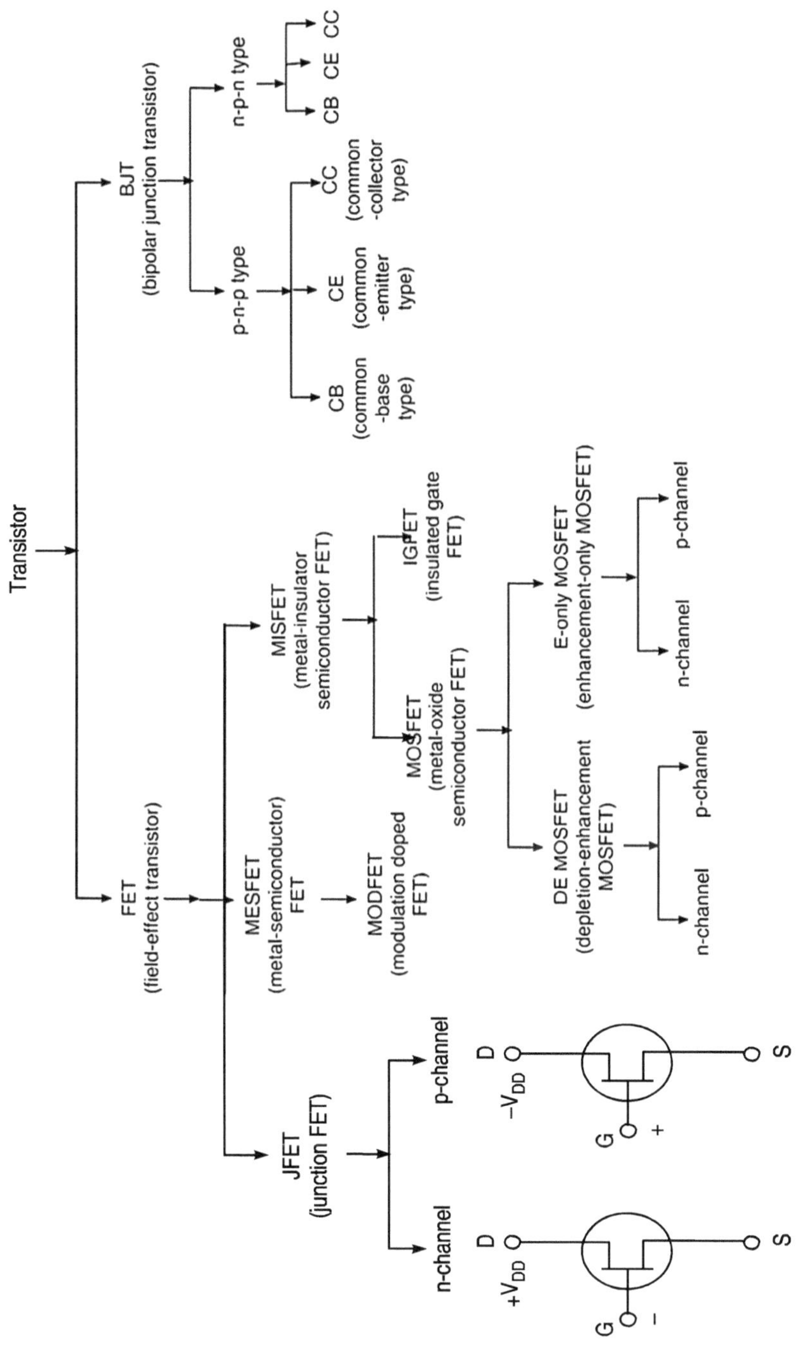

Fig. 10.1 Classification of transistors. Reproduced from [1]

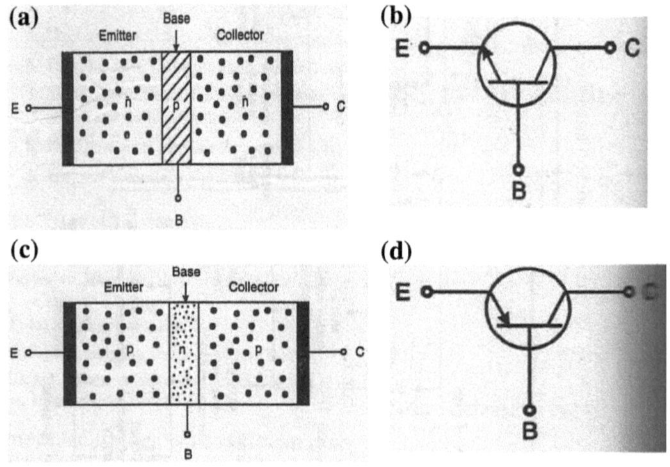

Fig. 10.2 Construction of a BJT **a**, **b** n-p-n type and its symbol, and **c**, **d** p-n-p type and its symbol. Reproduced from [1]

These are shown in Fig. 10.2a–d. Figure 10.2a, b show a n-p-n transistor and Fig. 10.2c, d depict a p-n-p transistor.

Salient features of their construction are the following.

- The middle region is called the base and the two outer regions are called the *emitter* and the *collector*. Although the two outer regions are of the same type (n-type), their function cannot be interchanged. They have different physical and electrical properties.
- In most transistors, the collector region is made physically larger than the emitter region since it is required to dissipate more heat.
- The base is very *lightly doped*, and is very thin. The emitter is *heavily doped*. Doping of the collector is intermediate between the heavy doping of emitter and light doping of the base.
- The function of the emitter is to emit or inject electrons (holes in case of a p-n-p transistor) into the base. The base passes most of these electrons (holes in case of p-n-p) on to the collector. The collector serves the function of collecting these electrons (holes in case of a p-n-p) from the base.

Other details of these transistors are as follows.

- A transistor has two p-n junctions. One junction lies between the emitter and the base, and is called the emitter-base junction or simply the emitter junction. The other junction is between the base and the collector, and is called collector-base junction or collector junction. Thus, a transistor is like two p-n junction diodes connected back-to-back.
- Figure 10.2b shows the symbol for an n-p-n transistor. In the symbol, the emitter (not the collector) has an arrowhead. The arrowhead points out in the direction of the conventional emitter current (i.e. from p-region to n-region).

- Figure 10.2c shows the structure of a p-n-p transistor and Fig. 6.2d shows its symbol. In a p-n-p transistor, the conventional emitter current flows from the emitter to the base. That is why the direction of arrowhead is inward (i.e. from p-region to n-region).
- Both types (p-n-p and a-p-n) of transistors are widely used, sometimes even together in the same circuit.

In this chapter we shall consider the transistor of n-p-n type. Since a p-n-p transistor is the complement of an n-p-n transistor, it is merely necessary to read hole for electron and electron for hole, negative for positive and positive for negative, for the corresponding operation of a p-it-p transistor.

- The choice of an n-p-n transistor is found to be more suitable because the major part of the current is transported by electrons. Since the mobility of electrons is greater than that of the holes, hence the action of it-p-n transistor is faster.

10.3 Fundamentals of BJT Operation

To understand the working of a BJT, we consider an n-p-n transistor as shown in Fig. 10.3. The emitter-base junction is forward-biased by V_{EE}, and the collector-base junction is reverse-biased by V_{CC}. When biased in this way, the transistor is said to work in *active region*. The directions of various currents that flow in the transistor are also indicated in Fig. 10.3. As per the usual convention, the direction of current flow has been taken opposite to the direction of electron movement. To understand the action of the transistor, some of the electrons and holes have been assigned numbers.

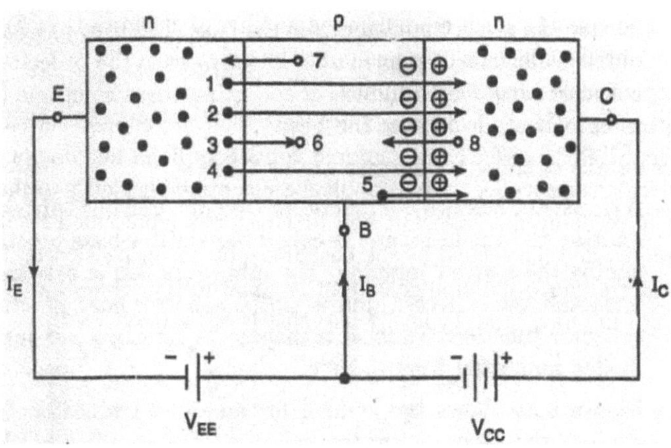

Fig. 10.3 Schematic arrangement of a n-p-n

Emitter injection ratio (or emitter efficiency γ). If the emitter junction is forward-biased (may be, by a fraction of a volt), the barrier potential is reduced. Then the space-charge region at the junction also becomes narrow. As such the majority charge carriers diffuse across the junction. The resulting current consists of electrons travelling from the emitter to the base, and holes passing from the base to the emitter. The electron current is made much larger than the hole current. This is done by doping the base region more lightly than the emitter region.

In Fig. 10.3, the electrons 1, 2, 3, and 4 are crossing from the emitter to the base, and hole 7 from the base towards the emitter. The total sum of these charge-carrier movements constitute the emitter current i_E. Only a portion of this current is due to the movement of electrons 1, 2, 3, and 4. These are the electrons injected by the emitter into the base. The ratio of the electron current to the total emitter current is known as *emitter injection ratio γ or the emitter efficiency*. Typically, γ is equal to 0.995. This means that only 0.5 % of the emitter current consists of the holes passing from the base to the emitter.

Base transportation factor β. Once the electrons are injected by the emitter into the base, they become minority carriers in the base region. These electrons do not have separate identities from those which are thermally generated in the base region itself (It is worth noting that these electrons are emitted by the emitter, and are in addition to the thermally generated minority carriers in the base region). The basic idea in transistor action is that the base is made very narrow (about 25 μm) and is very lightly doped. Because of this, most of the minority carriers (electrons) travelling from the emitter end of the base region to its collector end do not recombine with holes (like the electron numbered as 6). The ratio of the number of electrons arriving at collector to the number of emitted electrons is known as the *base transportation factor*. It is designated by symbol β'. Typically $\beta' = 0.995$.

Movement of hole 8 from the collector region and that of electron 5 from the base region constitute leakage current I_{CBO}. Movement of electron 3 and that of hole 7 constitutes a part of emitter current I_E. These two currents are not equal. Actually, the number of electrons (like 3) and holes (like 7) crossing the emitter-base junction is much more than the number of electrons (like 5) and holes (like 8) crossing the collector-base junction. The difference of these two currents in the base region makes the base current I_B.

DC alpha α_{dc}. The collector current is less than the emitter current, for two reasons. Firstly, a part of the emitter current consists of holes that do not contribute to the collector current. Secondly, not all the electrons injected into the base are successful in reaching the collector. The first factor is represented by the emitter injection ratio γ and the second by the base transport factor β', Hence, the ratio of the collector current to the emitter current is equal to $\beta'\gamma$. This ratio is called dc alpha (α_{dc}) of the transistor. Typically its value is $\alpha_{dc} = 0.99$.

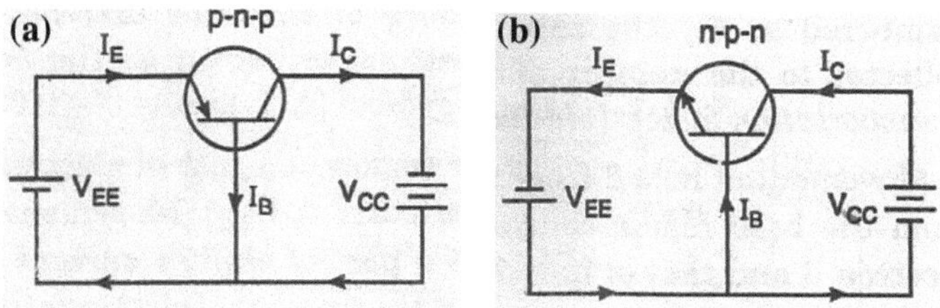

Fig. 10.4 Transistor biasing showing a **a** forward-biased emitter base junction, and **b** reverse-biased collector base junction

10.3.1 Transistor Biasing

Biasing means the system of connection of transistor to a battery. For smooth working of a transistor, it is essential to apply voltages of correct polarity across its two junctions. For that, it is connected to the battery in following two ways.

1. In forward-biasing, and
2. In reverse-biasing

The emitter-base junction is always forward-biased, and the collector-base junction is always reversed-biased.

Figure 10.4a, b show these arrangements. Here, the two batteries are used to provide dc emitter supply voltage V_{EE} and collector supply voltage V_{CC}. In Fig. 10.4a for *p-n-p* transistor, the positive terminal of V_{EE} is connected to a *p*-type emitter so as to repel (or push) the holes into the base. The negative terminal of V_{CC} is connected to the collector so as to attract (or pull) the holes through the base. Similar considerations apply to *n-p-n* transistor also. It is shown in Fig. 10.4b.

Leakage current. It is worth knowing that if the emitter-base junction is not forward-biased, the transistor will not conduct the current. It is because there would be no current due to majority charge carriers, although an extremely small current will flow due to majority charge carriers. This current is called as *leakage current* of the transistor.

Biasing considerations. Following criterions may be kept in mind to understand the assignment of positive and negative terminals in a transistor circuit.

(i) For a *p-n-p* transistor, the collector and base are *negative* (−) with respect to emitter, as shown in Fig. 10.5a. In it, the collector is *more negative* than base. It is therefore shown by a single '−' at the base, but by a double '–' at the collector.

(ii) For a *n-p-n* transistor, the collector and base are positive (+) with respect to emitter, as shown in Fig. 10.5b. In it, the collector is *more positive* than the base. That is why it is indicated by a single '+' at base and a double '++' at the collector.

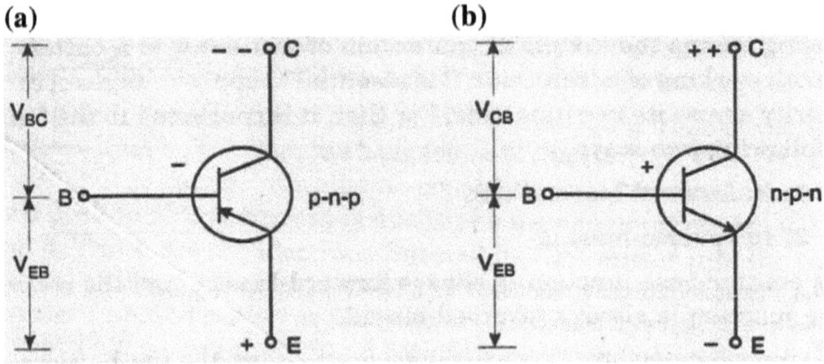

Fig. 10.5 Symbolic representation of **a** a *p-n-p* transistor, and **b** a *n-p-n* transistor

10.3.2 Transistor Currents

The three primary currents that flow in a properly biased transistor are the emitter current I_E, base current I_B, and collector current I_C. Their directions of flow along with the relative magnitudes of these currents are shown in Fig. 10.6a, b for two different modes of *p-n-p* transistor.

In Fig. 10.6a showing the common-base mode, the following current relations holds good.

$$I_E = I_B + I_C \tag{10.1}$$

Of the total emitter current, about 1–2 % goes to base and remaining current goes to collector. Figure 10.6b shows the flow of current when the transistor is connected in common-emitter mode. Again, it can be seen that $I_E = I_B + I_C$.

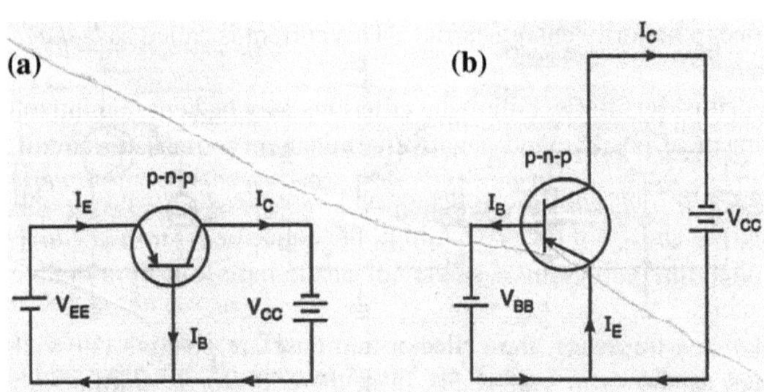

Fig. 10.6 Flow of transistor currents in a p-n-p transistor connected in **a** common-base mode, and **b** common-emitter mode

Notation for flow of current in transistor.

By normal convention, the flow of current into transistor is taken +ve while those flowing out are taken negative. Therefore in above circuits, I_E will be positive but I_B and I_C are negative. Hence on applying the Kirchhoff's current law, we can write

$$\because \quad \sum I = 0$$
$$\therefore \quad +I_E + (-I_B) + (-I_C) = 0 \quad \text{(on neglecting leakage current)}$$
$$\text{or} \quad I_E - I_B - I_C = 0$$
$$\text{i.e.} \quad I_E = I_B + I_C$$

10.4 Transistor Circuit Configurations and Their Characteristics

For operation of electronic devices, a transistor is connected in either of the following configurations.

1. Common-base (CB) configuration,
2. Common-either (CE) configuration, and
3. Common-collector (CC) configuration

The above configurations differ from each other in style of their connection with the external load. The term 'common' is used to denote the electrode which is common for input and output circuit, the common electrode is generally grounded. Hence, these configurations are also known as ground-base configuration, ground-emitter configuration, and ground-collector configuration. These are shown in Fig. 10.7a–c.

Transistor characteristics. Characteristics of bipolar junction transistors are expressed in terms of relations between the currents and voltages flowing in them. These relations are useful in studying the operational behaviour of transistors, when they are connected in a circuit. The characteristics are generally expressed in the form of curves. A transistor displays the following characteristics.

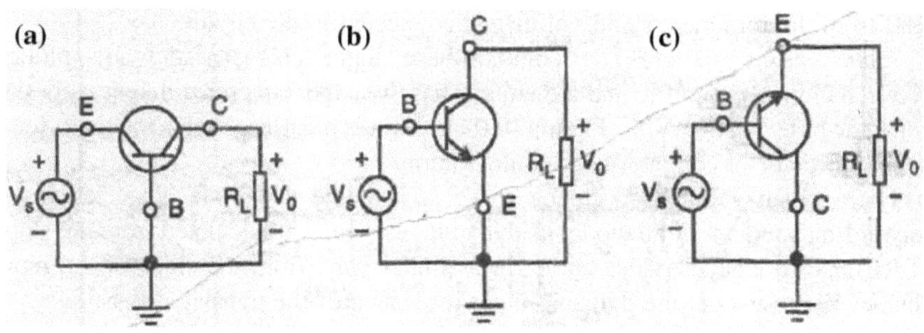

Fig. 10.7 Three configurations in which a transistor may be connected

1. **Static characteristics**

   (i) *input characteristics* (I/C) that relates input current i_{in} with input voltage v_i, for a given output voltage v_o.
   (ii) *output characteristics* (O/C) that relates the output current i_o with output voltage, for a given input current.
   (iii) *current transfer characteristic.*

2. **Dynamic characteristics.**

In this chapter, we shall study the static characteristics of all three configurations discussed above.

The common-collector configuration can be treated as a special case of common-emitter configuration (with feedback applied). The details of a CC amplifier can be known from the CE characteristics. The static characteristics of a transistor in CC configuration are therefore not required. Therefore the CC characteristics are not discussed here. Hence the following details will now be discussed in subsequent sections.

- For CB configuration: input characteristics, output characteristics, and property relationships.
- For CE configuration: input characteristics, output characteristics, and property relationships.

10.4.1 Common-Base (CB) Characteristics

The circuit arrangement for determining CB characteristics of a p-n-p transistor is shown in Fig. 10.8. In it the emitter-to-base voltage V_{EB} can be varied with the help of a potentiometer R_1. Since the voltage V_{EB} is quite low (less than one volt), we include a series resistor R_s (say, of 1 kΩ) in the emitter circuit. This helps in limiting the emitter current I_E to a low value. Without this resistor, the current I_E may change by a large amount even if the potentiometer R_1 setting is moved slightly. The collector voltage can be varied by adjusting the potentiometer R_2. The required currents and voltages for a particular setting of the potentiometers can be read from the millimeters and voltmeters connected in the circuit.

Input characteristics. The common-base input characteristics are plotted between emitter current I_E and the emitter-base voltage V_{EB}, for different values of collector-base voltage V_{CB}. Figure 10.9 shows a typical input characteristic for a p-n-p transistor in common-base configuration.

For a given value of V_{CB}, the curve is just like the diode characteristic in forward-bias region. For a diode, its dynamic resistance is calculated from the slope of its forward characteristics curve. In a similar way, from the slope of the input characteristics we get the dynamic input resistance of the transistor as below

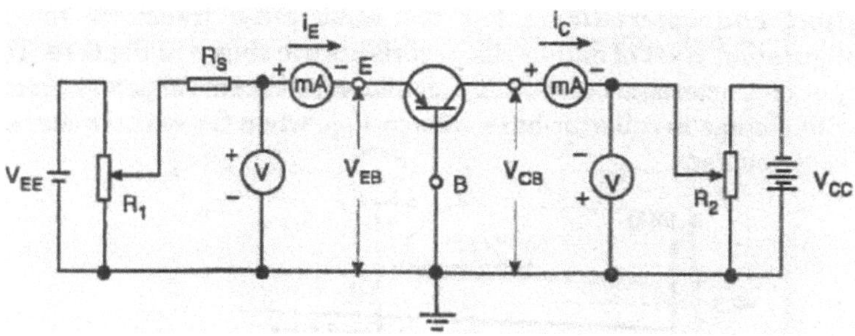

Fig. 10.8 Circuit arrangement for determining the static characteristics of a p-n-p transistor in CB configuration

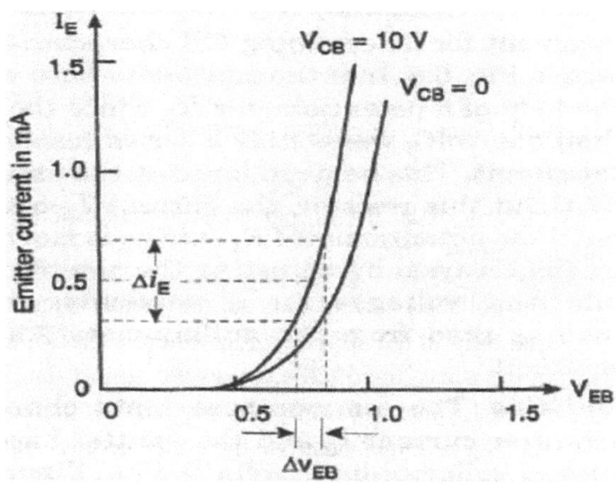

Fig. 10.9 Common-base input characteristics for a typical p-n-p silicon transistor

$$r_i = \frac{\Delta V_{EB}}{\Delta i_E} \bigg| V_{CB=constant} \quad (10.2)$$

The dynamic input resistance r_i of the transistor is very low (about 20–100 Ω). Since the curve is not linear, the value of r_i varies with the point of measurement. As the emitter-base voltage increases, the curve tends to become more vertical. It means that r_i decreases.

The input characteristics of an n-p-n transistor are similar to those as shown in Fig. 10.9. It differs only in I_E and V_{EB} which would be negative and V_{CB} would be positive.

Output characteristics: For the same p-n-p transistor in CB configuration, a set of output characteristics are shown in Fig. 10.10. The output characteristic curve indicates the variation in collector current i_C with change in collector-base voltage V_{CB}, when the emitter current I_E kept constant.

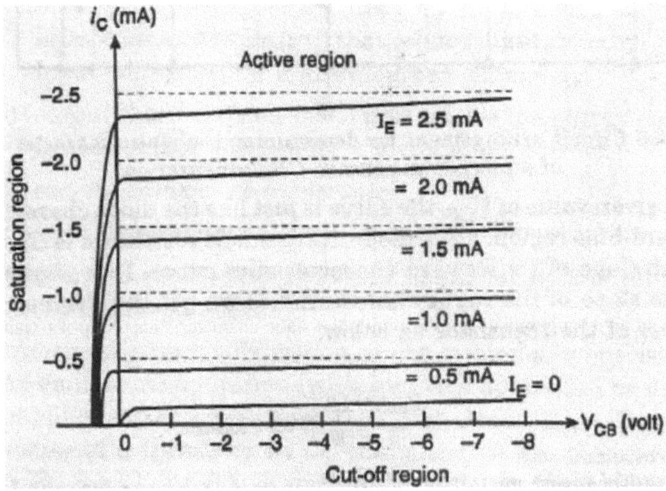

Fig. 10.10 Common-base output characteristics for a p-n-p transistor

A close look at the output characteristics of Fig. 10.10 reveals the following salient features.

(i) The collector current I_C is approximately equal to the emitter current I_E. This is true only in the active region, where collector-base junction is reverse-biased.

(ii) In the active region, the curves are almost flat. This indicates that the I_C (for a given I_E) increases only slightly as V_{CB} increase.

(iii) As V_{CB} becomes positive, the collector-base junction becomes forward biased. Hence the collector current I_C (for a given I_E) sharply decreases. This is the saturation region. In this region, the collector current does not depend much upon the emitter current.

(iv) The collector current is not zero when $I_E = 0$. It has a very small value. This is the reverse leakage current I_{CO}. The conditions that exist when $I_E = 0$ for CB configuration, is shown in Fig. 6.11. The notation most frequently used for I_{CO} is I_{CBO}, as indicated in the figure.

In this notation, the subscript CBO means the current between collector and base when the third terminal (the emitter) is open. The current I_{CBO} is temperature sensitive. At room temperature, its typical value ranges from

- 2–5 μA for germanium transistors, and
- 0.1–1 μA for silicon transistors (Fig. 10.11).

From the output characteristics of Fig. 10.10, we can determine a number of transistor parameters, such as given below.

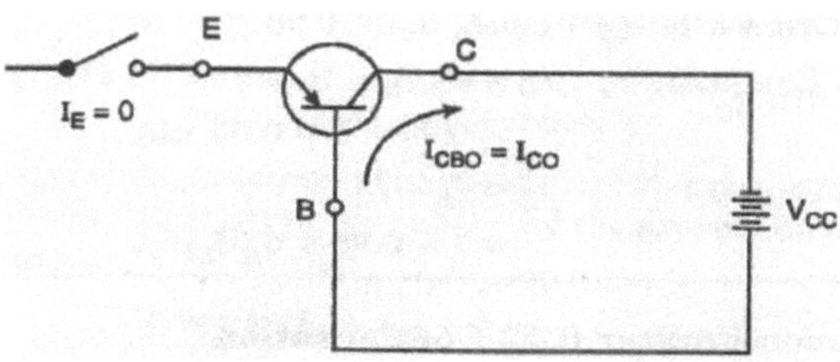

Fig. 10.11 Reverse leakage current in CB configuration

Dynamic output resistance is defined as

$$r_o = \frac{\Delta V_{CB}}{\Delta i_C} \,|\, I_{E=\text{constant}} \tag{10.3}$$

where ΔV_{CB} and Δi_c, are small changes in collector voltage and collector current around a given point on the characteristic curve (for a given I_E). Since the output curves are very flat; therefore for a given ΔV_{CB}, the Δi_c is very small. It means that the output resistance is very high (of the order of 1 M Ω).

- **dc alpha of the transistor** is defined as

$$\alpha_{dc} = \frac{I_C}{I_E} \tag{10.4}$$

It means that at any point in the active region on the curve, I_C is less than I_E, and the difference is very small. The value of α_{dc} is less than unity but very close to it. A typical value is $\alpha_{dc} = 0.98$.

- **ac current gain**. A transistor is also used as amplifier. The amplifier handles ac signals. Under such conditions, the small changes in voltages and currents are given by a parameter called ac current gain or ac alpha α_{ac}. It is defined as

$$\alpha_{ac} = \frac{\Delta i_C}{\Delta i_E} \,|\, V_{CB=\text{constant}} \tag{10.5}$$

The current gain is also called as short-circuit current gain because it is defined under short-circuit condition. The value of α_{ac}, is in the range of 0.95–0.995.

Example 10.1 In a CB configuration transistor, the emitter current is 1 mA and the ratio of collector current to emitter current is 0.95. Determine the value of (a) collector current, and (b) base current.

Solution. Given are: $I_E = 1$ mA, $\alpha_{dc} = 0.95$

(a) From Eq. 10.4 $I_C = \alpha_{dc} I_E$

$$= 0.95 \times 1 = 0.95 \text{ mA}$$

(b) From Eq. 10.1 $I_B = I_E - I_C$

$$= 1 - 0.95 = 0.05 \text{ mA}$$

10.4.2 Common-Emitter (CE) Configuration

In CE configuration, the emitter is made common to input and the output. The signal is applied between the base and emitter, and the output is developed between the collector and emitter. Whether the transistor works in CB or CE configuration, it is to be ensured that it works in the active region. It means that the emitter-base junction is forward-biased and the collector-base junction is reverse-biased. Such forward-reverse biasing (FR biasing) is achieved in CE configuration by connecting the batteries V_{BB} and V_{CC} as shown in Fig. 10.12. Here an n-p-n transistor has been used. The emitter-base junction is forward-biased by the battery V_{BB}.

This forward biasing needs a very small voltage (say, 0.6 V). The battery V_{CC} (say, of 9 V) is connected between the emitter and collector. Since the base is at $+V_{BB}$ potential with respect to the emitter and the collector is at $+V_{CC}$ potential with respect to the emitter, the net potential of the collector with respect to the base is '$V_{CC} - V_{BB}$'. The collector-base junction is reverse-biased by this potential. Since V_{CC} is much larger than V_{BB}, the reverse-bias may be taken only as V_{CC}.

Current relations. We have seen that in CB configuration, I_E is input current and I_C is output current. These currents are related as below.

$$I_E = I_C + I_B \tag{10.6a}$$

$$I_C = \alpha_{dc} I_E + I_{CBO} \tag{10.6b}$$

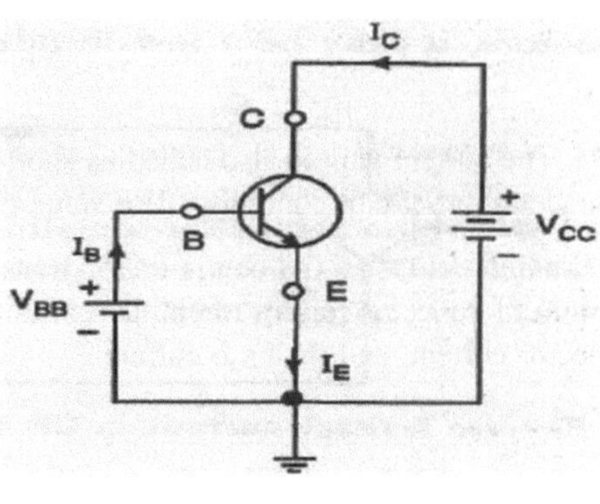

Fig. 10.12 FR biasing of an n-p-n transistor in common emitter (CE) configuration

In CE configuration, I_B becomes the input current and I_C is the output current. To obtain this relation, we substitute to the expression of I_E from Eqs. 10.6a into 10.6b, so that

$$I_C = \alpha_{dc}(I_C + I_B) + I_{CBO}$$

$$(1 - \alpha_{dc})I_C = \alpha_{dc}I_B + I_{CBO}$$

or

$$I_C = \left(\frac{\alpha_{dc}}{1 - \alpha_{dc}}\right).I_B + \left(\frac{1}{1 - \alpha_{dc}}\right).I_{CBO} \tag{10.7}$$

In this equation, I_C is given in terms of I_B. This can be simplified by defining

$$\beta_{dc} = \frac{\alpha_{dc}}{1 - \alpha_{dc}} \tag{10.8}$$

and

$$I_{CEO} = \frac{I_{CBO}}{1 - \alpha_{dc}} \tag{10.9}$$

Thus Eq. 10.7 becomes

$$I_C = \beta_{dc}I_B + I_{CEO} \tag{10.10}$$

This Equation states that this leakage current would flow between the collector and emitter, if the third terminal (base) were open. This is illustrated in Fig. 10.13. The magnitude of I_{CEO} is much larger than that of I_{CBO}, as indicated by Eq. 10.9. For example,

- if α_{de} = 0.98 the value of I_{CEO} is fifty times that of I_{CBO}.
- for Si transistor, I_{CEO} would typically be a few microamperes,
- for Ge transistors, it may be a few hundred microamperes.

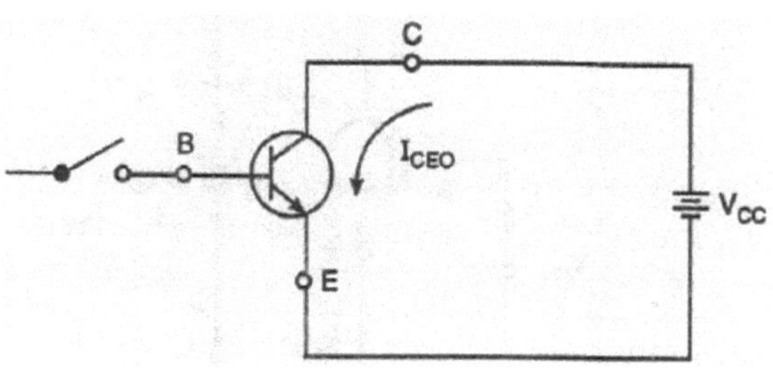

Fig. 10.13 Reverse leakage current in CE configuration

Relation between β_{dc} and I_{CEO}. The factor β_{dc} is called the common-emitter dc current gain. It relates the dc output current I_C to the input current I_B. Equation 10.8 indicates that β_{dc} can be very large. For example,

- if $\alpha_{dc} = 0.98$, the value of β_{dc} is

$$\beta_{dc} = \frac{0.98}{1 - 0.98} = 49$$

- typically β_{dc} can have values in the range of 20–300. If we solve Eq. 10.12 for β_{dc} we obtain

$$\beta_{dc} = \frac{I_{C-I_{CEO}}}{I_B} \tag{10.11a}$$

If I_{CEO} very small as compared to I_C (this is usual) then

$$\beta_{dc} = \frac{I_C}{I_B} \tag{10.11b}$$

Thus β_{dc} is the ratio of de collector current to dc base current.

Relation between α_{dc}, and β_{dc}, The dc current gain of a transistor when connected in CE configuration, is β_{dc}. It has been defined by Eq. 10.11b. The same transistor connected in CB configuration gives a de current gain of α_{dc}. Therefore, if the value of α_{dc} of a transistor is known, its β_{dc} can be calculated. Hence on manipulating Eq. 10.8, we get

$$\beta_{dc} = \frac{\alpha_{dc}}{1 - \alpha_{dc}}$$

or

$$\beta_{dc} - \beta_{dc} \cdot \alpha_{dc} = \alpha_{dc}$$

or

$$\beta_{dc} = \alpha_{dc}(1 + \beta_{dc})$$

or

$$\alpha_{dc} = \frac{\beta_{dc}}{1 + \beta_{dc}} \tag{10.12}$$

Relation between α_{ac}, and β_{ac}. When a transistor is used as an amplifier, the designers are more interested in knowing the ratio of small changes in collector and base currents, rather than the ratio of their absolute values. This ratio is called beta ac or dynamic beta (β_{ac}). Thus, the ac beta is related as

$$\beta_{ac} = \frac{\Delta i_C}{\Delta i_B} \mid V_{E=\text{constant}} \tag{10.13}$$

To a very close approximation, the value of β_{ac} is same as the $\beta_{dc.}$ Hence its typical value is also from 20 to 300.

Just as β_{dc} is related to α_{dc}, so is β_{ac} related to α_{ac}. We can establish this relation by considering

$$I_E = I_C + I_B$$

If I_C and I_B change by small amounts Δi_C and Δi_B, the I_E changes by Δi_E. We shall still have

$$\Delta i_E = \Delta i_C + \Delta i_B$$

On dividing the above equation by Δi_C and rearranging the terms, we get

$$\frac{\Delta i_E}{\Delta i_C} = 1 + \frac{\Delta i_B}{\Delta i_C}$$

or

$$\frac{1}{\alpha_{ac}} = 1 + \frac{1}{\beta_{ac}}$$

or

$$\beta_{ac} = \frac{\alpha_{ac}}{1 - \alpha_{ac}} \tag{10.14}$$

Example 10.2 When the emitter current of a transistor is changed by 1 mA, its collector current changes by 0.995 mA. Calculate (a) its common-base current gain α, and (b) its common-emitter short-circuit current gain β.

Solution. (a) Common-base short-circuit current gain is given by Eq. 10.5 as

$$\alpha_{ac} = \frac{\Delta i_C}{\Delta i_E} = \frac{0.995 \times 10^{-3}}{1 \times 10^{-3}} = 0.995$$

(b) Common-emitter short-circuit current gain is given by Eq. 6.14 as

$$\beta_{ac} = \frac{\alpha_{ac}}{1 - \alpha_{ac}} = \frac{0.995}{1 - 0.995} = 199$$

Example 10.3 The dc current gain of a transistor in common-emitter configuration is 100. Find its dc current gain in common-base configuration.

Solution. Using Eq. 10.12, we calculate the dc current gain in common-base configuration as

$$\alpha_{\mathrm{dc}} = \frac{\beta_{dc}}{1 + \beta_{dc}} = \frac{100}{1 + 100} = 0.99$$

10.5 Comparison Between CB, CE and CC Configurations

We have seen that a transistor behaves differently in different configurations. A configuration may be suitable for certain applications, but may not be suitable for others. Out of these configurations, the CC configuration has maximum dynamic input resistance. So we use this configuration where high input resistance is of prime importance, even though its voltage gain is less than unity. However, the decreased voltage gain can be compensated by subsequently using the CE configuration. A comparison between the CB and CE configurations is shown in Table 10.1. It gives the typical values of the important parameters also.

10.6 Amplification with BJTs

Though a transistor can be used to perform a number of functions, its main use lies in amplifying the electric signals. Hence, it is widely used as amplifier. Although there are various kinds of transistor amplifiers such as class A, B, AB, C types;

Table 10.1 Comparison between CB and CE configurations

S. no	Parameters	Common-base configuration	Common-emitter configuration
1.	Input dynamic resistance	Very low (about 20 Ω)	Low (about 1 kΩ)
2.	Output dynamic resistance	Very high (about 1 MΩ)	High (about 10 kΩ)
3.	Current gain	Less than unity (about 0.98)	High (about 100)
4.	Leakage current	Very small (about 500 nA for GE, 2 nA for Si)	Very large (about 50 μA for GE, 0.2 μA for Si)

single-stage and cascade types etc.; but we shall study the basic working of a single-stage amplifier. Such amplifiers have its one amplifying element connected in CB, CE and CC configurations. We shall describe each one of them now.

10.6.1 Amplification with CB Configuration

Circuit of a single-stage CB-amplifier using n-p-n transistor is shown in Fig. 10.14. Here, the input ac signal is injected into the emitter-base circuit and output is taken from the collector-base circuit. The E-B junction is forward-biased by V_{EE} whereas C-B junction is reverse-biased by V_{CC}. The Q-point[1] or dc working conditions are determined by the batteries along with resistors R_E and R_C.

When no signal is applied to the input circuit, the output just sits at the Q-point so that there is no output signal. When positive half-cycle of the signal is applied, then following occurrences take place.

(i) forward bias is decreased because V_{BE} is already negative with respect to the ground.
(ii) I_B is decreased.
(iii) I_E and hence I_C are decreased because they are both nearly β_{ac} times the base current.
(iv) $I_C R_C$ drop is decreased.
(v) V_{CB} is increased because $V_{CC} = V_{CB} + I_C R_C$.

It means that a positive output half-cycle is produced. Since a positive-going input signal produces a positive-going output signal, there is no phase reversal between the two.

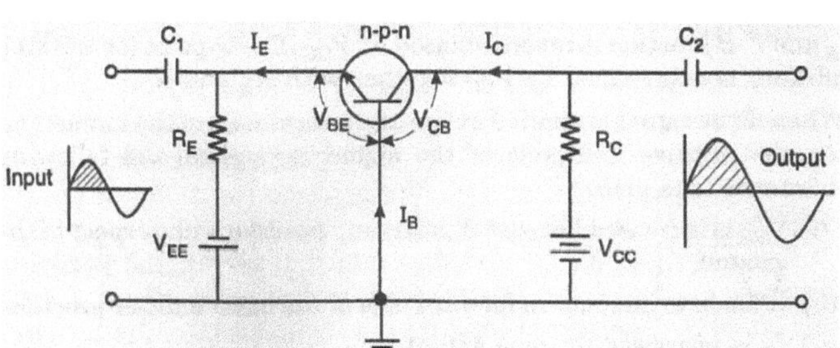

Fig. 10.14 Circuit of a single-stage CB amplifier using n-p-n transistors

[1]Q-point i.e. quiescent point is that point on de load line which represents the values of Ic and VCE that exists when no signal is applied.

Characteristics of a CB amplifier. It has

(i) a very low input resistance (about 30–150 Ω).
(ii) a very high output resistance (up to about 500 kΩ).
(iii) a current gain of about $\alpha < 1$.
(iv) a large voltage gain of about 1500.
(v) a power gain of up to 30 dB.
(vi) no phase reversal between input and output voltages.

Uses. One of the important uses of a CB amplifier is in matching a low-impedance circuit to a high- impedance circuit. It also has high stability of collector current with temperature changes.

10.6.2 Amplification with CE Configuration

Circuit of a single-stage CE amplifier using an n-p-n transistor is shown in Fig. 10.15. Here, base is the driven element. The input signal is injected into the base-emitter, whereas the output signal is taken out from the collector-emitter circuit. The E-B junction is forward-biased by V_{BB} and C-B junction is reverse-biased by V_{CC}. The Q-point (or working condition) is determined by V_{CC} together with R_B and R_C.

When an ac signal is applied at the input terminals of the circuit and when the positive half-cycle of the signal is applied, the following occurrences take place.

(i) V_{BE} is increased because it is already positive with respect to the ground.
(ii) It leads to increase in forward bias of the base-emitter junction.
(iii) I_B is increased to some extent.
(iv) I_C is increased by β_{ac} times the increase in I_B.
(v) I_CR_C drop is increased considerably.
(vi) V_{CE} which represents the output voltage, is decreased.

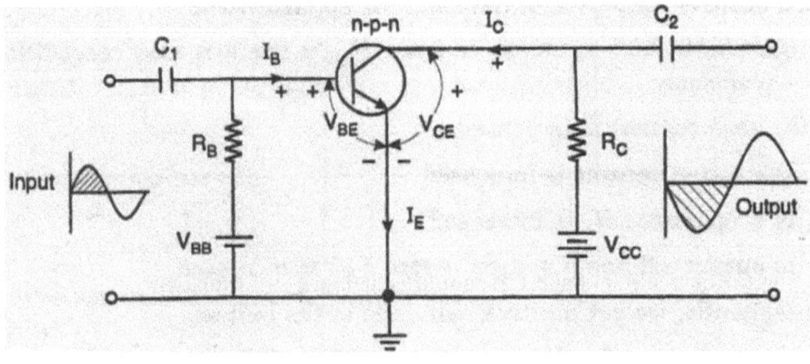

Fig. 10.15 Circuit of a single-stage CE amplifier using n-p-n transistor

As $V_{CC} = V_{CE} + I_C R_C$, hence a negative half-cycle of the output is obtained. It means that a positive-going input signal becomes a negative-going output signal as shown in Fig. 10.15.

Characteristics of a CE amplifier. A CE transistor amplifier has the following characteristics.

(i) It has moderately low input resistance (about 1–2 kΩ).
(ii) Its output resistance is moderately large (about 50 kΩ).
(iii) Its current gain β_{ac} is high (about 50–300).
(iv) It has a very high voltage gain of the order of about 1500.
(v) It produces very high power gain of the order of about 10,000.
(vi) It produces phase reversal of input signal i.e. input and output signals are 180° out of phase with each other.

Uses. Most of the transistor amplifiers are of CE type because of large gains in voltage, current and power. Moreover, its input and output impedance characteristics are suitable for many applications.

10.6.3 Amplification with CC Configuration

Circuit of a single-stage CC amplifier using an n-p-n transistor is shown in Fig. 10.16. The input signal is injected into the base-collector circuit and output signal is taken out from the emitter-collector circuit. The E-B junction is forward-biased by V_{EE} and C-B junction is reverse-biased by V_{CC}. When an ac signal is applied across the input circuit (i.e. when positive half-cycle of the signal is applied) then

(i) forward bias is increased since VEE is positive with respect to collector.
(ii) base current is increased.
(iii) emitter current is increased.
(iv) drop across RE is increased.
(v) output voltage (i.e. drop across R_E) is increased.

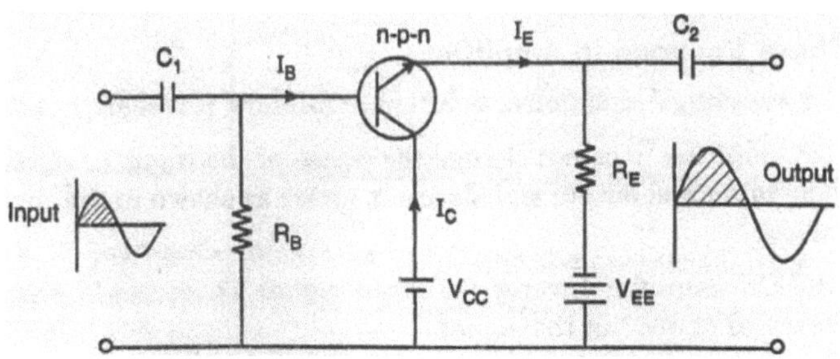

Fig. 10.16 Circuit of a single-stage CC amplifier using n-p-n transistor

Consequently, we get positive half-cycle of the output.

Characteristics of a CC amplifier. A CC amplifier has the following characteristics.

(i) High input impedance (about 20–500 kΩ).
(ii) Low output impedance (about 50–1 kΩ).
(iii) High current gain of about 50–500.
(iv) Voltage gain of less than 1.
(v) Power gain of about 10–20 dB.
(vi) No phase reversal of the input signal.

Uses. The CC amplifiers are used for the following purpose.

- For impedance matching i.e. for connecting a circuit having high output impedance to another circuit having a low input impedance.
- For circuit isolation.
- As a two-way amplifier, since it can pass a signal in either directions.
- For switching the circuits.

10.6.4 Phase Reversal in Amplifiers

From the foregoing discussion of different amplifiers, it is seen that the

(i) CB amplifier does not change the phase of the input ac signal. The input and output signals are in phase as shown in the chart below.
(ii) CE amplifier inverts the input signal i.e. causes a phase reversal of 180° in the signal.
(iii) CC amplifier does not change the phase of the input signal i.e. the input and output signals are in phase.

Type of amplifier	Input waveform	Output waveform
Common base (CB)	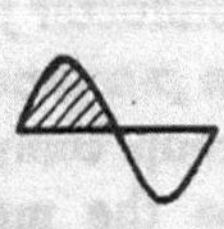	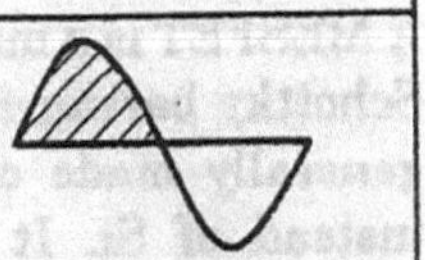
Common emitter (CE)	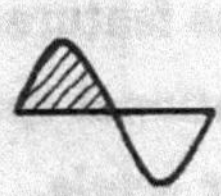	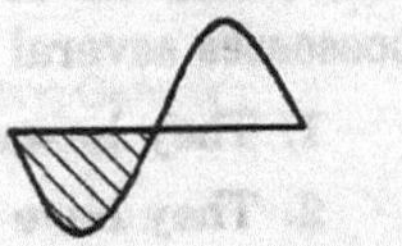
Common collector (CC)	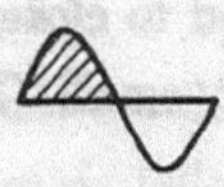	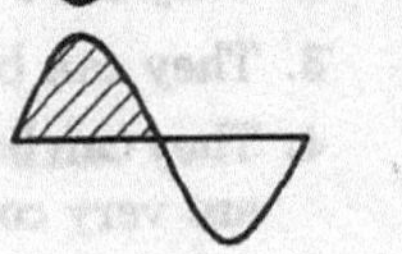

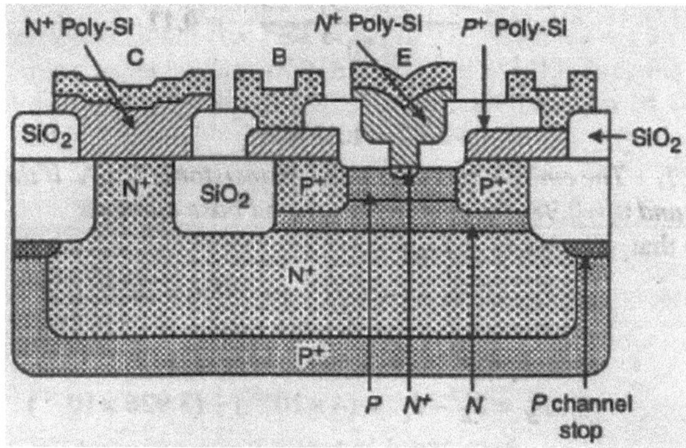

Fig. 10.17 Diffused junction transistors

10.7 BJT Fabrication

10.7.1 Diffused Junction Transistors

A double polysilicon self-aligned NPN Si-BJT transistor is shown in Fig. 10.17. A P-type Si substrate is oxidized and windows are defined using photolithography. The windows are then etched in the oxide. N^+ layer is formed in the open window using photoresist and oxide as an implant mask. Subsequently the photoresist and the oxide are removed, and a lightly doped N-type epitaxial layer is grown. During this high temperature growth, the implanted N^+ layer diffuses only slightly towards the surface and becomes a conductive buried collector. The N^+ sub-collector layer guarantees a low collector series resistance when it is connected to the collector ohmic contact. The lightly doped N-type collector region above the N^+ sub-collector in the part of the BJT, where the base and emitter are formed, ensures a high base-collector reverse breakdown voltage.

10.8 Solved Examples

Example 10.4 What are the specifications of a transistor?

Solution. Main specifications of a transistor are as follows.

- Polarity: whether p-n-p type or n-p-n type.
- Material: whether Si-transistor or Ge-transistor.
- Maximum power dissipation
- Frequency: at which the magnitude of CE short-circuit 'current gain' falls to unity.
- Maximum operation voltages: which can be applied between

Fig. 10.18 Typical transistor casing

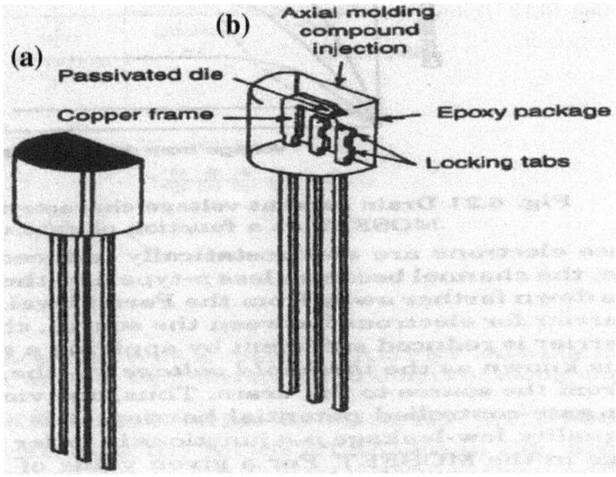

(i) emitter and base, with collector open.
(ii) collector and emitter, with base open.
(iii) collector and base with emitter open.

- Maximum collector current that can be allowed safely.
- Maximum junction temperature that can be allowed at the junction.
- Junction capacitance
- Case outline i.e. the type of packaging of transistors. A typical transistor casing is shown in Fig. 10.18.

Example 10.5 Use of n-p-n transistors is more popular than p-n-p transistors—why?

Solution. The n-p-n transistors are used more than the p-n-p transistors because the electrons have higher mobility than holes (i.e. $\mu_e > \mu_h$).

Example 10.6 How does a transistor work in switching operation?

Solution. In switching operation, a transistor is controlled in two different conduction states viz. ON-state and OFF-state. Ideally a switch appears as a short-circuit when turned-on, and as open circuit when turned off.

Example 10.7 Why is CE configuration used widely in amplifier circuits?

Solution. A transistor in CB configuration has a very low input resistance ($\cong$20 Ω). and very high output resistance ($\cong$1 MΩ). It is just the reverse of what we desire (high input resistance and low output resistance). That is why the CB configuration is unpopular. Comparatively, the CE configuration is much better, as regards its input and output resistances. Its input resistance is about 1 kΩ and output resistance about 10 kΩ. A transistor in the CE configuration makes a much better amplifier. Furthermore, the current gain, voltage gain and power gain of CE is much greater than those of CB.

From the point of view of cascading of amplifier stages, the CC configuration would have been the best. Its input resistance is very high ($\cong$150 kΩ). and output resistance is quite low ($\cong$800 Ω). However, unfortunately the voltage gain of the CC amplifier is low (less than unity). Therefore, we use CC amplifier only in such applications where the requirement of high input resistance is of prime importance.

Thus we see that the CE configuration is best suited for most of the amplifier circuits.

Example 10.8 What is a cascade amplifier? Why is it widely used?

Solution. Main utility of a transistor lies in its ability to amplify the signals. The transistor alone cannot perform this function. We have to connect some passive components (such as resistors and capacitors) and a biasing battery. Such circuit is then called an amplifier. Thus, an amplifier is an electronic circuit which is capable of amplifying (or increasing the level of) the signals.

Very often, a single-transistor amplifying stage is not sufficient. In almost all applications we use a number of amplifier stages, connected one after the other. The signal to be amplified is fed to the input. of the first stage. The output of the first stage is connected to the input of the second stage. The second stage feeds the third stage, and so on. Ultimately, the output appears across the load connected to the output of the final stage. Such a connection of amplifier stages is known as cascaded amplifier.

Figure 10.19 shows a cascaded amplifier having two stages. The first stage is energized by a signal source having voltage V_s and internal resistance R_i. The load is connected to the output of the second stage at terminals A_3 and B_3. If the cascaded amplifier is to work properly, certain conditions must be satisfied. The working of one stage should not adversely affect the performance of the other.

Example 10.9 The value of α_{dc}, for a certain transistor is 0.98 and the collector leakage current is 1 mA. Determine (a) the collector current, and (b) base current, when I_E = 1 mA.

Solution. Given are: α_{dc} = 0.98, I_C = 1μA = 1×10^{-6} A, and I_E = 1mA = 1×10^{-3} A

(a) The collector current is obtained from

$$\begin{aligned} I_C &= \alpha_{dc} I_E + I_{CO} \\ &= (0.98 \times 1 \times 10^{-3}) + (1 \times 10^{-6}) \\ &= 0.981 \times 10^{-3} = 0.981\ \mu A \end{aligned}$$

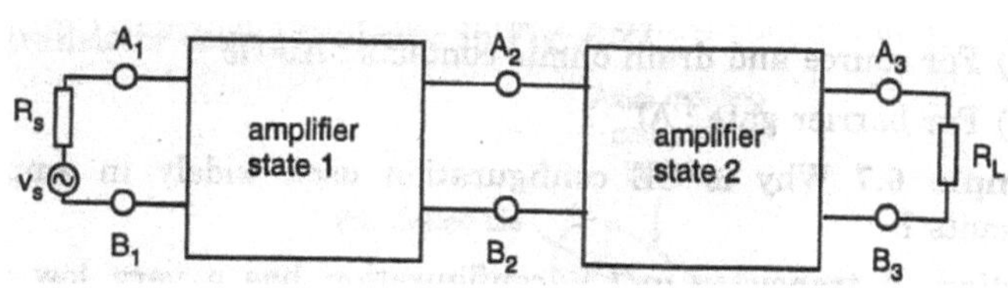

Fig. 10.19 Two amplifier stages cascaded to increase the amplifying action

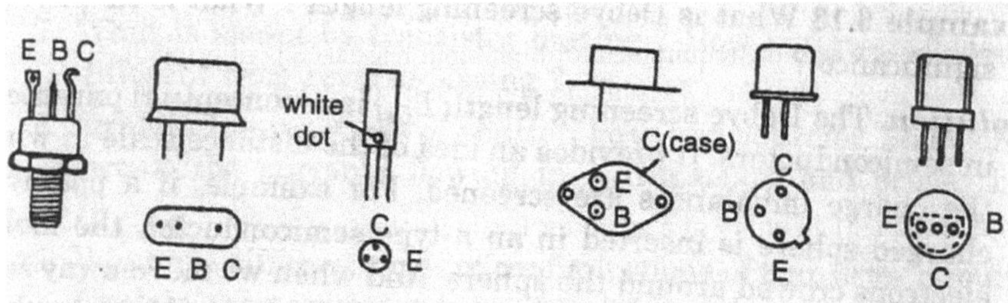

Fig. 10.20 Transistor terminal identification

(b) The base current is calculated from

$$\begin{aligned} I_B &= I_E - I_C \\ &= (1 \times 10^{-3}) - (0.981 \times 10^{-3}) \\ &= 0.019 \times 10^{-3} = 0.019 \text{ mA} = 19\ \mu\text{A}. \end{aligned}$$

Example 10.10 Describe the various terminal identifications of transistors.

Solution. These are shown in Fig. 10.20.

Example 10.11 What is 'Kirk effect'?

Solution. Kirk effect involves a phenomenon in which the current gain drops at high collector currents. It occurs on effective widening of neutral base due to modification of charge distribution in depletion space, at the reverse-biased base-collector junction.

Example 10.12 In a common base connection, $I_C = 0.96$ mA and $I_B = 0.05$ mA. What is the value of α?

Solution. We know that, $\alpha = I_C/I_E$ and $I_E = I_B + I_C$

so,

$$\begin{aligned} \alpha &= \frac{I_C}{I_C + I_B} \\ &= \frac{0.96 \times 10^{-3}}{0.96 \times 10^{-3} + 0.05 \times 10^{-3}} = \frac{0.96}{1.01} \\ &= 0.95 \end{aligned}$$

Example 10.13 In common base connection, the emitter current is 1 mA. If the the emitter side is open, the collector current is 60 μA. Calculate the total current if $\alpha = 0.93$.

Solution. We know that

$$I_T = \alpha I_E + I_{CO}$$
$$I_T = 0.93 \times (1 \times 10^{-3}) + (60 \times 10^{-6})$$
$$= 0.99 \text{ mA}$$

Review Questions

1. Define the following terms.
   (a) Bipolar junction
   (b) Field-effect transistor
   (c) Transistor saturation
   (d) Transistor cut-off
   (e) Base transportation factor
   (f) dc alpha
2. What are the advantages of transistor in electronic (semiconductor) devices? Bring out a broad classification of transistors.
3. Explain as to why an ordinary junction transistor is called `bipolar' ? How is it different from a `unipolar' transistor ?
4. Sketch and explain the construction and working of a BJT with the help of suitable diagrams.
5. Explain the fundamentals of a BJT operation.
6. What is meant by transistor biasing? How is forward-biasing different from reverse-biasing?
7. Explain the functions of emitter, base and collector. Show on a circuit the currents flowing in a transistor, and prove that $I_E = I_B + I_C$.
8. What are different types of configurations of transistor circuit ? Explain them with the help of suitable line diagrams.
9. Show the biasing arrangements for a p-n-p transistor in CB configuration so that it works in active region.
10. Explain the reason as to why the base current in a transistor is usually much smaller than I_E or I_C in active operation.

11. Sketch typical output characteristic curves for p-n-p transistor in CB configuration. Label all variable and indicate active, cut off and saturation regions.
12. Draw the circuit of a CE transistor configuration and explain all related information of its characteristics.
13. Sketch a typical CE input characteristics for an n-p-n transistor. Label all variables. Outline the procedure of calculating the input dynamic resistance of the transistor at a given point from these curves.
14. Derive the relationship between the beta and alpha parameters of a transistor.
15. How is a BJT useful in amplifier circuit ? Draw the circuit diagram of a CB amplifier and explain its salient features, gains and characteristics.
16. Draw the circuit diagram of a CE-amplifier and explain it. Also discuss its salient finding, gains, and characteristics.
17. Draw the circuit diagram of a CC-amplifier and explain the salient features, characteristics and gains.
18. Compare the CB, CE and CC amplifiers from different view points. Also draw their input and output waveforms.
19. Enlist the uses where each type of amplifier (CB, CE and CC) are used. Explain as to why the CE configuration is most popular in amplifier circuits.
20. Write the specifications of a transistor. Also mention the transistor terminal identification.

Numerical Problems

1. For a certain transistor, $7\alpha_{dc}= 0.98$ and emitter current $I_E= 2$ mA. Calculate the values of collector current I_C and base current I_B.

   [**Ans** : $I_C = 196$ mA, $I_B = 40$ mA]

2. The emitter current I_E in a transistor is 2 mA. If the leakage current I_{CBO} is 5 μA and $\mu_{dc} = 0.985$, calculate the collector and base currents.

   [**Ans** : $I_C= 0.975$ mA, $I_B = 25$ μA]

3. A transistor is supplied with dc voltages so that $I_B = 40$ μA. If $\beta_{dc} = 80$ and leakage current is 5 μA, what is the value of emitter current I_E?

   [**Ans** : $I_E = 3.645$ mA]

4. The collector current $I_C = 2.9$ mA in a certain circuit. If base current $I_B = 100$ μA calculate α_{dc} of the transistor.

[**Ans** : $\alpha_{dc} = 0.97$]

5. In a transistor circuit, $I_E = 5$ mA, $I_C = 4.95$ mA and $I_{CBO} = 200$ mA. Calculate β_{dc}, and leakage current I_{CBO}.

[**Ans** : $\beta_{dc} = 99$, $I_{CBO} = 2$ μA]

6. Collector current in a CB transistor is 5 mA. If $\beta_{dc} = 140$ and base current is 35 μA, calculate the leakage current I_{CO}.

[**Ans** : $I_{CO} = 0.71$ μA]

7. Table No. 10.7 gives values of the collector current and collector voltage for a series of base current values in a transistor in the CE configuration. Plot these characteristics and hence find (a) the current gain when the collector voltage is 6 V, (b) the output resistance for a base current of 45 μA.

Table 10.7

V_{CE} V)	Collector current (mA)			
	$I_B = 25$ μA	$I_B = 45$ μA	$I_B = 65$ μA	$I_B = 85$ μA
3	0.91	1.59	2.25	3.00
5	0.92	1.69	2.45	3.20
7	0.96	1.84	2.65	3.50
9	0.97	2.04	2.95	4.00

[**Ans** : (a) 40.25, (b) 13,33 kΩ]

8. A variation of 5 μA in the base current produces a change of 1.2 mA in the collector current. Collector-to-emitter voltage remains fixed during this variation. Calculate the current amplification factor β_{dc}.

[**Ans** : $\beta_{dc} = 240$]

9. A transistor is connected in CB configuration. When the emitter voltage is changed by 200 mV, the emitter current changes by 5 mA. During this variation, collector-to-base voltageis kept fixed. Calculate the dynamic input resistance of the transistor.

[**Ans** : $r_i = 40\ \Omega$]

10. In an n-p-n silicon transistor $\alpha_{dc} = 0.995$, $I_E = 10$ mA, leakage current $I_{CO} = 0.5$ μA. Determine I_C, I_B, β_{dc}, and I_{CEO}.

[**Ans** : $I_C = 9.9505$ mA, $I_B = 49.5$ μA, $\beta_{dc} = 199$, $I_{CEO} = 100$ μA]

Objective Questions

1. The current I_{CBO}
   (a) is generally greater in silicon than in germanium transistors
   (b) depends largely on the emitter-base junction bias
   (c) depends largely on the emitter doping
   (d) increase with an increase in temperature
2. The arrowhead on the transistor symbol always point in the direction of
   (a) hole flow in the emitter region
   (b) electron flow in the emitter region
   (c) minority-carrier flow in the emitter region
   (d) majority-carrier flow in the emitter region.
3. In CB configuration the output volt-ampere characteristics of the transistor may be shown by plots of
   (a) V_{CB} versus I_C for constant values of I_E
   (b) V_{CB} versus I_B for constant values of I_E
   (c) V_{CE} versus I_E for constant values of I_B
   (d) V_{CE} versus I_C for constant values of I_E
4. A transistor connected in common base configuration has
   (a) a low input resistance and high output resistance
   (b) a high input resistance and a low output resistance
   (c) a low input resistance and a low output resistance
   (d) a high input resistance and a high of output resistance
5. Compare to CB amplifier the CE amplifier has
   (a) lower input resistance
   (b) higher output resistance
   (c) lower current amplification
   (d) higher current amplification

6. Which of the following transistors is symmetrical in the sense that emitter and collector, source and drain terminals can be interchanged?
   (a) JFET (b) FET
   (c) NPN transistor (d) PNP transistor.
7. While using a bipolar junction transistor as an amplifier, the collector and emitter terminals got interchanged mistakenly. Assuming that the amplifier is a common emitter amplifier and the biasing is suitably adjusted, the interchange of terminals will result into which one of the following?
   (a) Zero gain (b) Infinite gain
   (c) Reduced gain (d) No change in gain.

Answers

1. (d) 2. (a) 3. (a) 4. (a) 5. (d) 6. (b)
7. (c)

Reference

1. Gupta, K.M., Gupta, N.: Advanced Electrical And Electronics Materials. Scrivener Wiley Publishing, USA (2015)

Chapter 11
Metal Semiconductor Field Effect Transistors, MOS Transistors, and Charge Coupled Device

Abstract Detailed classification of different types of transistors is presented. The field-effect transistors (FETs) are exhaustively sub-classified into MESFET, MISFET, MODFET, MOSFET, IGFET, DE-MOSFET, E-only MOSFET, n-channel and p-channel types. The advantageous features of FETs over BJTs and applications of FETs are given. MISFETs: their construction, basic types and performance are discussed. MESFET: its construction, operation and characteristics are explained. MOSFETs: their basic types, construction, operation and short-channel effects are elaborated. IGFET and HEMT, n-channel and p-channel FET are described. The charge coupled devices: its operation and salient uses are explained. Minute insight into the various topics are given through solved numerical and theoretical examples. Review questions, numerical problems and objective type questions are also given with their answers.

Keywords Field-effect transistor (FET) · Metal-semiconductor field-effect transistors (MESFETs) · Insulated-gate field-effect transistor (IGFET) · MOSFET · Enhancement type MOSFET · Depletion type MOSFET · n-channel MOSFET · p-channel MOSFET · Symbols for different types of MOSFETs · PMOS · Charge coupled devices

11.1 Introduction

A field-effect transistor is a semiconductor device. It is a junction device. Unlike a diode which has only one junction, a transistor has *two* junctions. It is one of the most important micro-electronic devices which have completely revolutionized the scenario of modern electronic industry. There has been a rapidly expanding effort to utilize and develop many types of semiconductor devices such as FET, MOSFET, UJT, SCR, etc.

K.M. Gupta and N. Gupta, *Advanced Semiconducting Materials and Devices*, Engineering Materials, DOI 10.1007/978-3-319-19758-6_11

Different types of transistors are used in vivid applications. They serve different functions in numerous applications such as in amplifiers, oscillators etc. A detailed classification is shown in Fig. 11.1, and their description follows in subsequent sections. The earlier Fig. 10.1 has been reproduced here as Fig. 11.1 for better understanding.

11.2 Field-Effect Transistor (FET)

FET is an abbreviation of field-effect transistor. It is a unipolar, 3-terminal solid-state device in which the current is controlled by an electric field. The FET was developed in early 1960s. There are several kinds of FETs viz. JFET, MOSFET, MISFET, IGFET etc. The FETs process many advantages over BJTs. These are listed below.

1. FETs have high current impedence.
2. FETs have high power gain.
3. FETs can be fabricated easily for different devices, particularly the ICs.
4. They are of very small size.

A FET differs from a junction transistor. Unlike BJT whose operation depends on both; the electrons and holes, the operation of FET depends on the flow of majority carriers only. It is, therefore, also known as *unipolar transistor* (UPT).

11.2.1 Applications of FETs

FETs are used in almost all those application where bipolar transistors are also used. Besides them, the FETs find exclusive uses in following applications.

1. In logic circuits where it is kept 'OFF' when the input is zero. It can be turned 'ON' with very little power input.
2. As voltage-variable resistor (VVR) in amplifiers and tone controls etc.
3. In large-scale integration (LSI) and computer memories.
4. As input amplifiers in the measuring and testing equipment, and also in oscilloscopes and electronic voltmeters, etc.
5. For mixer operation of FM and TV receivers.

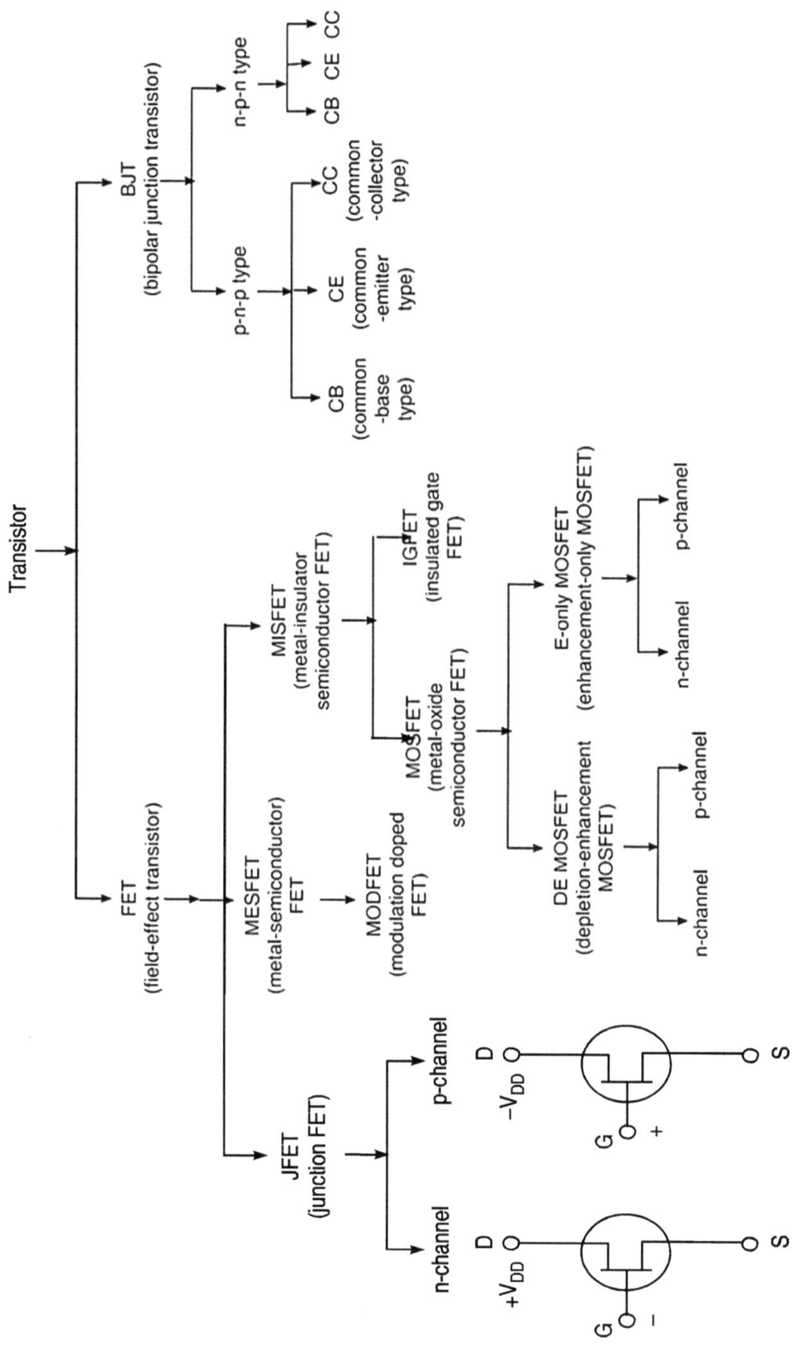

Fig. 11.1 Classification of transistors. Reproduced from [1]

11.3 Metal-Semiconductor Field-Effect Transistors (MESFET)

A MESFET is a metal-semiconductor junction. It uses a reverse-biased Schottky barrier instead of a *p-n* junction. The MESFET devices are generally made of III–V group compounds such as GaAs, InP etc. Instead of Si. It is because the metal-semiconductor FET device possesses several meritorious features such as given below.

1. They have high operational speed.
2. They have high power levels.
3. They can be fabricated to close geometrical tolerances.
4. They can be made very small. Gate lengths of less than 0.25 μm are very common.
5. They have higher mobility.
6. They have higher carrier drift velocities than Si.
7. They can be operated at higher temperatures.
8. They have better high frequency operations since the drift time and capacitance may be kept a minimum in them.

These devices are useful in following applications.

- High-speed digital circuits
- High-speed microwave circuits
- High-frequency services
- GaAs integrated circuits
- Very small size gates having lengths ≤0.25 μm.

11.4 Basic Construction of MESFETs

JFETs have low noise and relative insensitivity to cosmic radiation. Therefore, they are ideally suitable for amplification in GHz frequency range such as for the satellite communication systems. However, their main disadvantage is their low gain bandwidth product. Since these devices employ a Schottky barrier metal gate instead of the *p-n* junction, they are known as metal-semiconductor field-effect transistors (MESFETs).

Figure 11.2a–b shows the cross-section of a GaAs MESFET. The device is fabricated by growing an *n*-type epitaxial layer on a high-resistivity GaAs substrate. The substrate may be either of *n*-type or of *p*-type. The source and drain ohmic contacts are made by depositing a suitable metal by vacuum evaporation and by alloying. The metal gate offers a number of advantages viz. The formation of a Schottky barrier contact can be achieved at much lower temperatures than those

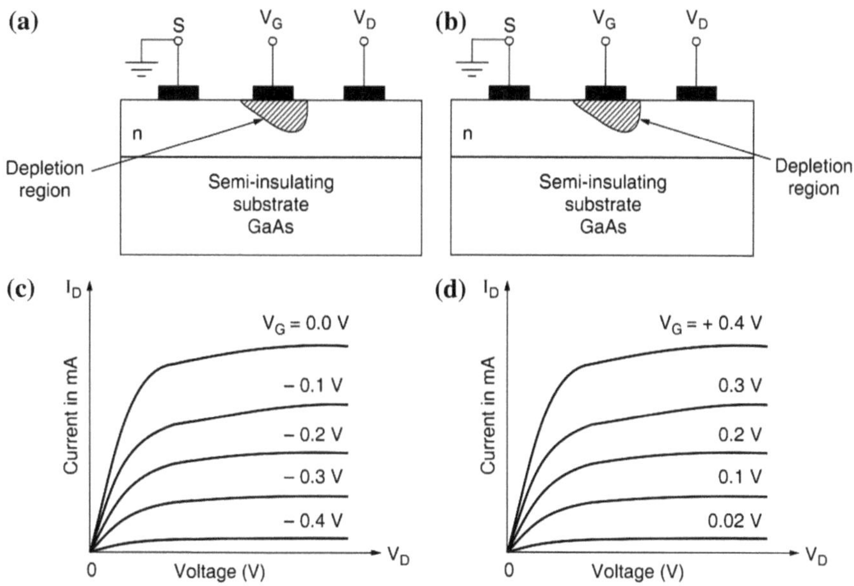

Fig. 11.2 Two basic types of MESFET: **a** depletion mode device and **b** enhancement mode device

required for *p-n* junction formation. And also, the gate length can be reduced to submicron dimensions by controlling the length of the metal electrodes.

11.4.1 Basic Types of MESFETs

Depending on the thickness of the epitaxial layer, a MESFET may be designed to operate either as a

1. depletion mode (normally on) device, or
2. enhancement mode (normally off) device.

In a *depletion mode device*, the thickness of the epitaxial layer is more than the zero-bias depletion region width of the Schottky barrier gate, and the transistor has a conducting channel at $V_G = 0$. Thus, the gate is biased negative to deplete the channel, as shown in Fig. 11.2a.

In the *enhancement-type device* (Fig. 11.2b), the epitaxial layer is kept thin, and the built- in voltage of the metal gate Schottky barrier junction is sufficient to deplete the channel completely at $V_G = 0$. Conduction in the channel occurs only for small positive values of V_G.

Enhancement-type devices are useful in high-speed, low-power applications; but the majority of MESFETs are the depletion type.

11.5 High Frequency Performance

The MESFET (or GaAs FET) is also known a high performance field effect transistor that is used for high performance microwave applications and in semiconductor RF amplifiers. The MESFET shares many features with FET or JFET, but the MESFET is able to offer superior performance in the region of RF microwave operation and for use within RF amplifiers. The small geometries and other aspects of the device make it ideal in this application. Typically a supply voltage of around 10 volts is used. Proper care must be taken when designing the bias arrangements because if the current flows in the gate junction, it will destroy the GaAs FET. Similarly care must be taken when handling the devices as they are 'static sensitive'. In addition to this, when used as an RF amplifier connected to an antenna, the device must be protected against during electrical storms. If these precautions are observed, the MESFET performs exceedingly well. It is an electronics component that is relatively cheap and performs well.

11.6 Models for I-V Characteristics of Short Channel MESFET

Since MESFETs have a very small channel length, typically less than 1 μm, therefore their operation is influenced by *short-channel effect.* These effects dominate the volt-current (*V-I*) characteristics of MESFET. For Example, the high-field effects occur when 1 V appears across a channel length of 1 μm (10^{-4} cm), giving an electric field of 10 kV/cm.

A simple approximation to the velocity-field curve assumes a constant mobility (linear) dependence up to some critical field ζ_c and constant saturation velocity v_s for higher fields. For Si, it is

$$v_d = \frac{\mu\zeta}{1 + \mu\zeta/v_s} \tag{11.1}$$

where μ is the low-field mobility. If we assume that the electrons passing through the channel drift with a constant saturation velocity v_s the current takes a simple form given as

$$I_D = env_sA = eN_dv_sZh \tag{11.2}$$

where h is a slow function of V_G. In this case the saturated current follows the velocity saturation, and does not require a true pinch-off in the sense of depletion region meeting at some point in the channel. As shown in Fig. 11.3, the $I_D - V_D$ curves are more evenly spaced if the constant saturation velocity dominates, compared with the V_G-dependent spacing for the long-channel constant-mobility case.

Fig. 11.3 Depiction of characteristic of MESFET

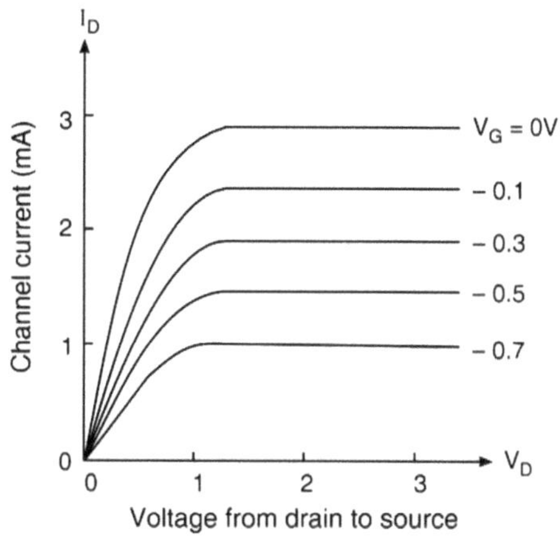

11.7 Operation of MESFET

Since increased doping causes an increased scattering by the ionized impurities, a high electron concentration is created in the channel of MESFET. It is done by growing a thin 'undoped well' bounded by wider band gap doped barriers (e.g. AlGaAs). This configuration is called 'modulation doping' and the resulting device is called *modulation doped field-effect transistor* (MODFET).

Figure 11.4a shows a conduction band view of modulation doping. Electrons in donor-doped AlGaAs fall into the GaAs potential well and become trapped. With this configuration, we can achieve a high electron concentration in the channel as well as retaining the high mobility. The mobile electrons generated by the donors in AlGaAs diffuse into small band gap layer of GaAs.

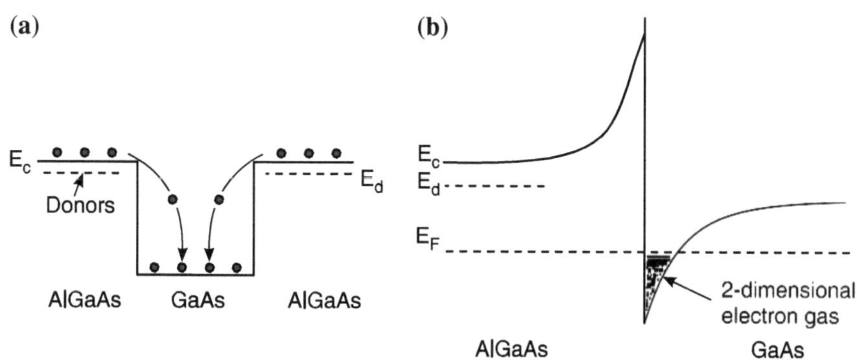

Fig. 11.4 Illustration of **a** modulation doping shows the electrons in donor doped AlGaAs falling into GaAs potential well and being trapped, and **b** a single heterojunction showing trapping of electrons in undoped GaAs

Figure 11.4b shows the diffusion of mobile electrons generated by the donors in AlGaAs into small band gap GaAs layer. They are prevented from returning to the AlGaAs by the potential barrier at the AlGaAs/GaAs interface. The electrons in almost triangular well form a two-dimensional electron gas.

11.8 Construction of MESFET Structure

Schematic construction of a MESFET is shown in Fig. 11.5. It comprises of the following components.

(i) A semi-insulating GaAs *substrate.*
(ii) A source *S*.
(iii) A drain contact *D* of Au-Ge alloy.
(iv) A Schottky gate *G* of Al, or alloys of Au, Ti or W.
(v) A *n*-type GaAs layer grown epitaxially on the substrate to form *channel region.*

The substrate may be doped or undoped. If doped, the doping is accomplished by chromium 'Cr' as its energy level is close to the centre of GaAs band gap. Its use as dopant results in producing a very high resistivity material, whose resistivity is of the order of about 10^8 Ω-cm. This is the reason due to which the substrate is referred to as semi-insulating GaAs. This is non-conducting.

Details of other components are as follows.

Source. It is the terminal through which the majority carriers enter into the system.

Drain. It is the terminal through which the majority carriers leave the system i.e. they are drained out from this terminal. The drain-to- source voltage V_{DS} drives the drain current I_D.

Gate. It is a heavily-doped impurity region. The gate-source voltage V_{GS} reverse-biases the gate. The gate is insulated from its conducting channel by an ultra-thin metal oxide insulating film.

Channel region. It is the space between two gates through which the majority carriers pass from source to drain, when V_{DS} is applied. This is formed of a thin

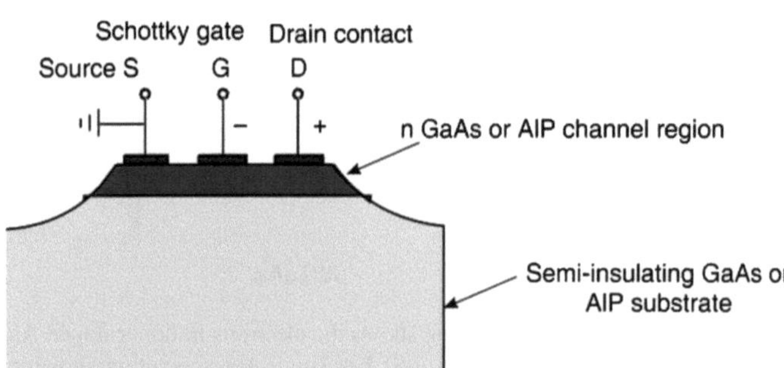

Fig. 11.5 Constructional details of a MESFET

layer of n-type GaAs on the substrate. The GaAs is lightly doped and is epitaxically grown. This epitaxial layer is also called *buffer zone* when grown between the two layers.

11.9 Insulated-Gate Field-Effect Transistor (IGFET)

A MISFET is a **m**etal-**i**nsulator-**s**emiconductor (MIS) transistor. This is one of the most used electronic devices. In this device the channel current is controlled by a voltage applied at gate electrode and an insulator is used to isolate this gate from the channel. Due to this reason, this device is also known as 'insulated-gate field-effect transistor (IGFET)'. The MISFET is used in following main applications.

- Digital ICs, in which it is switched from no conducting channel ('OFF' state) to conducting channel ('ON' state).

11.10 MOSFET

Metal Oxide Semiconductor Field Effect Transistor (MOSFET) is an important semiconductor device which is widely used in many circuit applications. The main advantage of MOSFEF over JFET is that the input impedance of a MOSFET is much more than that of a JFET, because of very small gate leakage current. Since the gate of MOSFET is insulated from the channel, it is also termed as Insulated Field Gate Effect Transistor (IGFET). MOSFETs are further classified into depletion type and enhancement type based on their basic mode of operation.

11.10.1 Basic Types of MOSFETs

MOSFETs may be categorized into two types viz.

1. enhancement type and
2. depletion type.

The enhancement type devices are normally off i.e. channel does not exist for $V_{gs} = 0$ and the applied $|V_{gs}|$ must be greater than V_T for the device to turn on. On the other hand, the depletion type devices are normally on i.e. channel does exist even for $V_{gs} = 0$ and the applied $|V_{gs}|$ must be reduced below V_T for the device to turn off. It implies that an n-channel enhancement type device has a positive V_T, whereas an n-channel depletion type device has a negative V_T. Similarly, a p-channel enhancement type device has a negative V_T, whereas a p-channel depletion type device has a positive V_T. The threshold voltage can be changed either by doping or by ion implantation. Since these are charged, they can change the charge state of the surface, hence the threshold voltage.

11.11 Construction of MOSFET

Since the materials used are Si, SiO_2 and metal; this transistor is commonly known as MOSFET. Fabrication details of a typical MISFET are shown in Fig. 11.6a. It is a *n*-channel construction comprising of the following components.

- *Source S* made of a metal contact.
- *Drain D* made of a metal contact.

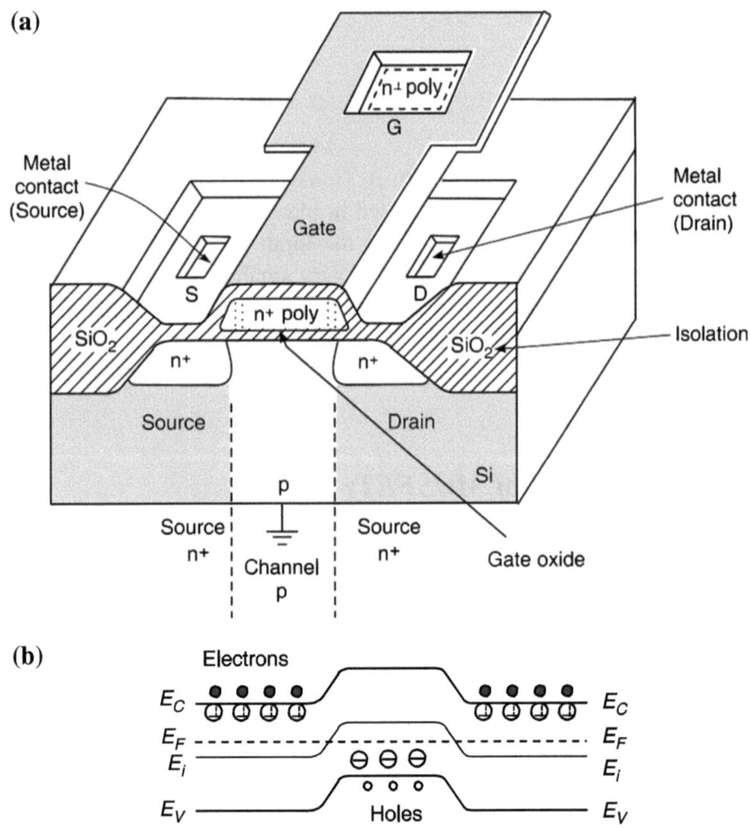

Fig. 11.6 An enhancement type *n*-channel MOSFET showing its **a** fabrication details, and **b** equilibrium band diagram along channel. Reproduced from [1]

- *Isolation* made of SiO_2.
- *Gate electrode* made of a heavily doped polysilicon.
- *Substrate* made of *p*-type Si.

The *n*+ source and drain regions are implanted (or diffused) into the substrate. Any current cannot flow from drain to source until a conducting *n*-channel is placed between them.

The MOSFET is surrounded on all sides by a thick SiO_2 layer. It is for the purpose of providing critical electrical isolation between the adjacent transistors on an *IC*. The equilibrium band diagram is shown in Fig. 11.6b.

11.12 Operating Principle and I-V Characteristics of Enhancement Type n-Channel MOSFET

To understand the operation of the MOSFET shown in Fig. 11.6a, we consider its band diagram shown in Fig. 11.6b. It depicts the equilibrium state along the channel. The Fermi level is flat in equilibrium. The conduction band is close to the Fermi level in *n*+ source/drain, while the valence band is closer to the Fermi level in *p*-type material. Hence, there is a potential barrier for an electron to go from the source to the drain, corresponding to the built-in potential of the back-to-back *p*-*n* junctions between the source and drain.

When a positive voltage is applied to the gate relative to the substrate (which is connected to the source in this case), positive charges are in effect deposited on the gate metal. In response, the negative charges are induced in the underlying Si, by the formation of depletion region and a thin surface region containing mobile electrons. These induced electrons form the channel of the FET, and allow current to flow from drain to source.

As Fig. 11.7 suggests, the effect of the gate voltage is to vary the conductance of this induced channel for low drain-to-source voltage, analogous to the JFET case.

Since electrons are electrostatically induced in the *p*-type channel region, the channel becomes less *p*-type and therefore the valence band moves down farther away from the Fermi level. This obviously reduces the barrier for electrons between the source, channel and the drain. If the barrier is reduced sufficient by applying a gate voltage in excess of what is known as the *threshold voltage* V_T, there is significant current flow from the source to the drain. Thus, one view of a MOSFET is that it is a gate-controlled potential barrier. It is very important to have high-quality, low-leakage *p-n* junctions in order to ensure a low off-state leakage in the MOSFET. For a given value of V_G, there will be some drain voltage V_D for which the current becomes saturated, after which it remains essenially constant.

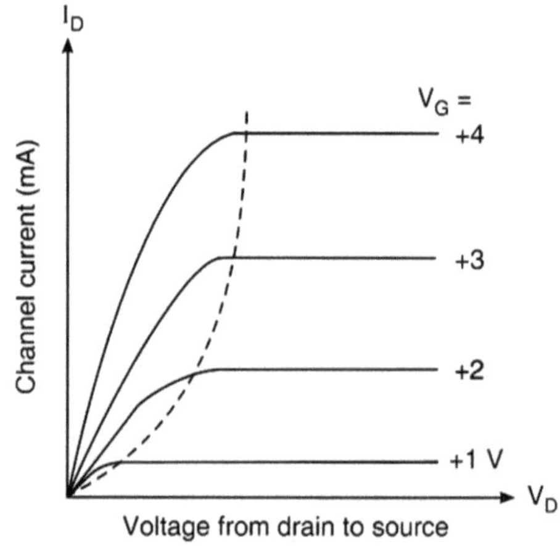

Fig. 11.7 Effect of gate voltage on channel current

11.13 p-Channel Enhancement MOSFET (PMOS)

Construction: Figure 11.8 shows the construction of p-channel enhancement type MOSFET. Its construction is exactly reverse of the n-channel enhancement type MOSFET. In this construction, n-type substrate and p-doped regions under the drain and source terminals. The terminals remain the same, but all the voltage polarities and the current directions are reversed.

11.13.1 Characteristics

The drain characteristics are shown in Fig. 11.9b. It can be noted that the drain current increases with increase in the negative gate to source voltages (V_{GS}). The transfer characteristics is shown in Fig. 11.9a. It can be further noted that it is exactly a mirror image of the transfer characteristics of an n-channel enhancement MOSFET. The drain current remains zero up to the voltage $V_{GS} < V_t$ and increases as negative V_{GS} becomes than V_t as shown in Fig. 11.9b.

11.13.2 Symbols for Enhancement MOSFET

Figure 11.10 shows the symbols for n-channel and p-channel enhancement type MOSFET. In the symbol, the dashed line between drain and source is chosen such that a channel does not exist between the two under no bias condition. It is the only difference between the symbols for the depletion type (see Fig. 11.17) and enhancement type MOSFETs.

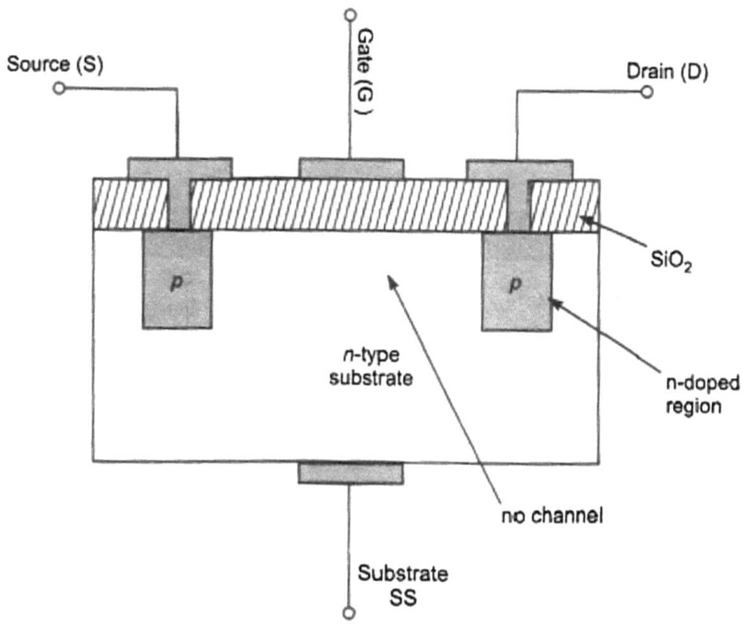

Fig. 11.8 Construction of p-channel DMOSFET

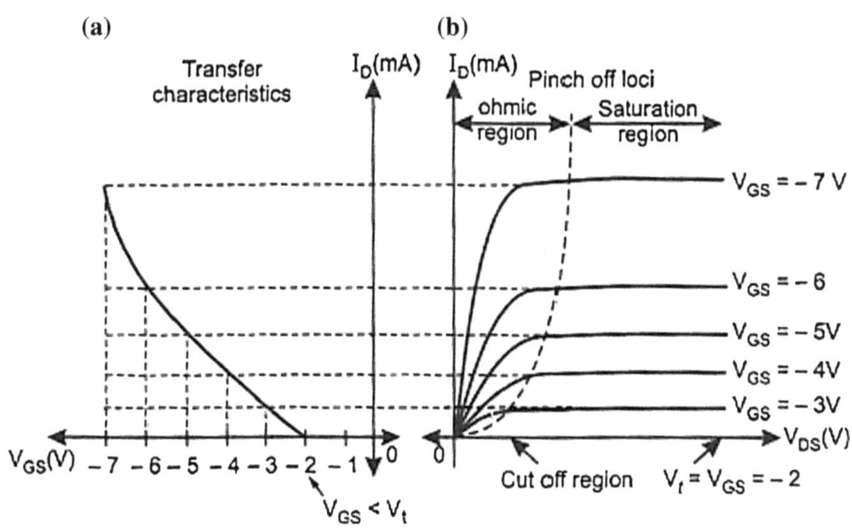

Fig. 11.9 Construction of p-channel EMOSFET

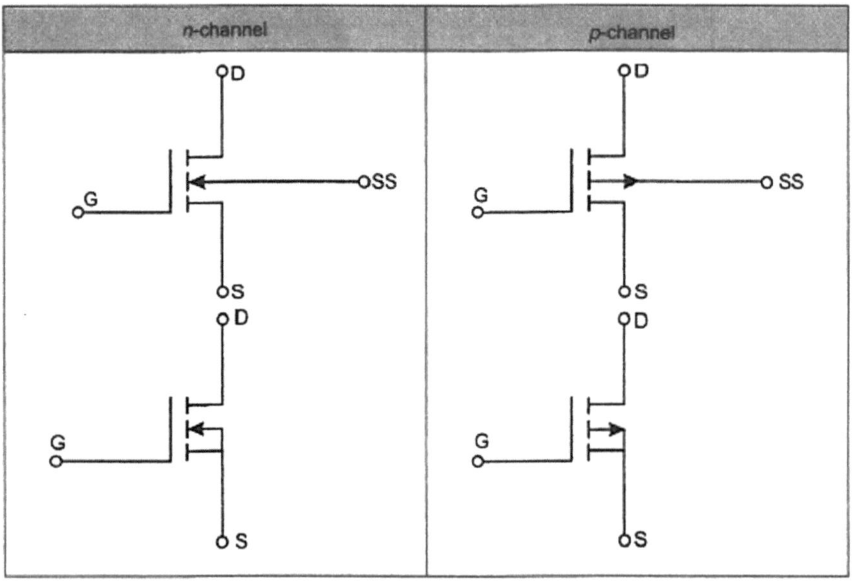

Fig. 11.10 Symbols for MOSFET

11.14 Depletion Type MOSFET

Construction The basic structure of an n-channel depletion type MOSFET is shown in Fig. 11.11. A slab of p-type material is formed from a silicon base as the substrate. However, many discrete devices provide an additional terminal labelled SS, resulting in a four terminal device. The source and drain terminals are connected through an *n*-channel as shown in Fig. 11.11a. The gate is also connected to a metal contact surface but remains insulated from the n-channel by a very thin silicon dioxide (SiO_2) layer. *There is no electrical connection between the gate terminal and the channel of MOSFET.* Due to the presence of the SiO_2 layer between gate terminal and n-type channel, the input impedance of MOSFET is very high.

For n-channel depletion type MOSFET there will be an n-type silicon region (channel) connecting the source and drain at the top of the p-type substrate as shown in Fig. 11.11a. Similarly for p-channel depletion type MOSFET there will be a p-type region channel connecting the source and the drain at the top of the N-type substrate as shown in Fig. 11.11b.

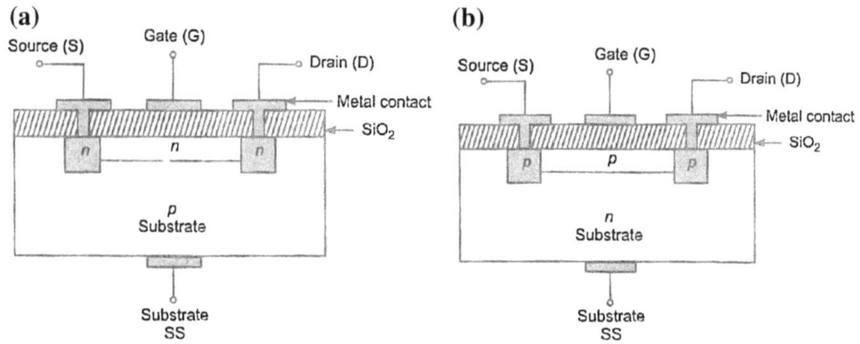

Fig. 11.11 Structure for depletion type MOSFET, **a** n-channel depletion type MOSFET, **b** p-channel depletion type MOSFET

11.14.1 Operation of n-Channel Depletion MOSFET

The gate to source voltage set to be zero and a voltage V_{DD} is applied across the drain to source terminals as shown in Fig. 11.12. This results in free electrons from the channel getting attracted towards positive potential of drain terminal. This establishes current flow through the channel to be denoted as I_{DSS} at $V_{GS} = 0$ V. The depletion type MOSFET can be operated in two different modes as given below.

1. Depletion mode: The device operates in this mode, when the gate voltage is negative
2. Enhancement mode: The device operates in this mode, when the gate voltage is positive.

11.14.2 Depletion Mode of Operation

Figure 11.13a shows a MOSFET with a negative gate to source voltage. The negative potential at the game will tend to pressure or repel electrons towards the p-type substrate and attracts the holes from the p-type substrate as shown in Fig. 11.13b. Depending on the magnitude of the negative bias established by V_{GS}, a level of recombination between repelled electrons and attracted holes will occur that will reduce the number of free electrons in the n-channel for conduction. The resulting drain current is reduced by increasing the negative gate voltage.

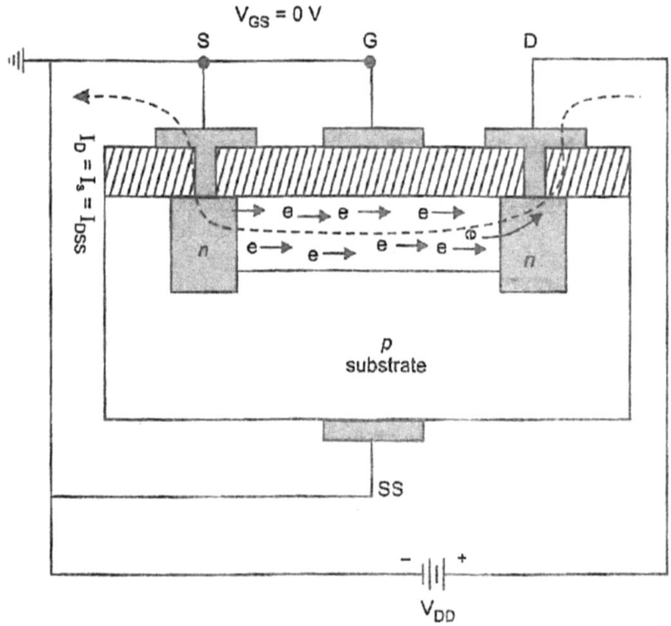

Fig. 11.12 n-channel depletion MOSFET operation when V_{GS}= 0 V

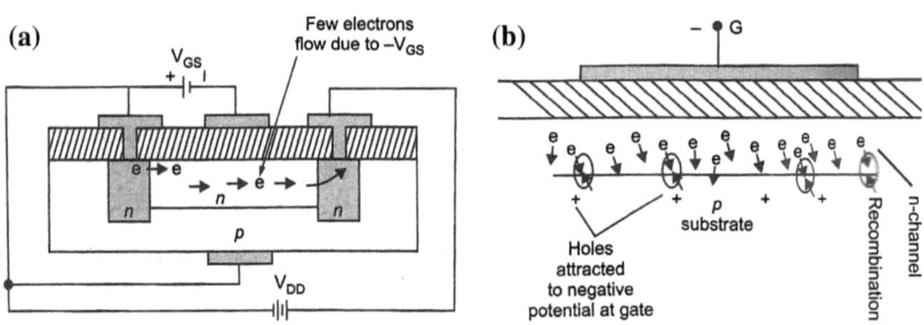

Fig. 11.13 Depletion mode of operation

The greater the negative voltage applied at the gate, the greater the depletion of n-channel electrons. The level of drain current will reduce with increasing negative bias for V_{GS}.

Fig. 11.14 Enhancement mode of operation

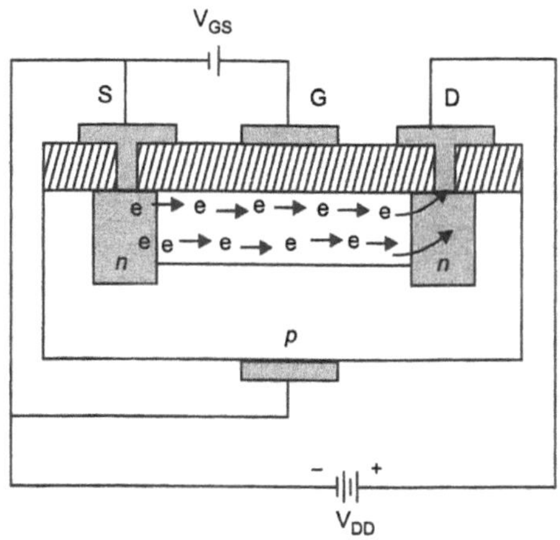

11.14.3 Enhancement Mode of Operation

Figure 11.14 shows the depletion MOSFET with a positive gate to source voltage. For positive values of V_{GS} the gate will draw additional electrons from the p-type substrate. As the gate to source voltage continues to increase, the drain current will increase at a rapid rate. Because of this, as gate to source voltage increases in positive direction, the drain current also increases. The application of a positive gate to source voltage has *enhanced* the level of free electrons in the channel as compared to that encountered with $V_{GS} = 0$ V. For this reason, this mode of operation is called as enhanced mode of operation.

11.14.4 Characteristics of n-Channel Depletion MOSFET

Figure 11.15 shows the drain and transfer characteristics for an n-channel depletion MOSFET. It may be note from this curve that the region 0AB of the characteristics is extends for the positive values of gate to source voltage V_{GS}. The value of I_{DSS} represents the current from drain to source with $V_{GS} = 0$, it is given by

$$I_D = I_{DSS}\left(\frac{V_{GS}}{V_{GS(off)}}\right)^2 \tag{11.3}$$

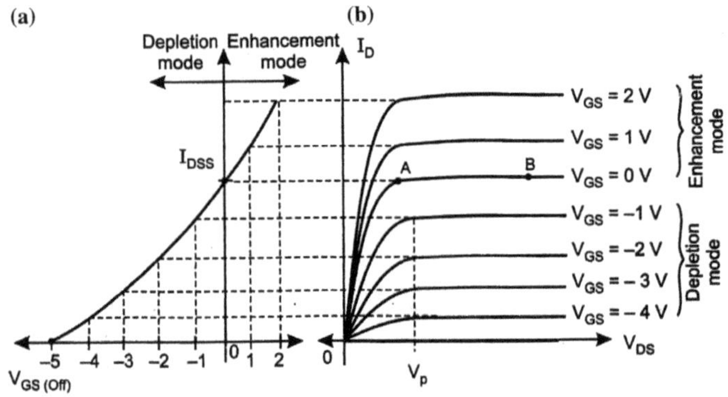

Fig. 11.15 Characteristics of depletion MOSFET, **a** transfer characteristic, **b** drain characteristic

The drain characteristics for an n-channel depletion type MOSFET is shown in Fig. 11.15b. These are tied for both negative and positive values of gate to source voltage (V_{GS}). When V_{GS} is zero and positive, the MOSFET operates in the enhancement mode. When V_{GS} is negative, the MOSFET operates the depletion mode.

Gate source cut off voltage [$V_{GS\{off)}$]. It is the gate source voltage where the channel is completely cut off and the drain current becomes zero. It can be easily understood if we refer to the transfer characteristics shown in Fig. 11.15a. As the reverse gate source voltage is increased, the cross sectional area of the channel decreases. This in turn decreases the drain current. At some gate-source voltage, the depletion layers extend completely across the channel. In this condition, the channel is off and drain current reduces to zero. The gate voltage at which the channel is cut off is called cut off voltage (V_{GSoff}). It is interesting that $V_{GS(off)}$ will always have the same magnitude value as V_P.

$$V_{GS(off)} = -V_P \tag{11.4}$$

11.14.5 *p-channel Depletion MOSFET and Its Characteristics*

The construction of a p-channel depletion type MOSFET is exactly reverse of n-channel as shown in Fig. 11.16b. The terminals are same but all the voltage

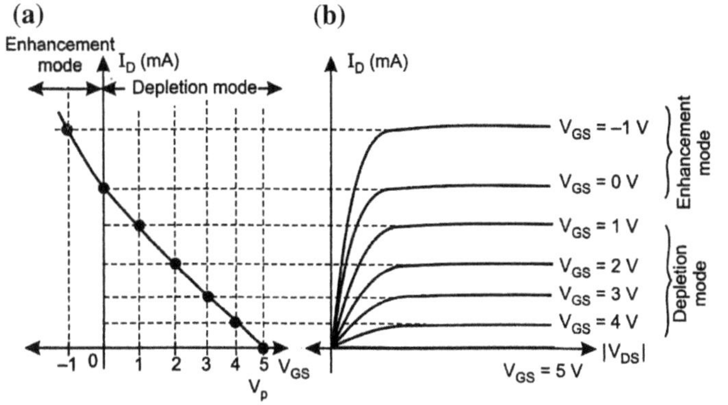

Fig. 11.16 **a** Transfer characteristic, **b** drain characteristic

polarities and the current directions are reversed. The drain characteristics would be same as n-channel depletion MOSFET, but with V_{DS} having negative values, I_D having positive values and V_{GS} having opposite polarities as shown in Fig. 11.16.

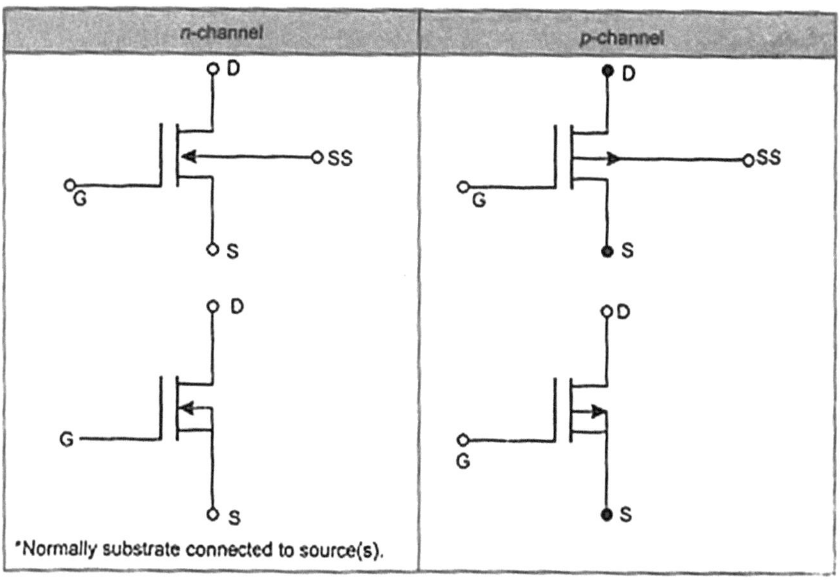

Fig. 11.17 Symbol for depletion MOSFET

11.15 Symbols of Depletion MOSFET

The symbol for n- and p-channel depletion MOSFET are shown in Fig. 11.17. This symbol is so chosen to reflect the actual construction of the device. There is no direct connection (due to the gate insulation) between the gate and channel. This is represented by a space between the gate and other terminal of the symbol. The vertical line representing the channel is connected between the drain and source and is supported by the substrate.

11.16 Comparison Between n-channel and p-channel MOSFET

S. no.	n-channel MOSFET	p-channel MOSFET
1.	It is costlier and somewhat difficult to fabricate than p-channel MOSFET	1. p-channel enhancement MOSFET is easier and cheaper to produce than n-channel MOSFET
2.	Since the hole mobility is nearly 2.5 times lower than the electron mobility, n-channel MOSFET occupy less area than p-channel MOSFET for same I_D rating	2. It occupies larger area than n-channel MOSFET for same I_D rating
3.	The drain resistance is lower than the identical p-channel MOSFET	3. The drain resistance is three times higher than of identical n-channel MOSFET
4.	It has higher packing density than p-channel MOSFET	4. Lower packing density than n-channel MOSFET
5.	Almost three times faster than n-channel MOSFET is switching application	5. It is slower than n-channel MOSFET

11.17 Comparison Between DMOSFETs and EMOSFETs

S. no.	DMOSFETs	EMOSFETs
1.	Channel between source and drain is present	Channel is absent between source and drain
2.	Drain current becomes zero when $V_{Gs} = 0$ V	Drain current becomes zero when $V_{GS} < V,$
3.	For an n-channel DMOSFET, I_D decreases as V_{Gs} becomes more and more negative	For an n-channel EMOSFET, I_D increases as V_{Gs} becomes more and more positive mode

(continued)

(continued)

S. no.	DMOSFETs	EMOSFETs
4.	It is operated in two modes 1. Deletion mode (V_{GS} is −Ve) 2. Enhancement mode (V_{Gs} is +Ve)	It is operated only in enhancement mode
5.	An insulating oxide layer (SiO_2) is present between gate and channel	The insulating oxide layer is present between gate and substrate

11.18 Comparison Between FETs and BJTs

Devices parameters	FETs	BJTs
Polarity of carrier	FET is a unipolar device. i.e. current device is carried either by electron or holes	BJT is a bipolar device. i.e. device current is carried by both electron and holes
Devices type	It is a voltage control device, i.e. applied voltage at the gate or drain terminal controls the amount of current flowing through the device	It is a current controlled device i.e. the base current controls the overall collector current
Input resistance	It's input resistance is very high of the order of several mega ohms	Its input resistance is very low as compared to FET and is of the order of few kΩ
Leakage current	This device does not suffer from reverse leakage current flow, because there is only one kind of carrier (either electron or holes)	It suffers from reverse leakage current flow which creates a lot of problem in device operation
Noise	It is less noisy than BJT or vaccum tubes that's why it is widely used in low-levei signal amplification	It is comparatively more noisy than FET
Switching speed	FET is much faster than BJT	BJT has a slow response as compared to FET
Fabrication	It's fabrication process is much easier than BJT	It is difficult to fabricate BJT devices
Packing density	FET has a high packing density as compared to BJT	BJT has a low packing density as compared to FET

11.19 Short-Channel Effects

A MOSFET device is considered to be a 'short device' when the channel length is of the same order of magnitude as the depletion-layer widths (x_{dD}, x_{dS}) of the source and drain junction. As the channel length L is reduced to increase both the operation speed and the number of components per chip, the effects of short-channel arise. Due to variations during the device fabrication, the geometry and doping parameters of a MOS transistor are subjected to uncertainties. Therefore, the electrical parameters of the transistor vary. Devices should have electrical parameters that are rather insensitive against the geometry fluctuations. Traditionally, the relationship between the threshold voltage and gate length, also called 'V_{th} roll-off', is used to measure the short-channel effect of a MOS transistor, as shown in Fig. 11.18 for the PCD device and the uniformly doped device. The high-drain bias threshold voltages are extracted using the constant current method which means that the gate-source voltage is taken at I_d = 100 nA. The PCD device also shows a much better V_{th} roll-off for shorter channel lengths.

The short-channel effects are attributed to two physical phenomena as given below.

(i) The limitation imposed on electron drift characteristics in the channel
(ii) The modification of the threshold voltage due to the shortening channel length.

Five different short-channel effects can be distinguished as follows.

1. Drain-induced barrier lowering and punch-through
2. Surface scattering

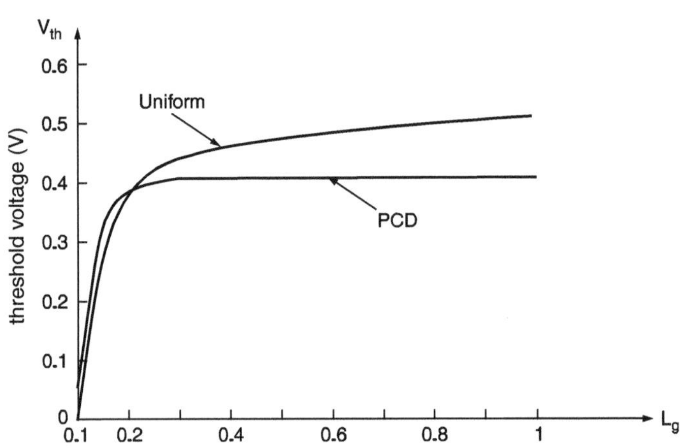

Fig. 11.18 Threshold voltage versus gate length characteristic of a MOS transistor to measure the short-channel effect

3. Velocity saturation
4. Impact ionization
5. Hot electrons.

Drain-induced barrier lowering and punch-through. When the depletion regions surrounding the drain and extends to the source so that the two depletion layers merge (i.e. $x_{dS} + x_{dD} = L$), the punch-through occurs. Here, x_{dS} and x_{dD} are the source junction width and drain junction width respectively. Punch-through can be minimized with the use of thinner oxides, larger substrate doping, shallower junctions, and longer channels.

In small-geometry MOSFETs, the potential barrier is controlled by both the gate-to-source voltage (V_{GS}) and the drain-to-source voltage (V_{DS}). If the drain voltage is increased, the potential barrier in the channel decreases; leading to *drain-induced barrier lowering* (DIBL). The reduction of the potential barrier eventually allows the electron flow between the source and the drain, even if the gate-to-source voltage is lower than the threshold voltage. The channel current that flows under this conditions ($V_{GS} < V_{TO}$) is called the sub-threshold current.

Surface scattering. As the channel length becomes smaller due to lateral extension of depletion layer into the channel region, the longitudinal electric field component increases and the surface mobility becomes field-dependent. Since the carrier transport in a MOSFET is confined with the narrow inversion layer, the *surface scattering* i.e. the collisions suffered by the electrons that are accelerated toward the interface, causes reduction of the mobility. Consequently, the electrons move with great difficulty parallel to the interface. Therefore the average surface mobility is about half as much as that of the bulk mobility.

Velocity saturation. The performance of short-channel devices are also affected by *velocity saturation*, which reduces the trans-conductance in saturation mode. At low electric field, the electron drift velocity v_{de} in the channel varies linearly with the electric field intensity. However as the electric field increases above 10^4 V/cm, the drift velocity tends to increase more slowly, and approaches a saturation value of $v_{de}(\text{sat}) = 107$ cm/s around 10^5 V/cm at 300 K.

Impact ionization. It is an undesirable short-channel effect, especially in NMOS. It occurs due to high velocity of electrons in presence of high longitudinal fields that generate electron-hole (*e-h*) pairs by *impact ionization* i.e. by impacting on silicon atoms and ionizing them. Normally most of the electrons are attracted by the drain, while the holes enter the substrate to form part of the parasitic substrate current.

Hot electrons. Another issue related to high electric fields is caused by so-called *hot electrons*. These high energy electrons enter the oxide where they get trapped and give rise to oxide charging. It accumulates with time and degrades the device performance by increasing *VT* and affects adversely the gate's control on the drain current.

11.20 A Charge Coupled Device

A charge coupled device (CCD) is an integrated circuit etched onto a silicon surface forming the light sensitive elements, called *pixels*. Photons incident on this surface generate charge that can be read and turned into a digital copy of the light patterns falling on the device. CCDs come in a wide variety of sizes and types. They are used in many applications such as

- cell phone cameras
- high-end scientific applications

A (CCD) stores and displays the data for an image in such a way that each pixel (picture element) in the image is converted into an electrical signal, the intensity of which is related to a colour in the colour spectrum such as for conversion into a digital value. This is achieved by '*shifting*' the signals between the stages, within the device, one at a time. CCDs move signals between capacitive *bins* in the device, with the shift allowing for the transfer of signals between bins. Figure 11.19 shows various CCDs of different sizes. The largest is mounted on a 15 cm wafer.

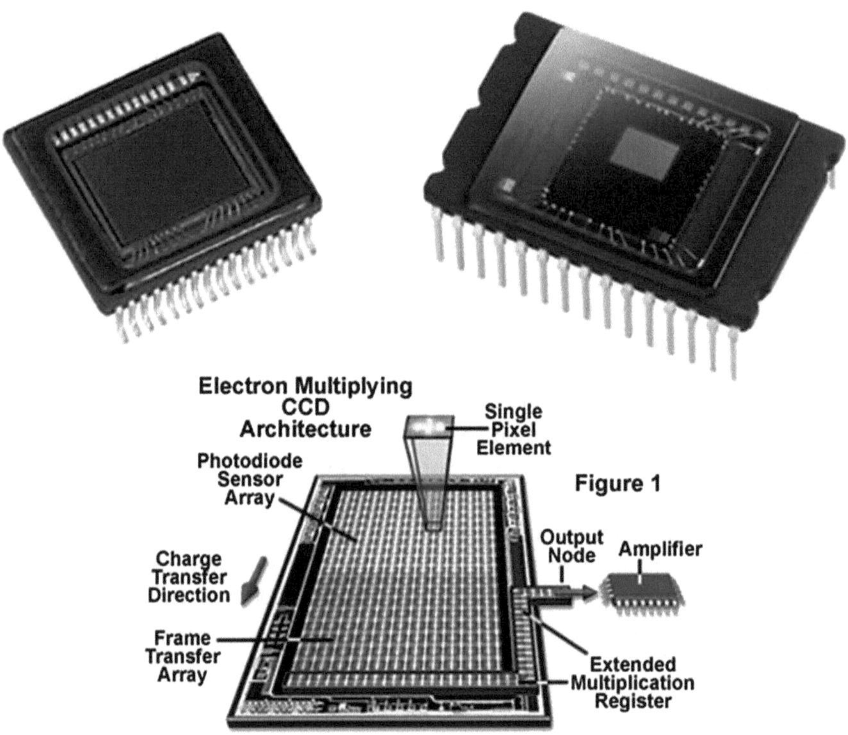

Fig. 11.19 Showing various types of CCDs [2, 3]

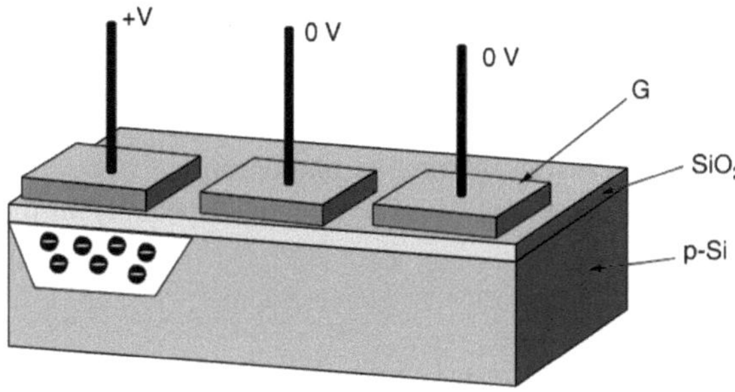

Fig. 11.20 Explanation of the basic principle of CCD

11.20.1 Operation

In a CCD, there is a photoactive region for capturing images, and a transmission region made out of a shift register (the CCD). The photoactive region is an epitaxial layer of silicon. An image is projected through a lens onto the array of capacitor. This is photoactive region causing each capacitor to accumulate an electric charge proportional to the light intensity at that location. A one-dimensional array used in line-scan cameras, captures a single slice of the image; while the other a two-dimensional array used in video and still cameras, captures a two-dimensional picture that corresponds to the scene projected onto the focal plane of the sensor. Once the array is exposed to the image, a control circuit causes each capacitor to transfer its contents to its neighbour which operates as a shift register. The last capacitor in the array dumps its charge into a charge amplifier, which converts the charge into a voltage. By repeating this process, the controlling circuit converts the entire contents of the array in the semiconductor to a sequence of voltages.

In a digital device, these voltages are sampled, digitized and stored in memory. In an analog device such as an analog video camera, they are processed into a continuous analog signal. It is done by feeding the output of the charge amplifier into a low-pass filter, which is then processed and fed to other circuits for transmission, recording, or other processing. Figure 11.20 shows the basic principle of CCD. In it, the charge packets (e.g. electrons, blue) are collected in potential wells created by applying the positive voltage at the gate electrodes (G). On applying the positive voltage to the gate electrode in correct sequence, the charge packets are transferred.

11.20.2 Salient Uses

CCDs are now-a-days commonly included in 'digital still' and 'video cameras.' They are also used in

- astronomical telescopes
- scanners
- bar code readers

The devices are also used in

- machine vision for robots
- in optical character recognition (OCR)
- in processing of satellite photographs
- in enhancement of radar images, especially in meteorology

11.21 Solved Examples

Example 11.1 What is the main difference between FET and BJT?

Solution. The FET is a majority carrier device and is therefore called as *unipolar transistor*; whereas the BJT operates by injection and collection of minority carriers. Since the electrons and holes both play role in this device, it is called as a *bipolar transistor.*

Example 11.2 Suggest the materials used to form (a) the source and drain ohmic contacts for MESFET, and (b) for Schottky barrier gate.

Solution. The following materials are suggested for these purposes.

(a) For source and drain ohmic contacts: Au-Ge
(b) For barrier gate: Al

Example 11.3 How many transistors are used in a semiconductor memory devices and microprocessors?

Solution. Million of MOS transistors are commonly used in them.

Example 11.4 Use of *n-p-n* transistors is more popular than *p-n-p* transistors—why?

Solution. The *n-p-n* transistors are used more than the *p-n-p* transistors because the electrons have higher mobility than holes (i.e. $\mu_e > \mu_h$).

Example 11.5 How does a transistor work in switching operation?

Solution. In switching operation, a transistor is controlled in two different conduction states viz. On-state and OFF-state. Ideally a switch appears as a short-circuit when turned-on, and as open circuit when turned off.

Example 11.6 What is HEMT?

Solution. This is high electron-mobility transistor. It is also known as MODFET.

Example 11.7 Define the transistor characteristics viz. dynamic drain resistance, mutual conductance, and amplification factor.

Solution. These characteristics are defined below.

- *Dynamic drain resistance* R_d. At an operating point, it is defined as the ratio of small change in drain voltage to small change in drain current, when the gate voltage is constant. That is

$$R_d = \left.\frac{\Delta V_D}{\Delta i_D}\right|_{V_{G=\text{constant}}} \tag{11.5}$$

Typically the value of R_d is about 400 kΩ.

- *Mutual conductance* or *transconductance* g_m. The mutual conductance at an operating point is defined as the ratio of small change in drain current to the small change in gate voltage, keeping the drain voltage constant. That is

$$g_m = \left.\frac{\Delta i_D}{\Delta V_G}\right|_{V_{\text{D=constant}}} \tag{11.6}$$

It is measured in the unit of Siemens S. Typically, its value ranges from 150 to 250 μS.

- *Amplification factor* μ. It is defined as the ratio of small change in drain voltage to the small change in gate voltage, when current I_D is kept constant. That is

$$\mu = \left.\frac{\Delta V_D}{\Delta V_G}\right|_{I_{D=\text{constant}}} \tag{11.7}$$

Since μ is a ratio of two voltages, it does not have any unit. The amplification factor of an FET can be as high as 100.

- *Inter-relation*. The above three parameters are related as

$$\mu = R_d g_m \tag{11.8}$$

Thus if any two parameters are known, the third can be computed.

Example 11.8 What is Debye screening length? What is its physical significance?

Solution. The Debye screening length L_{Dsl} is a conceptual parameter in semiconductors. It provides an idea of the distance scale in which the charge imbalances are screened. For example, if a positively charged sphere is inserted in an n-type semiconductor, the mobile electrons crowed around the sphere. And when we move away from the sphere by several Debye lengths, the positively charged sphere and negative electron cloud look like a neutral entity. The Debye length is expressed as

$$L_{Dsl} = \sqrt{\frac{\varepsilon_s KT}{e^2 p_o}} \quad \text{or} \quad \sqrt{\frac{\varepsilon_s KT}{e^2 n_o}} \tag{11.9}$$

where ε_s is surface permittivity, p_o and n_o are equilibrium hole and electron concentrations. The value of L_{Dsl} varies inversely with doping.

Review Questions

1. Answer the following terms.
   (a) Basic types of MESFETs (b) Field-effect transistor
   (c) Transistor saturation (d) Transistor cut-off
2. What are the advantages of field effect transistors in electronic (semiconductor) devices? Bring out a broad classification of transistors.
3. Differentiate between the following kinds of transistors.
   (a) BJT and FET (b) MESFET and MISFET
   (c) MODFET and MOSFET (d) JFET and IGFET
   (e) DE-MOSFET and E-only MOSFET (f) *n*-channel and *p*-channel
4. Explain the enhancement mode and depletion mode FET. How is n-channel MOSFET different from p-channel MOSFET?
5. Explain the meaning of MESFET. State its meritorious features and uses.
6. Discuss the basic types of MESFETs. How is the 'depletion mode' different from the'enhancement mode'?
7. What do you mean by high-frequency performance? Explain the models for I-V characteristics of a short-channel MESFET.
8. Sketch and explain the construction and operation of a MESFET. Discuss its characteristics and operation also.
9. Discuss the basic types of MOSFETs. Sketch and explain the construction, operating principleand I-V characteristicsof a MOSFET.
10. In respect of MESFET and MOSFET, give a brief account of the following:
    (a) Source and drain (b) Gate and channel
    (c) Symbols for *n* and *p* channels (d) Different parameters.
11. Explain the operation of a DE MOSFET under (a) depletion mode, and (b) enhancement mode, of *n*-channel
12. What is meant by MISFET? How isit different from MESFET and MOSFET? State its advantageous features and uses.
13. Explain the short-channel phenomenon, its causes and effects.
14. Sketch and explain the construction, operation and characteristics of a IGBT with the help of suitable diagrams.
15. Sketch and explain the construction and operation of a MISFET. Discuss its characteristics also.
16. Describe charge-coupled devices stating their operation and applications..
17. Write notes on the following.
    (a) Shockley diode (b) LASCR (c) Bilateral device

Objective Questions

1. The arrowhead on the transistor symbol always point inthe direction of
   (a) hole flow in the emitter region
   (b) electron flow in the emitter region
   (c) minority-carrier flow in the emitter region
   (d) majority-carrier flow in the emitter region.
2. A field-effect transistor (FET)
   (a) uses a high-concentration emitter junction
   (b) uses a forward-biased *p-n* junction
   (c) has a very high input resistance
   (d) depends on minority-carrier flow
3. Consider the following statements.
   A. Negative conductance is involved with instability of semiconductors.
   B. The Gunn diodes employ the combination of carrier injection and transit time.
   C. Atomic density of semiconductor is influenced due to heavy doping.
   D. At equilibrium, there is equal tunnelling from *n* to *p* and from *p* to *n*, in a tunnel diode.

   Of these, the correct statements are
   (a) A, B and C (b) A, C and D
   (c) A, B and D (d) B, C and D
4. Match list I with list II, and choose the correct answer from the codes given below the lists.

List I	List II
A. Region between E_v and E_F in degenerate *p*-type semiconductor	1. dV/dI is negative
B. On applying small forward bias to tunnel diode	2. GaAs
C. The Gunn diode	3. Almost completely filled with holes
D. IMPATT diode	4. p^+-n-i-n^+
	5. E_{Fn} moves up withrespect to E_{Fp}
	6. for only a part, filled with holes

Codes:

	A	B	C	D		A	B	C	D
(a)	1	6	4	2	(b)	1	5	2	4
(c)	3	5	2	4	(d)	3	6	4	2

5. Which one of the following statements is *not* correct?
   (a) Power MOSFETs are constructed so as to avoid punch through.
   (b) In power MOSFET, the channel length is relatively large and channel width is relatively small.
   (c) Power MOSFETs do not experience any minority carrier storage.
   (d) Power MOSFETs can be put in parallel to handle large currents.

6. In an n-channel JFET, the connections of the battery are such that
   (a) the drain is at zero potential MOSFETs with respect to source.
   (b) the drain is at positive potential MOSFETs with respect to source.
   (c) the drain is at negative potential MOSFETs with respect to source.
   (d) None of above.

7. Consider the following statements in respect of important features of IGBT:
   1. It has a high impedance gate as that of a MOSFET.
   2. It has a low ON state voltage as that of a BJT.
   3. It can block negative voltages as in the case of GTO.

   Which of these statements are correct?
   (a) 1 and 2. (b) 1 and 3.
   (c) 2 and 3. (d) 1, 2 and 3.

Answers

1. (a) 2. (c) 3. (b) 4. (c)

5. (b) 6. (d) 7. (a)

References

1. Gupta, K.M., Gupta, N.: Advanced Electrical and Electronics Materials. Scrivener Wiley Publishing, USA (2015)
2. http://www.topsurveillance.info/index.php/ccdvscmos.html
3. http://www.olympusmicro.com/primer/digitalimaging/concepts/emccds.html

Chapter 12
Power Semiconductor Devices

Abstract Many industrial applications require control of power such as in variable speed drives, light-intensity controllers, temperature regulators etc. They are controlled by Power devices such as four-layer devices, semiconductor controlled rectifiers (SCRs) etc. In this regard, a P-N-P-N diode has been explained stating its conduction mechanism, two-transistor analogy, triggering mechanisms etc. SCRs are also explained showing their various shapes and sizes, biasing, operation and triggering. Applications of SCR, half-wave and full-wave power control, and silicon controlled switch are also explained. Bilateral devices such as Diac and Triac are presented along with their working, characteristics and applications. Insulated gate bipolar transistor (IGBT), its characteristics and advantages are also elaborated. Meaning of uni-junction transistor (UJT) and gate controlled switch (GCS) are also briefed. Minute insight into the various topics are given through solved numerical and theoretical examples. Review questions, numerical problems and objective type questions are also given with their answers.

Keywords P-n-p-n diode · Silicon controlled rectifier · Silicon controlled switch · Bilateral devices · Diac · Triac · IGBT · Uni-junction transistor · Thyristors

12.1 Introduction

There are many industrial applications which require control of power. Some of such applications are variable speed drives, light-intensity controllers, temperature regulators etc. Control of power in them is accomplished by means of certain devices, which are known as 'power devices'. Basically these are electronic devices. Vivid kinds of power devices are used in different applications. One of the most common functions of these devices is in 'switching'. Switching requires a device to change its condition from an OFF (i.e. blocking) state to an ON (i.e. conducting) state. Transistors, diodes and other devices are used for switching operation. If transistor is used as switching device, its base current drives the device from 'cut-off state' to 'saturation'.

K.M. Gupta and N. Gupta, *Advanced Semiconducting Materials and Devices*, Engineering Materials, DOI 10.1007/978-3-319-19758-6_12

Switching requirements for different devices vary widely. Transistors, diodes and other devices are not suitable for all sorts of switching operations. They can be used to serve the purpose for certain types of switching only. There are a number of other important switching applications which require ‘a device such that the switching should remain in OFF state under forward bias, until it is switched to ON state by an external signal’.

1. Semiconductor controlled rectifier (SCR)
2. Insulated gate bipolar transistor (IGBT)

The semiconductor controlled rectifier is generally made of Si, therefore it is popularly known as silicon controlled rectifier (SCR). An IGBT is made of a combination of SCR and FET. Details of these devices are described in forthcoming sections.

12.1.1 *Different Types of Power Semiconducting Devices*

The power semiconducting devices are categorized as follows.

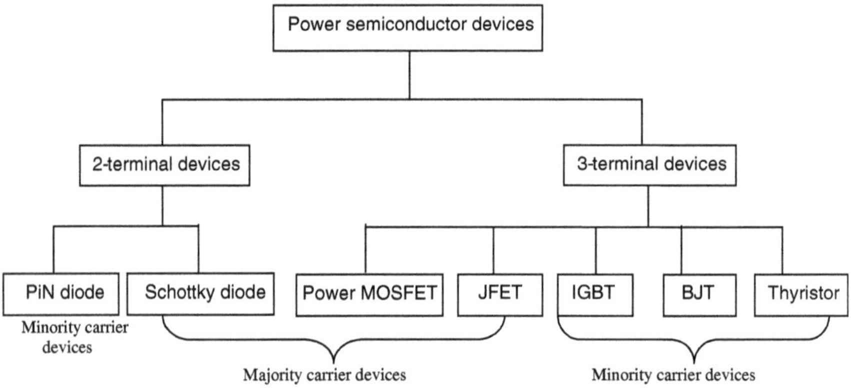

12.1.2 *Four Layer Devices*

A four-layer device is one which is formed of four different layers of p- and n-type semiconductors. The p- and n-semiconductors are arranged in alternate layers in such a way that they form a p-n-p-n structure. One such structure is shown in Fig. 12.1a. It is an SCR. There are many varieties of p-n-p-n structure to form a four-layer device. These devices effectively block the current through two terminals until the current is again turned on by a small signal, at the third terminal.

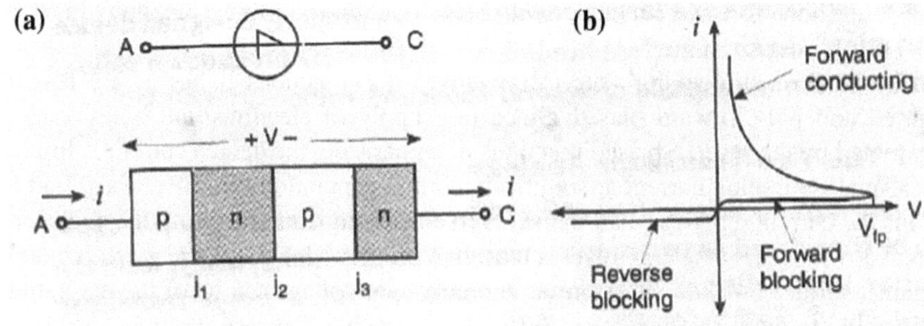

Fig. 12.1 A four-layer device showing **a** the two terminal p-n-p-n device, and **b** its I-V characteristic

The four-layer devices may be of different kinds. Basically, they may have

- four-layer structure with two terminals, and
- four-layer structure with three terminals

Normally, the third terminal is used for triggering (i.e. firing) the device.

12.2 P-N-P-N Diode

A p-n-p-n diode is a four-layer device, which can be considered for many purposes as the combination of a p-n-p and a n-p-n transistor. Its basic structure consists of an anode terminal A located outside the p-region with a cathode terminal K located outside the n-region (Fig. 12.1a). There are three junctions. The junction nearest to the anode is j_1, the center junction is j_2, and the junction nearest to the cathode (voltage positive), the device is forward biased. The forward-biased condition of this diode can be considered in two separate states viz.

1. the high-impedance or forward-blocking state, and
2. the low-impedance or forward-conducting state.

In this device the forward I-V characteristic (shown in Fig. 12.1b) switches from the blocking to the conducting states at a critical peak forward voltage V_{fp}.

12.2.1 Conduction Mechanisms

An initial positive voltage V places the junctions j_1 and j_3 under forward bias and the center junction j_2 under reverse bias. As V is increased, most of the forward voltage in the blocking state appears across the reverse-biased junction j_2. After switching to the conducting state, the voltage from A to K is very small (less than

1 V), and therefore we conclude that in this condition all three junctions must be forward biased.

In the reverse-blocking state (V negative), the junctions j_1 and j_3 are reverse biased and j_2 is forward biased. Since the supply of electrons and holes to j_2 is restricted by the reverse-biased junctions on either side, the device current is limited to a small saturation current arising from thermal generation of EHPs near j_1 and j_3. Thus the current remains small in the reverse-blocking condition until the avalanche breakdown occurs at a large reverse bias. In a properly designed device, with guards against surface breakdown, the reverse breakdown voltage can be of the magnitude of several thousand volts.

12.2.2 The Two-Transistor Analogy

The four-layer configuration of Fig. 12.1a suggests that the p-n-p-n diode[1] can be considered as two coupled transistors in which j_1 and j_2 form the emitter and collector junctions respectively of a p-n-p transistor. Similarly, j_2 and j_3 form the collector and emitter junctions of an n-p-n transistor. In this analogy, the collector region of n-p-n common with the base of p-n-p, and the base of n-p-n serves as the collector region of the p-n-p. The center junction j_2 serves as the collector junction for both transistors.

This two-transistor analogy is illustrated in Fig. 12.2. The collector current i_{c1} of p-n-p transistor drives the base of n-p-n, and the base current i_{B1} of p-n-p is dictated by collector current i_{C2} of the n-p-n. If we associate an emitter-to-collector current transfer ratio a with each transistor, we can solve for the current i. Hence using

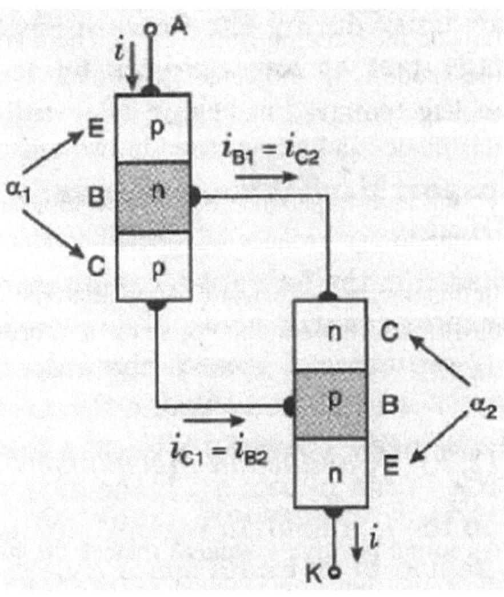

Fig. 12.2 Two-transistor analogy of p-n-p-n diode

[1]**Note:**The p-n-p-n diode is often called as Shockley diode.

$\alpha_1 = \alpha_N$ for the p-n-p, $\alpha_2 = \alpha_N$ for the n-p-n, and with I_{CO1} and I_{CO2} for the respective collector saturation currents, we have

$$i_{C1} = \alpha_1 i + I_{CO1} = i_{B2} \tag{12.1a}$$

and

$$i_{C1} = \alpha_2 i + I_{CO2} = i_{B1} \tag{12.1b}$$

But the sum of i_{C1} and i_{C2} is the total current through the devices; therefore

$$i_{C1} + i_{C2} = i \tag{12.2}$$

Taking this sum in Eqs. 12.1a, 12.1b we have

$$i\ (\alpha_1 + \alpha_2) + I_{CO1} + I_{CO2} = i$$

or

$$i = \frac{I_{CO1} + I_{CO2}}{1 - (\alpha_1 + \alpha_2)} \tag{12.3}$$

As Eq. 12.3 indicates, the current i through the device is small as long as the sum $\alpha_1 + \alpha_2$ is small in comparison with unity. As the sum of the alphas approaches unity, the current i increases rapidly. However, it does not increase without limit because the derivation is no longer valid as $\alpha_1 + \alpha_2$ approaches unity. Since the junction j_2 becomes forward biased in the forward-conducting state, both transistors become saturated after switching. The two transistors remain in saturation while the devices is in the forward-conducting state, being held in saturation by the device current.

12.2.3 Variation of a with Injection

Since the two-transistor analogy implies that the switching involves an increase in alphas to the point that $\alpha_1 + \alpha_2$ approaches unity, it may be helpful to review how alpha varies with injection for a transistor. The emitter-to-collector current transfer ratio α is the product of the emitter injection efficiency γ and the base transport factor B. Therefore an increase in α with injection can be caused by increases in either of these factors, or both. At very low currents (such as in the forward-blocking state of the p-n-p-n diodes), γ is usually dominated by recombination in the transition region of the emitter junction. As the current is increased, injection across the junction begins to dominate over recombination with the transition region and γ increases. There are several mechanisms by which the base transport factor B increases with injection, including the saturation of recombination centers as the excess carrier concentration becomes large. Whichever mechanism dominates, the increase in $\alpha_1 + \alpha_2$ required for switching of the p-n-p-n diode is automatically accomplished. In general, no special design is required to maintain $\alpha_1 + \alpha_2$ smaller than unity during the forward-blocking state. This requirement is usually met at low currents by the dominance of recombination within the transition region of j_1 and j_3.

12.2.4 Current Transport Mechanism in Forward-Blocking State of P-N-P-N Diode

When the device is biased in the forward-blocking state (Fig. 12.3a), the applied voltage V appears primarily across the reverse-biased junction j_2. Although j_1 and j_3 are forward biased, the current is small. The reason for this becomes clear if we consider the supply of electrons available to n_1 and holes to p_2. Focusing attention

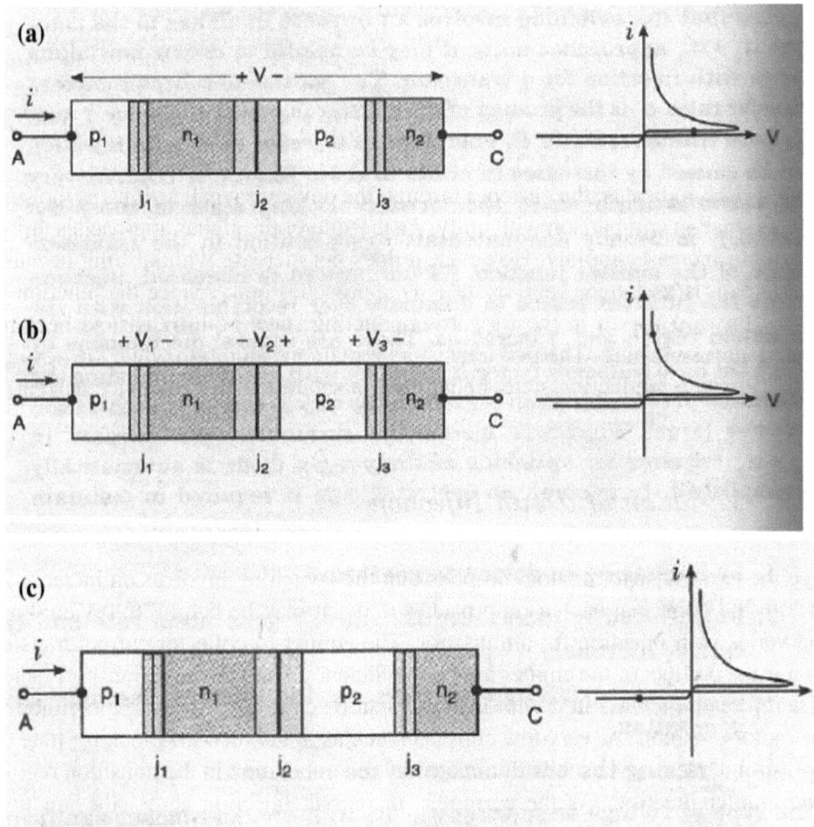

Fig. 12.3 Three bias states of a p-n-p-n diode: **a** the forward-blocking state; **b** the forward conducting sta, and **c** the reverse blocking state

first upon j_1, let us assume a hole is injected from p_i into n_1. If the hole recombines with an electron in n_1 (or in the j_1 transition region), that electron must be resupplied to the n_1 region to maintain space charge neutrality. The supply of electrons in this case is severely restricted. In a normal p-n diode the n region is terminated in an ohmic contact, so that the supply of electrons required to match recombination (and injection into p) is unlimited. In this case, however, the electron supply is restricted essentially to those electrons generated thermally within a diffusion length of j_2. As a result, the current passing through the j_1 junction is approximately the same as the reverse saturation current of j_2. A similar argument holds for the current through j_3. Holes required for injection into n_2 and to feed recombination in p_2 must originate in the saturation current of the center junction j_2. The applied voltage V divides approximately among the three junctions to accommodate this small current throughout the device.

12.2.5 Current Transport Mechanism in Forward Conducting State of P-N-P-N Diode

The charge transport mechanism changes dramatically when the transistor action begins. As $\alpha_1 + \alpha_2$ approaches unity, many holes injected at j_1 swept across j_2 into p_2. This helps to feed the recombination in p_2 and to support the injection of holes into n_2. Similarly, the transistor action of electron injected at j_3 and collected at j_2, supplies electrons for n_1. Obviously, the current through the device can be much larger once this mechanism begins. The transfer of injected carriers across j_2 is regenerative, in that a greater supply of electrons to n_1 allows greater injection of holes at j_1 while maintaining space charge neutrality. This greater injection of holes further feeds p_2 by transistor action, and the process continues to repeat itself.

If $\alpha_1 + \alpha_2$ is large enough, so that many electrons are collected in, n_1 and many holes are collected in p_2, the depletion region at j_2 begins to decrease. Finally the reverse bias disappears across j_2 and is replaced by a forward bias. When this occurs, the three small forward-bias voltages appear as shown in Fig. 12.3b. Two of these voltages essentially cancel in the overall V, so that the forward voltage drop from anode to cathode in the conduction state is not much greater than that of a single p-n junction. For Si this forward drop is less than 1 V, until ohmic losses become important at high current levels.

12.2.6 Triggering Mechanisms

For initiation of switching from one state to another, the basic requirement is that the carrier injection at j_1 and j_2 must be increased. It causes a significant transport of injected carriers across j_2. Once this transport begins, the regenerative nature of the process takes over and switching is completed. There are several means of switching (or triggering) a p-n-p-n diode from the forward-blocking state to the forward-conducting state. These are

1. by increasing in device temperature.
2. by sufficiently increasing the carrier generation rate and the carrier lifetimes.
3. by optical excitation to increase the current through EHP generation.
4. by raising the bias voltage to the peak value V_P.

This type of voltage triggering results in a breakdown (or significant leakage) of the reverse-biased junction j_2. The accompanying increase in current provides the injection at j_1 and j_3, and the transport required for switching to the conducting state. The breakdown mechanism commonly occurs by combination of base-width narrowing and avalanche multiplication.

12.3 Silicon Controlled Rectifier (SCR)

An SCR is a four-layered device, having three p-n. junctions j_1, j_2 and j_3 as shown in Fig. 12.4. It has three electrodes viz. cathode, anode and gate. It provides conduction of current in one direction only. Therefore also named as unidirection reverse-blocking thyristor. It can be switched from OFF to ON state by applying a positive trigger pulse to the gate. That is why it is also called a controlled rectifier.

Commercially, the SCRs are available in many shapes and sizes. Some of them are shown in Fig. 12.5. The SCRs are specified by the amount of anode current they can handle. Normally, they can handle the amount of anode currents from about 10 to 2000 A.

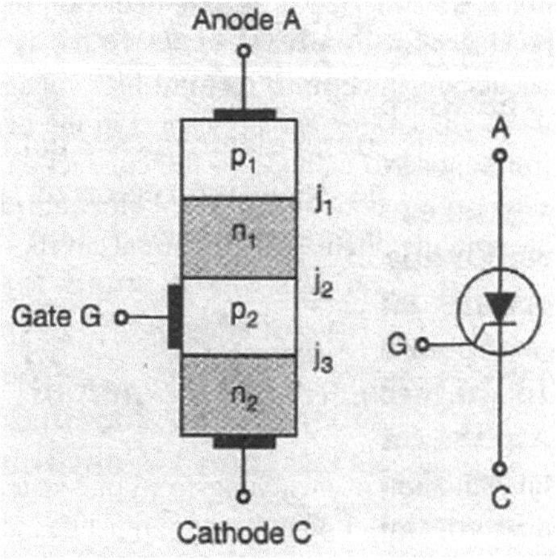

Fig. 12.4 A silicon controlled rectifier showing its **a** structure, and **b** representating symbol

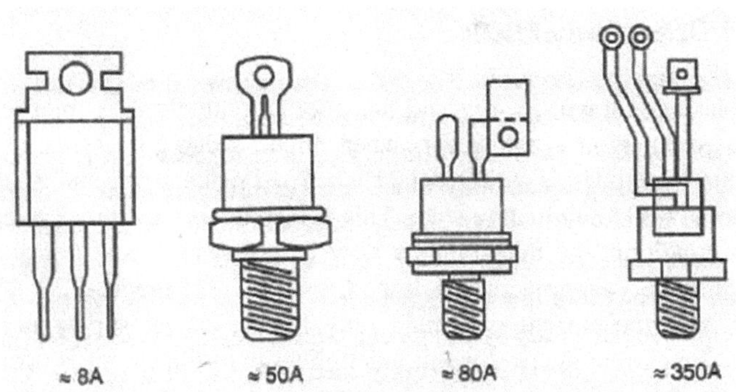

Fig. 12.5 Various shapes and sizes of commercial SCRs

12.3.1 Biasing of SCR

Similar to transistors, the SCRs are also biased in two different modes, namely

1. Forward biased, and
2. Reverse biased.

However, their biasing is junction-wise. For example, with polarity of applied voltage V as shown in Fig. 9.6a, the junctions

- j_1 and j_3 are forward-biased, while
- j_2 is reverse-biased.

In this case no current can flow through the SCR except the leakage current, if any.

However, with polarity of applied voltage V as shown in Fig. 12.6b, the junctions

- j_1 and j_3 are reverse-biased, while
- j_2 is forward-biased.

This is just a reverse condition from that given in Fig. 12.6a. In this case also, there is no flow of current through SCR.

Fig. 12.6 Biasing arrangement of an SCR showing **a** forward biasing, and **b** reverse biasing

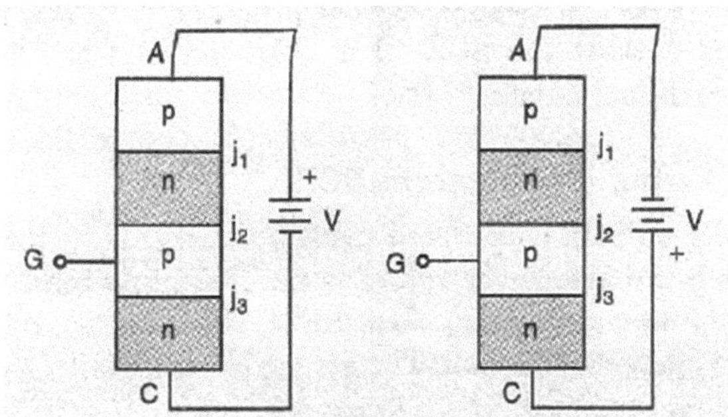

12.3.2 Operation of SCR

For the biasing shown in Fig. 12.6a, the current flow is blocked due to reverse-biased junction j_2. However, when anode voltage is increased, a certain critical value of voltage V_{BO} is reached when j_2 breaks down and SCR switches suddenly to a highly conducting state. This is called 'forward break over voltage'. Under this condition, the SCR offers a very little forward resistance about (0.01–1.0 Ω) so that the voltage across it drops to a low value (about 1 V) as shown in Fig. 12.7. Also the current is limited only by the power supply and the load resistance. The current keeps flowing indefinitely until the circuit is opened briefly or the current falls below the holding current I_H. It is also called latching current. The value of V_{BO} is determined by the magnitude of gate current. Higher will be the gate current, lower is the break over voltage.

With the type of biasing as shown in Fig. 12.6b, the current through the SCR is blocked by two reverse-biased junction J_1 and J_3. And when V is increased, a stage comes when Zener breakdown occurs, Fig. 12.7. It may destroy the SCR. Hence, it is observed that the SCR is a unidirectional device.

12.3.3 Firing (or Triggering) of an SCR

Usually an SCR is operated with an anode voltage slightly lower than the forward break over voltage V_{BO}. It is triggered into conduction by a low-power gate pulse. Once switched ON, the gate has no further control on device current. The gate signals can be of following two kinds.

1. dc firing signals as shown in Fig. 12.8a, and
2. Pulse signals as shown in Fig. 12.8b.

In case of push button switch S being open as shown in Fig. 12.8a, the SCR does not conduct and the lamp light is off. And when the switch S is closed momentarily,

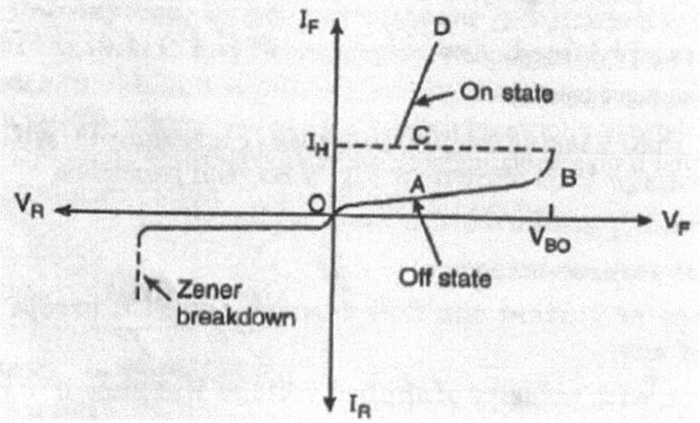

Fig. 12.7 V-I characteristic of an SCR

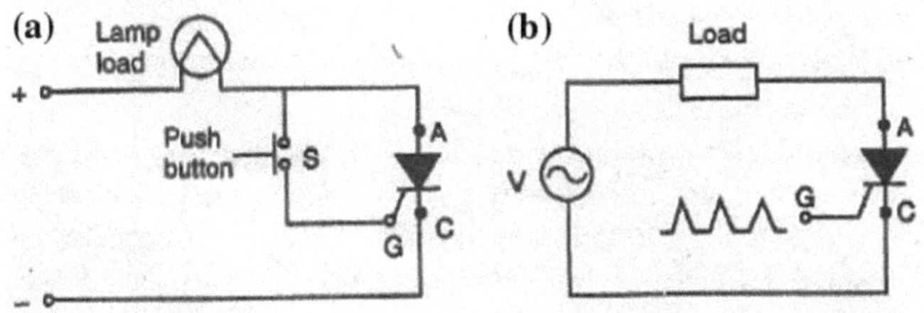

Fig. 12.8 Triggering of an SCR showing **a** d-c triggering, and **b** pulse signal triggering

a positive voltage is applied to the gate which forward-biases the p-n junction. As a result, the SCR is pulsed into conduction and the lamp 'lights up'. The SCR will remain in conducting state until the supply voltage is removed or reversed. Figure 12.8 shows triggering by 'timed pulses' obtained from a pulse source.

12.3.4 Turning OFF of SCR

As stated earlier that once 'fired', the SCR remains ON even when the triggering pulse is removed. This ability of the SCR to remain ON even when the gate current is removed, is referred to as latching. It is this fact that the SCR belongs to a class of devices known as latching devices. Thus, it is clear that an SCR cannot be turned OFF either by removing the gate pulse, or by applying a negative pulse. In fact, a number of techniques are employed to turn OFF an SCR. These are

1. by anode current interruption.
2. by reversing the polarity of anode-cathode voltage. It is done each half cycle by the voltage V as given in. Figure 12.7.
3. by reducing the current through SCR below the holding current I_H (Fig. 12.7). It is called low-current dropout.

12.4 Silicon Controlled Switch (SCS)

It is a four-layer, four-terminal p-a-p-n device. It has an anode A, cathode C, anode gate G_1 and cathode gate G_2 as shown in Fig. 12.9a. Its designating symbol is shown in Fig. 12.9b. An SCS is in fact a low-current SCR with two gate terminals. The two-transistor equivalent circuit is shown in Fig. 12.10.

Switching ON and OFF. An SCS may be switched ON or OFF by a suitable pulse applied at either gate. As seen from Fig. 12.10, a negative pulse is required at anode gate G_1 to turn the device ON, whereas a positive pulse is needed to turn it

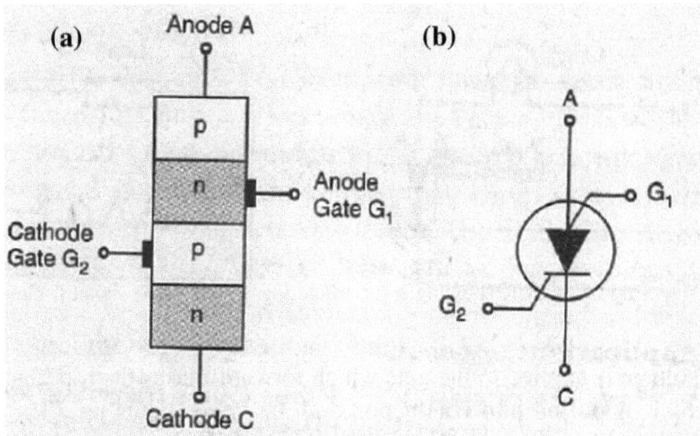

Fig. 12.9 A silicon controlled switch showing its **a** structure, and **b** representing symbols

Fig. 12.10 Circuit showing the working of a silicon controlled switch

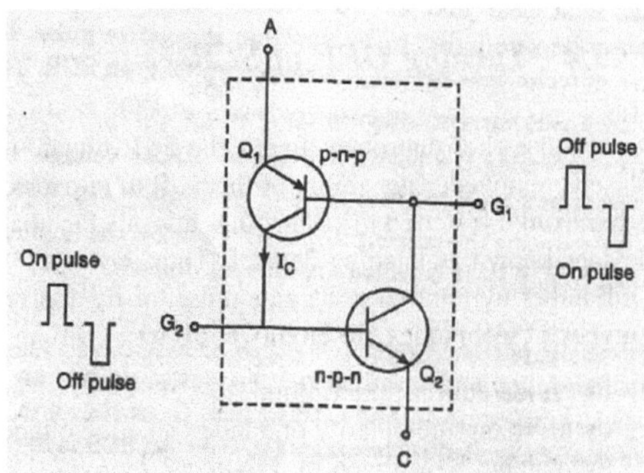

OFF. Similarly at cathode gate G_2, a negative pulse is required to switch the device OFF, and a positive pulse to turn it ON.

12.4.1 Mechanism

As seen from Fig. 12.10, when a negative pulse is applied to G_1, it forward-biases Q_1 (being p-n-p) which is turned ON. The resulting heavy collector current I_C being the base current of Q_2, turns it ON. Thus the SCS is switched ON. A positive pulse at G_1 will reverse-bias the emitter-base junction of Q_1, thereby switching the SCS OFF.

The SCCS is a versatile device because it can be operated in different ways. For example, it can be used as an SCR when its anode gate is not connected.

V-I Characteristics. The V-I characteristics of an SCS are essentially the same as those of the SCR as already shown in Fig. 12.7. As compared to an SCR, an SCS has much reduced' turn-OFF time. Moreover; it has a higher control and triggering sensitivity and a more predictable firing situation.

12.4.2 Applications

The more common areas of SCS application are as under. In

1. counters, registers and timing circuits of computers
2. pulse generators
3. voltage sensors
4. oscillators, etc.

12.5 Applications of SCR

Main application of an SCR is as a power control device. It has already been shown that when SCR is OFF, its current is negligible; and when it is ON, its voltage is negligible. Consequently, it never dissipates any appreciable amount of power even when controlling the substantial amount of load power. Other common areas of its applications include the following.

1. Relay controls
2. Static switches
3. Phase control
4. Regulated power supplies
5. Motor controls
6. Battery chargers
7. Inverters
8. Heater controls

SCRs are designed to control power up to 10 MW, including the individual ratings as high as 2000 A at 1.8 kV. Its frequency range application is up to about 50 kHz.

12.5.1 Half-Wave Power Control

SCRs are commonly used for the control of ac power such as for lamp dimmers, electric heaters, and motors etc. To understand it, a half-wave variable-resistance phase-control circuit is shown in Fig. 12.11. For this circuit, the gate triggering current is derived from the supply itself.

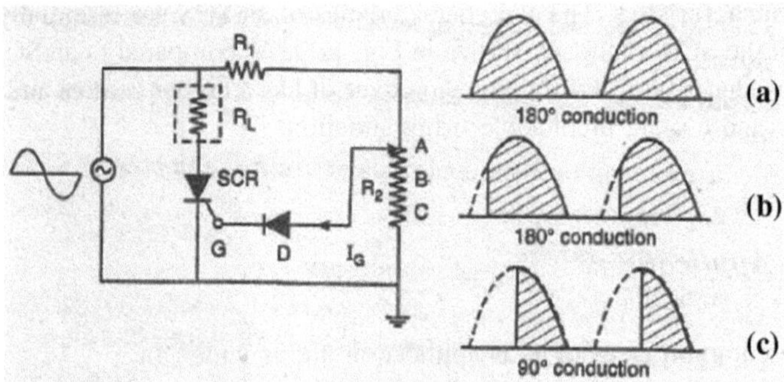

Fig. 12.11 Use of SCR in half-wave power control

Here, R_L is resistance of the load (say that of a heating element) and R_1 is current-limiting resistor. It limits the gate current during positive half-cycles of the supply. Potentiometer R_2 sets the trigger level for SCR. By adjusting R_2, the SCR can be made to trigger at any point on the positive half-cycle of ac waveform from 0° to 90° as shown in Fig. 12.11. If R_2 is adjusted to a very low value (at point A), the SCR will trigger almost immediately at the commencement of positive half-cycle of the input, thus giving a conduction angle of 180°. It means that the current will flow for the entire positive half-cycle, thereby delivering maximum power to the load.

However, if the value of R_2 is further increased by moving the contact to point B, the SCR triggering is delayed by 45° so that the conduction angle is reduced to 135°. Hence, relatively less power is delivered to R_L. If R_2 is further increased by moving the contact to point C, the SCR triggering is further delayed by 90° due to decrease in gate current I_G. In this case, still less power is delivered to the load. From the above description, it is clear that by adjusting R_2 the firing angle of the SCR can be changed and so can be the power delivered to the load.

12.5.2 Full-Wave Power Control

Figure 12.12 shows a full-wave, controlled rectifier arrangement which supplies higher average current. The two SCRs are fed by a centre-tapped transformer and supply the same to the load R_L. One SCR is triggered on each alternate half-cycle by the same trigger circuit. Each of the two SCRs remains conducting unless its current falls below the holding current I_H. The average dc voltage supplied by the above rectifier is expressed as

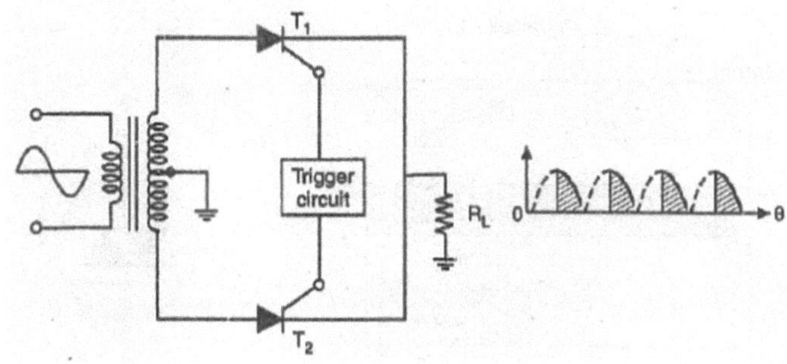

Fig. 12.12 Circuit showing the use of a SCR in full-wave power control

$$V_{dc} = \frac{V_m}{\pi}(1 + \cos\theta) \tag{12.4}$$

where V_m is maximum voltage across each thyristor, and θ is delay angle.

12.6 Bilateral Devices

The devices which can operate in both ways viz. in forward biasing mode as well as in reverse biasing mode, are called bilateral devices. They are useful in many applications requiring symmetrical switching. An example of this kind is switching with reverse and forward bias, such as switching ON and OFF of the sinusoidal signals in a-c circuits, during positive and negative parts of the cycle. There are a few bilateral devices in common use, but main among them are the following.

1. Bilateral diode switch (or diac)
2. Bilateral triode switch (or triac)

In a bilateral diode, the forward and reverse characteristics are symmetrical, which can be seen in Fig. 12.14 given in the next article. A bilateral triode can be triggered in a forward as well as reverse bias mode. Details of these devices are described in following sections.

12.7 Diac

A diac is a two-terminal semiconductor device that can conduct in either direction. It is, therefore, also known as bidirectional diode. Construction of a typical diac is shown in Fig. 12.13a. It has a configuration of the type of n_3-p_1-n_1-p_2-n_2. It has four junctions namely j_1, j_2, j_3 and j_4; anode A and cathode C. Although its basic configuration is of p-n-p-n type, but it is different from the p-n-p-n diode shown in

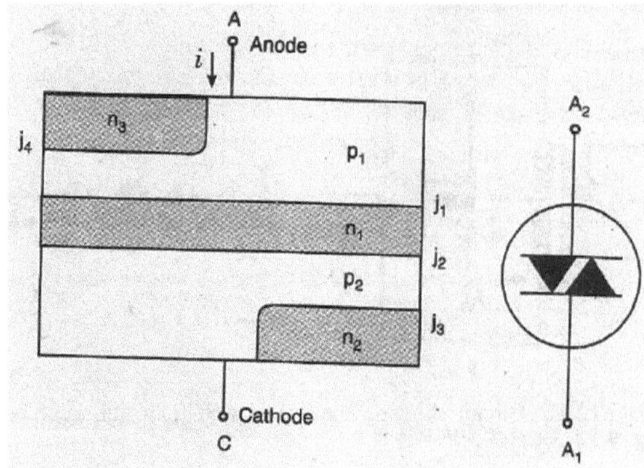

Fig. 12.13 Configuration of diac showing its **a** structure, and **b** representing symbol

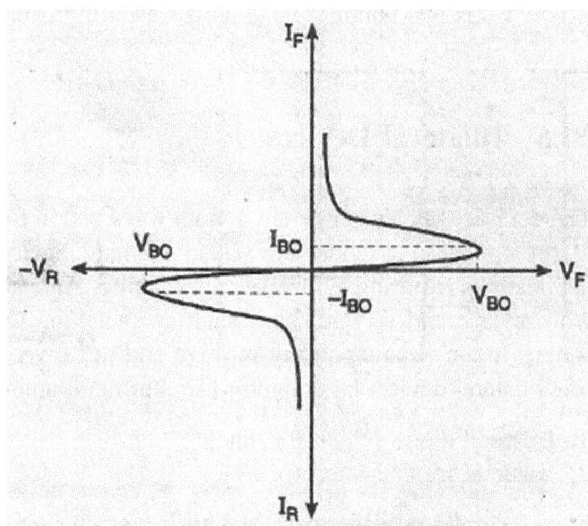

Fig. 12.14 VI characteristic of a iliac

Fig. 12.1. The difference lies in n_2 region which extends over only half of the width of the cathode, accompanied with another region n_3 which is diffused into half of the anode region.

The p-n-p-n configuration of Fig. 12.13a, in effect, can be considered to have two separate sections viz. p_1-n_1-p_2-n_2 and p_2-n_1-p_1-n_3. This is a symmetrical construction. Figure 12.13b shows the symbol for a diac.

12.7.1 Working

Since a diac in effect is equal to two diodes (i.e. two sections) connected in series, the voltage applied across it in either direction turns 'ON' one diode (i.e. one section) and reverse biasing the other. Hence, it can be switched from OFF to ON state for either polarity of the applied voltage. The operation of diac can be described as follows.

1. When the anode A is biased positive with respect to cathode C, the state of different junctions are as follows.
   - junction j_1 is forward biased
   - junction j_2 is reverse biased
   - junction j_3 is shorted by a metal contact (metal contact is not shown in figure).
   - junction j_4 is shorted by a metal contact on other end.

Now when j_2 is biased to breakdown state, a lateral current flows into p_2. It biasesthe left edge of junction j_3 into injection, and hence the device switches ON. Duringthis operation, j_4 remains dormant.

2. Now when the polarity is reversed i.e. the cathode C is made positive as compared to anode A, junction j_4 serves the function of a shorted emitter and j_1 is biased to breakdown in initiating the switching operation.

12.7.2 V-I Characteristics

The forward and reverse characteristics of a diac is shown in Fig. 12.14 on current-voltage diagram. As expected, the curve is symmetrical because of symmetrical configuration of the construction. The curve depicts the breakdown in either directions (i.e. forward and reverse biasing).

12.7.3 Applications

Some of the common uses of diacs are as triggering devices in triac phase control circuit used for the following.

- Light dimming
- Universal motor speed control
- Heat control

12.8 Triac

It is a three-terminal, five-layer, bi-directional device. It can be triggered into conduction by both positive and negative voltages at its anodei and with both positive and negative triggering pulses at its gate. A triac behaves like two SCRs connected in parallel, upside down with respect to each other. That is, the anode of one is tied to the cathode of the other and their gates are directly tied together. Hence, anode and gate voltages applied in either direction will fire a triac because they would fire at least one of the two SCRs which are in opposite directions.

12.8.1 Construction

As shown in Fig. 12.15a, a triac has three terminals A_1, A_2 and G. The gate G is closer to anode Al. Figure 12.15b shows that a triac is a two inverse parallel-connected SCRs with a common gate terminal. As seen, it has six doped regions. Figure 12.15c shows the schematic symbol which consists of two inverse-connected SCR symbols.

12.8.2 Working

When positive voltage is applied to A_2, the current flow follows p_1-n_2-p_2-n_2 path. The two junctions p_1-n_1 and p_2-n_2 are forward-biased whereas n_1-p_2 junction is blocked. The gate can be given either a positive or a negative voltage to turn ON the triac as explained below.

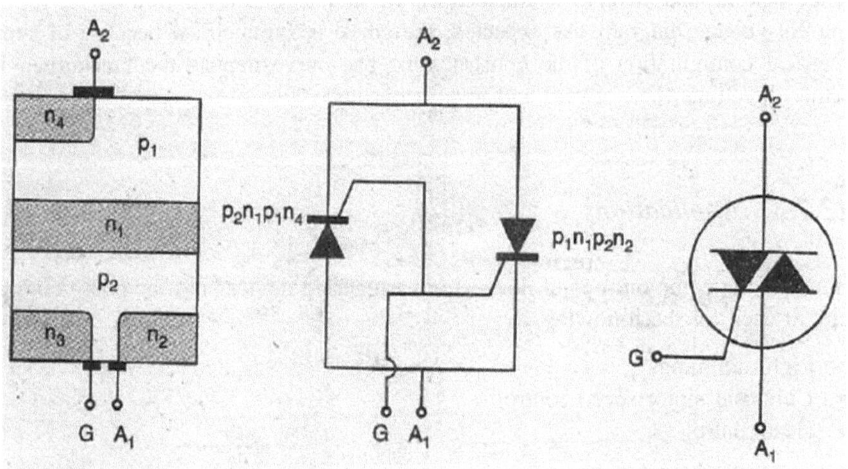

Fig. 12.15 A triac showing its **a** structure, **b** construction, and **c** representing symbol

1. **Positive gate**. A positive gate (with respect to A_1) forward-biases the p_2-n_2 junction and the break down occurs as in a normal SCR.
2. **Negative gate**. A negative gate forward-biases the p_2-n_2 junction and current carriers injected into p_2, turn on the triac.

When positive voltage is applied to anode A_1, the path of current flow is p_2-n_1-p_1-n_4. The two junctions p_2-n_1 and p_1-n_4 are forward-biased, whereas the junction n_1-p_1 is blocked. Conduction can be achieved by giving either a positive or a negative voltage to G as explained below.

1. **Positive gate**. A positive gate (with respect to A_1) injects current carriers by forward-biasing p_2-n_2 junction and thus initiates the conduction.
2. **Negative gate**. A negative gate injects current carriers by forward-biasing p_2-n_3 junction, thereby triggering the conduction.

The triac is said to be positively biased when the terminal A_1 is positive with respect to terminal A_2, and the operation occurs in quadrant I of V-I curve (see Fig. 12.16). Similarly, it is said to be negatively biased when terminal A_1 is negative with respect to terminal A_2, and the operation occurs in quadrant III of the same curve. The gate triggering occurs in any of the four operating modes given as follows.

(i) terminal A_1 to terminal A_1 voltage positive; gate. to terminal A_1 voltage positive: operating in quadrant I
(ii) terminal A_1 to terminal A_1 voltage positive; gate to terminal A_1 voltage negative: operating in quadrant I
(iii) terminal A_1 to terminal A_1 voltage negative; gate to terminal A_1 voltage negative: operating in quadrant III
(iv) terminal Al to terminal Al voltage negative; gate to terminal Al voltage positive: operating in quadrant III

It is seen that there are four triac-triggering modes, two each for the two anodes.

Fig. 12.16 V-I characteristic of a triac

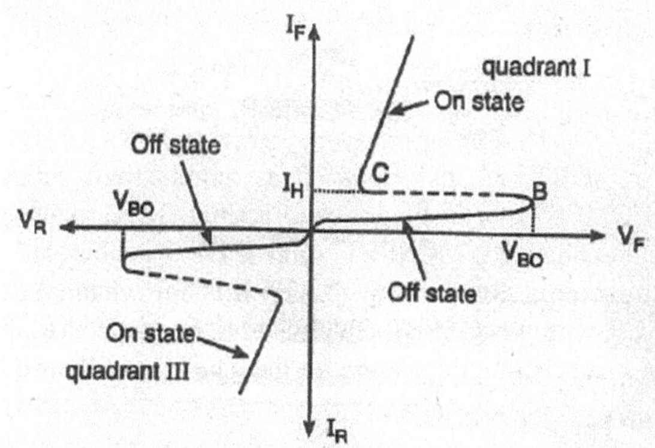

12.8.3 V-I Characteristics

Typical characteristics of a triac are shown in Fig. 12.16. As seen, the triac exhibits same forward blocking and forward conducting characteristics as an SCR, but for either polarity of voltage applied to the main terminal. Obviously, a triac has latch current in direction.

12.8.4 Applications

Main applications of a triac include the following.

1. For motor speed control
2. For light control
3. For minimizing the radio interference
4. As static switch to turn a-c power OFF and ON.

The only disadvantage of a triac is that it takes comparatively longer time to recover to OFF state. Hence, its use is limited to a-c supply frequencies of up to 400 Hz.

12.9 Insulated Gate Bipolar Transistor (IGBT)

An insulated-gate bipolar transistor is an improved device as compared to SCR. It can be turned-off from the conducting state to blocking state by the action of gate. This is an improvement over SCR which is difficult to be turned-off using the gate. An IGBT also avoids the use of additional expensive circuits, which is required in an SCR to reduce its anode to cathode current, below the holding current. This was a clumsy need in SCR, to change it from conducting state to blocking state.

12.9.1 Construction

An IGBT is also known as conductivity-modulated field effect transistor (COMFET), or insulated gate rectifier (IGR), or insulated gate transistor (IGT), or gain-enhanced MOSFET (GEMFET), or bipolar FET (i.e. BiFET). Its basic structure is shown in Fig. 12.17. It is a n^- channel device. It basically combines an SCR with a MOSFET which is able to connect or disconnect the n^+ cathode to the n^- base region, depending on the gate bias of the MOSFET. The MOSFET channel

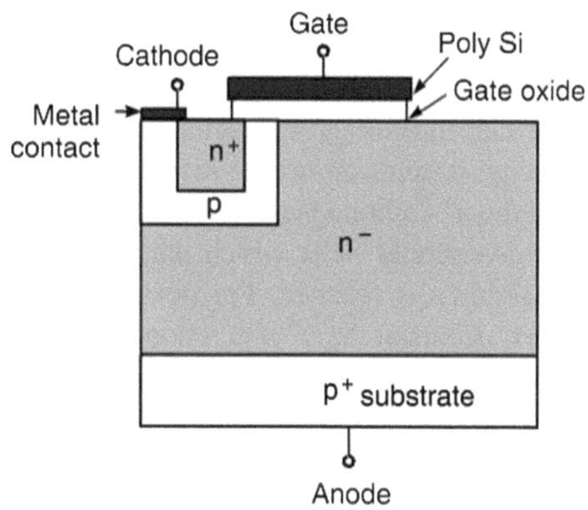

Fig. 12.17 Structure of an insulated gate bipolar transistor

length is determined by the *p*-region, which is formed by diffusion of the acceptors implanted in the same region as the n^+ cathode. Such a structure is known as a double-diffused MOSFET (DMOS), which is essentially an NMOSFET.

The main part of an IGBT is the n^-region, which acts as the drain of device. This is generally about 50 μm thick epitaxial regions with a low doping (≈1014 cm^{-3}). The doping is grown on a heavily p^+-doped substrate which forms the anode. This n^- region can therefore support a large blocking voltage in the OFF state. In the ON state, the conductivity of lightly doped region is modulated (increased) by the electrons injected from the n^+ cathode and the holes injected from the p^+ anode. That is why it is also known as conductivity modulated FET (COMFET). This increased conductivity allows the voltage drop across the device to be minimal in the ON state.

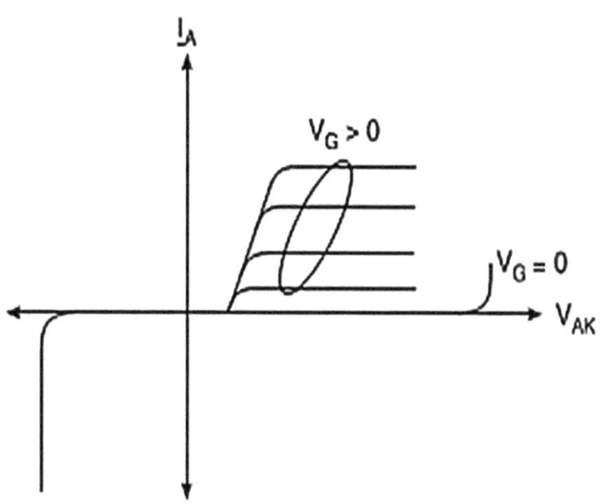

Fig. 12.18 Output current-voltage characteristics of an *n*-channel insulated gate bipolar transistor

12.9.2 Characteristics

The current-voltage characteristics are shown in Fig. 12.18. It depicts that if gate voltage is zero, an n-type inversion region is not formed in p-type channel region and the n^+ cathode is not shorted to the n base. The structure then looks exactly like a conventional SCR which allows minimum current flow in either polarity until breakdown is reached. For positive mode-to- cathode bias V_{AK}, avalanche breakdown occurs at the n^--p junction, while for negative V_{AK} avalanche occurs at the n^--p^+ junction.

When a gate bias is applied to the gate, there is significant current flow for positive V_{AK} as shown in Fig. 12.18. The characteristics look like that of a MOSFET, with one difference. Instead of the current to increase from the origin, there is an offset or cut-in voltage of about 0.7 V, just like for a diode. The reason for this can be understood by looking at the equivalent circuit shown in Fig. 12.18a. For small V_{AK} up to the offset voltage, the structure looks like a DMOS in series with a p-i-n diode, the n blocking region (base), and the n^+ cathode. In this regime, there is negligible voltage drop across the DMOS device and the p-i-n device are in forward bias. The current-voltage characteristics show an exponential behaviour, therefore

$$I_A \propto e^{eV}AK/2KT \tag{12.5}$$

On the other hand when V_{AK} is larger than the offset voltage of about 0.7 V, the characteristics look like that of a MOSFET, multiplied by a p-n-p bipolar junction transistor gain term. The equivalent circuit in this region is shown in Fig. 12.19b. In this regime, the current is given by

$$I_A = (1 + \beta_{pnp})I_{MOS} \tag{12.6}$$

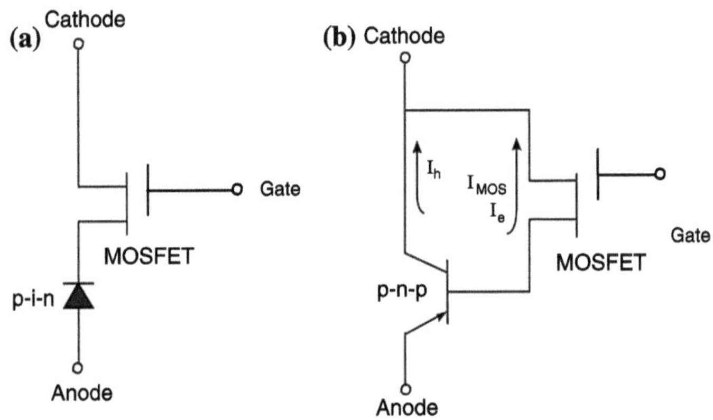

Fig. 12.19 IGBT equivalent circuit **a** *below* the offset voltage for low V_{AK}, and **b** *above* the offset voltage for high V_{AK}

12.9.3 Advantageous Features of IGBT

The construction of an IGBT incorporates the beneficial features of both: the MOSFETs and BJTs. These features are as follows.

1. Like MOSFET, it possesses high input impedance.
2. Like MOSFET, it has low input capacitance.
3. Like BJT, it has low resistance in ON state.
4. Like BJT, it has high current handling capacity in ON state.
5. Like BJT, it can be turned-off more easily than a SCR.

Because of above reasons, the IGBT is presently a favourable choice over the traditional SCR.

12.10 High Frequency Thyristors

These devices are used in applications requiring frequencies higher than the domestic AC mains supply (e.g. 50 or 60 Hz). The fast thyristors are made by diffusing heavy metals ions such as gold or platinum into the silicon, which act as charge combination centres. These thyristors can withstand very high reverse breakdown voltage and are also capable of carrying high current. However for switching circuits, once a high-frequency thyristor is 'latched-on' in the conducting state; it cannot be turned off by external control. The thyristor turn-off is passive, i.e., the power must be disconnected from the device.

12.10.1 Silion Carbide Thyristors

In recent years, some manufacturers have developed thyristors using Silicon carbide (SiC) as the semiconductor material. These have applications in high temperature environments, being capable of operating at temperatures up to 350 °C.

12.10.2 Sidac

It is a two leaded thyristor device that fires when its breakover voltage is met. However, it is usually used to carry the lead current directly in contrast to a diac which often riggers another power device such as a triac which in turn carries the lead current.

12.11 Solved Examples

Example 12.1 Discuss the uniqueness of property of any negative resistance device and mention its applications.

Solution. A uni-junction transistor (UJT) is a negative resistance device. An unique its property of this device is its triggering behaviour. It can be triggered by any one of three terminals. Once triggered, the emitter current increases regeneratively till it reaches a limiting value. The limiting value depends on the external power supply.

The UJT finds use in circuit meant for the following.

- Timing circuit
- Phase control
- Voltage regulation
- Sawtooth generator
- Switching
- Pulse generation
- Current regulation
- Sine wave generator

Example 12.2 Discuss a switching device which is suitable for power control applications?

Solution. A silicon controlled rectifier (SCR) is a suitable switching device for power control applications. The SCR is a rectifier with a control element. It can switch ON for any desired time duration and hence can deliver a selected amount of power to the load. An SCR possesses the advantages of both: a rheostat and a switch.

Example 12.3 What is a gate-controlled switch (GCS)? How is it different from an SCR?

Solution. A GCS is similar to an SCR in construction, but with a difference. It can be turned OFF by anegative gate pulse, whereas the SCR cannot be. That is why a GCS is also called as 'gate-turnoff switch' (GTO).

Review Questions

1. Define the following terms.
   (a) Power devices
   (b) Latching
   (c) Thyristor
   (d) Triggering
2. What do you mean by blocking and conducting states of power devices? Briefly explain a four layer device.
3. What is a p-n-p-n diode? Describe its construction and mechanism of conduction.
4. What is a two-transistor analogy. Discuss its physical significance.
5. Describe the variation of a with injection in a p-n-p-n. diode.
6. Explain the current transport mechanism in forward-blocking state of a p-n-p-n diode.
7. Explain the current transport mechanism in forward-conducting state of a p-n-p-it diode.
8. Discuss the necessity and importance of triggering mechanisms.
9. Explain the construction and working of an SCR under both modes of biasing.
10. Describe the mechanism of firing and turning-off of an SCR. Draw its V-I characteristics.
11. What is meant by holding current for an SCR? Why is only silicon and not germanium used in making SCR?
12. Describe the construction and working of a silicon-controlled switch. Discuss its V-I characteristics and applications .
13. Enumerate different applications of an SCR. Discuss the half-wave and full-wave power control mechanism of the SCR.
14. Explain the construction and working of a diac. Discuss its V-I characteristics and applications.
15. Is the diac a bidirectional diode? Explain as to how a diac is switched from OFF to ON state.

16. Explain the structure and operation of a triac. Discuss its V-I characteristics and applications.
17. Explain the structure and operation of an IGBT. Enlist its advantageous features over the MOSFETs and BJTs.
18. Distinguish between the following.
    (a) SCR and IGBT
    (b) Bilateral diode and bilateral triode
19. Write notes on the following.
    (a) Shockley diode (b) LASCR (c) Bilateral devices

Objective Questions

1. Which of the following statements is *not* correct?
   (a) A thyristor is equivalent to a unijunction transistor (UJT).
   (b) To turn a conducting SCR to OFF state, it is necessary to make the anode current less than the holding current.
   (c) If the gate current of an SCR is increased, there is a decrease in the break-down voltage.
   (d) A triac is a bidirectional device with three electrodes.
2. The anode current of a conducting SCR is 20A. If its gate current is reduced to one-fourth, its anode current will become
   (a) 10 A (b) 20A (c) 5A (d) zero
3. Consider the following statements.
   A. An SCR is specified by the anode current it can handle.
   B. An SCR is specified by the cathode current it can handle.
   C. An SCR is specified by the sum of anode and cathode currents it can handle.
   D. A SCS has a much reduced turn-off time as compared to an SCR.

   Of these, the correct statements are
   (a) A, B and D (b) B and C
   (c) A, B and C (d) A and D
4. Which of the following pairs does not match?

Name of device Representing symbol

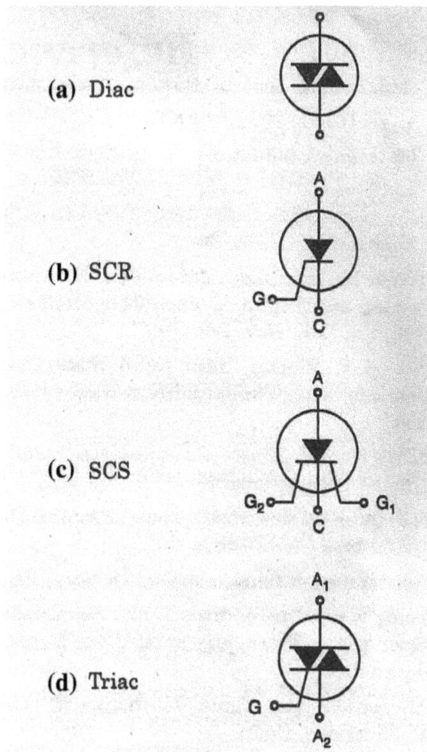

5. Which one is the most suitable power device for high frequency (> 100 kHz) switching application?

   (a) Power MOSFET (b) BJT
   (c) Schottky diode. (d) Micowave transistor

6. A thyristor equivalent of a thyratron tube is
   (a) SCR (b) UJT
   (c) Diac (d) Triac

7. Triacs cannot be used in ac voltage regulator for a
   (a) resistive load (b) back emf load
   (c) inductive load (d) resistive inductive load

Answers

1. (a)	2. (b)	3. (d)	4. (c)
5. (a)	6. (a)	7. (c)	

Part V
Fabrication Techniques

Chapter 13
Semiconductor Growth Techniques and Device Fabrication

Abstract This chapter presents a brief outline of the technology associated with semiconductor processing. The production of highly pure electronic-grade silicon from raw feedstock is presented. Different techniques of semiconductor crystal growth are given. Czochralski (CZ) method and Bridgman method are explained. Zone refining process is also described. Semiconductor fabrication technology, and thin film micro-electronic circuit fabrication are presented. Preparation of wafers, epitaxy, masking, photolithography, interconnection by metalizing are discussed. Methods of ion-implantation, molecular beam epitaxy (MBE), sputtering and etching are explained. Semiconductor p–n junction (diode) formed by different methods, are described. Stepwise procedure for the formation of a semiconductor p–n junction, and transistor manufacturing processes (of grown junction, alloy formation, diffusion) are briefly narrated. Minute insight into the various topics are given through solved numerical and theoretical examples. Review questions, numerical problems and objective type questions are also given with their answers.

Keywords Production of EGS · Crystal growth · Bridgman method · Czochralski method · Zone refining · Micro-electronic circuit construction · Wafer manufacturing · Ion-implantation · Epitaxy · Sputtering · Masking · Photo-lithography · Etching · Fabrication of p–n junction · Transistor manufacturing process

13.1 Introduction

A large number of processes are involved in the fabrication of semiconductor materials and devices. For example, the semiconductor to be used is prepared in single-crystal form and then is purified to electronic grade. The semiconductor crystal growth is accomplished by using different techniques. It is sliced into wafers that are further processed to obtain defect-free and highly polished surfaces. Subsequently, desired impurities are introduced into selected regions of a wafer, sometimes in conjunction with oxidation and masking, and sometimes without these to form appropriate junctions. This step is then followed by a metallization

K.M. Gupta and N. Gupta, *Advanced Semiconducting Materials and Devices*,
Engineering Materials, DOI 10.1007/978-3-319-19758-6_13

process to realize ohmic contacts, separation of devices into individual dies, lead attachment, and device encapsulation (packaging).

13.2 Production of Element Form of Silicon (Si)

The raw feedstock for Si crystal is silicon dioxide (SiO_2). SiO_2 is reacted with C in the form of coke in an arc furnace at very high temperatures (≈1800 °C) to reduce SiO_2 according to the following reaction:

$$SiO_2 + 2C \rightarrow Si + 2CO \tag{13.1}$$

This forms the metallurgical grade Si (MGS) which has impurities such as Fe, Al and heavy metals at levels of several hundred to several thousand parts per million (ppm). It is worth mentioning that 1 ppm of Si corresponds to an impurity level of $5 \times 10^{16}/cm^{+3}$. While the MGS is clean enough for metallurgical applications such as using Si to make stainless steel, it is not pure enough for electronic applications. It is also not a single crystal.

The MGS is refined further to yield semiconductor-grade or electronic-grade Si (EGS), in which the levels of impurities are reduced to parts per billion or ppb (1 ppb = $5 \times 10^{13}/cm^{+3}$). This involves reacting the MGS with dry HCl according to the following reaction to form trichlorosilane, $SiHCl_3$, which is a liquid having boiling point of 32 °C.

$$Si + 3HCl \rightarrow SiHCl_3 + H_2 \tag{13.2}$$

Along with $SiHCl_3$, chlorides of impurities such as $FeCl_3$ are also formed which fortunately have boiling points that are different from that of $SiHCl_3$. This allows a technique called fractional distillation to be used, in which the mixture of $SiHCl_3$ and the impurity chlorides is heated, and condense the vapours in different distillation towers held at appropriate temperatures. Thus the pure $SiHCl_3$ is separated from the impurities. $SiHCl_3$ is then converted to highly pure EGS by reaction with H_2,

$$2SiHCl_3 + 2H_2 \rightarrow 2Si + 6H \tag{13.3}$$

13.3 Semiconductor Bulk and Thin Films Growth Technologies

The semiconductor crystal growth using various techniques ranging from bulk crystal growth to the epitaxial growth of quantum dots include the following.

- Liquid-encapsulated Czochralski (LEC),
- Horizontal Bridgman (HB),
- Liquid-encapsulated Kyropoulos (LEK), and
- Vertical gradient freezing (VGF) methods.

There are also many improved methods available for the growth of bulk semiconductor crystals. For example,

- Magnetic LEC
- Direct synthesis LEC
- Pressure-controlled LEC, and
- Thermal baffle LEC methods are all variations of the original LEC technique, but with improved growth conditions.

Other bulk growth techniques include

- Dynamic gradient freezing,
- Horizontal gradient freezing,
- Magnetic LEK, and
- Vertical Bridgman methods.

The widely used epitaxial growth techniques are

- Molecular beam epitaxy (MBE)
- Metal-organic chemical vapour deposition (MOCVD), and
- Liquid-phase epitaxy (LPE).

The term epitaxy is of Greek origin and is composed of two words, epi (placed or resting on), and taxis (arrangement). Thus, epitaxy refers to the formation of single-crystal films on top of a substrate.

The growth techniques of bulk semiconductor crystals are designed to produce large-volume crystals under equilibrium conditions with almost no flexibility in the production of alloy composition. The ability of bulk semiconductor growth techniques to produce large single crystals that can be cut into submillimeter-thick wafers, subsequentially to be used for epitaxial growth, is invaluable.

Preparing wafers from the boules is called the wafering process and includes slicing, lapping, polishing, and cleaning. Since most wafers are used as substrates for epitaxial growth, the wafering process and the bulk crystal growth are very important for successful epitaxial growth. For many epitaxial growths, wafers can be used without any treatment prior to the growth. There are many technologies available to prepare ready-to-use wafers. For example,

(i) thermal oxidation and/or ultraviolet/ozone oxidation processes have been effective in producing thin oxide layers, which protect the wafer surface. These oxide layers can be removed by heating prior to epitaxial growth.
(ii) Packaging the wafers in nitrogen gas is an effective method used to reduce residual oxidation of polished surfaces during storage.

Semiconductor alloys, heterojunctions, and other quantum structures such as superlattices are currently grown by two main epitaxial growth techniques, namely, MBE and MOCVD. These growth techniques enable the synthesis of high-quality single-crystal thin film deposited layer by layer on suitable substrates.

13.4 Semiconductor Crystal Growth

Element form of Si and Ge are obtained by chemical decomposition of compounds such as $SiCl_4$, $SiHCl_3$ and GeO_2 etc. After isolation from these compounds, the Si and Ge are purified. They are melted (melting points of Si is 1410 °C and that of Ge is 937 °C) and then casted into ingots. When the casted ingots are cooled down, the Si and Ge are obtained in poly-crystalline form. It is now the turn of crystal growth. For growth of a single crystal, a very careful control has to be maintained at the boundary of 'solid form of material during cooling' and 'molten material'. Different methods are adopted for growth of crystals from molten material. These are listed below.

1. Bridgman method (or the method of growth from the melt)
2. Czochralski (CZ) method
3. Floating zone method

These three methods will be described now in subsequent sections.

13.4.1 Bridgman Method

Common technique for growing single crystal from the melt involves selective cooling of molten material so that the solidification occurs along a particular crystal direction as shown in Fig. 13.1. Here the silica crucible containing molten Si or Ge is pulled slowly through the furnace. Consequently, the solidification begins at one end and slowly proceeds down the length of the bar. To increase crystal growth, a

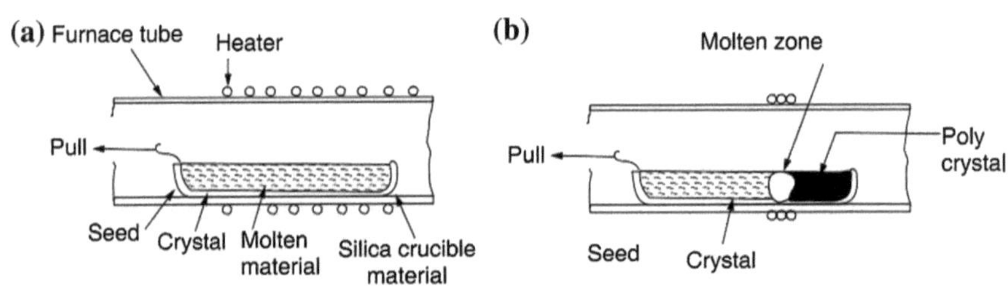

Fig. 13.1 Bridgman method of crystal growth from the melt in a crucible. Reproduced with permission from [1]

small seed crystal is placed at the end which is cooled first. Ge, GaAs and other semiconductor crystals are grown using this method.

As shown in Fig. 13.1b, a small region of polycrystalline material is melted and the molten zone is moved down the crucible at such a rate that a crystal is formed behind the zone, as it moves. This method is not suitable for growth of Si, as it has high melting point and sticks to the crucible.

13.4.2 Czochralski Method

In this method, a seed crystal is lowered down into molten material and is raised slowly, allowing the crystal to grow on to the seed. The crystal is rotated slowly as it grows to provide slight stirring of the melt and to correct any temperature variation. This arrangement is shown in Fig. 13.2.

The polysilicon material is kept in either a quartz or a graphite crucible and is melted by heating. The crystal growth is started by dipping a seed of single-crystal silicon into the melt. The seed is suspended over the crucible in a holder attached to a rotating arm and is inserted into the melt by lowering the arm. As the bottom of the seed begins to melt, the downward motion of the arm is reversed. The crystal is then pulled from the melt by slowly rotating and lifting the seed as the crystal grows at its lower end. The arm continues its upward movement forming a larger crystal. The crystal growth terminates when the melt in the crucible is depleted. During the pulling, provision is made to rotate the crucible. This provides stirring of the melt and avoids asymmetries in the heating. Material with desired impurity concentration is obtained by adding appropriate impurities to the melt prior to crystal growth. The atmosphere around the crystal puller is made inert by enclosing the entire assembly in a chamber that is flushed with an inert gas such as argon.

The CZ technique can also be used to grow single crystals of Ge, GaAs, and other compound semiconductors. In pulling crystal of a semiconductor such as GaAs from the melt, it is necessary to prevent volatile As atoms from vapourizing.

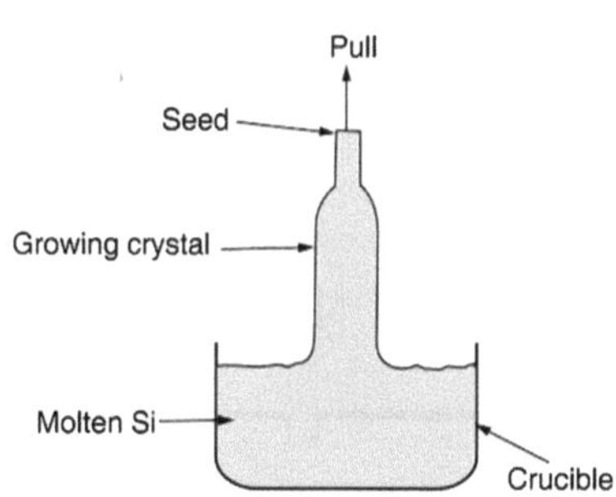

Fig. 13.2 Czochralski method of crystal growth. Reproduced with permission from [1]

13.4.3 *Float Zone Method*

The float zone (FZ) process shown in Fig. 13.3, uses a rod of ultrapure polycrystalline Si. The rod is maintained in a vertical position with the help of two chucks and is enclosed in a chamber in which an inert atmosphere is maintained by a flow of argon. An rf (radio frequency) heater coil is placed around the chamber. A single-crystal seed is clamped at the lower end of the rod, which is rotated around its axis during the growth process. The coil melts a small length of the rod starting with the seed crystal. The molten zone is then slowly moved upward along the length of the rod by moving the rf coil. As the coil is moved upward, recrystallization of the molten zone at the bottom occurs while the new material begins to melt at the top. The recrystallized region assumes the crystal structure of the seed. The molten zone is held together by surface tension of the liquid. The diameter of the growing crystal is controlled by the motion of the heater coil. The desired impurity level is obtained by starting with an appropriately doped polycrystalline material.

Comparison of FZ method with CZ process. The absence of a crucible in the FZ process offers many advantages over the CZ process. Not only is impurity contamination from the walls of the crucible eliminated, but also many volatile impurities can be removed by crystal growth in reduced atmospheric pressure. However, it is difficult to obtain large-diameter crystals using this method.

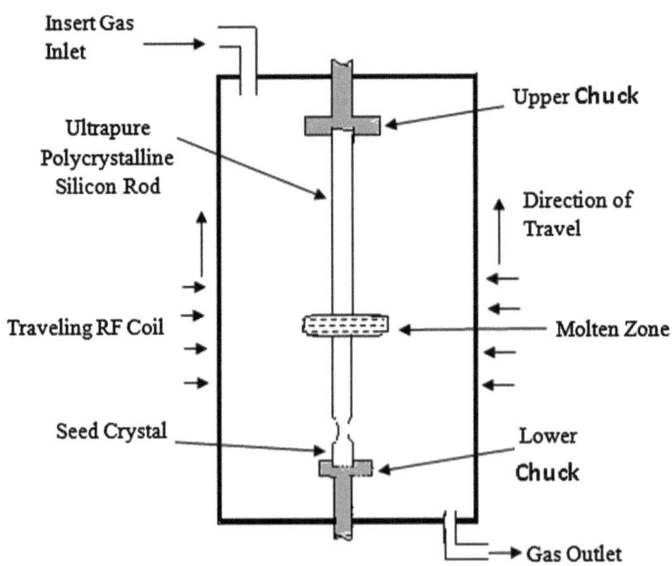

Fig. 13.3 Arrangement of a typical float zone system. Modified from [2] and self drawn

13.5 Processing of Semiconducting Materials

We have discussed earlier that a ppm (parts per million) addition of impurity (dopant) alters the electrical conductivity of semiconductors tremendously, hence this addition has to be minutely controlled to get a refined quality of extrinsic semiconductor. This necessitates an effective application of different refining techniques. The semiconductors can be refined by zone refining method. The *Zone refining* method is based on phase separation. Ultrapure single crystal silicon can be produced by this method.

13.5.1 Zone Refining Process

Although it is necessary to have acceptor or donor impurities, yet a practical semiconductor must be as far as possible chemically pure and crystallographically perfect. Thus, the *p*–*n* junction should occur in a single crystal because a grain boundary at the junction might contribute extra levels, thereby can complicate the matters. Recombination centres traps, and local distortion of the band structure may exist around a grain boundary. If a *p*–*n* junction occurs at a grain boundary, the operation of rectification is hindered or prevented.

Dislocations also have similar effects. In transistor materials, the lifetime of excess minority carriers should be as large as possible because, the operation of transistors depends on minority carrier injection across the base from the emitter base junction, to the base collector junction. If the excess minority carriers are arrested in their passage through the base by traps or recombination centres, the transistor operation is impossible. Impurities and imperfections are, therefore, highly undesirable. For practical purposes, it is also advantageous to control the resistivity of the material.

The level of undesirable impurities must, therefore, be of the order of parts per billion (ppb) before the donor is added, if real control of resistivity is so achieved. To attain the highest purity possible, zone refining is employed after chemical processing. Without zone refining, very little transistor grade material would be available. The zone refining process is based on the segregation of dissolved impurities during *non-equilibrium solidification.* Figure 13.4 describes the zone-refining process in nearly pure material where only one impurity is present.

Since the concentration of the impurity in solid and liquid are very small, the liquidus and solidus curves can be approximated by two straight lines. The ratio between the impurity concentration in the solid and that in the liquid is then a constant;

$$\frac{C_S}{C_l} = C_{seg} \tag{13.4}$$

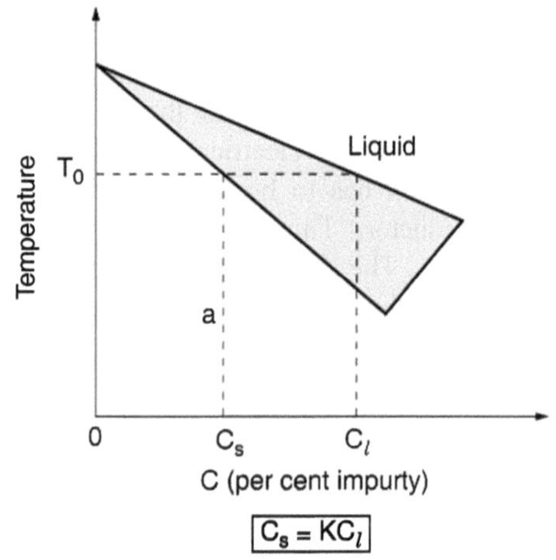

Fig. 13.4 The straight liquidus and solidus imply that the ratio of impurity concentration in the solid to that of the liquid is a constant C_{seg}, called segregation. Reproduced with permission from [1]

The constant C_{seg} is called the *distribution or segregation coefficient*. If the solidifying liquid is of the composition C_l, the first solid to emerge from the liquid has the composition $C_{seg}C_l$. If C_{seg} is less than one, as in Fig. 13.4, the solid is purer than the liquid. The material can be purified by partially solidifying and then pouring off the liquid. The first material to solidify is again the purest. A rod solidified from one end by moving the solid liquid interface continuously, leaves the purest and most useful material at the initial end.

13.5.2 Zone Refining Apparatus

Schematic representation of a zone-refining apparatus is shown in Fig. 13.5. Only a section of the bar of semiconductor material is melted at any one time. The source of heat is usually a high frequency induction coil. As the coil is moved along the bar, the molten zone moves with it. The liquid which no longer lies in the hot region, solidifies. The length of the zone remains constant if the power input of the induction coil and the environment of the bar remain constant.

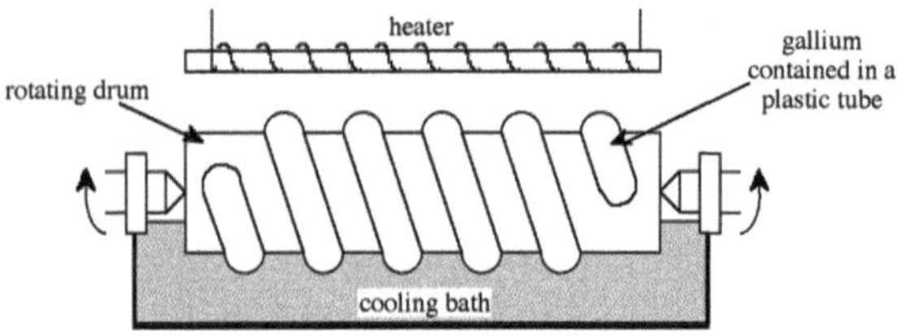

Fig. 13.5 Schematic set-up of a zone-refining apparatus. Reproduced from [3]

Example 13.1 A Si crystal is to be grown by the Czochralski method and it is desired that the ingot contain 10^{16} phosphorus atoms/cm^3. Determine the (a) concentration of phosphorus atoms that the melt should contain to give this impurity concentration in the crystal during the initial growth? For P in Si, $C_{seg} = 0.35$. (b) If the initial load of Si in the crucible is 5 kg, how many grams of phosphorus should be added? The atomic weight of phosphorus is 31.

Solution. (a) Using Eq. 13.4, the initial concentration of P in the melt is found as

$$\frac{10^{16}}{0.35} = 2.86 \times 10^{6}/\text{cm}^{3} \tag{i}$$

(b) The P concentration is so small that the volume of melt can be calculated from the weight of Si. The density of Si is 2.33 g/cm^3. On neglecting the difference in density between solid and molten Si, we write

$$\frac{5000\,\text{g of Si}}{2.33\,\text{g/cm}^{3}} = 2146\,\text{cm}^{3}\text{ of Si} \tag{ii}$$

Considering Eqs. (i) and (ii), we get

$$2.86 \times 10^{16}/\text{cm}^{+3} \times 2146\,\text{cm}^{3} = 6.14 \times 10^{19}\text{P atoms}$$

$$\therefore \quad \frac{6.14 \times 10^{19}\text{atoms} \times 31\,\text{g/mole}}{6.02 \times 10^{23}\text{ atoms/mole}} = 3.16 \times 10^{-3}\text{g of P}$$

13.6 Semiconductors Fabrication Technology [1]

The development of large-scale integrated circuits (LSICs) and decreasing size of electronic devices are remarkable achievements of semiconductor technology. A single piece of semiconductor can connect more than 1000 transistors on an area of about 1 cm^2. This has become possible due to advances in fabrication and processing techniques for semiconductors. In this regard, the processes of microelectronics may be grouped as follows.

1. Semiconductor microelectronics, and
2. Thin film microelectronics.

Diodes, *pnpn* switches, transistors and resistors belong to semiconductors of microelectronics group while the interconnections of various electronic circuits and capacitors are the product of thin *film microelectronics*.

13.6.1 Microelectronic Circuit Construction [1]

A semiconductor microelectronic circuit requires following sequential processes in its construction.

(i) Production of highly purified *n* or *p* type silicon.
(ii) Single crystal growth from the melt.
(iii) Manufacturing of 0.25–0.40 mm thick *wafer*.
(iv) Polishing of wafers to a thickness of 0.15–0.20 mm.
(v) Oxidation to isolate pockets in a wafer.
(vi) *Photoengraving* for cutting windows in the oxide layer.
(vii) Diffusion by doping via a vapour phase of B_2O_3, B_2H_6 or PH_3 in a stream of hydrogen, results in pn and *pnp* junctions.
(viii) *Epitaxy* for growing a new layer of same crystal orientation using hydrogen as carrier gas, and silane (SiH_4), dichlorosilane (SiH_2Cl_2) etc. as vapour sources.
(ix) Chemical processing such as etching to remove mask and oxide layers.
(x) *Masking i.e.* repeating the above operations a number of times on the surface of the wafer. The surface is then separated into chips. The circuit component per chip exceeds 1 million.
(xi) Interconnections by metalizing with aluminium which is deposited by vacuum evaporation.

13.6.2 Thin Film Circuit Fabrication

In fabricating thin film microelectronic circuits, glazed ceramics and glass substrates are used. The *substrate* is a polished surface on which many hundred I.Cs may be located. Deposition of I.Cs on thin films can be done by one or more of the following processes.

(i) Vacuum deposition,
(ii) Silk screening,
(iii) Vapour plating, decomposition,
(iv) Electron beam
(v) Plasma decomposition,
(vi) Sputtering, and
(vii) Anodization.

Thin film resistors employ metallic film while the thin film conductors employ film of gold or aluminium, and capacitors use a dielectric film. Ion implantation is a most modern method of doping microelectronic devices such as in MOSFET.

Readers are advised to refer particular references on specific topics as further information will be beyond the scope of this book.

13.7 Manufacturing of Wafers [1]

The single-crystal ingot is mechanically processed to manufacture wafers. It involves several steps, as listed below.

1. The first step involves mechanically grinding of the more-or-less cylindrical ingot into a perfect cylinder. This is important because in a modern integrated circuit fabrication, many processing tools and wafer handling robots require tight tolerances on the size of the wafers.
2. Then using X-ray crystallography, crystal planes in the ingot are identified. Most Si ingots are grown along the <100> direction.
3. For such ingots, a small notch is ground on one side of the cylinder to delineate a {110} face of the crystal. This is useful because for <100> Si wafers, the {110} cleavage planes are orthogonal to each other. This notch then allows the individual integrated circuit chips to be made oriented along {110} planes. Due to this; the chips are sawed apart, there is less chance of spurious cleavage of the crystal.
4. Next, the Si cylinder is sawed into individual wafers about 775 μm thick, by using a diamond-tipped inner-hole blade saw, or a wire saw.
5. The resulting wafers are mechanically lapped and ground on both sides to achieve a flat surface, and to remove the mechanical damage due to sawing. The flatness of the wafer is critical from the point of view of "depth of focus" or how sharp an image can be focussed on the wafer surface during photolithography.
6. The Si wafers are then rounded or chamfered along the edges to minimize the likelihood of chipping the wafers during processing.
7. Finally, the wafers undergo chemical-mechanical polishing using slurry of very fine SiO_2 particles in a basic NaOH solution to give the front surface of the wafer a mirror-like finish. The wafers are now ready for integrated circuit fabrication.

The economic value added in this process is impressive as from sand (SiO_2), we obtain Si wafers on which we can make hundreds of microprocessors. Thus the very cheap sand is converted into very costly products costing several hundred dollars.

13.8 Ion-Implantation

The process of ion-implantation is frequently used as an alternative to pre-deposition for introducing dopant atoms into the desired region of a semiconductor. The implanted impurities are then diffused using a drive-in step. The ion-implantation technique offers the advantage of precisely controlling the amount of dopant and its depth below the surface. And also, it is a low-temperature process that eliminates deformation of wafer caused at high temperatures.

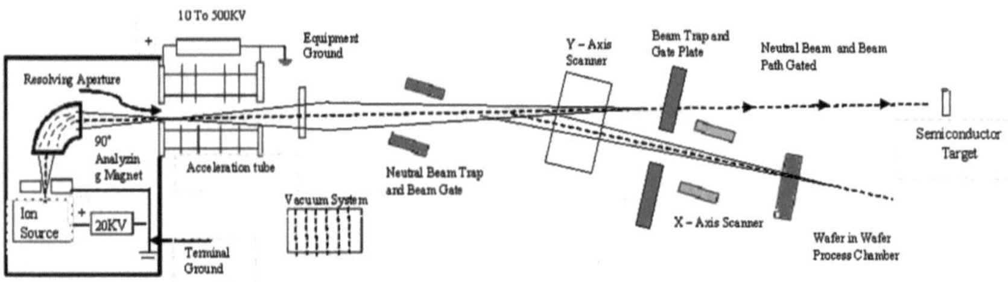

Fig. 13.6 Arrangement of an ion-implantation system. Modified from [2] and self drawn

In this technique, a beam of ionized atoms is accelerated through a desired potential (10–500 kV) and is made incident on a semiconductor target as shown in Fig. 13.6. A gas, containing the atoms of the desired impurity is introduced into a chamber where the atoms are ionized by collisions with high-energy electrons. The ion beam emerging from the chamber contains the desired dopant ions, as well as the ions of unwanted species. Now, the dopant ions are separated from the unwanted ions by passing the beam through a strong magnetic field that bends the desired ions through a 90°. The selected ions are then accelerated using an electric field and are made to strike the semiconductor target which is kept at ground potential. The beam is deflected horizontally and vertically so that it sweeps across the target to ensure homogeneous doping. Both the beam and the target are maintained under a high vacuum.

When the accelerated ions enter the semiconductor, they lose their kinetic energy through collisions with the electron cloud of the semiconductor atoms and the positively charged nuclei of the atoms. An ion comes to rest when its kinetic energy is reduced to zero. The total path length of the ion in the semiconductor is called its range, and the penetration depth of the ion in the target is called its projected range R_p.

13.9 Lithography

Lithography is a process in which the transfer of an image is done from the mask to the surface of the wafer. Lithography may be accomplished in different ways. These are:

1. Photolithography
2. Electron-beam lithography
3. X-ray lithography

Photolithographic process involves the transfer of an image from the mask to the surface of the wafer through the use of ultraviolet (UV) light and a photoresist. In *electron-beam lithography*, the E-beam systems may be either a scanning E-beam pattern generator or an E-beam projection system. The *X-ray lithography* uses the short wavelength X-rays instead of UV light.

13.9.1 Photoresists

Polymeric photoresist materials are generally used as a spin cast film as part of a photolithographic process. The film is modified by exposure to radiation such as visible light, ultraviolet light, X-rays of electrons. Exposure is usually conducted through a mask so that a pattern can be created in the photoresist layer and subsequently on the substrate through an etching or deposition process. Resists are either positive or negative depending on whether the radiation exposure weakens or strengthens the polymer. In the developer step, chemicals are used to remove the weaker material leaving a patterned photoresist layer behind. Important photoresist properties include resolution and sensitivity, particularly as feature sizes decrease.

13.9.2 Photolithography

Photolithography is a process of planar technology to fabricate an IC in which a desired pattern of doping is transferred onto the silicon wafer. The entire process involves:

(i) coating of SiO_2 layer with a film of photosensitive emulsion called *photoresist,*
(ii) blackening the portions to be doped on a transparent sheet, placing this sheet on photoresist layer and then exposing to ultraviolet light
(iii) removing the mask and dipping the wafer in trichlorothylene, and
(iv) *etching* the SiO_2 with hydrofluoric HF solution.

13.10 Epitaxy

Epitaxy means the processes which are used to grow a crystalline layer of semiconductor on a crystalline substrate in such a way that, the layer grown has the same lattice structure as the substrate. Epitaxial growth provides an alternative to the diffusion process for obtaining appropriately doped semiconductor regions. An added advantage of this process is that the impurity concentration in the grown layer can be adjusted independently of that in the substrate. The main processes of epitaxial growth are following.

- Vapour phase epitaxy (VPE)
- Liquid phase epitaxy (LPE), and
- Molecular beam epitaxy (MBE).

VPE has been used to grow Si and III–V compounds, whereas the LPE is mainly used with III–V compounds. MBE is more versatile and allows a precise control of doping profile.

13.10.1 Vapour Phase Epitaxy

Silicon epitaxy is accomplished using the chemical vapour deposition (CVD) technique which has been used to deposit Si from $SiCl_4$, silane (SiH_4), SiH_2Cl_2, SiH_3Cl, and some other compounds. Figure 13.7 shows a schematic of the growth system using $SiCl_4$ in which the liquid $SiCl_4$ is kept in a bubbler whose temperature is carefully controlled near 0 °C. H_2 gas is passed through the bubbler. The temperature of the bubbler and the flow rate of H_2 through it determine the concentration of the $SiCl_4$ in the mixture. The $SiCl_4 + H_2$ vapour is mixed with more H_2 near the mouth of the furnace and is passed through the quartz tube. The Si wafers are placed on a graphite susceptor kept in the quartz tube which is heated to a temperature in excess of 1100 °C by an rf induction coil. The following reaction takes place on the wafer surface:

$$SiCl_4 + 2H_2 \rightleftharpoons Si + 4HCl \tag{13.5}$$

The Si is deposited on the silicon substrate, whereas the HCl gas leaves through a suitable vent. The rf coil heats the graphite but not the quartz tube, therefore, there is no deposition on the walls of the tube. Since the reaction occurs in a chamber, the arrangement of Fig. 13.7 is called a reactor chamber or simply a reactor. The reaction represented by Eq. (13.5) is reversible. The reverse reaction etches Si and is used to clean the substrate prior to the start of deposition. This etching is performed by passing HCl through the reactor. The growth process is started by stopping the HCl flow and introducing the $SiCl_4 + H_2$ mixture into the reactor. In order to grow doped

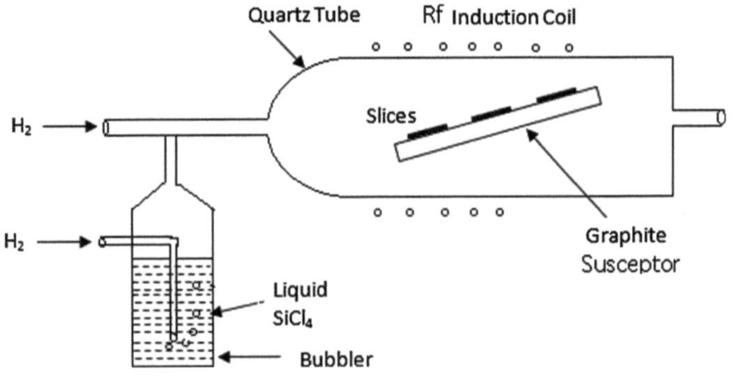

Fig. 13.6 Schemantic arrangement of an epitaxial growth system. Modified from [2] and self-drawn

Si with a desired resistivity and conductivity, dopant atoms are introduced into the gas system. Gases like PH_3, AsH_3, and SbH_3 are used for n-type doping; whereas diborane (B_2H_6) is used for p-type doping.

13.10.2 Liquid Phase Epitaxy

LPE is used to fabricate multilayer structures of ternary and quaternary compounds for use in LEDs and lasers. In this technique, a mixture of semiconductor with a second element may melt at a lower temperature than the semiconductor itself and thus may be used to grow a crystal from solution at the temperature of the mixture.

13.10.3 Molecular Beam Epitaxy (MBE)

MBE technique employs an evaporation method as a means of obtaining the epitaxial layer of desired constitution. The substrate is held in an ultrahigh vacuum, and the molecular or atomic beams of the constituents are made to impinge on its surface. Figure 13.8 depicts the arrangement for growth of AlGaAs epitaxial layers on a GaAs substrate. The components Al, Ga, and As of the compound and the dopants Sn and Be are heated in separate cells. Each of the source vessels has a controllable shutter port. Collimated beams of these substances are directed onto the substrate which is held at a temperature between 500 to 600 °C. The rate at which the atomic beams of the constituent atoms strike the substrate surface can be

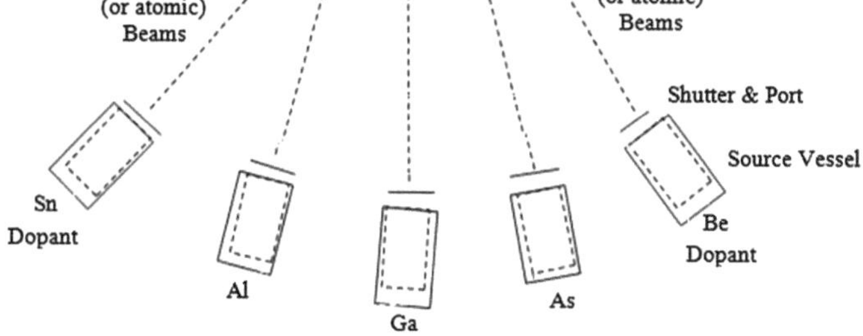

Fig. 13.8 Arrangement of molecular beam epitaxy method illustrating the growth of AlGaAs. Modified from [2] and self drawn

precisely controlled. High-quality films can be grown using this method. Because ultrahigh vacuum and close controls are involved, MBE setup remains a sophisticated one. However, its main advantages are

i. a low substrate temperature that minimizes the diffusion from the substrate, and
ii. a very precise control of dopant's profile.

13.11 Chemical Vapour Deposition (CVD)

In some cases, the thermal growth of SiO_2 is not possible such as when a substrate material other than Si is used. When thermal growth is not possible, a CVD technique offers an important alternative. For example, SiO_2 layers can be prepared by the reaction of SiH_4 and O_2 at temperatures between 250 and 500 °C.

$$SiH_4 + 2O_2 \rightarrow SiO_2 + 2H_2O \uparrow \tag{13.6}$$

The reaction may occur in a reactor similar to one as shown in Fig. 13.7 A gas mixture containing either N_2 or Ar with O_2 and about 1 % SiH_4 is used. The SiO_2 growth rate depends upon the substrate temperature and the rate of gas flow. This method also allows the deposition of SiO_2 layers doped with dopants such as As, P, and B. The doped oxide can then be used as a diffusion source.

13.12 Sputtering

In this method, the ions of a gas like Argon (Ar) produced in a glow discharge, are accelerated through a potential gradient and are then bombarded on a target. Atoms near the surface of the target become volatile and are transported as a vapour to the substrate where they condense and form a film of the material. A schematic diagram of a sputtering system is shown in Fig. 13.9. The material to be sputtered is made the cathode, while anode holds the substrate. Sputtering can be performed using either a D.C. or an rf power source. For SiO_2 deposition, a SiO_2 block is made the cathode, and the substrates are placed on the anode which is kept at the ground potential (Fig. 13.9). An rf voltage at about 10 MHz is applied between the cathode and the anode. The electrons in the plasma have a much higher mobility than the Ar-ions in the rf field. Consequently, electrons flow to the SiO_2 cathode, which then becomes negatively charged. Ar-ions bombard the cathode target, and SiO_2 is deposited on the substrates. These layers are very dense and can be used as insulating layers in various applications.

A problem common to all deposited SiO_2 layers is that the interface between the SiO_2 and the semiconductor is not as perfect as the Si–SiO_2 interface for thermally grown oxide.

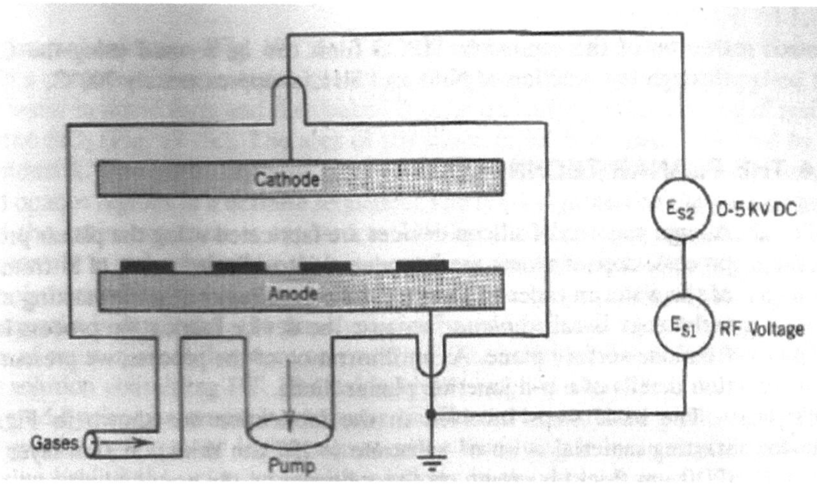

Fig. 13.9 Arrangement depicting an rf diode sputtering setup. Reproduced from [2]

13.13 Masking

The photomasking process can be divided into two distinct areas viz.

i. the generation of a mask whose image is to be transferred to the Si wafer; and
ii. the transfer of an image from the mask to the surface of the wafer.

For most present-day integrated circuits, mask making is done by using computer-controlled drawing boards and other equipment. In either case, copies of the circuit patterns are photographically reduced until they are ten times the ultimate size. The final mask is made from the 10× plate, using a step and repeat camera that has a reduction factor of 10. The step and repeat process results in rows and columns of identical images being transferred to a glass plate called a 'master'. The primary mask is used to make contact copies again using photosensitized glass plates. These copies are then used for the actual image transfer to a semiconductor wafer.

Another method utilizes the computer-controlled light flashes to generate the desired pattern on a photographic film by writing with a light pencil. The resulting pattern is then reduced and handled in a step and repeat system to create the production mask. Another method involves the use of an electron beam exposure system that can directly write the pattern in its final size onto an electron-sensitive photoresist in a hard surface mask.

13.14 Etching

Etching techniques are used for the selective removal of undesired dielectric and metallic layers. Different etching techniques are:

- Wet chemical etching
- Electrochemical etching,
- Sputter etching,
- Plasma etching, and
- Reactive ion etching.

A brief account of a few of these techniques is provided ahed.

Wet etching. Wet chemical etching is most widely used technique for selective removal of the regions of semiconductor material, metal, SiO_2, and Si_3N_4. For selective etching of different materials, the SiO_2 is etched in a buffered solution of HF + NH_4F; Si_3N_4 is etched in hot H_3PO_4; while Al is etched in either H_3PO_4, HNO_3, or acetic acid. A common method of etching is to immerse the wafer in the etching solution at a predetermined temperature. The etch time is determined by the etching rate of the solution.

Electrochemical etching. In electrochemical etching, a voltage is applied between the etchant and the material to be etched. Then the etching is performed at a controlled rate.

13.15 Metal Deposition Techniques

After a device has been fabricated on a semiconductor substrate, the ohmic contacts onto the various regions of the device are made by metal deposition. In case of an integrated circuit, all devices forming the circuit are suitably connected together with metallization to perform the intended circuit functions. The metal that is used to provide ohmic contacts for devices and interconnections for integrated circuits need to have high electrical conductivity and able to make good ohmic contacts with the semiconductor. Furthermore, it should have good adherence to the underlying surface and must not corrode under normal operating conditions. No single metal is known to perfectly meet all these requirements for a given semiconductor. However, in case of silicon, Al meets most of the requirements to a sufficient degree and is widely used for device interconnections, and in many cases for making ohmic contacts. Pure Al has a tendency to react with Si, hence to prevent this occurrence, some Si is added to Al-metallization.

Subsequent to metal deposition, the semiconductor wafer is placed in a furnace to alloy the metal to the semiconductor to ensure good electrical contact. It is then diced into separate pieces, each containing one or more semiconductor devices. Each chip is then soldered to a package, and wires from the package leads are connected to the metal on the semiconductor. Finally, the package is sealed with a ceramic or a metal cover.

13.16 Fabrication Techniques of P–N Junction

Based on the fabrication technique adopted, the semiconducting p–n junctions (diodes) may be of following types.

1. Grown junction.
2. Alloy type (or fused) junction.
3. Diffused junction.
4. Epitaxial grown (or planar diffused) junction.
5. Point contact diode junction

13.16.1 Grown Junction Diode

Diodes of this type are formed during the crystal pulling process. P- and N-type impurities can be alternately added to the molten semiconductor material in the crucible, which results in the formation of a P–N junction. After slicing, the larger area device can be cut into a large number (say in thousands) of smaller-area semiconductor diodes. Though such diodes, because of larger area, are capable of handling large currents but larger area also introduces more capacitive effects, which are undesirable. Such diodes are used for low frequencies.

13.16.2 Alloy Type (or Fused) Junction

Such a diode is formed by first placing a P-type impurity (a tiny pellet of aluminium or some other P-type impurity) into the surface of an N-type crystal and heating the two until liquefaction occurs where the two materials meet. An alloy will form that on cooling, will give a P–N junction at the boundary of the alloy substrate. Similarly, an N-type impurity may be placed into the surface of a P-type crystal. Alloy type diodes have a high current rating and large PTV (peak inverse voltage) rating. The junction capacitance is also large, due to the large junction area.

Alloyed junctions are made by heating a semiconductor slice in contact with an impurity that becomes liquid at the temperature used for heating and dissolves some of the semiconductor. As the liquid is cooled, the semiconductor recrystallizes with impurity atoms substituted in the semiconductor lattice. The fabrication process of a p–n junction formed by alloying Ge with In, is shown in Fig. 13.10a. A small pellet of In is placed on the surface of a slab of n-type Ge. The semiconductor slab is then

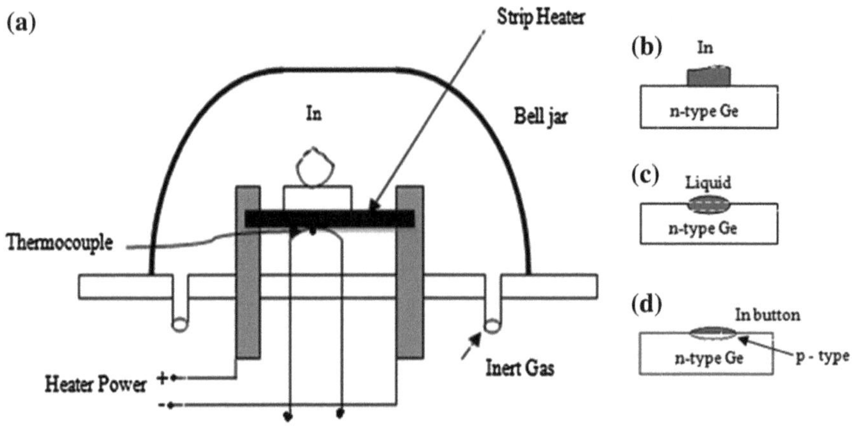

Fig. 13.10 Illustration of p–n junction formation by alloying. **a** Experimental setup, **b** Indium In–pellet on Ge before melting, **c** melting of into form a In–Ge mixture, and **d** the p–n junction formed after recrystallization. Modified from [2] and self-drawn

placed on a carbon strip heater that is covered with a glass bell jar. An inert atmosphere is maintained in the space covered by the bell jar by introducing an inert gas like argon or nitrogen. When the temperature of the strip heater is raised to about 500 °C, the In pellet on the surface of the Ge slab melts and dissolves some of the Ge (Fig. 13.10b) and forms a small puddle of a molten In–Ge mixture (Fig. 13.10c). When the temperature is lowered, the molten mass begins to solidify. The initial portion of the recrystallized material will be a single crystal of p-type Ge doped with In. As the solidification proceeds, the remaining mass becomes increasingly rich in In. Finally, when all the Ge is consumed, the material frozen at the outer surface of the recrystallized mass is pure In, which serves as the ohmic contact to the p-type Ge in Fig. 13.10d.

In the case of Si, p–n junctions can be made in a similar way by alloying Al to n-type Si. Alternatively, Au with about 0.1 % Sb can be prepared in the form of a thin disk which may then be alloyed to p-type Si to produce the n-region of the p–n junction. The alloy process is a simple and efficient method of making p–n junctions. However, it does not permit tight control on the area and the depth of the junction and thus has found only limited use.

13.16.3 Diffused Junction

Diffusion is a process in which a heavy concentration of particles diffuse into surrounding region of lower concentration. The main difference between the diffusion and alloy process is the fact that the liquefaction is not reached in the diffusion process. In the diffusion process, heat is applied only to increase the activity of elements involved.

For the formation of such junctions, either the solid or gaseous diffusion process can be employed. The process of solid diffusion starts with formation of layer of an acceptor impurity on an N-type substrate and heating the two until the impurity diffuses into the substrate to form the P-type layer. A large P–N junction is divided into parts by cutting process. Metallic contacts are made for connecting anode and cathode leads. Solid-state impurity diffusion is the most widely used technique for making p–n junctions, especially in silicon. The diffusion of dopant impurity atoms into a semiconductor is governed by the same basic laws as the diffusion of free carriers.

In the process of gaseous diffusion instead of layer formation of an acceptor impurity, an N-type substrate is placed in a gaseous atmosphere of acceptor impurities and then heated. The impurity diffuses into the substrate to form P-type layer on the N-type substrate. The process is illustrated in Fig. 13.11a, b. Though, the diffusion process requires more time than the alloy process; but it is relatively inexpensive, and can be very accurately controlled.

The diffusion technique leads itself to the simultaneous fabrication of many hundreds of diodes on one small disc of semiconductor material. It is most commonly used in the manufacture of semiconductor diodes. This technique is also used for the production of transistors and ICs.

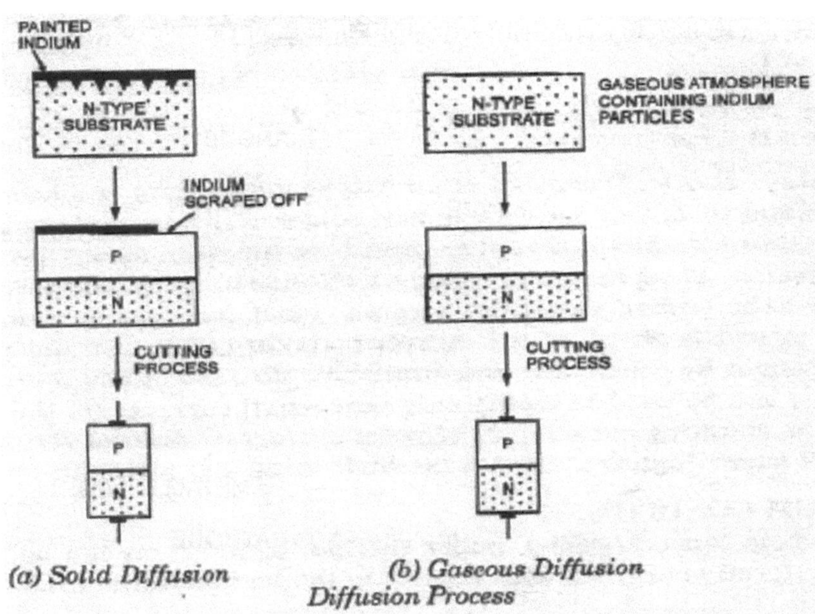

Fig. 13.11 Schemantic illustration of the formation of diffused junction by **a** solid diffusion process, and **b** gaseous diffusion process

13.16.4 Epitaxial Growth (or Planar Diffused) Junction (Fig. 13.12)

To construct an epitaxially grown junction diode, a very thin (single crystal) high impurity layer of semiconductor material (silicon or germanium) is grown on a heavily doped substrate (base) of the same material. This complete structure then forms the N-region on which the P-region is diffused. SiO_2 layer is thermally grown on the top surface, photo-etched and then aluminium contact is made on to the P-region. A metallic layer at the bottom of the substrate forms the cathode to which lead is attached. This process is usually employed in the fabrication of IC chips.

13.16.5 Point Contact Junction (Fig. 13.13)

It consists of an N-type germanium or silicon wafer about 12.5 mm × 12.5 mm square and 0.5 mm thick, one face of which is soldered to a metal base by radio-frequency (rf) heating. The other face has a phosphor bronze or tungsten spring pressed against it. A barrier layer is formed around the point contact by a pulsating current forming process. This causes a P-region to be formed round the wire and since pure germanium is N-type, a very small P–N junction in the shape of a hemisphere is formed round the point contact. This forming process cannot be controlled with precision. Because of small area of the junction, the point contact diode can be used to rectify only very small currents, of the order of mA.

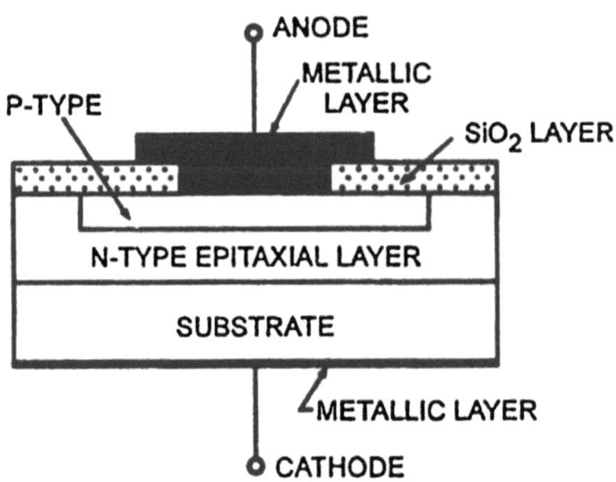

Fig. 13.12 Epitaxially grown (or planar diffused) diode

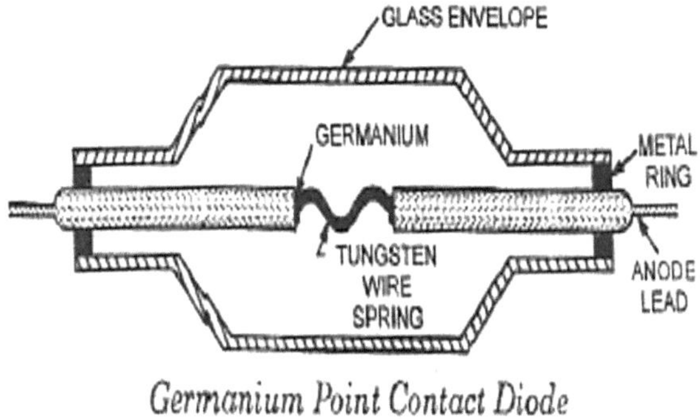

Fig. 13.13 Schematic arrangement of a point contact diode junction

13.17 Summary of the Fabrication of a Semiconductor P–N Junction

The main steps involved in the fabrication of a *p–n* junction are given below in Fig. 13.14. The diagrams show a typical case in which only four diodes per wafer are shown for simplicity. Also the relative thicknesses of the oxide, photoresist, and Al-layers are exaggerated.

13.18 Transistor Manufacturing Processes

Majority of the methods employed in transistor fabrication are the extensions of the methods employed in the manufacture of semiconductor diodes. The methods most commonly used for fabrication of transistors include point-contact, alloy junction, grown junction and diffusion. Various techniques adopted to manufacture a transistor, are displayed stepwise in Fig. 13.15. The details are self-explanatory.

13.19 Solved Examples

Example 13.2 Semiconductor materials are the natural choices for electronic devices—why?

Solution. Semiconductor materials are the natural choices because their conductivity and other electrical properties can be easily varied to suit the requirements of desired electronic devices. For example, their conductivity can be varied by following means.

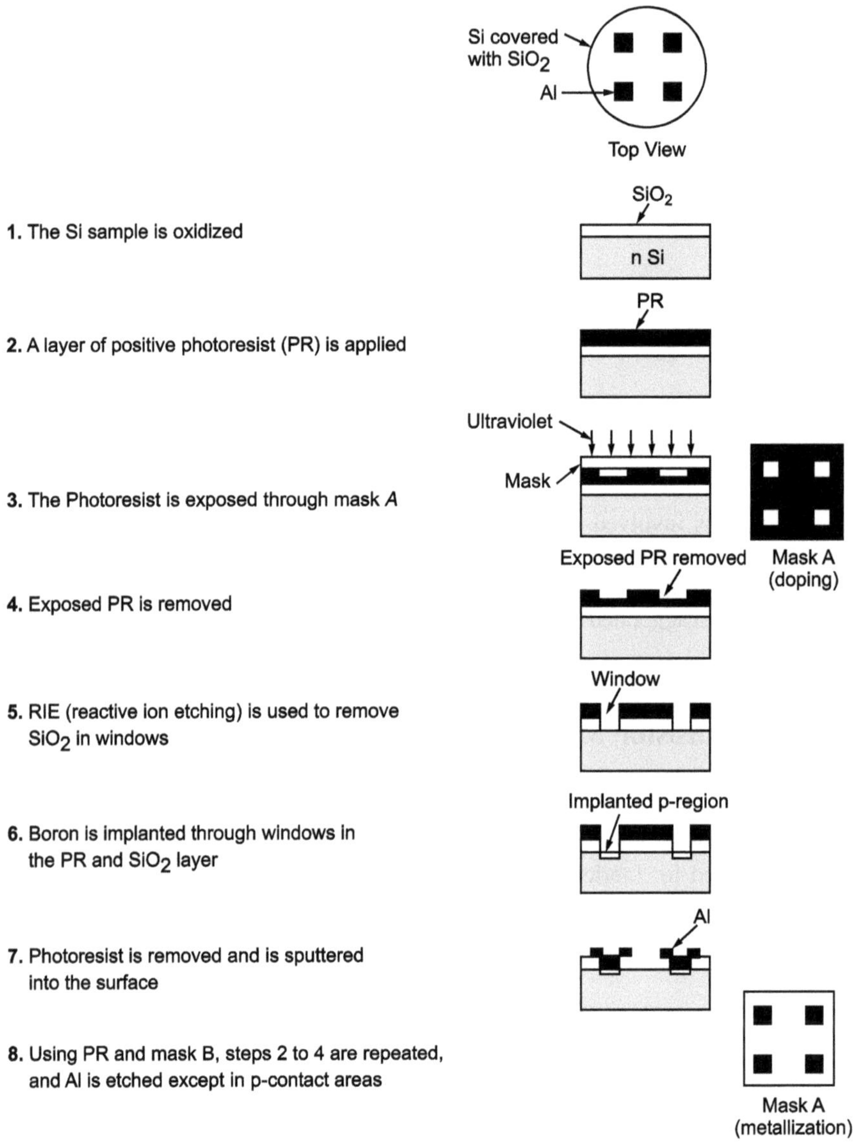

Fig. 13.14 Simplified description of steps in the fabrication of *p–n* junctions. Reproduced with permission from [1]

- By controlling the impurity content
- By optical excitation
- By change in temperature

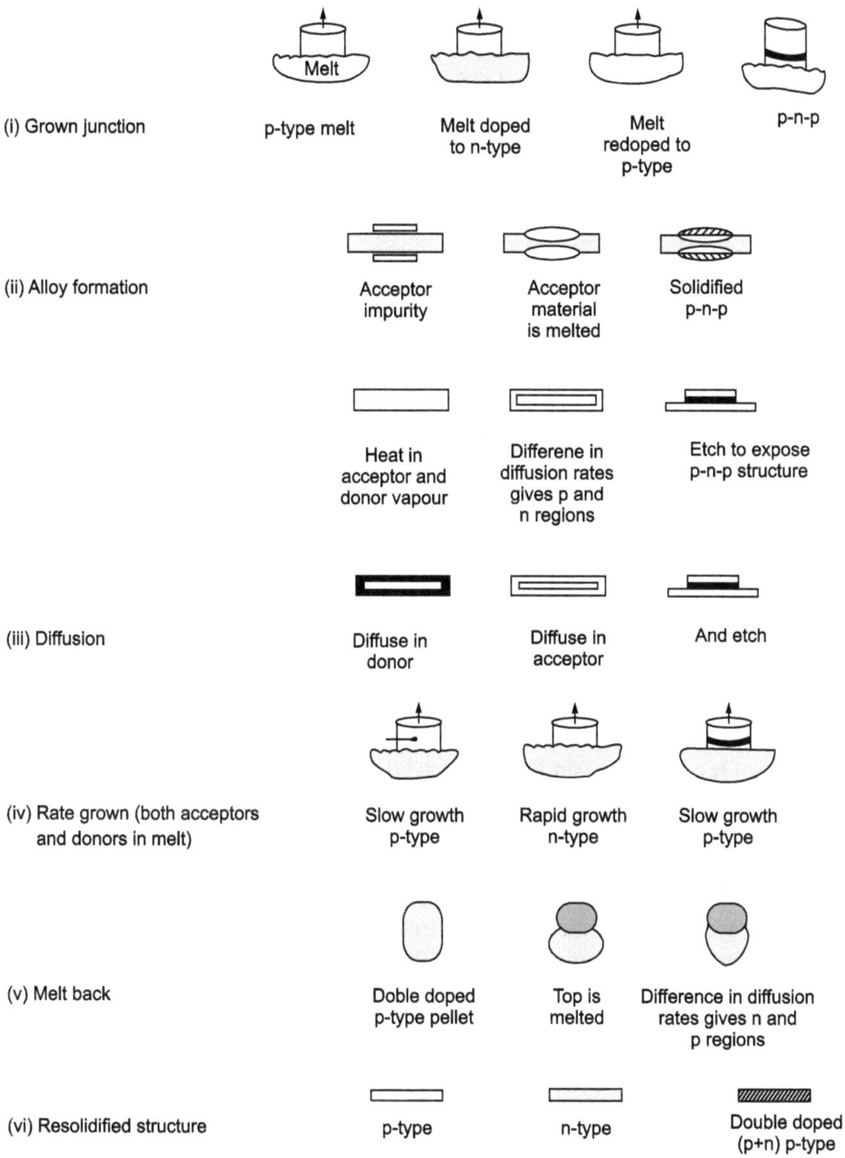

Fig. 13.15 Schematic arrangement of transistor manufacturing technique. Reproduced with permission from [1]

Example 13.3 How does a metallurgical grade Si (MGS) differ from an electronic grade Si (EGS)? Which grade is more suitable for making semiconductor devices?

Solution. A MGS is an output product obtained from the reaction of SiO_2. The MGS contains impurities such as Al, Fe and other heavy metals of the order of

100–1000 ppm. The MGS is an input product to obtain the EGS. EGS is highly pure semiconductor containing impurities of the order of a few ppb (1 ppb[1] = 10^{-3} ppm). Due to its high level purity, the EGS grade is more suitable for making semiconductor devices.

Example 13.4 Differentiate between the Bridgman's method and Czochralski's method of recrystallization.

Solution. In *Bridgman's recrystallization method*, the material is melted in a vertical cylinder that tapers conically to a point at the bottom. The vessel is gradually lowered in the cold zone where crystallization begins at the tip and the growth continues from the nucleus. In *Czochralski's method*, a single crystal is introduced on the surface of the melt and is drawn slowly upward, in the cold zone. The grown crystal usually acquires the form of container through which it has been drawn.

Example 13.5 Why the impurity concentration in extrinsic semiconductor is kept very small?

Solution. The electrical conductivity of pure semiconductors is primarily of electron carrier type. The hole conductivity is comparatively weaker. This makes the semiconductors extremely sensitive to the presence of any impurity. Hence impurity concentration is kept very small, about $1{:}10^6$ to $1{:}10^9$.

Example 13.6 Give a brief account of, as to how the p–n junction is formed?

Solution. In practice, the P–N junction is formed from a single monocrystalline structure by adding carefully controlled amounts of donor and acceptor impurities. The first and foremost requirement is to obtain an extremely pure germanium or silicon. Impurity of less than one part in ten billion (10^{10}) is required for most semiconductor device fabrication to-day. For obtaining pure semiconductor material, it is first purified chemically. For reducing the impurities further, and to ensure the formation of a monocrystalline structure, technique of floating zone is quite often employed. Once a pure monocrystalline semi-conductor is produced, carefully controlled amount of donor and acceptor impurities are added to the semiconductor without disturbing the orderly monocrystalline structure. Thus a P–N junction is formed.

Review Questions

1. What do you mean by metallurgical grade silicon and electric grade silicon? How is the element form of silicon produced?

[1] 1 ppb = 1 part per billion.

2. What are different processing techniques of semiconducting materials? Briefly describe them.
3. Explain the meaning of crystal growth. What are its different techniques? Explain any one of them.
4. Explain the Bridgman method of growing single crystal with the help of suitable diagram.
5. Describe the working of 'Czochralski method' crystal growth in semiconductor material with the help of suitable diagram.
6. Describe the float zone method of crystal growth with the help of suitable diagram.. Write its merits and demerits as compared to CZ method.
7. What is meant by zone refining? Why is it necessary for semiconducting materials?
8. Describe the technique of zone refining with the help of a suitable sketch and apparatus. How is this process accomplished?
9. Discuss any one technique of 'zone refining' in semiconductor material.
10. Describe in brief, the stepwise methodology to construct a semiconductor microelectronic circuit.
11. Explain the methodology of wafer manufacturing and the process of lithography.
12. What is the process of ion- implantation? With the help of suitable diagram, explain its process.
13. What is lithography? What are its different kinds? Discuss the 'photoresist' and photolithography.
14. What is/are the function(s) of epitaxy? Name its different processes, and explain the various phase epitaxy.
15. With the help of suitable diagram, explain the technique of molecular beam epitaxy.
16. What is sputtering? What is its importance in semiconductor fabrication? With the help of suitable diagram, explain this process.
17. Explain the following.
    (a) Chemical vapour deposition
    (b) Masking
    (c) Etching
    (d) Metal deposition techniques

18. Explain the procedure of manufacturing of a p-n junction.
19. Explain the procedure of manufacturing of a p-n-p or n-p-n junction.
20. Discuss in detail the various fabrication techniques of fabrication for p-n junctions. Compare the merits and demerits of each technique.
21. Write notes on the following:
    (a) Grown junction (b) Fused junction
    (c) Diffused junction (d) Planer diffused junction

Objective Questions

1. The silicon used for electrical purposes have silicon percentage of
   (a) 0.5%. (b) 2.5%.
   (c) 3.5%. (d) 4%
2. Consider the following functions:
   1. To mask against diffusion or ion implant.
   2. To act as a component in MOS devices.
   3. To provide low resistivity paths.
   4. To facilitate the entry of dopants.
   The function of an oxide layer on a silicon wafer would include
   (a) 1 and 2. (b) 2 and 3.
   (c) 3 and 4. (d) 1 and 4.
3. The correct sequence of processes involved in fabrication of a p-n junction diode is (out of many, only 4 to 5 processes are written)
   (a) thermal oxidization, photolithography, diffusion, etching
   (b) chemical vapour deposition, ion implantation, rapid thermal processing, diffusion
   (c) metallization, etching, photolithography, oxidization
   (d) oxidization, rapid thermal processing, ion implantation, etching, metallization

Answers

1. (c) 2. (a) 3.(d)

References

1. Gupta, K.M., Gupta, N.: Advanced electrical and electronics materials. Scrivener Wiley Publishing, USA (2015)
2. Tyagi, M.S.: Introduction to semiconductor materials and devices. Wiley India Pvt. Ltd., Delhi (2008). (Reprint)
3. http://cnx.org/contents/bbe0373c-eb45-4d40-8abd-3e6e077547c3@2/Electronic-Grade-Gallium-Arsen

Part VI
Special Purpose and Nano-Structured Semiconductors, and Recent Advances

Chapter 14
Special Semiconducting Materials in Vivid Fields (for Thermoelectrics, Integrated Circuits, Photocatalytics, Spintronic Devices, etc.), Plasmonic Solar Cell, and Photonics

Abstract Semiconducting materials are used in widely varying fields for specific applications. Therefore, they are special purpose semiconductors. Such semiconductors are described in this chapter, for example; three-dimensional semiconductor solar cell, optical semiconductor, thermoelectric semiconductors, integrated circuit purpose semiconductors. The photocatalytic semiconductors which are used to make the environment free of pollution, are also discussed. Transparent oxide semiconductors and semiconductor based spintronic devices are also explained. Spin-charge converter, spin injection materials, plasmonic solar cells, and photonic semiconductors are also elaborated. Plasmonic and photovoltaic kinds of semiconductors are introduced. Applications of plasmonic (photovoltaic) solar cells, photonic processor, photonic crystals, and superlattices are given. Future of photonics is also incorporated. Minute insight into the various topics are given through solved numerical and theoretical examples. Review questions, numerical problems and objective type questions are also given with their answers.

Keywords Solar cells · Bi_2Te_3 thermoelectric semiconductor · Chalcogenides · Skutterudites · SiC · SiGe · Cu_2O · BiOX (X = Cl, Br, I) photocatalytic semiconductors · Plasmonics · Photonics · Exiton

14.1 Semiconductor Nanoparticles in Solar Cell Construction

The ideal material used in solar cell must have high surface area for light absorption and charge separation. Nanoparticles, having a comparable surface area to volume ratio, provide the same. Titanium dioxide (TiO_2) nanoparticles are used to make nanoporous thin film supported upon a glass substrate. The material, thus, obtained has optical transparency, excellent stability and good electrical conductivity. The benefit of these novel photoelectrical solar cells is that they can be fabricated from cheap, low purity materials by simple and low cost procedures. The size-tunable bandgap of the semiconductor nanoparticles, due to size quantization, also means

K.M. Gupta and N. Gupta, *Advanced Semiconducting Materials and Devices*,
Engineering Materials, DOI 10.1007/978-3-319-19758-6_14

that, more efficient solar cells can be produced for photovoltaics (electricity production) and water splitting (hydrogen production) processes.

Today the environment needs renewable energy. Among the most important primary sources of clean energy is solar energy. The basis of it is called the photovoltaic effect, whereby the energy of sunlight is converted directly into electrical energy due to the interaction of photons and electrons in a semiconductor material. So far, the conversion of sunlight into electricity has been conducted almost exclusively with photovoltaic devices, but the situation could change with the use of nanotechnology. The use of nanomaterials in solar cells will create more efficient solar cells, smaller and with reduced production costs.

14.2 Three-Dimensional (3D) Semiconductor Solar Cells

The scientists have developed new solar cells from nanoscale crystals of (CdSe and CdTe) in the form of nanorods. However, current research lead us to think about the development of three-dimensional solar cells, which produce high performance with a significant reduction in size. 3D solar cell is based on the deposition of a semiconductor material on a three-dimensional nanostructure. The first 3D solar cells of low cost were based on a heterojunction based on the deposition of $CuInSe_2$ (p-type semiconductor) on TiO_2 (n-type semiconductor). But these are not the only materials that can be used.

The provision of 3-D allows the solar cells absorb light received from many different positions, so it is not necessary that the cell receives light directly. In addition, once the cell has absorbed the radiation, it is trapped inside the structure, can be absorbed by different parts of it. As a result of absorbing more photons, the cell coating may be thinner. Therefore, the electrons spend less time in semiconductor, reducing the possibility of recombination. These characteristics make the 3D solar cells to have important applications especially in the space field by eliminating the need for systems that maintain a specific orientation as satellite systems. It also reduces the weight and complexity, and increases the reliability and performance. However, these types of solar cells have to demonstrate their capability of withstanding the loads that support launch and operation of spacecraft, and to ensure a minimum of life in the harsh vacuum and thermal cycles.

14.3 Semiconductor (ZnS) as Optical Material Applications [1–6]

ZnS is used as detector because it emits light on excitation by X-rays or electron beam, making it useful for X-ray screens and cathode ray tubes [1]. It also exhibits phosphorescence due to impurities on illumination with blue or ultraviolet light [2].

Zinc sulfide, with addition of few ppm of suitable activator, is used as phosphor in many applications, from cathode ray tubes through X-ray screens to glow in the dark products [3]. When silver is used as activator, the resulting colour is bright blue, with maximum at 450 nm. Manganese yields an orange-red colour at around 590 nm. Copper provides long glow time and the familiar glow-in-the-dark greenish colour. Copper doped zinc sulfide (ZnS + Cu) is used also in electroluminescent panels [4].

Zinc sulfide is also used as an infrared optical material, transmitting from visible wavelengths to over 12 μm. It can be used planar as an optical window or shaped into a lens [5]. It is made as microcrystalline sheets by the synthesis from hydrogen sulfide gas and zinc vapour and sold as FLIR (Forward Looking IR) grade ZnS a pale milky yellow visibly opaque form [6].

14.4 Scope of ZnS Nanoparticles Semiconductor [7–20][1]

ZnS has wide direct bandgap (3.54 eV). Therefore, it's most common potential applications are in laser diodes and light emitting diodes (LEDs). Some optoelectronic applications of ZnS overlap with that of GaN which has a similar bandgap ($\sim$3.4 eV at room temperature). Compared to GaN, the ZnS has a larger exciton binding energy ($\sim$60 meV, 2.4 times of the room-temperature thermal energy), which results in bright room-temperature emission from ZnS. Other properties of ZnS favourable for electronic applications include its stability to high-energy radiation and to wet chemical etching. Radiation resistance makes ZnS a suitable candidate for space applications. ZnS is currently the most promising candidate in the field of random lasers to produce an electronically pumped UV laser source.

The pointed tips of ZnS nanorods result in a strong enhancement of an electric field. Therefore, they can be used as field emitters. Aluminium-doped ZnS layers are used as transparent electrodes. The constituents Zn and Al are much cheaper and less poisonous compared to the generally used indium tin oxide (ITO). One application which has begun to be commercially available is the use of ZnS as the front contact for solar cells or of liquid crystal displays.

Transparent thin-film transistors (TTFT) can be produced with ZnS. As field-effect transistors, they even may not need a p–n junction, thus avoiding the p-type doping problem of ZnS. Some of the field-effect transistors even use ZnS nanorods as conducting channels. The future scope of ZnS nanoparticles lies in the following applications.

Zinc sulfide nanorod sensor. Zinc sulfide nanorod sensors are devices detecting changes in electrical current passing through zinc sulfide nanowires due to adsorption of gas molecules appears to be effective in the catalytic dissociation of

[1]References 8–20 appear inside the matter of [7].

hydrogen molecules into atomic. The sensor detects hydrogen concentrations down to 10 ppm at room temperature, whereas there is no response to sulphur.

Spintronics. ZnS has also been considered for spintronics applications. If doped with 1–10 % of magnetic ions (Mn, Fe, Co, V, etc.), ZnS could become ferromagnetic, even at room temperature.

Piezoelectricity. The piezoelectricity in textile fibres coated with ZnS has shown capability of fabricating "self-powered nanosystems" with everyday mechanical stress from wind or body movements. Scientists have reported producing electricity generating device (called flexible charge pump generator) delivering alternating current by stretching and releasing zinc oxide nanowires. This mini-generator creates an oscillating voltage up to 45 mV, converting close to 7 % of the applied mechanical energy into electricity. Researchers used wires with lengths of 0.2–0.3 mm and diameters of 3–5 μm, but the device could be scaled down to smaller size.

Biosensors. ZnS has high biocompatibility and fast electron transfer kinetics. Such features advocate the use of this material as a biomimic membrane to immobilize and modify biomolecules.

14.5 Semiconductors in Thermoelectric (TE) Applications [21]

Within the framework of simple electronic band structure of solids, the metals are poor TE materials. Hence, most of the TE work put much emphasis on semiconductors. In order to have a maximum ratio of electrical to thermal conductivity, the material should have a low carrier concentration with very high mobilities. Crystal structure and bonding strongly influence the mobility. Materials with diamond or zinc-blende structures with a high degree of covalent bonding frequently have high mobilities (e.g., Si, Ge, InSb), but also exhibit high thermal conductivity. On the other hand, the low lattice thermal conductivities are found in conjunction with low Debye temperatures and high anharmonic lattice vibrations. These conditions are best satisfied by highly covalent intermetallic compounds and alloys of the heavy elements such as Pb, Hg, Bi, Tl, or Sb, and S, Se, or Te. Once a material system selected with a favourable electrical-to-thermal conductivity ratio, the composition optimized to enhance the ZT (Z is figure-of-merit of the thermoelectric material and T is the temperature. A higher value of Z is required) by doping the material, to maximize the density of states at the Fermi level and achieve a high Seebeck coefficient. Crystal structure of the most studied TE material, Bi_2Te_3 is shown in Fig. 14.1.

It crystallizes in a layer structure with rhombohedral–hexagonal symmetry. The hexagonal unit cell dimensions at room temperature are a = 3.8 Å and c = 30.5 Å. The Bi and Te layers are held together by strong covalent bonds, whereas the bonding between adjacent Te layers is of the van der Waals type. This weak

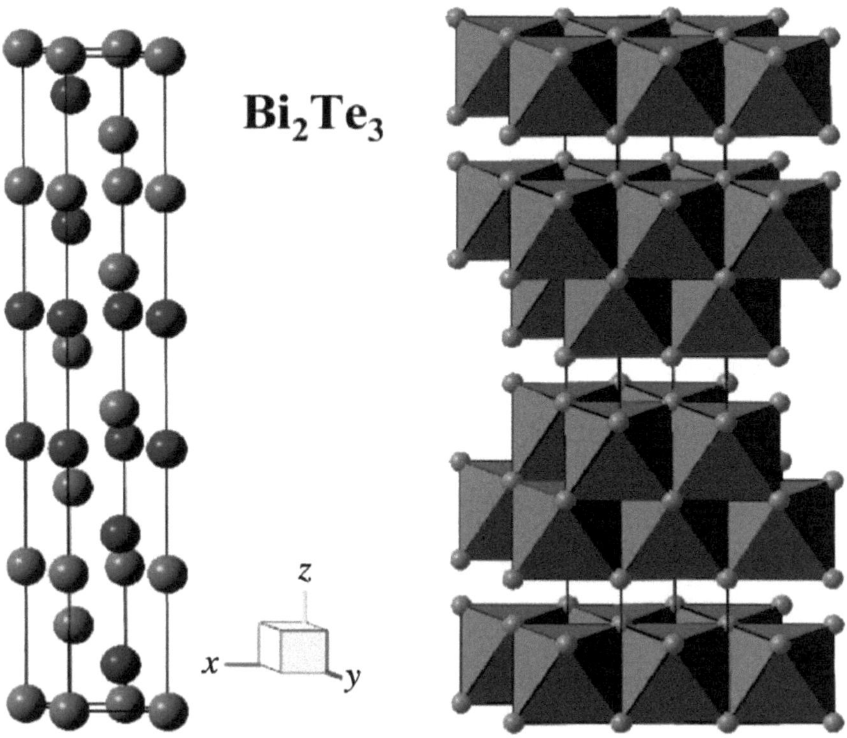

Fig. 14.1 Crystal structure of the thermoelectric material Bi_2Te_3.The *blue* atoms are Bi and *pink* atoms are Te. Reproduced with permission from [21]

binding between the Te layers accounts for the ease of cleavage along the plane perpendicular to the *c*-axis and the anisotropic thermal and electrical transport properties of Bi_2Te_3. For example, the thermal conductivity along the plane perpendicular to the to *c*-axis is nearly twice that of the value along the *c*-axis direction.

When grown from a melt or by zone refining, the Bi_2Te_3 crystals are always non-stoichiometric and show *p*-type behaviour. On the other hand, the *n*-type materials could be grown from the melt containing excess Te, iodine, or bromine. The thermal conductivity of both *p*- and *n*-type Bi_2Te_3 are $\sim 1.9\ \mathrm{W\,M^{-1}K^{-1}}$, giving a *ZT* of about 0.6 near room temperature. The alloying could further reduce the lattice thermal conductivity of Bi_2Te_3 through the scattering of short-wavelength acoustic phonons. The optimum compositions for thermoelectric cooling devices are normally $Bi_2Te_{2.7}Se_{0.3}$ (*n*-type) and $Bi_{0.5}Sb_{1.5}Te_{0.3}$ (*p*-type) with $ZT \approx 1$ near room temperature.

14.6 Semiconductors as High-ZT Thermoelectric Materials [22]

Z is the figure-of-merit of a thermoelectric material. A higher Z shows a higher performance at a specific temperature range. Performing higher Z over a wide temperature range is essential to obtain higher thermoelectric density. Hence, high ZT thermoelectric materials are desired. Important and emerging semiconductors for this purpose are

1. Ternary chalcogenides
2. Quarternary chalcogenides
3. Pentatellurides
4. Skutterudites
5. Oxide thermoelectrics

14.6.1 Chalcogenides and Pentatellurides: The Complex Inorganic Structures [21]

Recent approaches to high-performance bulk TE materials focus on ternary and quaternary chalcogenides containing heavy atoms with low dimensional or isotropic complex structures to take advantage of the large carrier effective masses and low lattice thermal conductivity associated with such systems. Along these lines, Cs Bi_4Te_6 possessing the layered structure has been identified in September 2012 by a team of researchers of MIT as a material showing a *ZT* of 0.8 at 225 K, which is 40 % greater than that of the $Bi_{2-x}Sb_xTe_{3-y}Se_y$ alloys [21].

Other potential low-temperature TE materials are low dimensional semiconducting or semimetallic doped layered 'pentatellurides' $ZrTe_5$ and $HfZrTe_5$. These compounds have a structure similar to Bi_2Te_3, with van der Waals gaps between the individual layers. Doped pentatellurides exhibit very high power factors (exceeding the optimally doped Bi_2Te_3 solid solutions) in the low-temperature range ($\sim$250 K), their thermal conductivity is relatively high ($\sim$4–8 W M^{-1} K^{-1}) and the materials need to be compositionally tuned further to make them useful as thermoelectrics.

14.6.2 Skutterudutes: The Crystal Structures with "Rattlers" [21]

The method of lowering the lattice thermal conductivity through mixed-crystal or solid-solution formation does not always produce enough phonon scattering to lower the lattice thermal conductivity. This approach especially works well in highly covalent semiconductor materials based on clathrates (Si, Ge, or Sn) and void structures formed by heavy elements of low electronegativity differences (e.g., $CoSb_3$-based skutterudites) Some doped skutterudites show exceptionally high *ZT* values at elevated temperatures (*ZT* 1.5 at 600–800 K).

14.6.3 Oxide Thermoelectrics [22]

There are numerous oxides with metal atoms in their common oxidation states that are stable at elevated temperatures and show electrical properties ranging from insulating to superconducting. Nevertheless, oxides have received very little attention for TE applications. This is due to their strong ionic character, with narrow conduction bandwidths arising from weak orbital overlap, leading to localized electrons with low carrier mobilities. This situation changed with the unexpected discovery of good TE properties in a strongly correlated layered oxide, $NaCo_2O_4$. This oxide attains $ZT \sim 0.7$–0.8 at 1000 K. Inspired by the striking TE performance of $NaCo_2O_4$, most of the current studies are focused on Co-based layered oxides, such as $Ca_3Co_4O_9$ and $Bi_2Sr_3Co_2O_y$, crystallizing in "misfit" (lattice-mismatched) layered structures. Among the *n*-type oxides, Al-doped ZnO ($Al_{0.02}Zn_{0.98}$) shows reasonably good TE performance (*ZT* 0.3 at 1000 K).

14.7 Semiconducting Materials for Integrated Circuits [23]

An integrated circuit, commonly referred to as an IC, is a microscopic array of electronic circuits and components that has been diffused or implanted onto the surface of a single crystal (or chip) of semiconducting material such as silicon Fig. 14.2. It is called an integrated circuit because the components, circuits, and base material are all made together or integrated, out of a single piece of silicon, as opposed to a discrete circuit in which the components are made separately from different materials and assembled later. ICs range in complexity from simple logic modules and amplifiers to complete microcomputers containing millions of elements. ICs have become the principal components of almost all electronic devices. These miniature circuits have demonstrated low cost, high reliability, low power requirements, and high processing speeds compared to the vacuum tubes and transistors which preceded them. Integrated circuit microcomputers are now used as controllers in equipment such as machine tools, vehicle operating systems, and other applications where hydraulic, pneumatic, or mechanical controls were previously used. IC microcomputers are smaller and more versatile than previous control mechanisms. They allow the equipment to respond to a wider range of input and produce a wider range of output. They can also be reprogrammed without

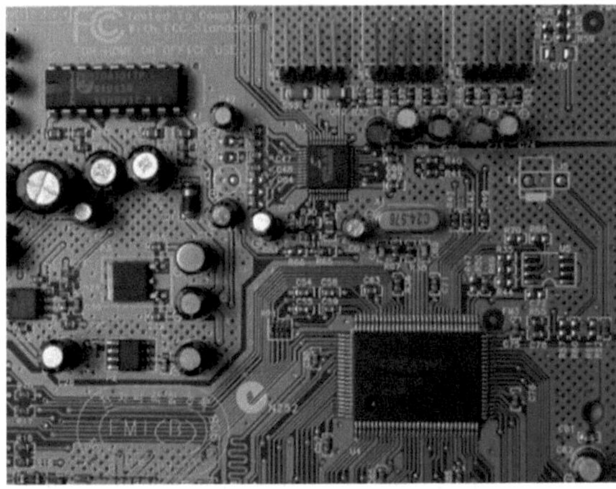

Fig. 14.2 Integrated circuit. Reproduced from [24]

having to redesign the control circuitry. Integrated circuit microcomputers are so inexpensive they are even found in children's electronic toys.

14.7.1 Silicon Carbide (SiC) for High Temperature ICs [25]

Conventional silicon circuits break down beyond 350 °C, but the ability to get sensor information from high-temperature environments such as inside a jet engine or in a deep oil well, can improve efficiency. It also provides capabilities for space missions to extreme locations. Research into silicon carbide logic ICs could move signal processing and control closer to the point of sensing, thereby removing the need for long wires which could fail, or complicate the systems.

Silicon carbide allows for high-temperature devices because of its wide bandgap. In ordinary silicon, high temperatures can kick electrons into the conduction band, causing errant currents to flow and fouling logic operations. Silicon carbide's bandgap is so much wider that heat does not cause much conduction. Simple SiC integrated circuits have been demonstrated reliable at 500 °C. Junction field effect transistor (JFET) technology avoids MOS oxide reliability issues, especially at high temperature limits to integration because of power consumption (1 mW per gate).

The high-temperature SiC integrated circuits are used in following applications.

- Analog differential amplifier
- Digital NAND, NOR gates
- Operational Amplifier
- A/D Converter
- Multiplexer
- Microcontrollers

- In-package amplifier for SiC dynamic pressure sensor
- Digital electronics for transmission of sensor signals on power cable
- Wireless capacitive pressure sensor for propulsion health monitoring

The Enabling Technologies for SiC Electronics are the following.

- Metal-to-SiC contacts
- High temperature packaging
- SiC transistor (JFET) design
- Multilevel interconnect structure
- Conformal metal layers
- High temperature insulating layers
- Increase circuit density
- Increase reliability and yield
- Reduce crystalline defects

14.7.2 Gallium Nitride (GaN) [26]

Gallium nitride (GaN) is an option for MMICs (Monolithic Microwave Integrated Circuit). Because GaN transistors can operate at much higher temperatures and work at much higher voltages than GaAs transistors, they make ideal power amplifiers at microwave frequencies. A MMIC is a type of integrated circuit (IC) device that operates at microwave frequencies (300 MHz–300 GHz). These devices typically perform the functions such as:

- microwave mixing,
- power amplification,
- low-noise amplification, and
- high-frequency switching.

Inputs and outputs on MMIC devices are frequently matched to a characteristic impedance of 50 Ω. This makes them easier to use, as the cascading of MMICs does not then require an external matching network.

14.7.3 Indium Phosphide (InP) [27]

Indium phosphide is used to make photonic integrated circuit (PIC) or integrated optical circuit which is a device that integrates multiple photonic functions. It is analogous to an electronic integrated circuit. The major difference between the two is that a photonic integrated circuit provides functionality for information signals imposed on optical wavelengths, typically in the visible spectrum or near infrared (850–1650 nm) range. The most commercially utilized material for photonic integrated circuits is indium phosphide, which allows for the integration of various optically active and passive functions on the same chip.

14.7.4 Silicon Germanium (SiGe) [28]

The advents of 200 GHz SiGe:C technologies pave the way for realizing the millimeter-wave integrated circuits with their lower cost and excellent performance. This work is focused on the design and implementation of circuits in IHP's low-cost SiGe:C technology at W-band and frequencies beyond. Different types of high-speed frequency dividers as benchmarking circuits are designed and measured to show the speed and power performance of the SiGe technology.

14.7.5 Gallium Arsenide (GaAs) [29]

Gallium Arsenide (GaAs) is used for building integrated circuits that operates better than Si. GaAs has five times higher electron mobility than Si and has high peak (threshold) velocity. GaAs can be used both for analog microwave applications and for high speed digital devices up to a frequency of 20 GHz. The short minority lifetime and high saturated drift velocity of GaAs is useful for applications such as planar schottky mixer diodes, transferred electron devices, planar Gunn and PIN diodes. In fact, both: the digital and analog functions can be integrated on the same chip together with lasers, producing integrated optical devices.

Applications of GaAs Semiconductor. GaAs technology can be used above 3 GHz with good performance. Several GaAs monolithic chip functions are already available from dc to 35 GHz such as listed below.

- Receivers
- Transmitters (1 W or greater) phase shifters
- Switches
- Broadband amplifiers
- Active filter oscillators
- Sample and hold circuits
- Flash A to D converter
- PLL (phase locked loop) synthesizers
- Log amplifiers mixers etc.

Monolithic-GaAs microwave IC's are useful in several military applications including communication, radar, ECM and weapon systems like phased array antenna systems, spread spectrum systems, global satellite TDMA communications etc. They are cost effective and hence useful in consumer applications such as

- Direct satellite TV
- Telecommunication
- High speed data transmission
- Instrumentation, and
- Microwave test equipment.

These could employ mixed analog and digital subsystems on the same chip. GaAs IC's enable system designers to have high performance microwave components at low cost with advantage of batch processing, automatic assembly and testing.

14.7.6 Comparative Study of Different Semiconducting Materials for Integrated Circuits

A comparison of various properties of different semiconducting materials used for making ICs is illustrated in Table 14.1.

14.8 Semiconductor Photocatalytic Materials [30]

The steadily worsening environmental pollution and energy shortages have raised awareness of a potential global crisis. For the sustainable development of human society, the development of both: the pollution-free technologies for environmental remediation and alternative clean energy supplies is an urgent task. Among the wide variety of green earth and renewable energy projects underway, semiconductor photocatalysis has emerged as one of the most promising technologies because it represents an easy way to utilize the energy of either natural sunlight or

Table 14.1 Comparative study of different semiconducting materials for integrated circuits

Comparison of RF semiconductor material properties						
	Si	GaAs	InP	SiGe	SiC	GaN
Lattice constant (Å)	5.4	5.7	5.9	5.5	3.1	3.2
Saturation velocity (cm/s)	1×10^7	0.8×10^7	2.2×10^7	–	2×10^7	2.5×10^7
e-mobility (cm^2/Vs)	1350	8000	10000	3000	900	1500
Eg bandgap (eV)	1.1	1.4	1.3	0.7–1.1	3.3 (4H)	3.4
F_t (GHz) FET	20	150	300	50	20	150
Power density (W/mm)	0.2	0.5	–	0.3	10	>30
Thermal conductivity (W/cm K)	1.5	0.5	–	–	4.9	~2.0

Reproduced from [29]

artificial indoor illumination, and is thus abundantly available everywhere. The potential applications of photocatalysis are found mainly in the following fields:

1. Photolysis of water to yield hydrogen fuel
2. Photo-decomposition or photo-oxidization of hazardous substances;
3. Artificial photosynthesis;
4. Photo-induced super-hydrophilicity;
5. Photoelectrochemical conversion, etc.

14.8.1 ZnO Semiconductor as Photocatalytic Material [31]

ZnO is considered a valid photocatalytic materials because of its good optoelectronic, catalytic and photochemical properties along with its low cost. ZnO has a bandgap of 3.0 eV. Due to the position of the valence band of ZnO, the photogenerated holes have strong oxidizing power to decompose most organic compounds. ZnO has been tested to decompose aqueous solutions of several dyes and many other environmental pollutants. In many cases, ZnO has been reported to be more efficient than TiO_2, but the occurrence of photocorrosion and the susceptibility of ZnO to facile dissolution at extreme pH values, have significantly limited its application in photocatalysis.

Several studies have shown that ZnO is quite active under visible light illumination for the photodegradation of some organic compounds in aqueous solution. Methyl green was successfully decolourized and degraded by ZnO under visible light irradiation at low watt irradiation and the addition of an oxidant ($Na_2S_2O_8$ or H_2O_2) enhanced the degradation rate of the pollutant. Nanostructured ZnO obtained by thermal decomposition of zinc oxalate without using any additives or solvents was more efficient than commercial ZnO for the mineralization under solar irradiation.

14.8.2 Cu_2O Semiconductor as Photocatalytic Material [32]

Cu_2O is a p-type semiconductor with a direct bandgap of 2.0–2.2 eV. It has been studied for application in solar energy conversion. The photocatalytic properties of Cu_2O are strongly dependent on the shape of the crystals that determines the number of atoms located at the edges, corners, or surfaces. Cu_2O octahedra show better photocatalytic activity than cubes, because the {1 1 1} faces are more active than {1 0 0} faces due to the dangling bonds of {1 1 1} surfaces; whilst {1 0 0} faces have saturated chemical bonds and no dangling bonds exist.

14.8.3 Iron Oxides Semiconductor as Photocatalytic Material [33]

Iron (III) hydroxides absorb light up to 600 nm. Most of them have semiconductor properties and have been studied as photocatalysts even though a very efficient electron/hole recombination generally takes place.

14.8.4 Sulfides Semiconductor as Photocatalytic Material

Some transition metal sulfides have been considered as photocatalysts for the removal of organic pollutants because they have narrow bandgap and proper band potentials, which match with the visible light photon energy for the degradation of many compounds. CdS, with a direct band-gap energy of 2.42 eV, has been extensively studied for the photocatalytic production of hydrogen from water with visible light and for photovoltaic devices. However, it is not suitable for the photocatalytic removal of organic pollutants because it is unstable under irradiation and suffers photocorrosion that not only destroys the photocatalyst but more importantly, releases toxic cadmium ions in solution.

Other heavy metal sulfides as Bi_2S_3, ZnS and MoS_2 have been tested for the photocatalytic degradation of organic pollutants. ZnS nanoporous nanoparticles comprising hexagonal wurtzite, ZnS nanocrystals of several nanometers in diameter were prepared by a solution-phase thermal decomposition route in the presence of poly(N-vinyl-2-pyrrolidone). The ZnS nanoporous nanoparticles showed much greater activity for the photodegradation than that of ZnS nanocrystallites. Crystalline Sb_2S_3 synthesized by solid-state reaction had a very narrow bandgap of about 1.55 eV. The photocatalytic activity of Sb_2S_3 for the degradation of MO under visible light was much higher than that of Bi_2S_3, and CdS. The high efficiency of Sb_2S_3 was ascribed to the broad spectrum response and the suitable valence band position.

14.8.5 Semiconductor Chalcogenide as Photocatalytic Material [34]

$ZnIn_2S_4$ is a ternary semiconductor chalcogenide that belongs to the AB_2X_4 family of the ternary compounds. $ZnIn_2S_4$, with a bandgap of 2.2 eV, is an interesting potential photocatalyst for solar-driven applications.

$SnIn_4S_8$ is another ternary sulfide that has been studied for the visible light-induced decomposition of organic pollutants. In particular, porous $SnIn_4S_8$ microspheres were highly efficient and stable towards the photodegradation of MO, RhB and MB. The photocatalytic activity of the $SnIn_4S_8$ microspheres is much higher than that of the binary constituent In_2S_3 and SnS_2 and the ternary chalcogenide $ZnIn_2S_4$.

14.8.6 Bismuth Oxyhalides Semiconductors as Photocatalytic Material [35]

Bismuth Oxyhalides (BiOX where X = Cl, Br, I) are other ternary semiconductors that are potential candidates for photocatalytic applications. All BiOX compounds crystallize in the tetragonal matlockite structure, a layer structure characterized by (Bi_2O_2) slabs interleaved by double slabs of halogen atoms. Among the bismuth oxyhalides, BiOI has the smallest bandgap (1.8 eV) whilst BiOF has the largest value (3.4 eV). BiOI nanosheets exhibited a high photocatalytic performance to remove sodium pentachlorophenate from an aqueous solution under simulated solar light irradiation. BiOF is an effective photocatalyst for degradation of many organic pollutants such as phenol, MO, salicylic acid and RhB in aqueous media.

14.9 Transparent Thin Film Transistors [36]

The thin film transistor (TFT) was the first solid-state amplifier. The field-effect transistor (FET) was probably the first successful solidstate amplifier. Later a new field-effect transistor (called MISFET) was patented where the thickness of the dielectric layer, which insulates the metal control electrode from the copper sulfide channel, was about 10^{-5} cm. The year 1990 marked the debut of a new class of TFT, based upon the organic semiconductor active layer material, with electron mobilities similar to that of a-Si:H. This new class of TFTs are very promising candidates for integration onto flexible plastic substrates for a future generation of rugged, lightweight displays than can be rolled up like a map.

More recently, a new generation of oxide semiconductors are being studied and are applied as the active material to TFT, in special zinc oxide (ZnO). The transparent electronics is emerging technology for the next generation of optoelectronic devices. Oxide semiconductors are very interesting materials because they combine simultaneously high/low conductivity with high visual transparency and have been widely used in a variety of applications, such as given below.

- Antistatic coatings
- Touch display panels
- Solar cells
- Flat panel displays
- Heaters
- Defrosters
- Optical coatings

14.9.1 Advances in Transparent Oxide Semiconductor Based Transistors [37]

Transparent oxide semiconductor based transistors have recently been proposed using as active channel non-doped ZnO. One of the main advantages exhibited by these transistors lies on the magnitude of the electron channel mobility leading to higher drive currents and faster device operating speeds. The mobility reported in the literature is ranging from 0.2 to 7 cm^2/V s with an on/off current ratio from 105 to 107 and a threshold voltage between −1 and 15 V. To date, ZnO channel layers have been deposited using substrate heating or submitted to post-thermal annealing in order to increase, mainly the crystallinity of the ZnO layer and so the film's mobility. The main advantage of using ZnO deals with the fact that it is possible to growth at/near room temperature high quality polycrystalline zinc oxide, which is a particular advantage for electronic drivers, where the response speed is of major importance. Besides that, since ZnO is a wide bandgap material (3.4 eV), it is transparent in the visible region of the spectra and therefore, also less light sensitive.

14.10 Semiconductor Based Spintronic Devices [38]

Spintronics (or spin electronics) refers to the study of the role played by electron spin in solid, and possible devices that specifically exploit the spin properties instead of/or in addition to charge degrees of freedom. The spin relaxation and spin transport in metals and semiconductors are of fundamental research today for the potential these phenomena have in electronic technology. The prototype device that is already in use in industry as a 'read head' and a 'memory-storage cell' is the giant-magnetoresistive (GMR) sandwich structure which consists of alternating ferromagnetic and non-magnetic metal layers. Depending on the relative orientation of the magnetizations in the magnetic layers, the device resistance changes from small (parallel magnetizations) to large (antiparallel magnetizations). This change in resistance (also called magnetoresistance) is used to sense changes in magnetic fields. Recent efforts in GMR technology have also involved the magnetic tunnel junction (MTJ) devices where the tunneling current depends on spin orientations of the electrodes [38].

Current efforts in designing and manufacturing spintronic devices involve two different approaches. The first is perfecting the existing GMR-based technology by either developing new materials with larger spin polarization of electrons, or making improvements or variations in the existing devices that allow for better spin filtering. The second effort focuses on finding the novel ways of both generation and utilization of spin-polarized currents. These include investigation of spin transport in semiconductors and looking for ways in which semiconductors can function as spin polarizers and spin valves. The importance of this effort lies in the fact that the existing metal-based devices do not amplify signals (although they are successful switches or valves), whereas semiconductor based spintronic devices could in principle provide amplification and serve, in general, as multi-functional devices. Perhaps even more importantly, it would be much easier for semiconductor-based devices to be integrated with traditional semiconductor technology.

While there are clear advantages for introducing semiconductors in novel spintronic applications, many basic questions pertaining to combining semiconductors with other materials to produce a viable spintronic technology remain open. For example, whether placing a semiconductor in contact with another material would impede spin transport across the interface is far from well-understood. In the past, one of the strategies to advance understanding of spin transport in 'hybrid semiconductor structures' was to directly borrow knowledge obtained from studies of more traditional magnetic materials. However, there is also an alternative approach involving the direct investigation of spin transport in all-semiconductor device geometries. In such a scenario a combination of optical manipulation (for example, shining circularly polarized light to create net spin polarization) and material in homogeneities (e.g. by suitable doping as in the recently discovered $Ga_{1-x}Mn_xAs$ type ferromagnetic materials where Mn impurities act as dopants) could be employed to tailor spin transport properties.

14.10.1 GaAs Spin—Charge Converter [39]

'Spin-charge converters' are important devices in spintronics. Spin-charge converters enable the transformation of electric into magnetic signals and vice versa. Recently, the research group of Professor Jairo Sinova [39] has realised a new efficient spin-charge converter based on the common semiconductor material GaAs. Spintronics does not only make use of the electron's charge to transmit and store information but it takes also the advantage of electron's spin. The spin can be regarded as a rotation of the electron around its own axis, and generates a magnetic field like a small magnet. To make real use of the electron spin, it has to be manipulated precisely. It has to be aligned, transmitted and detected. The work [39] shows that it is possible to do so using electric fields rather than magnetic ones. Thus, the very efficient, simple and precise mechanisms of charge manipulation already established in semiconductor electronics can be transferred to the world of spintronic and thereby combine semiconductor physics with magnetism.

Spin-charge converters are essential tools for that. They can transform charge currents into spin currents, and vice versa. The main principle behind these converters is the so called 'spin-Hall effect'. The spin-Hall effect appears when an electric field drives the electrons through a (semi-) conductor plate. The interaction of moving electrons and an external magnetic field forces the electrons to move to one side of the plate, perpendicular to their original direction. This leads to the so called Hall voltage between both sides of the plate. For the spin-Hall effect, electron-spins are generated by irradiating the sample with circularly polarised light. To make practical use of this effect, it is essential to get a highly efficient spin separation.

Up to now, platinum has been the most efficient spin-charge converter material, as it is a heavy metal, and the spin-orbit coupling of heavy metals is known to be especially strong due to the large amount of protons (positive charge) in their core. Sinova and his colleagues [39] have shown that the gallium-arsenide (GaAs), a very common and widely used semiconductor material, can be an as efficient spin-charge converter as platinum, even at room temperature, which is important for practical applications. The reason for this lies in the existence of certain valleys in the conduction band of the semiconductor material. By taking into account the valleys in the conduction band, new ways are open to find highly efficient materials for spintronics. Especially, since current semiconductor growth technologies are capable of engineering the energy levels of the valleys and the strength of spin-orbit coupling, e.g. by substituting Ga or As with other materials like Aluminium.

14.11 Heusler Alloy Search for the New Spintronic Materials [40]

The search for material combining properties of the ferromagnet and the semiconductor has been a long-standing goal but an elusive one because of differences in crystal structure and chemical bonding. The advantages of ferromagnetic semiconductors (FS) is their potential as spin-polarized carrier sources and easy integration into semiconductor devices. The ideal FS would have Curie temperatures above room temperature and would be able to incorporate not only p-type, but also n-type dopants. The Eu chalcogenides, the early magnetic semiconductors, in which the magnetic species (Eu21) resides on every lattice site, failed in the practical sense because their ferromagnetic transition temperatures, T_{fm}, were much lower than the room temperature.

The discovery of ferromagnetic order temperatures as high as 110 K in III–V—based diluted magnetic semiconductors (DMS), alloys in which some atoms are randomly replaced by magnetic atoms, such as Mn has generated much attention. To achieve large spin polarization in semiconductors, the Zeeman splitting of the conduction (valence) band must be greater than the Fermi energy, E_F, of the electrons (holes). In concentrated materials, this occurs easily because the net magnetization is proportional to the concentration of magnetic species. Externally applied magnetic fields may be necessary to produce large polarization in DMS where the Mn content is generally low (5 %) and may be limited by phase separation while the carrier concentrations are high (10^{20} cm^3). Another approach is to search for new materials that exhibit large carrier spin polarization. These include the ferromagnetic oxides and related compounds, many of which are predicted to be "half-metallic". There are efforts to produce magnetic Heussler alloys such as NiMnSb (T_{fm} 5728 K), which has been used as an electrode in tunneling junctions.

14.11.1 Spin Injection Materials [41]

The purpose of the spin injection material is to inject a highly spin polarized current into a semiconductor. This can be accomplished either through high intrinsic spin polarization, or band symmetry matching with the adjacent semiconductor and/or tunnel barrier. The material to be used for a spin injecting contact should have several attributes, as given below.

(i) It must be ferromagnetic, with a Curie temperature of over 400 °K;
(ii) It must have significant easy axis remnant magnetization, i.e., zero field magnetization, of at least 50 % of the saturation magnetization;
(iii) It should provide high spin polarization of the injected current, producing high spin polarization in the semiconductor; and
(iv) It should be thermally stable against intermixing with adjacent layers, and degradation of its FM properties with processing.

Requirements (i) and (ii) provide the non-volatile reprogrammable characteristics, which are highly desirable for applications such as field programmable gate arrays, logic elements, or memory. In general, the spin injection contact material needs to be selected and tailored for a particular semiconductor or tunnel barrier.

Three broad families of materials could be used for polarized spin injection, namely: ferromagnetic metals, half metals, and ferromagnetic semiconductors, but each has different challenges. Traditional FMMs such as Fe, Co, Ni and alloys are well-known to the magnetic recording industry and readily meet criteria (i) and (ii) above. Because of the large conductivity mismatch between a FMM and a semiconductor, an intervening tunnel barrier is required to enable efficient spin injection. This may take the form of a tailored reverse-biased Schottky contact or a discrete metal oxide layer (e.g., Al_2O_3, MgO, etc.). Several FMMs have been shown to meet criteria for selected semiconductors and/or tunnel barriers.

14.11.2 Spin Transport in Semiconductors and Their Nanostructures [41]

For several devices, once the spin is injected into a semiconductor, it is important that the spin not loose coherence in the time that transport, manipulation, and detection occur. Most experimental work on spin transport in semiconductors typically focuses on III–V direct gap materials such as GaAs, because polarization dependent optical absorption/emission spectroscopies provides easy, direct and quantitative insight into carrier spin polarization and dynamics. The long spin lifetimes expected for the low-Z (weak spin orbit) Group IV semiconductors make spin angular momentum especially attractive. Spin transport, via electrical injection and detection of spin polarized carriers from FM metal contacts (e.g., Fe, CoFe) into Si, has been demonstrated, with reported electron spin polarizations of 30 % or more.

Magnetic field induced coherent precession of the pure spin current and spin polarized charge current has been demonstrated in lateral and vertical transport geometries, respectively. These results collectively show that information can be fed in, processed and read out using spin rather than charge as the state variable. However, these results have previously been limited to low temperature, due to thermal noise generated by the contact resistance. Options to reduce this contact resistance, by controlling the depletion width in the Si, have been identified. Recent work has demonstrated successful electrical injection, detection and manipulation of spin accumulation in Si using ferromagnetic metal/SiO_2 tunnel barrier contacts at temperature to 500 K, easily exceeding the operating temperature requirements for commercial application.

14.11.3 Graphene, a Challenge to Semiconductors [41]

Graphene exhibits *spin* transport characteristics that surpass those of any other semiconductor studied to date, demonstrating large magnetoresistance (MR) at room temperature. Such large MR has not been seen in any other semiconductor materials or nanostructures, including InAs to GaN to Si.

For low dimensional materials, CNTs are attractive as spin transport materials because their low dimensionality results in a suppression of certain spin orbit scattering mechanisms at higher temperatures (>70 K), leading to longer spin lifetimes. Limited successes have been reported for spin injection into CNTs from magnetic metal contacts at low temperature. However, obtaining reliable contacts and reproducible results continue to be challenges.

14.12 Plasmonic Solar Cells

Plasmonic Solar Cells (PSC) as shown in Fig. 14.3 are known as photovoltaic devices. They have great potential in driving down the cost of solar power. For the solar cell (SC) technology to be a viable energy source to compete with fossil fuels, the price needs to be reduced by 2–5 times. Approximately 90 % of the market for solar cells use silicon wafers. These wafers are typically around 200–300 μm thick and constitute about 40 % of the cost. Plasmonic solar cells are a type of thin-film solar cells which are typically 1–2 μm thick. They also use substrates such as glass, plastic or steel which are cheaper than silicon. The major problem for thin film solar cells is that they don't absorb as much light as the current solar cells. Methods for trapping light on the surface, or in the solar cell, are crucial in order to make thin film solar cells viable. An important point about plasmonic solar cells is that they are applicable to any kind of solar cell. This includes the standard silicon or newer thin-film types.

Fig. 14.3 Plasmonic solar cells. Reproduced from [42]

Current solar cells have ways of trapping light. However, their method is to create pyramids on the surface which have dimensions bigger than most thin film solar cells. Making the surface of the substrate rough, typically by growing SnO_2 or ZnO on surface, with dimensions of the order of incoming wavelengths and depositing the solar cell on top, has also been explored. This method increases the photocurrent, but the thin film solar cells would then have poor material quality. Other methods have been explored for increasing the absorption of light in thin film solar cells. One method which has been explored over the past few years is to scatter light using metal nanoparticles excited at their surface plasmon resonance. Different approaches using these metal nanoparticles and the basic principles which govern as to how the scattering/absorption is increased, will be the focus of the following information along with a detailed application of plasmonic solar cells.

There have been quite a few pioneer working with plasmonic solar cells. One of the main focuses has been on improving the thin film solar cells through the use of metal nanoparticles distributed on the surface. It has been found that the Raman scattering can be increased by order of magnitude when using metal nanoparticles. The increased Raman scattering provides more photons to become available to excite surface plasmons which cause electrons to be excited and travel through the thin film solar cell to create a current.

The promising field of plasmonics has yielded methods for guiding and localizing light at the nanoscale. Now focus of plasmonics research is turning to photovoltaics, where design approaches based on plasmonics can be used to improve absorption in photovoltaic devices, permitting a considerable reduction in the physical thickness of solar photovoltaic absorber layers, and yielding new options for solar cell design. Due to a combination of the resonant plasmonic properties of metallic nanoparticles with thin-film photovoltaic technology, a new generation of plasmonic solar cell has evolved with similar performance to silicon cells but at potentially a fraction of the cost. Today, plasmonic solar cells are emerging as promising candidates amongst many solar energy technologies, spurring continuing research to improve device performance.

14.12.1 Nonlinear Plasmonic Antennas

Contrary to traditional optical elements, plasmonic antennas made from nano-structured metals permit the localization of electromagnetic fields on length scales much smaller than the wavelength of light. This results in huge amplitudes for the electromagnetic field close to the antenna being conducive for the observation of nonlinear effects already at moderate pump powers. Thus, these antennas exhibit a promising potential to achieve optical frequency conversion and all-optical control of light at the nano-scale. This opens unprecedented opportunities for ultrafast nonlinear spectroscopy, sensing devices, on-chip optical frequency conversion, nonlinear optical metamaterials, and novel photon sources. Here, the review of some of the recent advances in exploiting the potential of plasmonic antennas has been presented to realize robust nonlinear applications.

Surface plasmon polaritons (SPPs) refer to the oscillation of charge density waves resonantly coupled to the electromagnetic field at a metal–dielectric interface. SPPs stand out in comparison with light in conventional all-dielectric optical elements for their confinement is not restricted by the conventional diffraction limit. Therefore, SPPs can be localized and guided deeply sub-wavelength, which promises a viable route toward the realization of ultra-compact waveguides and antennas for integrated optics.

14.12.2 Scope and Applications of Plasmonic Solar Cells

The applications for plasmonic solar cells are endless. The need for cheaper and more efficient solar cells is too much. In order for solar cells to be considered cost effective, they need to provide energy at a smaller price than that of traditional power sources such as coal and gasoline. The movement toward a greener world has helped to spark research in the area of plasmonic solar cells. Currently, solar cells cannot exceed efficiencies of about 30 % (First Generation). With new technologies (Third Generation), efficiencies of up to 40–60 % can be expected. With a reduction of materials through the use of thin film technology (Second Generation), prices can be driven lower. In such perspective, plasmonic solar cells carry enormous scope and extensive applications in driving high efficiency at low cost. Some of the applications are:

Space exploration: Certain applications for plasmonic solar cells would be for space exploration vehicles. A main contribution for this would be the reduced weight of the solar cells. An external fuel source would also not be needed if enough power could be generated from the solar cells. This would drastically help to reduce the weight as well.

Rural electrification: Solar cells have a great potential to help rural electrification. An estimated two million villages near the equator have limited access to electricity and fossil fuels and that approximately 25 % of people in the world do not have access to electricity. When the cost of extending power grids, running rural electricity and using diesel generators is compared with the cost of solar cells, many times the solar cells win. If the efficiency and cost of the current solar cell technology is decreased even further then many rural communities and villages around the world could obtain electricity when current methods are out of the question. Specific applications for rural communities would be

- water pumping systems,
- residential electric supply and
- street lights.

A particularly interesting application would be for health systems in countries where motorized vehicles are not overly abundant.

Marine industry: Solar cells could be used to provide the power to refrigerate medications in coolers during transport. Solar cells could also provide power to lighthouses, buoys, or even battleships out in the ocean. Industrial companies could use them to power telecommunications systems or monitoring and control systems along pipelines or other system.

High power: If the solar cells could be produced on a large scale and be cost effective then entire power stations could be built in order to provide power to the electrical grids. With a reduction in size, they could be implemented on both commercial and residential buildings with a much smaller footprint. They might not even seem like an eyesore.

Hybrid system: Other areas are in hybrid systems. The solar cells could help to power high consumption devices such as automobiles in order to reduce the amount of fossil fuels used and to help improve the environmental conditions of the earth.

Low power: One application which has not been mentioned is consumer electronics. Essentially, solar cells could be used to replace batteries for low power electronics. This would save everyone a lot of money and it would also help to reduce the amount of waste going into landfills.

14.13 Photonic Materials

The term 'Photonics' comes from photons which is the smallest unit of light, similar to an electron which is the smallest unit of electricity. Photonics is the generation, processes and manipulation of photon to achieve a certain function. Photonics is related to generation, guiding and detection of light. The study of photonic materials is interdisciplinary of chemistry, physics and materials science. It focuses on glasses, optical fibres and sol-gel based materials.

Glass is Isotropic in nature, with optical properties easily tuned for specific applications. Glasses play a fundamental role in photonics;

Optical fibres prepared from optical glasses, are the choice for guiding media for applications in communications, sensing and light transmission. Its newer applications in new areas are such as plasmonics and biophotonics.

Sol-Gel based materials and processes are more popular in preparation of colloidal suspensions and special optical thin films.

The photonic crystals had been studied in one-dimensional crystal form of periodic multi-layer dielectric stacks, such as the Bragg mirror. Today, such structures are used in a diverse range of applications from reflective coatings to enhancing the efficiency of LEDs to highly reflective mirrors in certain laser cavities e.g. VCSEL (Vertical Cavity Surface Emitting Laser).

14.13.1 Need of Photonics Instead of Electronics

Photonics possesses following favourable features than electronics.

1. Uninhibited light travels thousands of times faster than electrons in computer chips.
2. Optical computers will compute thousands of times faster than any electronics computer can ever achieve due to the physical limitation i.e. the differences between light and electricity.
3. It can pack more wavelengths into a optical fibre so that the transmission bandwidth is increased than conventional copper wires.
4. Light encounters no electromagnetic interference as compared to that of electron in copper wires.

14.14 Possible Applications of Photonic Materials [43]

Some of the exciting potential applications of nonlinear optics (photonic materials) include the following.

- Quantum computing,
- Quantum communications,
- Quantum imaging,
- All-optical switching,
- Optical power limiting, and
- Nonlinear optical image processing.

However, the implementation of these applications has historically been held back by the limited availability of materials with the required properties of large optical nonlinearity combined with high optical transparency and high resistance to laser damage. Several strategies have traditionally been exploited in an attempt to create new materials with more desirable optical properties. These strategies include the synthesis of new chemical compounds with intrinsically large optical response, the use of quantum coherence techniques to more anciently utilize the nonlinear response present in a given material, and the creation of composite materials that

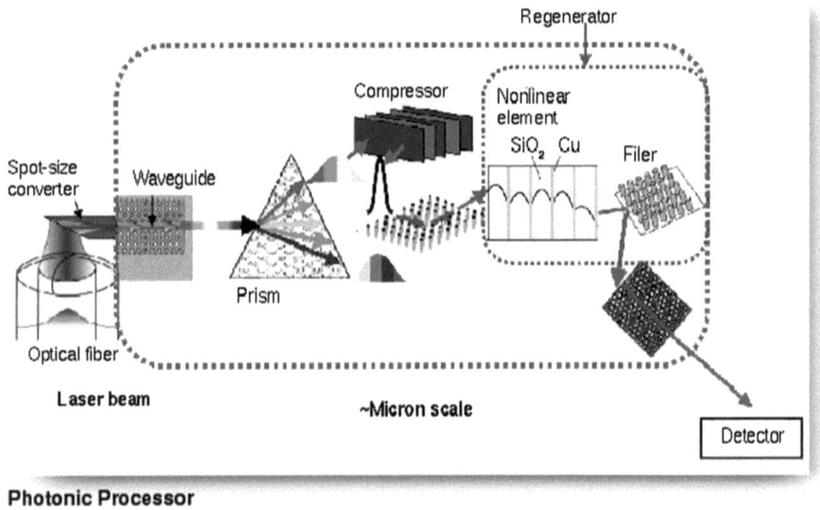

Fig. 14.4 Photonic processor

can combine the desirable characteristics of two or more constituent materials to create a new material with tailored optical properties. Some of the photonic applications are described in subsequent sections.

14.14.1 Photonic Processor (Fig. 14.4)

There are materials that reflect the frequency range of interest, with essentially no loss at all. Such materials are widely available all the way from the ultraviolet regime to the microwave. Nonlinear effects using non-linear properties of materials for construction of photonic crystal lattices open new possibilities for molding the flow of light. In this case the dielectric constant is additionally depending on intensity of incident electromagnetic radiation and any non-linear optics phenomena can appear.

14.14.2 Photonic Crystals [44]

Photonic chips go 3D. The dream of building computer chips that use light signals rather than electricity has entered the realm of serious research in recent years with the advent of photonic crystal, a material that blocks and channels light within extremely small spaces. This image (Fig. 14.5) shows the microscopic structure of a three-dimensional photonic crystal that is capable of channeling and emitting light in the visible and telecommunications ranges.

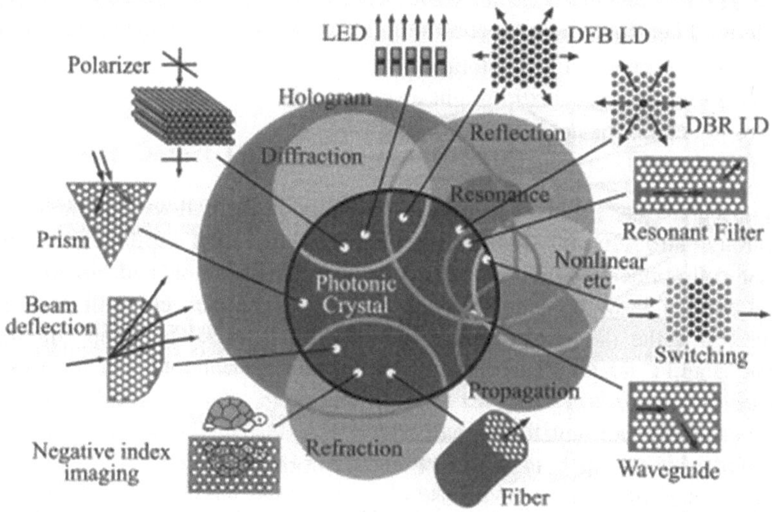

Fig. 14.5 Photonic crystals. Reproduced from [45]

14.14.3 *Photonic Integrated Circuits*

The key driving force of photonic research is the miniaturization and increase of the functionality of photonic circuits. The application of photonic crystal technology allows us to design waveguides, couplers, and routers on a much smaller scale than previously possible, thus allowing the design of high-density integrated circuits Applications of this work are in telecommunications, e.g. for devices that manage high-speed, high volume data (e.g. internet) traffic.

14.14.4 *The Future of Photonics [46]*

Photonic materials and devices play an important role in following.

- Telecommunications
- Information processing and storage
- Chemical sensors
- Solar cells
- Light emitting diodes
- Magneto-optic memories
- Video systems
- Colour imaging, etc.

They encompass a wide variety of materials including

1. wide band-gap semiconductors
2. materials for magnetic data storage
3. diamond materials for extra-bright display screens
4. micro-materials for next-generation displays
5. semiconductor laser materials and more.

Recent progress has led to spectacular advances in light flow and light-matter interaction control through photonic crystal technology, light trapping in thin films for solar cell applications, photonic crystal fibres; the control, understanding and application of light at the nano-scale through nano-photonics, controlling the flow of light below the diffraction limit, plasmonics and optical materials. In future, photonic material can play important role in information technology and processing signal technology.

The discovery and development of new photonic materials has been and continues to play a vital role in the development of information technology. It has led to the development of semiconductor lasers, optical fibres and amplifiers, and to high bandwidth telecommunications networks. The exploitation of new optical phenomena and the development of optical devices create opportunities for advances in

- optical memory storage
- holographic, and
- magneto-optic storage.

14.15 Where Photovoltaic Meets Microelectronics [47]

The PV industry has to systematically reduce its manufacturing cost in order to reach grid parity for the main markets in the coming decade. Crystalline Si solar cells are dominating presently the PV-market and this dominance is expected to continue for at least the next decade. Industrial production at a cost of 1 $/Wp for crystalline Si solar cells has been announced for 2011. Increasing cell efficiencies is a prominent pathway in view of the strong leverage on costs of materials in the module fabrication sequence. In order to further reduce $/Wp costs, eventually down to the 0.5 $/Wp level, it is clear that the dominant position of crystalline Si solar cells on the market was partially achieved thanks to the existing knowledge and equipment base within the context of micro (nano)-electronics, although the cost drivers in both cases are principally different: cost/functionality reduction for microelectronics versus cost/Wp for photovoltaics. In order to achieve the ambitious goals stated higher, it makes sense to have a closer look how the process and analysis toolbox available in the microelectronics area can be used at the benefit of the further development of crystalline Si-based photovoltaic devices. The use of a nanotechnology toolbox is however not limited to crystalline Si solar cells. Also fields like concentrator PV (CPV) might take profit from developments in the domain of photonics enabled by the technological capabilities developed within micro (nano) electronics.

Continued advances will depend on continued reduction in the cost of photonic components, increased functionality and increased levels of integration, novel trends in the fields of photonic materials and devices, novel materials, material structures and the development of advanced fabrication techniques, characterization methods and applications of photonic materials.

14.16 Solved Examples

Example 14.1 What is photonic integrated circuit?

Solution. By definition an integrated circuit is a micro-electronic device that houses multiple circuits on a chip. For example, an IC is built by lithographic fabrication of numerous transistors on a silicon chip. Similarly a photonic IC is a device that houses integrated photonic functions on a chip. The development of PICs promises to increase functionality, density, and significantly reduced cost when compared to optical components assembled from discrete devices.

Example 14.2 What is superlattice?

Solution. A semiconductor superlattice is a periodic structure that can be used to illustrate the behaviour of a periodic potential. This class of systems is composed of a number of semiconductor quantum wells of thickness L separated by barriers of thickness h. Generally speaking, the number of periods ranges between 10 and 50, but for the analysis the number of periods is approaching infinity. The superlattice means that the quantum wells are close to each other such that an electron can tunnel through the barriers and exists in any of the wells with a non-zero probability.

Example 14.3 What are intraband transition and interband transition?

Solution. The optical properties of any material are the result of photon interaction with the constituents of the material. The photon interactions with semiconductor materials, that lead to effects that are the basis for technologies such as detectors, emitters, optical communications, display panels, and optical oscillators. The interaction of photons with electrons in semiconductor materials is most important and gives rise to many phenomena. Electrons in semiconductor materials can absorb photons and be excited from the valence band to the conduction band. This is called the *interband transition*.

The inverse of this process occurs when electrons decay from a higher energy level, such as a conduction band to a lower energy level, such as a valence band, and photons are emitted. This is the basis for light-emitting diodes (LEDs) and laser diodes.

Electrons can absorb photons and be excited from one state to another within a particular band, such as a conduction band. This transition is called an *intraband transition*. When the electrons are excited from a bound state to another bound state in the conduction band, the transition is called an 'inter-subband transition'. These terminologies are also applied to heavy or light holes in semiconductors.

Example 14.4 What is optical bandgap of a bulk semiconducting material?

Solution. The band-to-band transition in a bulk material is usually referred to as the optical bandgap. In the case of quantum structure, the conventional optical bandgap is no longer allowed, and the effective bandgap is referred to as the transition from the ground state in the valence band to the ground state in the conduction band.

Example 14.5 What is Excition?

Solution. The effective bandgap in quantum structures is larger than the conventional optical bandgap in bulk materials. If an electron is excited from the valence band to the conduction band of a semiconductor, it leaves behind a positively charged hole. This process is called electron–hole pair generation. When the electron and hole interact with each other due to Coulomb interaction, the result is called an *exciton*. The excitonic energy levels are usually formed in the fundamental bandgap. If the exciton is trapped by an impurity or an atom in the crystal, it is called a *bound exciton* or a *Frenkel exciton*. The binding energy of a 'free exciton' is usually smaller than that of a 'bound exciton'.

Example 14.6 What is meant by excitons in bulk semiconductors?

Solution. Excitons are quasi-particles used to describe electron-hole pairs coupled by Coulomb interaction in a manner similar to the hydrogen atom. There are two types of excitons viz. free and bound excitons. Excitons in semiconductors are stable so long as their binding energy is smaller than the thermal energy (kT). The optical absorption and photoluminescence emission of excitons affect the optical properties of the band edge of semiconductors and their heterojunction. Exciton absorption is profound at low temperatures in most direct bandgap semiconductor materials. It can even be observed at room temperature.

Review Questions

1. Discuss the role of semiconductor nanoparticles in construction of solar cells. Describe the 3-D semiconductor solar cells.
2. Explain the various uses of ZnS as optical materials; as nanoparticles in nanorod sensor, spintronics, piezoelectricity etc.

3. Describe the thermoelectric purpose semiconductors. Also write the high ZT-thermoelectric materials.
4. Write notes on the following.
   (a) Chalcogenides and pentatellurides
   (b) Skutterudutes
   (c) Oxide thermoelectric materials
5. Enlist the various semiconducting materials used for making ICs. Write their properties and characteristics.
6. Discuss the SiC as high-temperature IC purpose semiconductor. Write its properties and applications also.
7. Describe the characteristics and properties of GaAs and write its applications.
8. Compare the various IC purpose semiconducting materials from different aspects.
9. What are semiconducting photocatalytic materials? What purpose do they serve? Enlist their examples and discuss their properties and specialities.
10. Describe the following with all related etails.
    (a) Transparent thin-film transistors
    (b) Transparent oxide-semiconductor based transistors
    (c) Semiconductor based spin-charge converter
    (d) Spin-injection semiconducting materials
11. Sketch and explain the construction and working of plasmonic solar cells. Write their applications also.
12. What are photonic materials? How are they useful as compared to electronics? Write the various possible applications of photonic materials. Discuss the future of photonics.
13. Describe the photonic processor, photonic crystals, and photonic ICs with all related details.
14. Write notes on the following
    (a) Graphene, a challenge to semiconductors
    (b) Superlattice
    (c) Excition
    (d) Bulk semiconductors

References

1. Burs, L.E.: Appl. Phys. A **53**, 465 (1991)
2. Alivisatos, A.P.: J. Phys. Chem. **100**, 13226–13239 (1996). As quoted in [1]
3. Gunasekaran, M., Ichimura, M.: Jpn. J. Appl. Phys. **44**, 7345–7350 (1999). As quoted in [1]
4. Nien, Y.T., Chen, P.W., Chen, I.G.: J. Alloy Compd. **462**, 398–403 (2008). As quoted in [1]
5. Drezner, Y., Berger, S., Hefetz, M.: Math. Sci. Eng. B **18**, 59 (2001). As quoted in [1]
6. Klein, C.A., Dibenedetto, B., Pappis, J.: Opt. Eng. **25**, 519 (1986). As quoted in [1]
7. http://en.wikipedia.org/wiki/Zinc_oxide
8. Bakin, A. et al.: ZnO-GaN hybrid heterostructures as potential cost efficient LED technology. Proc IEEE **98**(7), 1281–1287 (2010) [IEEE (Invited Paper for a special volume)]. doi:10.1109/JPROC.2009.2037444. As quoted in [7]
9. Look, D.: Recent advances in ZnO materials and devices. Mater. Sci. Eng. B **80**, 383 (2001). doi:10.1016/S0921-5107(00)00604-8. As quoted in [7]
10. Kucheyev, S.O. et al.: Ion-beam-produced structural defects in ZnO. Phys. Rev. B **67**(9), 094115 (2003) (Bibcode:2003PhRvB..67i4115K). doi:10.1103/PhysRevB.67.094115. As quoted in [7]
11. Li, Y.B., Bando, Y., Golberg, D.: ZnO nanoneedles with tip surface perturbations: excellent field emitters. Appl. Phys. Lett. **84**(18), 3603 (2004) (Bibcode:2004ApPhL..84.3603L). doi:10.1063/1.1738174. As quoted in [7]
12. Oh, B.-Y., Jeong, M.-C., Moon, T.-H., Lee, W., Myoung, J.-M., Hwang, J.-Y., Seo, D.-S.: Transparent conductive Al-doped ZnO films for liquid crystal displays. J. Appl. Phys. **99**(12), 124505 (2006) (Bibcode:2006JAP….99l4505O). doi:10.1063/1.2206417. As quoted in [7]
13. Nomura, K., Ohta, H., Ueda, K., Kamiya, T., Hirano, M., Hosono, H.: Thin-film transistor fabricated in single-crystalline transparent oxide semiconductor. Science **300**(5623), 1269–72 (2003) (Bibcode:2003Sci…300.1269N). doi:10.1126/science.1083212. PMID 12764192. As quoted in [7]
14. Heo, Y.W., Tien, L.C., Kwon, Y., Norton, D.P., Pearton, S.J., Kang, B.S., Ren, F.: Depletion-mode ZnO nanowire field-effect transistor. Appl. Phys. Lett. **85**(12), 2274 (2004) (Bibcode:2004ApPhL..85.2274H). doi:10.1063/1.1794351. As quoted in [7]
15. Wang, H.T., Kang, B.S., Ren, F., Tien, L.C., Sadik, P.W., Norton, D.P., Pearton, S.J., Lin, J.: Hydrogen-selective sensing at room temperature with ZnO nanorods. Appl. Phys. Lett. **86**(24), 243503 (2005) (Bibcode:2005ApPhL..86x3503W). doi:10.1063/1.1949707, as quoted in [7]
16. Tien, L.C., Sadik, P.W., Norton, D.P., Voss, L.F., Pearton, S.J., Wang, H.T., Kang, B.S., Ren, F., Jun, J. et al.: Hydrogen sensing at room temperature with Pt-Coated ZnO thin films and nanorods. Appl. Phys. Lett. **87**(22), 222106 (2005) (Bibcode:2005ApPhL..87v2106T). doi:10.1063/1.2136070. As quoted in [7]
17. Che, M.A., El-Shaer, A., Bakin, A., Waag, A., Ahlers, H., Siegner, U., Sievers, S., Albrecht, M., Schoch, W. et al.: Magnetic property investigations on Mn-doped ZnO layers on sapphire. Appl. Phys. Lett. **87**(6), 062501 (2005) (Bibcode:2005ApPhL..87f2501M). doi:10.1063/1.2007864. As quoted in [7]
18. Brandon, K.: Piezoelectric nanowires turn fabric into power source. Wired News (CondéNet) (2008). As quoted in [7]
19. Qin, Y., Wang, X., Wang, Z.L.: Editor's summary: nanomaterial: power dresser. Nature **451** (7180), 809–813 (2008) (Bibcode:2008Natur.451..809Q). doi:10.1038/nature06601. PMID 18273015. As quoted in [7]
20. New small-scale generator produces alternating current by stretching zinc oxide wires. Science Daily. Accessed 10 Nov 2008
21. Tritt, T.M., Subramanian, M.A.: Thermoelectric materials, phenomena, and applications: a bird's eye view. MRS Bull. **31**, 188–198 (2006)
22. Gupta K.M.: Engineering Materials: Research, Applications and Advances. CRC Press, Taylor and Francis Group (2014)
23. http://www.madehow.com/Volume-2/Integrated-Circuit.html

24. https://articlefind.wordpress.com/2012/01/26/integrated-circuit/
25. http://spectrum.ieee.org/semiconductors/devices/silicon-carbide-logic-circuits-work-at-blistering-temperatures
26. http://en.wikipedia.org/wiki/Monolithic_microwave_integrated_circuit
27. http://en.wikipedia.org/wiki/Photonic_integrated_circuit
28. Wang, L.: Millimeter-wave integrated circuits in SiGe:C technology (2008)
29. http://www.microwavejournal.com/articles/17674-future-rf-market-opportunitiesforgan?v=preview
30. Jiang, G., Chen, T., Yang, Q.: Photocatalytic materials. Adv. Mater. Sci. Eng. **2012**, 2 (2012). Article ID 186948. http://dx.doi.org/10.1155/2012/186948
31. Gowthaman, P., Saroja M., Venkatachalam, M., Deenathayalan, J., Shankar, S.: Doping effects of strontium on Zn nonanorods and their photocatalytic properties. J. Nanosci. Nano Technol. **2**(4), 377–383. ISSN 2279-0381
32. Zhang, Y., Deng, B., Zhang, T., Gao, D., Xu, A.-W.: Shape effects of Cu_2O polyhedral microcrystals on photocatalytic activity. J. Phys. Chem. C **114**(11), 5073–5079 (2010). doi:10.1021/jp9110037
33. Belattar, S., Mameri, Y., Seraghni, N., Debbache, N., Sehili, T.: Catalytic degradation of 3,5-Dimethylphenol with goethite and hydrogen peroxide. J. Environ. Eng. Technol. **1**(3), 21–28 (2012)
34. Banica, R., Ursu, D., Racu, A.V., Vaszilcsin, N.: Hydrothermal synthesis of $ZnIn_2S_4$ microspheres under controlled pressure. Nanocon 2014, Nov 5–7, Brno, Czech Republic, EU
35. Di Paola, A., García-López, E., Marcì G., Palmisano, L.: A survey of photocatalytic materials for environmental remediation. J. Hazard. Mater. **211–212**, 3–29 (2012)
36. Fortunato E., Barquinha, P., Goncalves, A., Marques, A., Pereira, L., Martins, R.: Recent advances in ZnO transparent thin film transistors. Thin Solid Films **487**, 205–211 (2005)
37. Fortunato, E. et al.: Recent advances in ZnO transparent thin film transistors. Thin Solid Films **9**, 205–211 (2005). doi:10.1016/j.tsf.2005.01.066
38. http://www.physics.umd.edu/rgroups/spin/intro.html
39. Universität Mainz. A new, tunable device for spintronics. ScienceDaily. Accessed 28 Aug 2014
40. Wolf, S.A., Awschalom, D.D., Buhrman, R.A., Daughton, J.M., von Molna, S., Roukes, M.L., Chtchelkanova, A.Y., Treger, D.M.: Spintronics: a spin-based electronics vision for the future. Magnet. Mater. **294**, 1488–1495 (2001)
41. International Technology Roadmap for Semiconductors, Edition 2011
42. http://www.erbium.nl/research.html
43. Boyd, R.W.: Novel photonic materials for advanced imaging applications. J. Korean Phys. Soc. **43**(4), 603–606 (2003)
44. http://www.trnmag.com/Stories/2004/072804/Photonic_chips_go_3D_072804.html
45. Inoue, K. et al.: Photonic crystal optics and various applications (2004)
46. http://www.mdpi.com/journal/materials/special_issues/materials_applications
47. Poortmans, J., Baert, K., Hozel, J., Posthuma, N., John, J., Flamand, G., Gordon, I., Bruynseels, F., Mertens, R.: Where photovoltaics meets microelectronics. In: International Conference on Materials for Advanced Technologies, 2011

Chapter 15
Nano-Structured Semiconducting Materials and Devices

Abstract The present era is of nanotechnology and nano-materials. Therefore, a large number of investigations are being carried out in nano-structured semiconductors also. Consequently, the newer nano-based semiconducting materials and devices are being researched. Many recent developments in this field are included briefly in this chapter by incorporating the published research papers. The main emerging nano-semiconducting materials and devices included here are: semiconductor nanowires, nano crystals, nanogap devices, nanoparticle- based organic photovoltaic devices, nanoscale semiconductor, artificial electronic skin, semiconductor nanomembranes, nano-structured tin-dioxide materials, nano-structures based solar energy applications, metal-semiconductor hybrid nanostructures, inorganic semiconductor nanostructures etc.

Keywords Semiconductor nanowires · Nanogap devices · Nano-scale semiconductor · Semiconductor nanomembranes · Nanostructure based photonic switching devices · Nanostructure based solar energy applications · Nanostructure based environmental applications · Hybrid nanostructures

15.1 Surface Science of Free Standing Semiconductor Nanowires [1]

One dimensional free-standing semiconductor nanostructures, in the form of nanowires and nanotubes, have the potential to become central components in future electronics and photonics applications within information technology, renewable energy and life-sciences. These 1D nanostructures have been grown in virtually all types of materials, ranging from one component carbon and silicon to compound semiconductors and complex oxides. One serious challenge, though, is the control and characterization of their surface structure, chemistry and morphology. 1D nanostructures will have very large surface to bulk ratios, ultimately reaching 100 % for the single walled-carbon nanotube. Even for a 100 nm diameter

K.M. Gupta and N. Gupta, *Advanced Semiconducting Materials and Devices*,
Engineering Materials, DOI 10.1007/978-3-319-19758-6_15

Si wire, the outermost atom layer will make out only ~4 % of the total number of atoms. Compared to typical doping levels, this is a very significant amount of material. For semiconducting wires, especially important are effects such as surface band bending (which can completely dominate electrical properties), electron/hole traps at the surface (which will strongly affect optical performance) and the severe problem of proper (Ohmic) electrical contacts for nanowire devices. The large surface areas also have positive implications, as wires can be used as extremely delicate sensors, and electric field control in transistors is possible to a much more extensive degree. In addition, the contact area to other functional components can be high and varying the surface chemistry along the wire can be used in the design of electronic devices. Due to the wide range of possible applications, there is a strong current interest in semiconductor nanowires. Surfaces govern important processes for nanowire growth, physical properties and the ability of nanowires to interact with their surroundings. In this prospective, recent results obtained using surface sensitive electron microscopy/spectroscopy and scanning tunneling microscopy are highlighted and future directions discussed, in this paper.

15.2 Application of Semiconductor as Nanowires Sheathed Inside Nanotubes: Manipulation, Properties and Applications [2]

Since the discovery of metals encapsulated into multi-walled carbon nanotubes (CNTs), such sheathed structures attracted extensive interest with respect to the development of various synthetic strategies for producing the unique structure of nanowires sheathed inside nanotubes. The nanowire materials varied from metals to alloys, from semiconductors to insulators, and even metal–semiconductor heterojunctions were tried. In recent years, the studies on these nanostructures have been mainly focused on in situ manipulation, property analysis and applications. In this review, the significant developments and achievements in regards of manipulation, property measurements and device applications of inorganic nanowires sheathed inside nanotubes according to different categories of the filling materials, i.e., metals, alloys, compounds and semiconductor–metal heterojunction nanowires, are summarized.

15.3 Copper Phthalocyanine Nanocrystals Embedded into Polymer Host: Preparation and Structural Characterization [3]

In the last years, considerable research efforts have been devoted to the fabrication and characterization of various organic semiconducting thin films. Metal phthalocyanine is one of the most important organic compounds which can be applied in electro-optic devices, photoconducting agents, photovoltaic cell elements nonlinear optics, electrocatalysis, and other photoelectronic devices. Among the metal substituted phthalocyanine, copper II phthalocyanine (Cu-Pc) was found to have superior properties. It is well known that copper-phthalocyanine (Cu-Pc) is an organic semiconductor. They are chemically and thermally very stable; most of them can easily form ordered thin films. Cu-Pc has a lower mobility of about 10^{-4} Cm^2/V s. High mobility of Cu-Pc thin films could be achieved by employing an elevated substrate heating temperature, which directly affects the crystal structure and morphology of the thin films.

15.4 Electrical Nanogap Devices for Biosensing [4]

For detecting substances that are invisible to the human eye or nose, and particularly those biomolecules, the devices must have very small feature sizes, be compact and provide a sufficient level of sensitivity, often to a small number of biomolecules that are just a few nanometres in size. Electrical nanogap devices for biosensing have emerged as a powerful technique for detecting very small quantities of biomolecules. The most charming feature of the devices is to directly transduce events of biomolecules specific binding into useful electrical signals such as resistance/impedance, capacitance/dielectric, or field-effect.

Nanogap devices in electrical biosensing have become a busy area of research which is continually expanding. A wealth of research is available discussing planar and vertical nanogap devices for biosensing. Planar nanogap devices including label-free, gold nanoparticle-labeled, nanoparticles enhanced, nanogapped gold particle film, and carbon nanotube nanogap devices as well as vertical nanogap devices with two and three terminals for biosensing are carefully reviewed.

15.5 Nanoparticle-Based Plasmonic Organic Photovoltaic Devices [5]

Plasmonic metallic nanoparticles (NPs) have recently been identified as a breakthrough route for enhancing the efficiency of organic photovoltaic (OPV) devices. The present article highlights the different strategies of incorporating plasmonic NPs for light trapping into either the active or the buffer layer or at various interfaces within the OPV cell architecture. In addition, the different enhancement mechanisms that indicates future trends in the development of NPs-based solution processable OPVs are summarized.

The recent events in countries with large fossil fuel reserves has clearly demonstrated the need for the rapid development of a low cost, large scale and 'green' energy technology. In this concept, solar energy through photovoltaic energy conversion is the most promising candidate for long term sustainable energy production. Organic photovoltaics (OPVs) hold the promise for a cost-effective, lightweight solar energy conversion platform particularly for the provision of off-grid electricity and portable consumer electronics. One highly successful OPV design concept is based on the polymer–fullerene bulk heterojunction (BHJ) type of devices, which is characterized by a complex interpenetrating network of the two phases, where power conversion efficiencies (PCEs) of over 8 % have been reported. The commercialization of OPVs can spark into life once the efficiencies for single junction devices exceed 10 %. One promising plasmon enhanced light-trapping approach is the utilization of plasmonic metallic nanoparticles (NPs) either between interfaces, or inside the buffer or the active layers of OPV devices in order to promote absorption, thereby increasing the optical thickness of OPV materials for light harvesting. Noble metallic NPs are known to exhibit a strong absorption band in the UV–Vis region, which lies within the optical absorption band of the conjugated polymers used in the active layer of OPVs.

Electrical biosensors rely solely on the measurement of currents and/or voltages to detect binding. Owing to its inherent superiorities of electrical transduction methods, such as excellent compatibility with advanced semiconductor technology, miniaturization, and low cost, biosensors based on electrical detection are capable of behaving at high performance with a simple miniaturized readout. Nanogap electrodes are defined as a pair of electrodes with a nanometer gap. Nano sized biomolecules can be trapped into a gap between two electrodes and connecting the electrodes; the biomolecules are therefore detected by observing their electrical behaviour (resistance/impedance, capacitance/dielectric, or field-effect).

15.6 Nanoscale Semiconductor "X" on Substrate "Y"—Processes, Devices, and Applications [6]

Recent advancements in the integration of nanoscale, single-crystalline semiconductor 'X' on substrate 'Y' (XoY) for use in transistor and sensor applications are presented. XoY is a generic materials framework for enabling the fabrication of various novel devices, without the constraints of the original growth substrates. Two specific XoY process schemes, along with their associated materials, device and applications are presented. In one example, the layer transfer of ultrathin III–V semiconductors with thicknesses of just a few nanometers on Si substrates is explored for use as energy-efficient electronics, with the fabricated devices exhibiting excellent electrical properties. In the second example, contact printing of nanowire-arrays on thin, bendable substrates for use as artificial electronic-skin is presented. Here, the devices are capable of conformably covering any surface, and providing a real-time, two-dimensional mapping of external stimuli for the realization of smart functional surfaces. This work is an example of the emerging field of "translational nanotechnology" as it bridges basic science of nanomaterials with practical applications.

15.6.1 Introduction

In this Progress Report, recent advancements in the heterogeneous integration of single-crystalline semiconductors on non-conventional and often non-crystalline substrates for various applications are presented, The approach decouples the choice of the active semiconductor layer from the support substrate, and thus allows for integration of any functional semiconductor 'X' on any support substrate 'Y'. This concept, abbreviated as X-on-Y or XoY, presents a versatile material, processing and device technology platform for enabling new applications such as artificial electronic skin (e-skin), paper-like displays, or energy-efficient electronics (Fig. 15.1). XoY architecture has been commonly utilized for synthetic nanoscale electronic materials, such as carbon nanotubes, graphene, etc. More recently, the concept has been applied to epitaxial thin films with nano- or micrometer-scale dimensions. Besides the enabled technologies, XoY facilitates the exploration of the fundamental science associated with the interfaces of highly dissimilar materials which often have drastic effects on the device properties. While the approach is generic for both micro- and nano-scale semiconductors, films and wires with nano-scale dimensions are discussed, where the physical and chemical properties can be drastically altered by various size-effects.

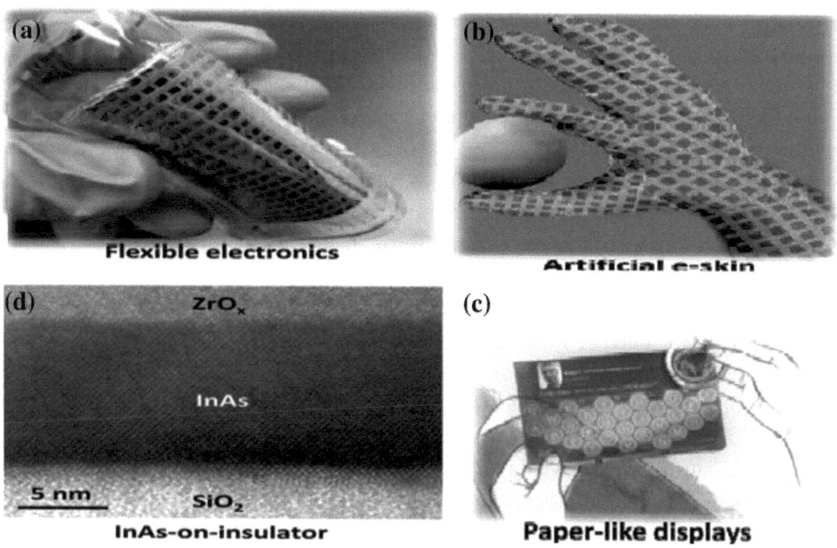

Fig. 15.1 Examples of various applications enabled by the XoY concept, **a** Flexible electronics, "Reproduced from [6] " **b** Artificial electronic-skin, "Reproduced from [6] " **c** Paper-like displays (copyright Nokia, Morph concept; reproduced with permission from Nokia)," Reproduced with permission from [6] " and **d** Ultrathin III–V compound semiconductors on Si/SiO_2 substrates for low-power electronics. "Reproduced from [6] ". Original source for **a**, **b** K. Takei et al., Nat. Mater. 2010, 9, 821; **c** T. Ryhänen et al., Nanotechnologies for Future Mobile Devices, Cambridge University Press, 2010; **d** H. Ko et al., Nature. 2010, 468, 286

15.6.2 Large-Scale Nanowire-Array XoY [6]

In recent years, integration of inorganic nanowires has been demonstrated on various rigid and flexible substrates, which have led to an increasing exploration of their unique properties within many different applications. First of all, the electrical and optical properties of NWs can be tuned through the precise synthesis of composition, size and shape. Furthermore, they are compatible with large scale integration methods, such as contact printing, which has already been demonstrated as a viable route for obtaining high-throughput fabrication of NW based electronics and sensors. One of the key advantages of the NW-based device technology is that single-crystalline NWs can be grown non-epitaxially on low cost amorphous substrates first, followed by their eventual transfer, for instance, by contact printing, as parallel arrays on a desired device substrate using room-temperature processing. This makes the technology compatible with virtually all support substrates and readily enables for heterogeneous integration of NWs with different composition and functionality on substrates through a multi-step assembly process.

15.6.3 Artificial Electronic Skin [6]

The large-scale assembly of NWs ('X') on plastic substrates ('Y') for use as the active back-plane of artificial electronic-skin (e-skin) is discussed. As an example of a large-scale NW-based electronic system, an artificial e-skin device, capable of touch mapping is demonstrated. Here, Ge/Si core/shell NW-arrays were contact printed on polyimide substrates and used as the active channel material for the transistor back-plane. A pressure sensitive rubber (PSR) was then laminated on top to function as the sensing element. Each pixel consists of a single NW-array FET with the drain connected to the ground through the PSR. PSR consists of carbon black nanoparticles that are embedded in an insulating silicone rubber matrix. In a relaxed state, PSR is insulating, however, when pressed, the nanoparticle spacing is reduced such that current flows by tunneling between the particles, resulting in a drastic change of the conductance.

The e-skin devices were fabricated by spincoating a polyimide layer (~24 μm in thickness) on a silicon handling wafer followed by electron-beam evapouration of SiO_x as a surface adhesive layer for the subsequent NW printing step. Next, NW-arrays were contact printed on photolithographically pre-patterned regions. Subsequently, source (S) and drain (D) Ni electrodes (~50 nm in thickness), Al_2O_3 gate dielectric (~50 nm in thickness by atomic layer deposition) and Al top gate electrodes (~100 nm in thickness) were fabricated by photolithography and lift-off processes. Parylene (~500 nm in thickness) was then vapour deposited as the insulating spacer layer. Patterned oxygen plasma etching of the parylene layer was performed to fabricate via-holes for electrical connection between the drain of each

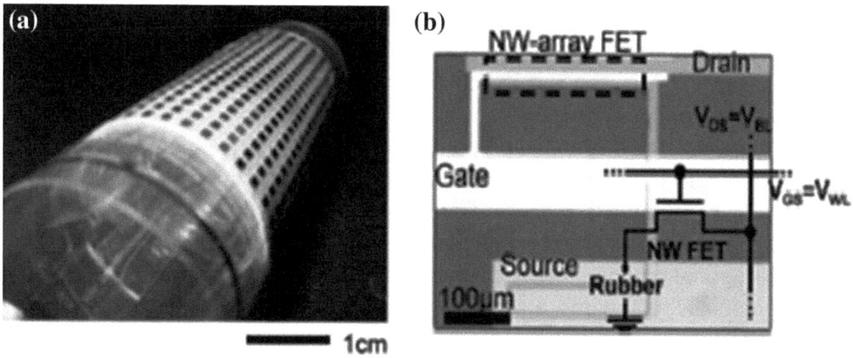

Fig. 15.2 E-skin device with NW active matrix circuitry. **a** A photograph of a rolled 7 cm × 7 cm e-skin device, and **b** An optical image of a pixel, consisting of a NW-array FET. "Reproduced with permission from [6] "

transistor and the PSR. PSR was then laminated on the top surface followed by the peel-off of the entire e-skin device from the handling wafer, resulting in the highly flexible integrated sensors and electronics.

Figure 15.2a displays a photograph of a fully fabricated 19 × 18 pixel-array e-skin device. The device is truly macroscale (7 × 7 cm in area) which presents the largest-scale demonstration of ordered NW array electronics for a fully functional system to date. An optical image of a single pixel is shown in Fig. 15.2b, highlighting the active and passive components.

15.7 Recent Advances in Semiconductor Nanowire Heterostructures [7]

Semiconductor nanowires have made a deep impact on materials science related research, and are being explored for applications in several disciplines. Many of these applications require heterostructures, which can be defined as the combination of two or more materials within the same nanowire structure. In this paper, the current state-of the epitaxial nanowire heterostructures, growth, understanding, and promising applications of such structures are discussed. In addition, interesting applications relying on heterostructures in nanowires, illustrating that nanowires and their heterostructures have been grown in a plethora of materials. Examples from a wide range of semiconductor materials have been picked up.

Semiconductor nanowires can be described as quasi-one-dimensional pieces of semiconductor material with diameters in the 10–100 nm range and lengths of the order of micrometres. Nanowires have made a strong impact in electronics as a promising route to further downscaling and extensions of Moore's law, in photovoltaics as promising solar cell materials, and in optoelectronics as highly efficient light emitting diodes. Additionally, many more applications can be anticipated in life-sciences and in sensor structures. There are several methods to fabricate semiconductor nanowires. In this highlight, the growth of vertically aligned nanowires with epitaxial orientation to the substrate, is focussed. This is the most technologically interesting class of nanowires, which offers the highest degree of control over the growth process. The favoured growth direction for nanowires is {111}. For III–V semiconductors, the group V terminated {111} direction, {111}B (or its hexagonal counterpart) is favourable. Thus, to ensure growth of vertically aligned, epitaxial nanowires, a substrate with the same orientation as the nanowire growth direction should be used.

There are essentially two methods to grow vertically aligned epitaxial nanowires: (i) metal particle seeded growth, and (ii) selective area growth. In metal particle seeded growth, the first step is to pattern a (piece of a) semiconductor wafer (the substrate) with metal particles. There are different ways to do this. The most prominent methods are to deposit colloidal or aerosol fabricated particles or to define the metal particles by means of lithography, metal evaporation, and lift-off. The former methods result in randomly distributed metal particles and the latter in position controlled arrays of metal particles.

Metal particle seeded nanowire growth competes with growth from the vapour phase, which results in 2D film growth on the substrate and the nanowire side walls, leading to tapering. Thus, conditions to maximize nanowire growth over film growth have to be realized. A common strategy is to grow nanowires at temperatures where film growth is suppressed, often a couple of hundred degrees lower than optimum film growth temperatures.

In selective area growth, the idea is to partially mask the substrate so that material only grows at specific areas where the substrate is exposed. The first step is to fabricate a mask. This is often achieved by depositing an oxide or nitride layer on the substrate and then by lithographically opening holes in this mask. After masking, the substrate is transferred to an epitaxy reactor (most often MOVPE or MBE).

It is commonly claimed that one of the major advantages of nanowires is the seemingly endless possibilities to form heterostructures in nanowires as compared to in epitaxial films. In heterostructures, two or more materials are grown on top of each other; schematic illustrations of the three relevant categories of heterostructures are shown in Fig. 15.3. In Fig. 15.3a, a heterocombination where the nanowire and the substrate are of different materials is shown, in Fig. 15.3b an axial or longitudinal, heterostructures is shown, and in Fig. 15.3, a radial, or core/shell, heterostructure is shown. The major advantage with heterostructures is that different materials have different bandgaps and this is something that is heavily utilized in modern electronics and optoelectronics applications. Design and fabrication based on combinations of different materials with different bandgaps is known as bandgap engineering.

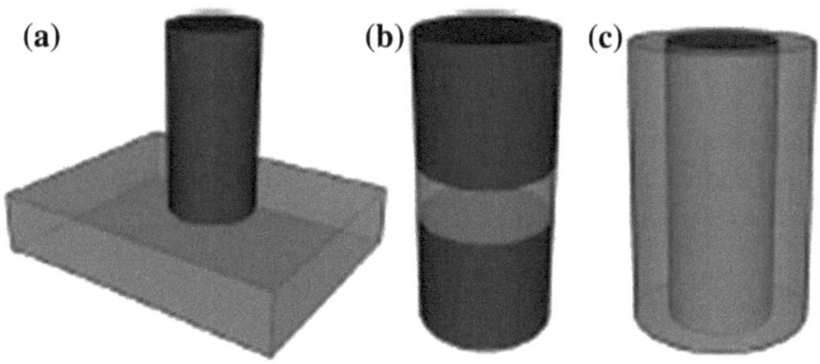

Fig. 15.3 Simple schematic illustrations of nanowire heterostructures. **a** Nanowire–substrate heterostructure, **b** Example of an axial heterostructure, **c** Example of a radial heterostructure. The substrate is only shown in (**a**), where it is part of the heterostructure. "Reproduced from [7] "

15.8 Synthesis, Assembly and Applications of Semiconductor Nanomembranes [8]

Research in electronic nanomaterials, historically dominated by studies of nanocrystals/fullerenes and nanowires/nanotubes, now incorporates a growing focus on sheets with nanoscale thicknesses, referred to as nanomembranes. Such materials have practical appeal because their two-dimensional geometries facilitate integration into devices, with realistic pathways to manufacturing. Recent advances in synthesis provide access to nanomembranes with extraordinary properties in a variety of configurations, some of which exploit quantum and other size-dependent effects. This progress, together with emerging methods for deterministic assembly, leads to compelling opportunities for research, from basic studies of two-dimensional physics to the development of applications of heterogeneous electronics.

Semiconductor nanomembranes (NMs) are monocrystalline structures with thicknesses of less than a few hundred nanometres and with minimum lateral dimensions at least two orders of magnitude larger than the thickness. They differ from thin films in that they exist as free-standing, isolated forms at some critical stage in their growth or processing, or in their final, device-integrated forms. Because NMs offer many features that cannot be reproduced in other material formats, they are of central importance to a rapidly expanding frontier in nano-science and technology. Studies of these and other 'zero-dimensional' materials evolved to include nanowires and carbon nanotubes, partly because it is comparatively easy (although still difficult) to form electrical contacts to such 'one-dimensional' structures. Although diverse types of semiconductor device are possible with individual wires/tubes, their practical application in high-yield, scalable systems faces formidable engineering challenges in assembly and other aspects of manufacturing. NMs also have finite-size and quantum characteristics in their electronic, phononic and optical properties, and have unique mechanical features, with effects related to shape distortions and folding of sheets, not found in zero- and one-dimensional materials. NMs can be made uniformly and repeatably (in size, shape, surface orientation, thickness and surface roughness) by 'top-down" methods used in semiconductor device manufacture.

15.9 Inorganic Nanomembranes [8]

Single-crystalline inorganic semiconductor NMs with thicknesses that match length scales of important physical processes (a few hundred nanometres or less) offer opportunities in basic and applied research, as well as in technology. Figure 15.4 illustrates representative examples of NM. Their properties in mechanics, electronics, thermoelectrics and photonics are given in [8]. The extremely small thicknesses of NMs (down to, 2 nm for silicon) lead to flexural rigidities that can be more than fifteen orders of magnitude smaller than those of bulkwafers ($\sim$200 μm) of the same materials. The resulting values are so small that they qualitatively change the nature of the material to allow otherwise impossible, non-planar geometries and multilayer integration options. As a consequence, NMs conform and bond robustly to nearly any surface, thereby enabling them to be stacked onto one another or onto foreign hosts to yield unusual, heterogeneous systems that cannot be achieved with wafer-bonding technologies or epitaxy. Such stacking can lead to unusual electronic, electromechanical, thermoelectric, optoelectronic, optomechanical and photonic behaviour.

Fig. 15.4 Inorganic nanomembranes of extremely small thicknesses. Reproduced from [8]

15.10 III-V Compound Semiconductor Nanostructures on Silicon: Epitaxial Growth, Properties, and Applications in Light Emitting Diodes and Lasers [9]

Significant developments have occurred in the area of III-V compound semiconductor nanostructures. The scope of developments includes quantum dots and nanowires epitaxially grown on Si substrates, as well as their applications in light emitting diodes and lasers. Such nanoscale heterostructures exhibit remarkable structural, electrical, and optical properties. The highly effective lateral stress relaxation, due to the presence of facet edges and free surfaces, enables the achievement of nearly defect-free III-V nanoscale heterostructures directly on Si, in spite of the large lattice mismatches and the surface incompatibility. With the incorporation of multiple quantum dot layers as highly effective three-dimensional dislocation filters, self-organized quantum dot lasers monolithically grown on Si exhibit, for the first time, relatively low threshold current (J_{th} = 900 A/cm^2) and very high temperature stability (T_0 = 244 K). III-V semiconductor nanowire light emitting diodes on Si, with emission wavelengths from UV to near-infrared, have also been demonstrated.

There has been a growing need for the monolithic integration of electronic and photonic components on the same chip for future high-speed and multifunctional systems. One critical, technology is a coherent light source on Si. Although advances have been made in Si-optoelectronics, the achievement of an electrically injected all-Si laser has been elusive. Alternatively, the integration of III-V semiconductor materials and devices on Si, that involves the use of direct epitaxial growth, wafer bonding, or compliant substrates has been explored. The resulting devices, however, generally exhibit extremely short lifetime and poor characteristics, due to the presence of a high density of dislocations associated with the large differences in lattice constants and thermal expansion coefficients between III-V materials and Si, as well as their polar/non-polar surface incompatibility. To overcome such fundamental challenges, III-V compound semiconductor nanostructures, including quantum dots and nanowires epitaxially grown on Si have been extensively investigated. This has been motivated by the effective lateral stress relaxation of such nanoscale heterostructures related to the presence of facet edges and sidewalls that can minimize, or eliminate the formation of dislocations, thereby leading to nearly defect-free III–V semiconductor nanostructures on Si. Additionally, semiconductor lasers, that use quantum dots or nanowires as gain regions, are expected to exhibit superior performance, due to the strong carrier confinement and the resulting near-discrete density of states.

Recently, significant progress has also been made in the development of III-V quantum dot and nanowire heterostructures on Si. 1.0–1.3 μm In(Ga)As/GaAs quantum dots grown directly on Si exhibit superior optical quality. With the incorporation of multiple layers of InAs quantum dots as effective three-dimensional dislocation filters, room temperature operational quantum dot lasers on Si have been achieved, which are characterized by relatively low threshold current (Jth = 900 A/cm^2) and very high temperature stability (T_0 = 244 K). High quality arsenide-, nitride-, and phosphide-based III-V nanowires on Si are also being investigated.

15.11 Nanostructured Tin Dioxide Materials for Gas Sensor Applications [10]

Detection of pollutant, toxic, refining, combustible and process gases is important for system and process control, safety monitoring and environmental protection. Various methods can be used to accomplish gas sensing including gas chromatography, Fourier-transform infrared spectroscopy, chemiluminescence detectors, mass spectrometry, semiconductor gas sensors, and others. Gas sensors based on solids tate semiconductor materials offer considerable advantages in comparison to other gas sensing methods. Semiconductor sensors are inexpensive to produce, easy to miniaturize, rugged, reliable and can be designed to operate over a range of conditions including high temperatures. Semiconductor sensors can be produced in arrays to allow sensing of multiple species simultaneously and with advances in sensitivity; detectivity limits are approaching part-per-million (ppm) levels for some species.

Tin dioxide (also called stannic oxide or tin oxide) semiconductor gas sensors were first proposed and patented in 1962, and since then stannic oxide gas sensors have undergone extensive research and development. Tin dioxide (SnO_2) is the most important material for use in gas sensing applications. It is the dominant choice for solidstate gas detectors in domestic, commercial and industrial application.

15.12 Quantum Transport in Semiconductor Nanostructures [11]

In recent years, semiconductor nanostructures have become the choice for investigations of electrical conduction on short length scales. This development was made possible by the availability of semiconducting materials of unprecedented purity and crystalline perfection. Such materials can be structured to contain a thin layer of highly mobile electrons. Motion perpendicular to the layer is quantized, so that the electrons are constrained to move in a plane. As a model system, this two-dimensional electron gas (2DEG) combines a number of desirable properties, not shared by thin metal films. It has a low electron density, which may be readily varied by means of an electric field (because of the large screening length). The low density implies a large Fermi wavelength (typically 40 nm), comparable to the dimensions of the smallest structures (nanostructures) that can be fabricated today. The electron mean free path can be quite large (exceeding 10 μm). Finally, the reduced dimensionality of the motion and the circular Fermi surface form simplifying factors.

Quantum transport is conveniently studied in a 2DEG because of the combination of a large Fermi wavelength and large mean free path. The quantum mechanical phase coherence characteristic of a microscopic object can be maintained at low temperatures (below 1 K) over distances of several microns, which one would otherwise have classified as macroscopic. The physics of these systems has been referred to as mesoscopic, a word borrowed from statistical mechanics.

Semiconductor nanostructures are unique in offering the possibility of studying quantum transport in an artificial potential landscape. This is the regime of ballistic transport, in which scattering with impurities can be neglected. The transport properties can then be tailored by varying the geometry of the conductor, in much the same way as one would tailor the transmission properties of a waveguide. The physics of this transport regime could be called electron optics in the solid state. The formal relation between conduction and transmission, known as the Landauer formula, has demonstrated its real power in this context. For example, the quantization of the conductance of a quantum point contact (a short and narrow constriction in the 2DEG) can be understood using the Landauer formula as resulting from the discreteness of the number of propagating modes in a waveguide.

15.13 Photonic Switching Devices Based on Semiconductor Nanostructures [12]

The global communication revolution that has been enabled through massive data transmission within optical fibre networks has had a tremendous impact on people's lifestyle and modern industry. Over the past two decades, high-speed optical links have been successfully applied to various systems ranging from longhaul transmission lines to short-haul communications within buildings. However, the research focus of photonic integrated circuits (ICs) is moving towards the optical manipulation of information at shorter distances, ultimately on circuit boards or IC chips. Current research in optical interconnection aims to provide an ambitious solution based on optical networks to the physical limitations existing in today's electronic systems, which is believed to be a critical issue for realizing future faster information transfer and processing.

One of the key devices to establish a high-bit-rate photonic signal processing system is a photonic switch which functions with an ultrashort delay time. The idea of using optical networks to replace present-day electronic switching fabrics was initially driven by the power hungry nature of the optical to electronic (O/E) and electronic to optical (E/O) conversion. Additionally, there is a speed limitation for the electronic interconnection which typically employs millions of closely spaced metal wires in the state-of-the-art computer systems. Despite impressive advancements in the material and device technology, the current record of the on chip global interconnection lies at 5 Gb/s per channel whilst the speed record of the chip-to-chip interconnection is at 10 Gb/s per channel. A photonic switch that controls optical signals directly by another light beam with potential recovery times in the pico- or femto-second regimes has the capability for terahertz switching speed. The availability of such a component, combined with the use of low-loss optical waveguides and fibres would provide a promising step towards high-speed and low-cost photonic networks at short distances.

Semiconductor nanostructures such as epitaxial quantum dots (QDs) offer the possibility to meet these system goals as a natural consequence of their small volume and atom-like density of states which lead to a high differential gain/absorption. QD based nano-structures are anticipated to generate high optical nonlinearity with ultralow energy consumption. From the viewpoint of integrating devices to realize a compact, energy-efficient photonic system, major advantages of QD-based photonic devices include:

- High energy efficiency.
- High thermal stability.
- Broad bandwidth.
- Compatibility with an Si-platform.

15.14 Zinc Oxide Nanostructures: Growth, Properties and Applications [13]

Zinc oxide is a unique material that exhibits semiconducting and piezoelectric dual properties. Using a solid–vapour phase thermal sublimation technique, nanocombs, nanorings, nanohelixes/nanosprings, nanobelts, nanowires and nanocages of ZnO have been synthesized under specific growth conditions. These unique nanostructures unambiguously demonstrate that ZnO probably has the richest family of nanostructures among all materials, both in structures and in properties. The nanostructures could have novel applications in optoelectronics, sensors, transducers and biomedical sciences. This article reviews the various nanostructures of ZnO grown by the solid–vapour phase technique and their corresponding growth mechanisms. The application of ZnO nanobelts as nanosensors, nanocantilevers, field effect transistors and nanoresonators is demonstrated.

15.15 Recent Trends on Nanostructures Based Solar Energy Applications: A Review [14]

Nanostructure based solar energy is attracting significant attention as possible candidate for achieving drastic improvement in photovoltaic energy conversion efficiency. Although such solar energy expected to be more expensive, there is a growing need for the efficient and light-weight solar cells in aero-space and related industries. It is required to rule the energy sector when the breakeven of high performance is achieved and its cost becomes comparable with other energy sources. Various approaches have been intended to enhance the efficiency of solar cells. Applications of nanotechnology help us to solar devices more economically. Nano photovoltaic cells are used to improve the efficiency to create effective systems for conversion cost, efficient solar energy storage systems or solar energy on a large scale.

A solar cell performs two major functions: photogeneration of charge carriers in a light absorbing material and separation of the charge carriers to a conductive contact that will transmit the electricity. Solar cells are electronic devices used for the direct conversion of solar energy to electricity, using the photovoltaic (PV) effect. Fundamental properties of nanostructured materials are currently extensively studied because of their potential application in numerous fields which includes electronic devices, opto electronics, optics, tribology, biotechnology, human medicine and others. Nanostructured materials contain structures with dimensions in the nanometer length scale which includes polycrystalline materials with nanometer sized crystallites, materials with surface protrusions spatially separated by few nanometers granular or porous materials with grain sizes in the nanometer range or nanometer sized metallic clusters embedded in a dielectric matrix. The motivation for using nanostructured materials emerges from their specific physical and chemical properties. Enhancing the regular crystalline structure using nanocrystalline materials can increase the absorbance of all incident solar spectra in the form of thin films or multilayered solar cells.

15.16 Design, Fabrication, and Modification of Nanostructured Semiconductor Materials for Environmental and Energy Applications [15]

The environment and energy are the biggest challenges of the 21st century. Ironically, the solution to these large problems may lie in something very small. Nanomaterials, with attractive chemical and physical properties, are being explored for potential uses in energy and environmental applications. During the past decade, rapid advances in materials science have led to significant progress in environmental remediation and renewable energy technologies such as photocatalytic oxidation, adsorption/separation processing, solar cells, fuel cells, and biofuels. The design and creation of new materials and substances chemically modified from the molecular and atomic levels to sizes on the nanoscale promise significantly enhanced functions for environmental and energy applications. Meanwhile, the development of advanced characterization techniques has facilitated a fundamental molecular-level understanding of structure-performance relationships, which are strongly related to grain size and size distribution, shape, chemical composition, presence of interfaces (grain boundaries and free surfaces), and interactions between the constituent domains. This knowledge, together with effective synthesis strategies, has inspired the design and fabrication of novel nanostructured materials for a wide variety of applications.

This article reviews the recent progress in the design, fabrication, and modification of semiconductor nanostructured materials. It also highlights their environmental and energy applications, including photocatalytic treatment, environmental monitoring, water splitting, and hydrogen storage. The synthesis of nanostructured materials is a very active research field. The ability to fabricate and process nanostructured materials lies at the heart of nanotechnology, paving the way for understanding novel properties and realizing their potential applications. To date, many technologies have been explored to synthesize nanostructured materials. These technical approaches can be essentially grouped in two paradigms: top-down and bottom-up. In particular, versatile bottom-up methods based on chemistry have attracted considerable attention because of their relatively low cost and high throughput. Bottom-up approaches refer to the buildup of a material from the bottom: atom-by-atom, molecule-by-molecule, or cluster-by-cluster.

Growth species such as atoms, ions, and molecules, after impinging on the growth surface, assemble into crystal structures one after another. In recent years, a number of techniques, including co-precipitation, sol-gel processes, micro-emulsions, freeze drying, hydrothermal processes, laser pyrolysis, ultrasound and microwave irradiation, templates, and chemical vapour deposition, have been developed to control the size, morphology, and uniformity of nanostructures simultaneously. The successful implementation of the bottom-up strategy requires, in the end, the controlled growth of nanostructures. Among various media for crystal growth, the solution based method offers significant advantages, including (i) low reaction temperatures, (ii) size-selective growth, (iii) morphological control, and (iv) large-scale production. The liquid-phase approach to the synthesis of inorganic nanostructures has been recently reviewed.

The scope of this article covers a variety of semiconductor materials, focusing particularly on TiO_2-based nanostructures (e.g., pure, doped, coupled, nanoporous, mesoporous, hierarchically porous, and ordered mesoporous TiO_2). The preparation of nanoparticles, hierarchical nanoarchitectures, thin films, and single crystals by sol-gel, microemulsion, hydrothermal, sonochemical, microwave, photochemical, and nanocasting methods is discussed.

15.17 Application of Semiconductor and Metal Nanostructures in Biology and Medicine [16]

Advances in nanotechnology research have led to the creation of new generation of contrast agents, therapeutics, and delivery systems. These applications are expected to significantly improve the diagnosis and treatment of a variety of diseases. Two nanotechnologies—semiconductor and metallic nanostructures—are the most advanced in this field and have been extensively investigated for clinical use. These nanostructures are currently the "model" for the developments of many novel nanostructures.

Advancements in the synthesis, characterization, and surface modifications of nanoscale structures have provided a foundation for their utility in medical applications. The unique aspect of these nanostructures is their tunable optical, electronic, magnetic, and biological properties. Due to these properties, the incorporation of nanostructures in detection schemes or the use the nanostructures as contrast agents could improve the specificity and accuracy of current diagnostics. Furthermore, nanostructures can improve the effectiveness of drugs in the treatment of a variety of diseases when they are used as delivery vehicles because of the increase in therapeutic payload. In the last fifteen years, researchers have characterized the tunable properties by altering the nanostructure size, shape, and chemical composition and have developed reproducible strategies to make nanostructures of desired properties. Further, researchers have also developed strategies to render nanostructures biocompatible and capable of being coated with biological molecules. For example, semiconductor nanocrystals synthesized via organometallic methods contain the hydrophobic ligand tri-n-octylphosphine oxide (TOPO). These nanocrystals can subsequently be made water-soluble by covering the TOPO with the amphiphillic polymer polyacrylic acid grafted with octylamine. The hydrophylic carboxylic acid groups of the polymer coat render the nanocrystal stable in an aqueous environment. Furthermore, these groups can be used for conjugation to polymers and/or biological molecules such as antibodies and oligonucleotides via the formation of peptide bonds.

15.18 Semiconductor Nanocrystals [16]

Semiconductor nanocrystals are commonly used as probes for labeling cells and tissues owing to their intense, tunable fluorescence. The origin of this unique physical property is quite interesting. A semiconductor crystal's band-gap energy becomes strongly size-dependent as its physical dimensions approach the exciton Bohr radius (the natural radius of an electron-hole pair in bulk). This dependence is a result of a phenomenon known as "quantum confinement," in which the allowed electronic states of the exciton are broadened when its position is restricted by the dimensions of its parent crystal, in a manner analogous to the pedagogical "particle in a box." In this size range, semiconductor nanocrystals are considered electronically zero-dimensional and are referred to as quantum dots. The dimension at which quantum confinement becomes significant depends on the exciton Bohr radius in bulk and thus the particular material in question. For the well-studied II-VI direct band-gap semiconducting material cadmium selenide (CdSe), strong confinement occurs in the range of 1-10 nm. This results in a shift in fluorescent band-edge.

15.19 Metal/Semiconductor Hybrid Nanostructures for Plasmon-Enhanced Applications [17]

Hybrid nanostructures composed of semiconductor and plasmonic metal components are receiving extensive attention. They display extraordinary optical characteristics that are derived from the simultaneous existence and close conjunction of localized surface plasmon resonance and semi-conduction, as well as the synergistic interactions between the two components. They have been widely studied for photocatalysis, plasmon-enhanced spectroscopy, biotechnology, and solar cells. The preparation of the hybrid nanostructures is first presented according to the semiconductor type, as well as the nanostructure morphology.

Nanomaterials have been long researched for exploring their unique physical and chemical properties that are different from those of the corresponding bulk materials or constituting atoms. The properties that result by shrinking down materials to the nanoscale, in many cases, are caused by the quantum confinement effect and/or the large surface-to volume ratio. A vast number of studies have been devoted to the preparations of various nanostructures. Tremendous progresses have been made in the synthetic control of the size and shape of different nanostructures. Various metal and semiconductor nanocrystals, especially Au and Ag nanocrystals, which have intriguing plasmonic properties in the visible-to-near-infrared regions, have been prepared using wet chemistry methods. Precise control of the shape and size of nanocrystals allows for the tailoring of their properties nanorods can be synthetically tuned from the visible to near-infrared region by control of the length-to-diameter aspect ratio of the nanorods. High catalytic activities can be obtained by controlling the shape, size, and morphology of nanocrystals. Moreover, various hybrid nanostructures, including metal/metal, metal/semiconductor, metal/magnet, and magnet/semiconductor ones, have been prepared. Compared with single component ones, hybrid nanostructures can exhibit superior properties or new functionalities. In bimetallic nanostructures, for instance, coating thin Ag shell on Au nanocrystals results in plasmonic properties that are similar to those of pure Ag nanocrystals and improved chemical stabilities. Au/Pd bimetallic nanostructures possess simultaneously LSPR and high catalytic activities. They have been utilized for plasmon-enhanced photocatalysis. (Co core)/(Ag shell) nanostructures exhibit a strong enhancement in magneto-optical Faraday rotation caused by the LSPR of the Ag shell.

15.19.1 Semiconductor Nanostructures

Semiconductors are the foundation of modern electronic and optoelectronic devices, including field-effect transistors, solar cells, light-emitting diodes, photodetectors, digital and analog integrated circuits. Because of their extreme importance, semiconductor nanostructures are one type of the earliest explored nanomaterials. The well-known examples are semiconductor nanocrystals and nanowires, the former of which are also called quantum dots. However, semiconductor nanostructures generally have relatively small optical cross-sections in the visible region, especially wide-band-gap semiconductors. Most of semiconductor nanostructures therefore interact weakly with visible light, which accounts for a substantial energy fraction (44 %) of the entire solar spectrum. Tremendous effort has been made to increase the visible-light response of wide bandgap semiconductor nanostructures. A number of studies have shown that the visible-light response of oxide semiconductor nanostructures, for example, TiO_2, can be enhanced by several methods, including doping with metal or non-metal elements, tuning the morphology, forming heterostructures, and integrating semiconductors with plasmonic metals. Integration of semiconductors with plasmonic metals to form (plasmonic metal)/semiconductor hybrid nanostructures has recently attracted much attention because of the unique properties of LSPR. Various (plasmonic metal)/semiconductor

15.20 Application of Nd:Yag Laser in Semiconductors' Nanotechnology [18]

The main tendency in development of modern electronics and optoelectronics is the use of functional objects of small size. Unique properties of the nano-objects are mainly determined by their atomic and electronic processes occurring in the structure, which has a quantum character. Surface nanostructures and their unique properties play a significant role in such objects as highly dispersed systems—adsorbents and catalysts, fillers, composite materials, film and membrane systems. Formation on the surfaces of ordered and disordered ensembles of nanoparticles allow creating materials with new unique physical properties. Nanostructuring of the surface leads to improvement of optical, electrical, thermal, mechanical and field electron emission properties of materials, for example, reducing of the work function of electron from silicon, enhancing biocompatibility with implants in living tissue and prosthetic devices used in orthopaedics and dentistry. Such materials find application in selective nanocatalyse, microelectronics, nanophotonics, photovoltaic, spectroscopy, and optics. On their base devices are created for recording and storing information with ultra-high density, as well as light-emitting devices.

Lasers have been used for more than 50 years in diverse fields of application starting from simple laser micromachining processes for micro-electro-mechanical systems (MEMs), such as, cutting, drilling, to more smarter ones, e.g.: restoration of art works (laser stone cleaning), pulse laser deposition of coatings and films; local defect annealing after ion implantation; formation the precipitation areas of impurities in Si in medicine, medical diagnosis, treatment, or therapy etc. Among solid state lasers, Nd:YAG laser has an important role due to its high efficiency, possibility to tune it in different wavelengths from infrared (λ = 1064 nm) till ultraviolet (λ = 213 nm) and change pulse duration from milliseconds down to picoseconds.

15.21 Inorganic Semiconductor Nanostructures and Their Field-Emission Applications [19]

Nanomaterials and nanostructures are not only at the forefront of the hottest fundamental materials research nowadays, but they are also gradually intruding into our daily life. Nanostructures can be divided into zero-dimensional (0D when they are uniform), one-dimensional (1D when they are elongated), and two-dimensional (2D when they are planar) based on their shapes. The recent emphasis in the nanomaterials research is put on 1D nanostructures at the expense of 0D and 2D ones, perhaps due to the intriguing possibility of using them in a majority of short-term future applications.

Inorganic semiconductor nanostructures are ideal systems for exploring a large number of novel phenomena at the nanoscale and investigating the size and dimensionality dependence of their properties for potential applications. The use of such nanostructures with tailored geometries as building blocks is also expected to play crucial roles in future nanodevices. Since the discovery of carbon nanotubes much attention has been paid to exploring the usage of inorganic semiconductor nanostructures as field-emitters due to their low work functions, high aspect ratios and mechanical stabilities, and high electrical and thermal conductivities. This article mainly focuses on the most widely studied inorganic nanostructures, such as ZnO, ZnS, Si, WO_3, AlN, SiC, and their field-emission properties.

15.22 Benchmarking Nanotechnology for High-Performance and Low-Power Logic Transistor Applications [20]

Recently there has been tremendous progress made in the research of novel nanotechnology for future nanoelectronic applications. In particular, several emerging nanoelectronic devices such as carbon-nanotube, field-effect transistors (FETs), Si nanowire FETs, and planar III–V compound semiconductor (e.g., InSb, InAs) FETs, all hold promise as potential device candidates to be integrated onto the silicon platform for enhancing circuit functionality and also for extending Moore's Law. For high-performance and low-power logic transistor applications, it is important that these research devices are frequently benchmarked against the existing Si logic transistor data in order to gauge the progress of research. In this paper, we use four key device metrics to compare these emerging nanoelectronic devices to the state-of-the-art planar and nonplanar Si logic transistors. These four metrics include: (1) or intrinsic gate delay versus physical gate length; (2) energy-delay product versus; (3) subthreshold slope versus; and (4) versus on-to-off-state current ratio ON OFF. The results of this benchmarking exercise indicate that while these novel nanoelectronic devices show promise and opportunities for future logic applications, there still remain shortcomings in the device characteristics and electrostatics that need to be overcome. We believe that benchmarking is a key element in accelerating the progress of nanotechnology research for logic transistor applications.

15.23 Nanomaterials for Solar Energy Conversion [21]

There is an urgent need for cheap and efficient solar cell technology in order for solar energy to become more economically viable. Research employing nanomaterials for solar energy conversion either focuses on improving the energy conversion through better energy/photon management and/or reducing the processing cost through high-defect tolerance conversion approaches, but a deeper understanding of the photon-to-carrier conversion mechanisms and charge transport in nanoscale morphologies is needed. The following researches are going on in this field.

- Nanostructures for solar energy conversion (nanocrystals, nanowires, quantum dots);
- Carbon nanotubes in solar cell devices;
- Organic solar cells and bulk heterostructure devices;
- Dye-sensitized nanomaterials, Gratzel cells;
- Optical and electronic processes in nanomaterials for solar energy conversion;
- Advanced concepts for solar energy conversion; multiple exciton generation, up and down conversion, intermediate band, etc.;
- Exciton and charge-transfer dynamics in nanomaterials;
- Charge transport in nanoscale architectures;
- Modeling and optimizing solar cell efficiency

References

1. Mikkelsen, A., Lundgren, E.: Surface science of free standing semiconductor nanowires. Surf. Sci. **607**, 97–105 (2013)
2. Liu, Qian, Zou, Rujia, Bando, Yoshio, Golberg, Dmitri, Hu, Junqing: Nanowires sheathed inside nanotubes: Manipulation, properties and applications. Prog. Mater. Sci. **70**, 1–49 (2015)

3. Touka, N., Benelmadjat, H., Boudine, B., Halimi, O., Sebais, M.: Copper phthalocyanine nanocrystals embedded into polymer host: Preparation and structural characterization. J. Assoc. Arab Univ. Basic Appl. Sci. **13**, 52–56 (2013)
4. Chen, X., Guo, Z., Yang, G.-M., Li, J., Li, M.-Q., Liu, J.-H., Huang, X.-J.: Electrical nanogap devices for biosensing. Mater. Today. **13**(11), (2010)
5. Stratakis, E., Kymakis, E.: Nanoparticle-based plasmonic organic photovoltaic devices, Mater. Today. **16**(4), (2013)
6. Madsen, M., Takei, K., Kapadia, R., Fang, H., Ko, H., Takahashi, T., Ford, A.C., Lee, M.H., Javey, A.: Nanoscale semiconductor "X" on substrate "Y"—processes, devices, and applications. Adv. Mater **23**, 3115–3127 (2011) (Wiley-Vch Verlag GmbH & Co. KGaA, Weinheim)
7. Johansson, J., Dick, K. A.: Recent advances in semiconductor nanowire heterostructures. Royal Soc. Chem. Cryst. Eng. Comm. **13**, 7175–7184, 7175 (2011)
8. Rogers, J.A., Lagally, M.G., Nuzzo R.G.: Synthesis, assembly and applications of semiconductor nanomembranes. Macmillan Publishers Limited, **477** (2011)
9. Mia, Z., Chang, Y.-L.: III-V compound semiconductor nanostructures on silicon: Epitaxial growth, properties, and applications in light emitting diodes and lasers. J. Nanophotonics **3**, 031602 (23 January 2009)
10. Miller, T.A., Bakrania, S.D., Perez, C., Wooldridge, M.S.: Nanostructured tin dioxide materials for gas sensor applications, American Scientific Publishers, (2006)
11. Beenakker, W.J., van Houten, H.: Quantum transport in semiconductor nanostructures, **23** (2004)
12. Jin, C.-Y., Wada, O.: Photonic switching devices based on semiconductor nanostructures
13. Wang, Z.L., Phys, J.: Zinc oxide nanostructures: growth, properties and applications. Condens. Matter **16** R829–R858 (2004)
14. Sagadevan, Suresh: Recent trends on nanostructures based solar energy applications: a review. Rev. Adv. Mater. Sci. **34**, 44–61 (2013)
15. Hu, X., Li, G., Yu, J. C., Langmuir I.: Design, fabrication, and modification of nanostructured semiconductor materials for environmental and energy applications. **26**(5), 3031–3039 (2010)
16. Walkey, C., Sykes, E.A., Chan, W.C.W.: Application of semiconductor and metal nanostructures in biology and medicine
17. Jiang, Ruibin, Li, Benxia, Fang, Caihong, Wang, Jianfang: Metal/semiconductor hybrid nanostructures for plasmon-enhanced applications. Adv. Mater. **26**, 5274–5309 (2014)
18. Medvid, A., Mycko, A., Onufrijevs, P., Dauksta, E.: Application of Nd:YAG laser in semiconductors' nanotechnology. www.intechopen.com
19. Fang, Xiaosheng, Bando, Yoshio, Gautam, Ujjal K., Ye, Changhui, Golberg, Dmitri: Inorganic semiconductor nanostructures and their field-emission applications. J. Mater Chem. **18**(509–522), 509 (2008)
20. Chau, R., Datta, S., Doczy, M., Doyle, B., Jin, B., Kavalieros, J., Majumdar, A., Metz, M., Radosavljevic, M.: Benchmarking nanotechnology for high-performance and low-power logic transistor applications, IEEE Trans. Nanotechnol. **4** (2), (2005)
21. http://nanoandgiga.com/ngc2014/nanosolar.shtml

Chapter 16
Recent Advances in Semiconducting Materials and Devices

Abstract Advancement in the knowledge of science and technology and newer developments has always shown the path of research. The present researches in the field of semiconducting materials and devices are numerous. To get the readers acquainted with those researches, the emerging trends in their investigations and developments are presented in this chapter. Consequently, the leading developments included here are: semiconductor disk lasers, bandgap engineering and gain mirror technology, optically pumped semiconductor lasers, polymer fullerene solar cells, polymer semiconductor crystals, plastic solar cells, skin-inspired electronic devices, dye-sensitized solar cells from photoanodes, Piezoelectric oxide semiconductor FET, oxide semiconductor thin-film semiconductors, SiC power devices etc.

Keywords Semiconductor disk lasers (SDLs) · Gain mirror technology · Optically pumped semiconductor lasers · Laser diodes · Polymer semiconductor crystals · Graphene for radio frequency electronics · Green and biodegradable electronics · Skin-inspired electronic devices · Dye-sensitized solar cells · POSFET touch sensing device · Oxide semiconductor transistors · Semiconductor devices for power electronics · SiC and GaN power devices · Semiconductor for magnetoelectronics · Semiconductor with magnetic materials · Tunable left-handed metamaterial

16.1 Semiconductor Disk Lasers: Recent Advances in Generation of Yellow-Orange and Mid-IR Radiation [1]

The recent advances in the development of semiconductor disk lasers (SDLs) producing yellow-orange and mid-IR radiation is focused in this article. High-power GaInNAs- and GaSb based gain mirrors material systems have recently sparked a new wave of interest in developing SDLs for high-impact applications in medicine, spectroscopy, or astronomy. The dilute nitride (GaInNAs) gain mirrors enable emission of more than 11 W of output power at a wavelength range of 1180–

K.M. Gupta and N. Gupta, *Advanced Semiconducting Materials and Devices*, Engineering Materials, DOI 10.1007/978-3-319-19758-6_16

1200 nm and subsequent intracavity frequency doubling to generate yellow–orange radiation with power exceeding 7 W. The GaSb gain mirrors have been used to leverage the advantages offered by SDLs to the 2–3 μm wavelength range. Most recently, GaSb-based SDLs incorporating semiconductor saturable absorber mirrors were used to generate optical pulses as short as 384 fs at 2 μm, the shortest pulses obtained from a semiconductor laser at this wavelength range.

16.1.1 Introduction

Conceptually, the idea of an optically pumped semiconductor disk laser (OP-SDLs) was suggested already in 1966 by Basov et al. in a paper describing lasers with radiating mirrors [1]. However, it was not until the 1990s that the concept was acknowledged and the first working devices were reported [2–6]. In its essence, the concept of an OP-SDL is based on using an optically pumped semiconductor gain structure (i.e., gain mirror) with vertical emission. We note here that in addition to OP-SDL, also acronyms like OP-VECSEL (optically pumped vertical external-cavity surface-emitting laser) and OPSL (optically pumped semiconductor laser) are commonly used in literature to describe the same type of laser. The laser resonator is typically formed between the gain mirror and one or more external-cavity mirrors. In many ways, this laser architecture is similar to that of traditional solid state disk lasers. An essential difference is that in traditional solid state lasers the emission wavelength is dependent on certain fixed atomic transitions in a host material, whereas in an SDL the wavelength can be specifically tailored in a wide range by engineering the composition of the semiconductor material. This added wavelength versatility is one of the key factors that have made SDLs successful also commercially.

16.1.2 Cavity Designs

Structurally the SDL gain mirror resembles a half-VCSEL design that comprises a high reflectivity mirror and a semiconductor gain region. The gain region usually includes several quantum well (QW) or quantum dot (QD) layers separated by spacer/barrier layers. A typical mirror structure consists of a stack of quarter-wavelength semiconductor layers, forming a distributed Bragg reflector (DBR), although metallic, dielectric, or hybrid mirror structures can be used in some cases as well. While in VCSELs the single transverse mode operation is achieved by confining the laser mode to a very small gain area, in SDLs the same functionality is achieved by controlling the fundamental mode size via cavity design to have it match with the pumped area on the gain. Figure 16.1 shows various cavity configurations of SDLs.

The simplest conventional SDL cavity has an I-shape that is formed between the gain mirror and a single external output coupler (OC) mirror. However, in practice it is often easier to use a V-shaped cavity formed between the gain mirror, one

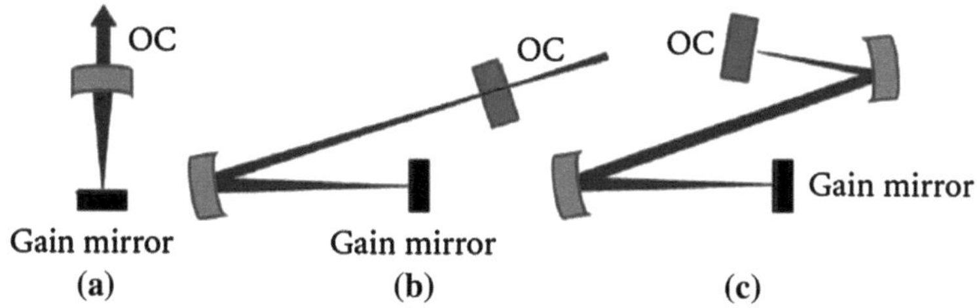

Fig. 16.1 Typical SDL cavities. **a** I-shaped cavity; **b** V-shaped cavity; **c** Z-shaped cavity. *OC* output coupler. Reprinted from [1] with the written permission from Hindawi Publishing Corporation

curved folding mirror, and a planar output coupler. The advantage of the V-cavity is that planar output couplers with various coupling ratios are often cheaper and more widely available on stock than equivalent curved couplers. Another practical advantage of the V-shaped cavity is related to the alignment of the laser; if the final alignment is done by monitoring the output of a photodiode placed behind the output coupler, the folding mirror in a V-shaped cavity collects the light efficiently to the photodiode enhancing the available signal while in I-shaped cavity the spontaneous emission from the gain is rapidly dispersed to all direction. More complex cavity configurations are often used for frequency conversion and mode-locking. For efficient frequency conversion the nonlinear crystal is often placed at a location near or at the mode waist. This is usually easier to do in a V-shaped or Z-shaped cavity than in an I-shaped cavity. More complicated Z-shaped cavities are typically used in mode-locking SDLs to accommodate also a semiconductor saturable absorber mirrors (SESAMs).

16.1.3 Thermal Management of SDLs

It is very important to conduct the heat away from the gain region with minimal thermal resistance between the heat sink and the active region. Generally speaking, thermal resistance is dependent on the thermal conductance of the materials used and on the distance that heat needs to be transferred. In short, one should aim to minimize the distance between the heat sink and the gain and at the same time use materials that have high thermal conductance. Using planar gain mirror geometry, the pumping is concentrated on an area that has typically a diameter of some tens or hundreds of micrometers, whereas the overall thickness of the semiconductor layer structure is only a few microns (e.g., 5–6 μm). In other words, the heated area is very large compared to the thickness of the layers. Thermal simulations show that in such a structure the heat flow is essentially one dimensional and is directed normal to the sample surface. We should point out that the epitaxial layers are grown on a semiconductor substrate that is typically some 200–600 μm thick and presents a major obstacle for the heat flow. Two assembling techniques of the gain mirror to

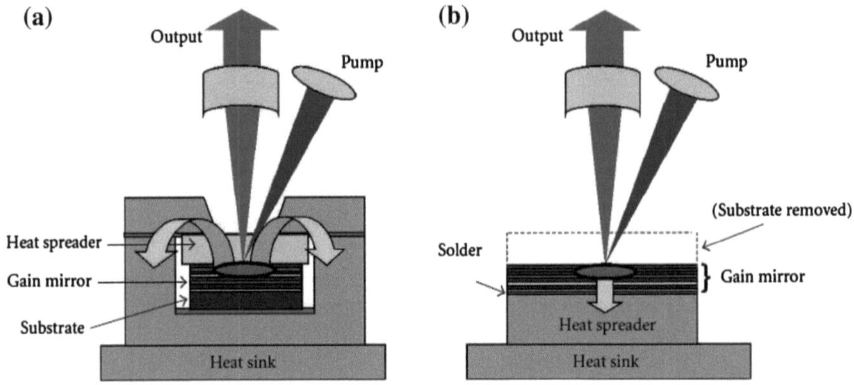

Fig. 16.2 Description of two heat extraction strategies. **a** Laser equipped with intracavity heat spreader. **b** Flip-chip mounted thin-device from which the semiconductor substrate is removed. Reprinted from [1] with the written permission from Hindawi Publishing Corporation

the heat sink are typically employed to overcome this issue. The so-called "intracavity heat spreader" method, is conceptually simple and involves contacting a transparent heat spreader element onto the gain mirror. See Fig. 16.2 for general description.

To summarize, efficient heat removal is highly important for high-power operation of SDLs. Use of high thermal conductance heat spreader materials, such as diamond, greatly improves the heat extraction from the point source. The distance from the gain region to the heat spreader can be minimized by optically contacting the heat spreader onto the sample or by flip-chip bonding the component on a heat spreader/heat sink. The application and the type of gain material determines which process is more suitable. Flip-chip processing suits well for mode-locking, continuous spectral tuning, and single-frequency operation since the laser spectrum is not affected by the intracavity heat spreader element. The intracavity heat spreader approach suits particularly well for long wavelength (GaSb and InP) lasers and applications that are not spectrally sensitive.

16.1.4 Wavelength Coverage

The gain mirror is essentially a stack of epitaxially grown semiconductor thin-films, fabricated on a GaAs, InP, GaSb, or other suitable semiconductor substrate by epitaxial growth. It is quite essential that one is able to grow high-quality gain material (QW, QD, or bulk) with desired bandgap energy, while keeping the material strain within reasonable limits. Secondly, the DBR should provide sufficient reflectance with a reasonable stack thickness and level of strain. Excessive

Table 16.1 Different technologies used for fabricating SDLs with emission at 1150–1300 nm

S. No.	Strategies for wavelength extension to 1150–1300 nm	Challenges
1.	GaAsSb/GaAs QW gain material	Low confinement of carriers in the QWs. Poor temperature behaviour
2.	InP-based gain with InP-based Bragg reflector	Compromised reflectivity, low thermal conductance of the DBR, Increased stack thickness
3.	Hybrid mirrors with InP-based gain	Compromised thermal conductance
4.	Wafer fusion of different gain and active regions	More expensive processing. Two growths required for one component
5.	InAs/GaAs QDs	Reduced design flexibility and low modal gain
6.	Strain compensated high indium content InGaAs QWs	Strain-related lifetime issues
7.	Dilute nitride GaInNAs/GaAs QWs	Formation of nitrogen-related defects

Reproduced from [1]

material strain, arising from the difference between the lattice constants of the semiconductor layers, can lead to formation of crystalline defects and ultimately to relaxation of the layered structure.

The 1150–1300 nm wavelength range has previously been very challenging for the growth of SDLs for two main reasons. First, for conventional InGaAs/GaAs QW material a relatively large content of indium must be used to reduce the bandgap energy to the desired value and the high indium content increases the lattice constant of the material causing buildup of strain in the layer structure. Alternatively, one could also resort to the use of InP-based QWs which work at 1.2–1.6 μm, but unlike with GaAs, the DBR materials lattice matched to InP have very low index contrast. Therefore, the thickness of the Bragg reflector must be increased significantly in order to achieve high reflectance. A number of techniques have been proposed to extend the emission wavelength of GaAs-based structures beyond the typical InGaAs spectral window near 1 μm or to enable the use of InP-based gain regions in surface normal lasers; the main techniques have been listed in Table 16.1 with related challenges.

16.2 High-Power Yellow-Orange SDLS Based on Dilute Nitride Gain Mirrors [1]

To produce yellow emission by second harmonic emission, the indium content of the conventional InGaAs/GaAs QWs needs to be relatively high ($x > 35$ %). The high indium content increases the compressive lattice strain close to the point where misfit dislocations start to appear. The high lattice strain, together with high operation temperatures, can strongly deteriorate the lifetime of a device based on such QWs. By adding a small amount of N (typically less than 3 %) to InGaAs, one can reduce at the same time the lattice constant and the band-gap of the material. This opens up great opportunities for GaAs-based technology. For example, the compressive strain of InGaAs/GaAs material system can be compensated by N incorporation enabling emission at wavelengths up to 1.55 μm. These dilute nitride compounds can be in fact lattice matched to GaAs; $Ga_{1-x}In_xN_yAs_{1-y}$ with $x \approx 2.8\ y$ is lattice matched to GaAs whereas compositions with $x > 2.8\ y$ and $x < 2.8\ y$ lead to compressively and tensile strained compounds, respectively. Furthermore, GaNAs layers exhibit a tensile strain that can be used for balancing the compressive strain of GaInAs layers.

The dramatic effect of nitrogen on the band gap is generally explained as being caused by the small size and large electronegativity of N atoms (radius ~0.068 Å, electronegativity ~3.04 in units of Pauling scale) as compared to As atoms (radius ~0.121 Å, electronegativity ~2.18) of the host crystal. Such impurity atoms create localized energy levels close to the conduction band edge and, as a result, modify the conduction band structure of the alloy. The interaction between the localized states and the conduction band is usually modeled using a so-called band anti-crossing model (BAC). BAC has been very successful in explaining anomalous properties of the dilute nitrides, especially the conduction band structure and the related electron effective mass. The theoretical dependence of the GaInNAs band-gap as a function of N and In composition is shown in Fig. 16.3.

The band gap decreases strongly by incorporating only a few percent of nitrogen and the 1200 nm wavelength range is readily achievable by using GaInNAs with relatively low N content. We should also note that nitrogen incorporation is associated with an increase of the nonradiative recombination centers. Incorporation of

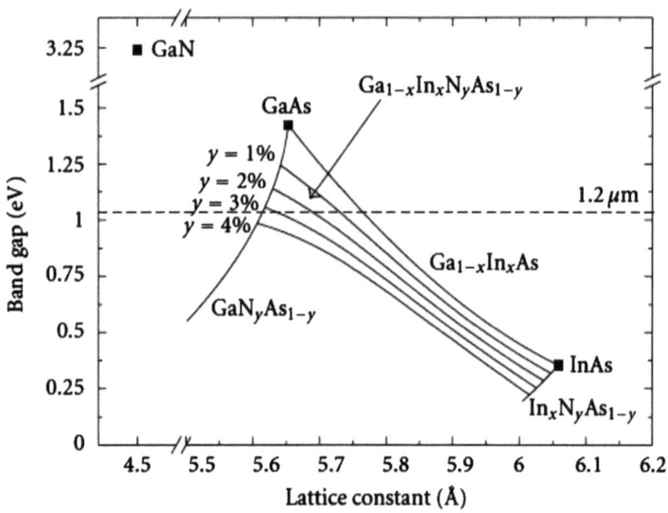

Fig. 16.3 Band gap and lattice constant of dilute nitride GaInNAs. Reprinted from [1] with the permission from Hindawi Publishing Corporation

higher amounts of nitrogen can cause clustering and phase separation having a detrimental effect on the optical quality of the material. To some extent, this effect can be alleviated by rapid thermal annealing (RTA) which, however, leads to a considerable blue shift of the PL wavelength, an effect that should be taken into account in order to achieve the desired laser performance.

In general, the control and understanding of epitaxial processes used to fabricate dilute nitrides is rather challenging. For example, the range of suitable growth temperatures for fabricating high-quality dilute nitrides is narrower than that for growing GaInAs. The typical growth temperature for GaInNAs is in the range of ~460 °C, while GaInAs QWs are grown typically at ~520 °C. The highest performance InGaAsN-based heterostructures are routinely fabricated by molecular beam epitaxy. The standard technique used to incorporate N is dissociation of atomic nitrogen from molecular nitrogen using a radio-frequency (RF) plasma source attached to the MBE growth chamber. Optimization of the plasma operation is one of the key issues that need to be addressed in order to fabricate high-quality dilute nitride heterostructures. The state of the nitrogen plasma depends on the RF power, the flow of N_2, and pressure. The main constituents of the plasma are the molecular nitrogen, atomic nitrogen, and nitrogen ions, each of them having a specific spectral signature that can be used for optimizing the plasma operation. Although the energy of the ions is small, they can cause significant degradation of the optical quality as they impinge on the semiconductor structure during the formation of the QWs. Another important growth parameter affecting the quality of dilute nitrides is the As pressure.

Figure 16.4 displays the structure of a typical dilute nitride gain mirror comprising 10 $Ga_{0.33}In_{0.67}N_{0.006}As_{0.994}$ QWs placed in five pairs. The $GaN_{0.006}As_{0.994}$ layers surrounding the QWs shift their ground state to lower energy and compensate for the compressive strain. For achieving lasing at around 1180 nm, the room temperature emission wavelength of the QWs was designed to be ~1145 nm. The

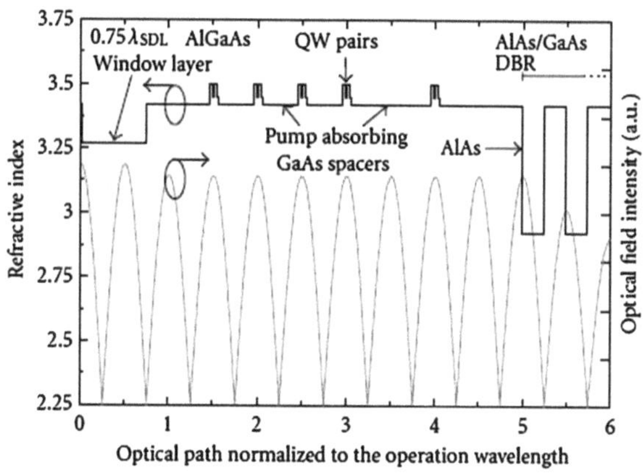

Fig. 16.4 Structure of a typical dilute nitride gain mirror. "Reprinted from [1] with the permission from Hindawi Publishing Corporation"

first four QW pairs were equally spaced at one half wavelength distance apart from each other. The last QW pair was located a full wavelength distance apart from the fourth pair in order to compensate for the pump intensity drop along the gain structure. A 0.75-λ $Al_{0.25}Ga_{0.75}As$ window layer was grown on top of the active region. The active region was grown on top of a 25.5-pair AlAs/GaAs DBR. The growth rate was 0.95 μm/h and the As/III beam equivalent ratio was 25. After the growth, the sample was kept in the growth chamber under As pressure for a 7 min in situ anneal at 680 °C to improve the luminescence properties.

16.3 GaSb-Based SDLs for 2–3 μm Wavelength Range [1]

Another commercially and scientifically very interesting spectral domain is that located between 2 and 3 μm. This spectral range can be accessed using GaSb material system. GaSb-based SDLs with high-power (>1 W) and widely tunable (up to ≈160 nm) operations have been reported by several groups. In addition to continuous wave lasers also ultrashort pulse SDLs in this wavelength range are of interest, as they could be used as seed sources for mid-IR supercontinuum sources or for pumping of mid-IR optical parametric oscillators. However, because of limited availability of some essential components, such as like semiconductor saturable absorber mirrors (SESAMs), the first passively mode-locked 2 μm GaSb-based SDLs were reported very recently.

The development of GaSb-based (AlGaln)(AsSb) heterostructures designed for 2–3 μm wavelength range has struggled with many obstacles, such as increased Auger recombination, typical in narrow bandgap semiconductors, and reduced carrier confinement leading to type-II band alignment in QWs instead of preferred type-I. Regardless, electrically pumped, edge-emitting lasers based on GaSb epitaxy have demonstrated CW operation even beyond 3 μm with careful band-gap engineering and utilization of quinternary AlGaInAsSb waveguides. For optical pumping, the decreased thermal properties of GaSb compared to conventional GaAs make thermal management more demanding for mid-IR SDLs. Typically, the optical pumping is based on commercially available diode pumps at 780–980 nm wavelength range, causing excessive heating due to a large quantum defect. The power scalability of GaSb-based SDLs is therefore limited by the effectiveness of thermal management. To reduce the thermal load, different methods have been investigated, such as in-well pumping, a flip-chip process with GaSb substrate removal, and the use of high thermal conductivity substrate, such as Si or GaAs, in combination with metamorphic growth.

The benefits brought by GaSb-based material system to SDLs are the high index contrast (Δn ~ 0.6) of lattice matched $AlAs_{0.08}Sb_{0.92}$/GaSb DBR layers, which enables to achieve high reflectivity in exceptionally broad band (~300 nm) with a relatively small number of layer pairs. This makes GaSb SDLs very attractive for spectroscopic application where broad tunability of the laser is needed. High-quality AlAsSb/GaSb DBR can also be used for SESAMs. GaAs-based 1 μm

SESAMs have been exploited extensively and their properties can be nowadays tailored to produce ultrashort pulses in various laser types. However, investigation of GaSb-based SESAMs has received far less attention. SESAMs operating at wavelengths around 2 μm and above would have a significant impact on the development of practical ultrafast lasers required in medical applications and time-resolved molecular spectroscopy, or as seeders for optical amplifiers and mid-IR supercontinuum lasers.

16.3.1 Continuous Wave GaSb Disk Laser

The 2 μm range SDLs employed a V-shaped laser cavity. The output characteristics obtained with 99–97 % reflective couplers are shown in Fig. 16.5. Here the cooling water temperature was set to 3.5 °C and the pump spot diameter was about 350 μm. The emission wavelength was about 1990 nm (Fig. 16.5), slightly depending on the power and output coupler. We observed a general tendency for a spectrum shift towards longer wavelengths with increased coupler reflectance, which could be caused by different heat loads on the gain.

16.3.2 Femtosecond Pulse Generation

As discussed earlier, ultrashort optical pulses have been generated in GaAs- and InP-based disk lasers in various configurations using both active and passive mode-locking schemes. On the contrast, the development of ultrafast GaSb disk lasers has been much slower, possibly due to lack of SESAMs and more demanding SESAM characterization. We have shown only very recently that also GaSb-based disk laser can generate sub-picoseconds pulses at 2 μm. The development of low-nonlinearity GaSb-based SESAMs was instrumental for demonstrating ultrafast 2 μm SDLs. The right combination of the dynamic properties of the SESAM (saturation fluence, absorption recovery time, nonlinear reflectivity, and nonsaturable losses) is quite essential for achieving stable mode-locking. An interesting finding was made that as-grown 2 μm GaSb saturable absorber mirrors had much faster recovery time than typical as-grown GaAs based components operating in the 1 μm regime.

16.3.3 Future Outlook

SDLs (or VECSELs, or OPSLs) combine a small footprint, multiwatt output power capability, high beam quality, and the capability to fill spectral gaps that cannot be reached by traditional solid state disk lasers. In terms of semiconductor technology,

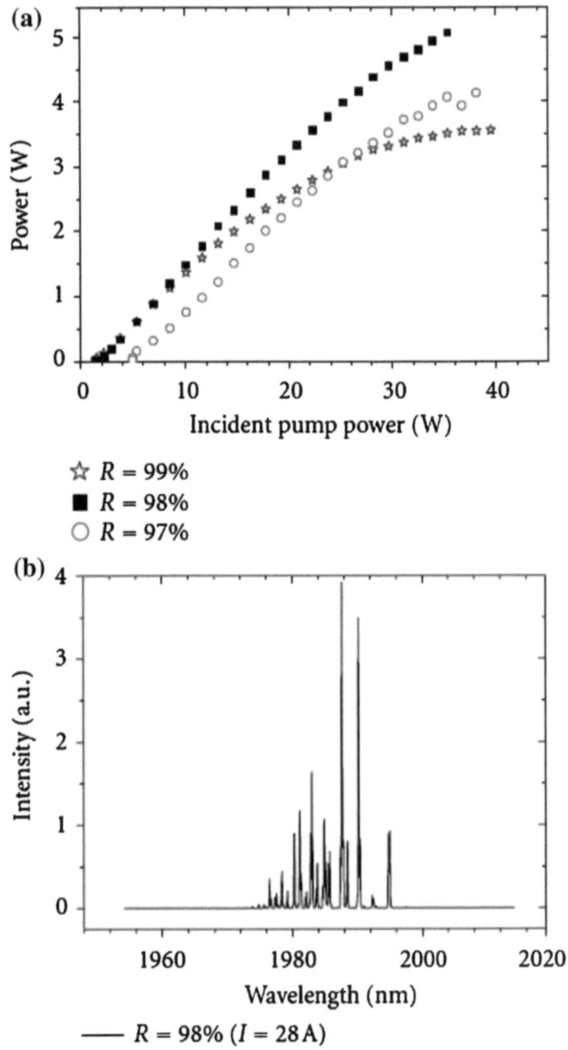

Fig. 16.5 **a** Laser output for different output couplers. T = 3.5 °C (cooling water temperature). Pump spot ∅ = 350 μm. *R* output coupler reflectivity. **b** Laser spectrum for a pump power of 35.2 W. Reprinted from [1] with the permission from Hindawi Publishing Corporation

OP-SDLs are in many aspects simpler than VCSELs; they do not require doping of the mirrors, usually make use of only one semiconductor mirror, and their processing is simpler. Owing to these advantageous features and intense developments efforts in the last decade, SDLs have reached a development stage that makes them very attractive for application deployments. While SDLs with emission at green or blue have been commercialized for several years by Coherent Inc., there are other wavelength regions where SDLs could have a tremendous impact on the development of new applications. Leveraging the advantages of SDLs technology to broader wavelength domains is inherently linked to the development of new semiconductor structures enabling wavelength tailoring and increased functionality.

16.4 Recent Advances in Optically Pumped Semiconductor Lasers [2]

Optically pumped semiconductor lasers offer significant advantages with respect to all traditional diode-pumped solid state lasers (including fibre lasers) in regards to wavelength flexibility, broad pump tolerance, efficient spectral and spatial brightness conversion and high power scaling. In this section, recent progress in the lab and applying this technology to commercial systems are described. Results include diversified wavelengths from 460 to 570 nm, power scaling to >60 W of CW 532 nm, and the launch of a low cost 5 W CW visible source for forensic applications.

16.4.1 Laser Diodes

Semiconductors are in many ways ideal laser materials. They are suitable for both optical pumping and direct electrical pumping. They are able to produce high optical gain and their quantum efficiency is high. Their emission is not limited to discrete lines set by atomic levels but can instead be chosen by design. Similarly multi-layered structures with varying index of refraction can be fabricated with great precision. The same physical processes that are responsible for such advantages impose limitations on the kind of devices that can be realized. In order to achieve optical gain in semiconductor, the active material cannot have arbitrary shape: one of the dimensions must be small (a few microns in the case of optical pumping, around one micron or less for electrical pumping), thus defining a plane that dictates the geometry of the laser devices.

To date, the typical (and enormously successful) implementation of the semiconductor laser is what is commonly known as a "laser diode", variations of which can be found everywhere. Low power devices are used in CD drives and carry phone conversations and data through optical communication lines, high power diodes pump solid state laser materials and are used in material processing applications. Laser diodes are electrically pumped monolithic devices, contained within a "chip" of semiconductor material with typical dimensions around 1 mm. In spite of its many applications, this particular implementation of the semiconductor laser severely limits the performance of semiconductors as laser materials.

If laser emission is in a direction contained in the plane, the most common case known as "edge emitters", the laser is optically just a waveguide with gain, the modal characteristics of the laser defined by the waveguide. In order to obtain single mode operation the mode size has to be small (typically 1×3 μm) and the output power is limited by optica damage at the facets. High power can be obtained of course, but at the expense of brightness. Even single mode edge emitters are astigmatic and highly divergent, in order to obtain round beams with lower divergence the active region can be placed between two highly reflective mirrors, commonly made out of semiconductor materials. In that case the light is emitted perpendicular to the plane of the device. The resonator Fresnel number for this configuration (known as VCSEL, i.e. Vertical Cavity Surface Emitting Laser) is very high, and again there is a severe limitation to the power generated in a single mode. Contrary to the case for edge emitters, VCSELs cannot be practically scaled to high power by increasing the emission area. To bring the pump current to the center of the active region highly doped, or very thick layers would be needed right in the optical path, where they would cause unacceptable absorption loss.

16.5 Optimization and Simplification of Polymerefullerene Solar Cells Through Polymer and Active Layer Design [3]

As sources of alternative energy are increasingly recognized to be more important in the 21st century, solar energy holds a special place as the only energy source that could single-handedly meet the ever-growing world energy demand. Interest in thin-film photovoltaic technologies has grown out of the desire to find inexpensive and readily deployable solar technologies. As a potential thin-film solar technology, organic photovoltaics (OPV) have been the subject of an enormous research effort. A great number of approaches to OPV are currently being investigated Polymer-fullerene bulk heterojunction (BHJ) solar cells have consistently been at the forefront of the growing field of organic photovoltaics (OPV). The enduring vision of OPV is the promise of combining a simple, low-cost approach with an efficient, flexible, lightweight platform. While a great number of specific device architectures have been explored, a loose classification divides OPV into single-junction and multi-junction (tandem) solar cells Among this class, the polymer-fullerene bulk heterojunction (BHJ) solar cells are the most explored and most successful to date, with a number of examples of efficiencies exceeding 8 % and even 9 % now reported and certified.

16.6 Polymer Semiconductor Crystals [4]

From a fundamental viewpoint, polymer semiconductor single crystals are critical for understanding the physics of polymer crystallization. They are also important tools for elucidating macromolecular interactions and solid-state packing in benchmark materials, i.e., poly(3-hexylthiophene)(P3HT), and in newly synthesized polymer semiconductors. Polymer single crystals are also important for studying intrinsic charge transport and determining the performance limitation of the material in question. These highly organized solids will 1 day help answer the age-long question regarding the mechanism of charge transport in conjugated polymers. From a technical point of view, polymer crystals may play an important role in consumer electronic applications such as in flexible displays based on high mobility transistors, solar cells, gadgets and toys, or possibly in technical applications not yet realized by us. Organic single crystals grown from small-molecule semiconductors have attracted much attention in recent years because of their utilization in both fundamental and applied science. One of the long-standing challenges in the field of polymer semiconductors is to figure out how long interpenetrating and entangled polymer chains self-assemble into single crystals from the solution phase or melt. The ability to produce these crystalline solids has fascinated scientists from a broad range of backgrounds including physicists, chemists, and engineers. Scientists are still on the hunt for determining the mechanism of crystallization in these information-rich materials. Understanding the theory and concept of crystallization of polymer semiconductors will undoubtedly transform this area from an art to an area that will host a bandwagon of scientists and engineers.

16.7 Graphene for Radio Frequency Electronics [5]

Past several decades have witnessed dramatic advances in electronics that found widespread application in computing, communications, automation, and other areas to affect every aspect of our lives. To a large extent, these advances have been made possible by the continued miniaturization or 'down-scaling' of electronic devices, particularly the silicon-based transistors; leading to denser, faster, and more power-efficient circuitry. Today, processors containing over two billion metal oxide semiconductor field-effect transistors (MOSFETs), many with gate lengths of just 30 nm, are in mass production. By itself, miniaturization has allowed for the continued performance improvements required for successive generations of integrated circuits. However, the continued increase in device speed and computation power through the evolutionary miniaturization of silicon devices will soon reach a technological limit. To ensure the continued progress of electronic technology and fulfill the future demands of society, transformative technologies that employ alternative materials and/or fundamentally different operating principles are necessary.

Carbon based materials, such as carbon nanotubes and graphene, are emerging as an exciting class of new materials for future electronics due to their superior electronic properties. Carbon nanotubes and graphene share a lot of common features in terms of their fundamental electronic properties. Graphene is emerging as an attractive electronic material for future electronics. With the highest carrier mobility, high saturation velocity, high critical current densities, and single atomic thickness, graphene has great potential for ultra-high speed transistors, with the highest projected cutoff frequency exceeding 1 THz. However, the fabrication of high speed graphene transistors is of significant challenge, since conventional electronic fabrication processes often introduce undesirable defects into grapheme lattices. Significant efforts have made to mitigate these challenges. In this paper, the review has been made of the opportunities, challenges, as well as the recent advances in the development of high speed graphene transistors and circuits.

16.8 Green and Biodegradable Electronics [6]

Sustainability is the ability of generation to ensure its needs for the present without compromising the ability of the future generations to meet their own needs. Consumption is the human transformation of materials and energy (along the production-consumption chain) that makes the transformed materials or energy less available for future use, or negatively impact biophysical systems in such a way to threaten human health, welfare, or other things people value. Plastic consumption and waste are two of the major concerns in the modern world. Polyethylene is currently the leading plastic material, mostly for use in plastic bag production. Due to the increased demand in countries with emerging economies, plastics consumption is projected to increase approximately by a factor of three during the current decade. The outcome of the constant demand for plastics is the buildup of non-biodegradable solid waste and plastic litter (estimated at 25 million tons/year in the year 2000) with negative consequences on our environment.

Plastics normally biodegrade very slowly, with full degradation occurring after 500 or 1000 years. We live in a world where the lifetime of electronics is becoming shorter, now approaching an average of several months. This poses a growing ecological problem. This article presents some steps taken to address the issue of electronic waste with biodegradable organic electronic materials. Many organic materials have been shown to be biodegradable, safe, and nontoxic, including compounds of natural origin. Additionally, the unique features of such organic materials suggest they will be useful in bio-functional electronics; demonstrating functions that would be inaccessible for traditional inorganic compounds. Such materials may lead to fully biodegradable and even biocompatible/bio-metabolizable electronics for many low-cost applications. This article highlights recent progress in these classes of material, covering substrates and insulators, semiconductors, and finally conductors.

16.9 Modern Plastic Solar Cells: Materials, Mechanisms and Modeling [7]

The sharp rise in societal interest in carbon-neutral, green energy technology in the 21st century has spurred a commensurate rise in research into OPV materials. Focus is given on two broad categories of OPV devices, ‘plastic’ solar cells comprising a conjugated polymer and a fullerene acceptor and hybrid organic–inorganic solar cells that blend inorganic semiconductors into organic materials. Photovoltaic cells use the energy of absorbed photons to generate free charge carriers (holes and electrons) which can do electrical work. Organic photovoltaic cells are photovoltaic devices that accomplish this conversion of energy using organic materials—either entirely or as part of a blend. In all cases, at least one of the charge carriers traverses a bulk organic material. Silicon-based (and many other inorganic-based) photovoltaic devices generate free carriers directly from the absorption of a photon, but presently known organic materials do not screen charges as effectively as silicon (because of their lower dielectric constant), In modern OPV devices, the absorption of a photon creates an exciton (i.e. a bound electron-hole pair) which must then be split into free carriers before it decays back to the ground-state. Thus OPV devices are 'excitonic' solar cells. This scission process takes place at the interface between the two constituent materials, a donor and an acceptor.

These materials are chosen such that energy offsets between the energy levels exist, thus making charge transfer energetically favoured. These materials can be layered or intimately mixed, provided there is a continuous pathway for both holes and electrons to traverse toward the electrodes. A photon is absorbed by either the donor or the acceptor, creating an exciton. The exciton then diffuses to the interface between the donor and the acceptor.

16.10 Miniature Wire-Shaped Solar Cells, Electrochemical Capacitors and Lithium-Ion Batteries [8]

It is critically important to develop miniature energy harvesting and storage devices in modern electronics, for example, for portable and foldable electronic facilities. In this review article, novel miniature solar cells, electrochemical capacitors and lithium-ion batteries as well as their integrated devices are carefully summarized. Particular emphasis has been paid to wire-shape energy devices that exhibit unique and promising advantages such as being lightweight and weaveable compared with the conventional planar architecture. Recent new materials and attractive designs are highlighted for these wire-shaped energy devices.

Electronic products ranging from cell phones and laptops to the electronic systems in spacecraft must be small, lightweight and flexible. The conventional silicon wafer-based photovoltaic technologies cannot fully satisfy these requirements. Therefore, there is increasing interest in developing next-generation photovoltaic devices, including dye-sensitized and polymer solar cells that can be made into flexible structures. In the case of the storage systems, electrochemical capacitors and lithium-ion batteries have been the most extensively explored. In this article, the main areas of progress in these miniatures wire-shaped energy devices, including dye-sensitized solar cells (DSCs), polymer solar cells (PSCs), electrochemical capacitors, and lithium-ion batteries, as well as integrated cells are briefly described that may convert solar energy to electric energy and simultaneously store it.

16.10.1 Miniature Inorganic Solar Cells

Over decades of the photovoltaic industry, crystalline silicon has remained the primary photovoltaic materials in commercial modules due to high efficiencies and natural abundance; though the high cost in producing Si wafers has largely limited their use in the future. Therefore, a lot of effort has been made toward developing an ultrathin silicon film (thicknesses of 5–50 μm) as an active photovoltaic layer. For instance, Rogers and co-workers fabricated microcells from single-crystalline silicon (Fig. 16.6a), which could be further printed onto flexible substrates such as poly (dimethylsiloxane) (Fig. 16.6b). Here, an individual microcell was composed of phosphorus-doped, boron-doped and un-doped regions (Fig. 16.6c). The energy conversion efficiency achieved 11.6 % with a white backside reflector under AM 1.5 illumination (Fig. 16.6d).

The main limitation of these ultrathin silicon cells was that the absorption of photons was reduced greatly due to the very thin active material, which lowered the device performance. To this end, some modifications were made to improve the

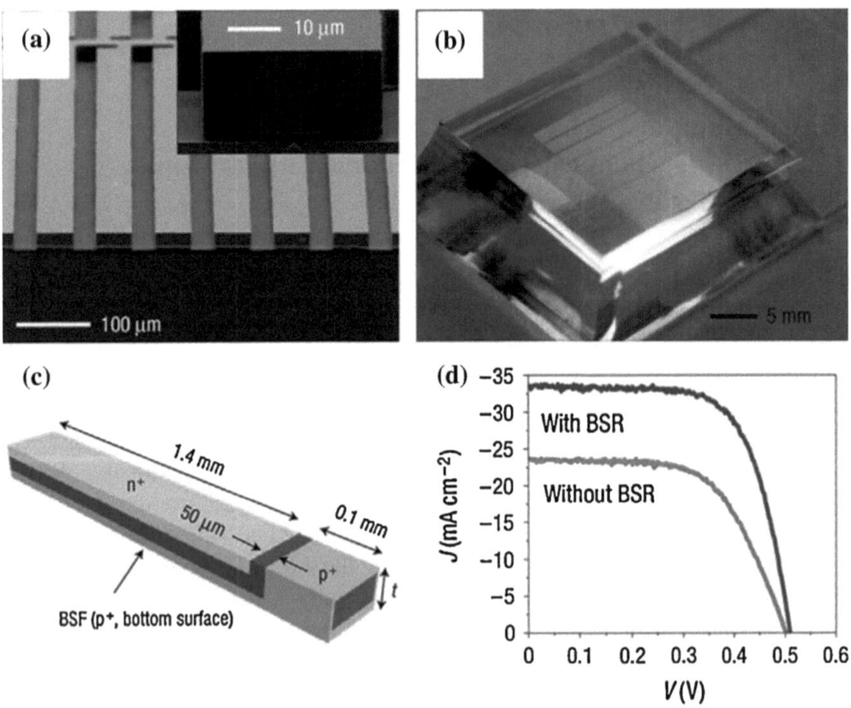

Fig. 16.6 Silicon-based solar microcells. **a** SEM image of an array of microcells on a source wafer. **b** Optical image of an array of microcells on an elastomeric polydimethylsiloxane substrate. **c** Schematic illustration of a microcell, BSF, back-surface field. **d** Representative J–V curves of a microcell with and without a white diffusing backside reflector (BSR) under AM 1.5 illumination of 1000 W/m^2 condition. Reproduced with permission from Ref. [8]. Copyright 2014, Elsevier

light trapping in ultrathin silicon solar microcells. For example, nanostructured relief features were introduced to the surface of the device to enhance the light absorption through scattering or diffraction and passive concentration through the luminescent waveguide. Both incident and diffusing photons from the bottom and sidewall could be harvested.

16.11 Skin-Inspired Electronic Devices [9]

Electronic devices that mimic the properties of skin have potential important applications in advanced robotics, prosthetics, and health monitoring technologies. Methods for measuring tactile and temperature signals have progressed rapidly due to innovations in materials and processing methods. Imparting skin-like stretchability to electronic devices can be accomplished by patterning traditional electronic materials or developing new materials that are intrinsically stretchable. The incorporation of sensing methods with transistors facilitates large-area sensor arrays. While sensor arrays have surpassed the properties of human skin in terms of sensitivity, time response, and device density, many opportunities remain for future development.

Human skin exhibits a remarkable range of properties. It is sensitive enough to detect a gentle breeze and distinguish subtle differences in the texture of materials, while simultaneously being robust enough to protect our bodies against damage from external objects and harmful biological entities. As our interface with the outside world, skin provides complex tactile sensing functionality that facilitates many of our essential activities. Skin provides temperature sensing capabilities that help us understand our surroundings and avoid damaging temperatures. While an electronic version of skin (e-skin) provides the opportunity to incorporate additional functionalities such as chemical sensors and energy generation, the strategies used to mimic the mechanical properties and tactile- and temperature-sensing properties of human skin is focussed here.

E-skin provides the promise of facilitating important advancements in the fields of robotics and medical devices. Since their inception, robots have been restricted to highly repetitive tasks in a structured environment. However, with the ability to collect complex information about their surroundings using e-skins, the robots could succeed in more dynamic and variable tasks, such as rescue missions or caring for the elderly. Used in prosthetics, synthetic skin could provide the ability to sense touch and temperature for amputees and individuals with nerve damage. Sensor skins are also capable of monitoring health parameters such as pulse waveforms and temperature distributions.

16.12 Recent Advances in Dye-Sensitized Solar Cells: From Photoanodes, Sensitizers and Electrolytes to Counter Electrodes [10]

Dye-sensitized solar cells (DSSCs), as low-cost photovoltaic devices compared to conventional silicon solar cells, have received widespread attention in recent years; although much work is required to reach optimal device efficiencies. This article highlights recent developments in DSSCs and their key components, including the photoanode, sensitizer, electrolyte and counter electrode. Motivated by continuously growing global energy demands and the depletion of readily accessible fossil fuels; the search for alternative energy sources, particularly renewable solar energy, has become vital. Despite the clear advantages associated with the adoption of solar cells, they need to be cost-effective and priced competitively in comparison to conventional energy resources, as any technological or performance improvements must be balanced against the associated cost.

The most attractive properties of DSSCs are their low-cost and simple manufacturing processes together with their advantageous attributes (e.g. lightweight, flexible, low toxic, and good performance in diverse light conditions). A DSSC typically consists of several micron thick semiconductor (e.g. TiO_2, ZnO and SnO_2) film served as a photo-anode that is coated or grown on a conductive substrate, a sensitizer (i.e. dye; e.g. N719, N3 or organic dyes by monolayer adsorption or quantum dots (e.g. CdS, CdSe and PbS)), an electrolyte (e.g. I_3^-/I^- and Co^{2+}/Co^{3+} redox couples) injected between the sensitizer and counter electrode, and a counter electrode (e.g. Pt and carbon materials) deposited on another conductive substrate. Illumination by visible light irradiation on the photoanode causes photo-excitation of the absorbed dye molecules to generate excited electrons which are subsequently injected into the conduction band of the semiconductor and quickly shuttled to the external circuit through the conductive substrate, producing an electric current. The original state of the dye is subsequently restored by electron donation from the redox electrolyte. The counter electrode returns charge from the external circuit back to the cycling circuit in the cell.

Areas of interest have included the construction of nanostructured semiconductor photoanodes with effective architectures for high dye loading and fast electron trans- port, the exploitation of versatile sensitizers with strong visible light harvesting ability, the utilization of redox electrolytes with useful compositions for efficient hole transport, the optimization of the Pt counter electrode as well as the development of other equiva-lent alternatives at lower costs. After various attempts, a recorded power conversion efficiency (PCE) of 12.3 % for liquid- electrolyte DSSCs was obtained in 2011 by using a zinc porphyrin dye YD2-o-C8 co-sensitized with another organic dye Y123, and CoII/IIItris(bipyridyl)-based redox electrolyte at AM 1.5 global full sun. In 2013, a new milestone for solid-state mesoscopic TiO_2 solar cells sensitized with lead iodide perovskite ($CH_3NH_3PbX_3$) was reported to achieve an exciting power conversion efficiency of more than 15 %, with expected future values as high as 20 %.

16.13 Investigation of the Optical Absorption of a–Si:H Solar Cells on Micro-and Nano-Textured Surfaces [11]

Light trapping is essential for thin-film silicon solar cells due to their thin absorber layers. Textured surfaces are widely used to scatter light, thereby increasing the pathlength of light in these solar cells. In a–Si:H solar cells different interfaces can be textured, e.g. the glass/TCO (transparent conductive oxide) interface, the TCO/a–Si:H interface, or the rear TCO/Al (rear reflector) interface. Nano-textured TCOs are widely used for scattering light at the TCO/a–Si:H interface. Additionally, the micro-texturing of the glass/TCO interface is effective for light scattering, especially in the long-wavelength range. Aluminium induced texturing (AIT) is a method to texture glass surfaces. Previous studies have shown that AIT glass increases the quantum efficiency of micromorph thin-film solar cells in the long-wavelength range. In this contribution, the results suggest that the nano-texturing increases the parasitic absorption in the TCO and the p-doped silicon layer, thus decreasing the blue response of the solar cell. However, absorption in the intrinsic layer is significantly enhanced in the long-wavelength region. Particularly in the double-junction a–Si:H solar cells on AIT micro-textured superstrates, the current is enhanced in both the top and the bottom cell.

16.14 POSFET Tactile Sensing Arrays Using CMOS Technology [12]

This work presents fabrication and evaluation of novel POSFET (Piezoelectric Oxide Semiconductor Field Effect Transistor) devices based tactile sensing chip. In the newer version presented here, the tactile sensing chip has been fabricated using CMOS (Complementary Metal Oxide Semiconductor) technology. The chip consists of 4 × 4 POSFET touch sensing devices (or taxels) and both, the individual taxels and the array are designed to match spatiotemporal performance of the human fingertips. To detect contact events, the taxels utilize the contact forces induced change in the polarization level of piezoelectric polymer (and hence change in the induced channel current of MOS). The POSFET device on the chip have linear response in the tested dynamic contact forces range of 0.01–3 N and the sensitivity (without amplification) is 102.4 mV/N.

Touch sensing plays an important role in various application domains such as robotics, electrotextiles and medical prosthesis. A large amount of efforts has been devoted on the development of tactile sensors and over the years, touch sensing technology has improved. Many new touch sensors using different materials and transduction methods, viz: Resistive, Piezoresistive, Quantum Tunneling, Capacitive, Optical, Ultrasonic, Magnetic, Piezoelectric etc. have been developed. Suitability of these sensors to a particular application depends on number of

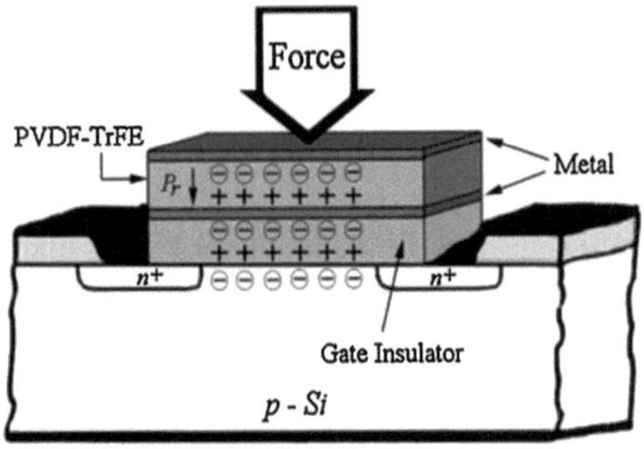

Fig. 16.7 The structure and working principle of a POSFET touch sensing device. Reproduced with permission from [12]

constraints, which include size of the sensors, response time of the sensors, and physical features such as bendability or conformability etc. Many times the sensors are big in size and considering also the associated electronic circuitry, they are unsuitable for body sites like robot's fingertips, where large numbers of sensors with high density are needed. For this reason, MEMs (Microelectromechanical Systems) based miniaturized touch sensors with on-chip electronics have been explored.

16.14.1 Working Principle of a POSFET Touch Sensing Device

The structure of a POSFET touch sensing devices is shown in Fig. 16.7. The piezoelectric polymer film is present over the gate area of the MOS device. The transducer material is thus an integral part of the device. The remnant polarization (Pr) of the polarized polymer results in an intrinsic electrical dipole equivalent to fixed charges ±Q, as shown in Fig. 16.7.

16.15 Oxide Semiconductor Thin-Film Transistors: A Review [13]

Transparent electronics is today one of the most advanced topics for a wide range of device applications. The key components are wide bandgap semiconductors, where oxides of different origins play an important role, not only as passive component but also as active component, similar to what is observed in conventional semiconductors like silicon. Transparent electronics has gained special attention during the last few years and is today established as one of the most promising technologies for leading the next generation of fl at panel display due to its excellent electronic performance. Recent progress in n- and p-type oxide based thin-film transistors (TFT) and n-type TFTs processed by physical vapour deposition methods is discussed: The p-type oxide TFTs, mainly centred on two of the most promising semiconductor candidates: copper oxide and tin oxide, and complementary metal oxide semiconductor (CMOS) devices based on n- and p-type oxide TFT is also presented.

16.15.1 Amorphous Oxide Semiconductor [13]

Oxide semiconductors, especially the amorphous ones, are a promising class of TFT materials that have made an impressive progress particularly in display applications challenging silicon not only in conventional applications but opening doors to completely new and disruptive areas like paper electronics. This new class of semiconductor materials, amorphous oxides semiconductors (AOS) exhibits a stimulating combination of high optical transparency, high electron mobility and amorphous microstructure. Besides that, AOSs do not have grain boundaries, thereby obviating the primary limitation of mobility in polycrystalline semiconductors, which is a huge advantage for process integration. Other advantages include low temperature deposition routes and ultrasmooth surfaces for suppressing interface traps and scattering centres. Besides the advantages of this new technology like the capability for providing real solutions and overcoming many obstacles and limitations of the conventional silicon technology, The used materials as well as the technology itself are environmental friendly and much less expensive.

A significant worldwide interest appeared, especially for 'active matrix for organic light emitting diodes (AMOLED)' technology. Samsung has released at the end of year 2010 what they boast proudly as one of the world's finest and largest 3D Ultra Definition (UD) TV with 70 in. (175 cm) diagonal and a sports resolution of 240 Hz that will meet the demands of 3D capability. Samsung has claimed that the making of this device is the first of the kind on oxide semiconductor TFT technology that supports high pixel density, with a resolution of 3840 × 2160, which is equivalent to 8 MP. Although the first transparent display has been done using the conventional a–Si:H technology in 2005.

16.16 Wide Band Gap Semiconductor Devices for Power Electronics [14]

It is worldwide accepted today that a real breakthrough in the Power Electronics field may mainly come from the development and use of Wide Band Gap (WBG) semiconductor devices. WBG semiconductors such as SiC, GaN, and diamond show superior material properties, which allow operation at high-switching speed, high-voltage and high-temperature. These unique performances provide a qualitative change in their application to energy processing. From energy generation (carbon, oil, gas or any renewable) to the end-user (domestic, transport, industry, etc.), the electric energy undergoes a number of conversions. These conversions are currently highly inefficient to the point that it is estimated that only 20 % of the whole energy involved in energy generation reaches the end-user. WGB semiconductors increase the conversion efficiency due to their outstanding material properties. The recent progress in the development of high-voltage WBG power semiconductor devices, especially SiC and GaN, and the performances of various rectifiers and switches, is discussed. Material and process technologies of these WBG semiconductor devices and future trends in device development and industrialization are also addressed.

16.16.1 Introduction

Power Electronics play a key role in the generation-storage-distribution cycle of the electric energy. This is because the main portion of the generated electric energy is consumed after undergoing several transformations, many of them carried out by power electronic converters. Examples of this can be found in all ranges of power levels (from a few W to MW), and they include many types of different equipments (power supplies for computers, industrial and telecom systems, domestic appliances, motor drives, industrial converters, etc.). The largest portion of the power losses in these power electronic converters are dissipated in their power semiconductor devices. Nowadays, these devices are based on the mature and very well established Silicon technology. However, Si exhibits some important limitations regarding its voltage blocking capability, operation temperature and switching frequency. Therefore, a new generation of power devices is required for power converters in applications where electronic systems based on traditional Si power devices cannot operate. The use of these new power semiconductor devices will allow increasing the efficiency of the electric energy transformations achieving a more rational usage of the electric energy.

Novel and innovative power devices based on WBG semiconductors can play a main role in energy efficient systems. Among the possible candidates to be the base materials for these new power devices, SiC and GaN present the better trade-off between theoretical character-istics (high-voltage blocking capability, high-temperature operation and high switching frequencies), and real commercial availability of the starting material (wafers) and maturity of their technological processes. Table 16.2 summarizes the main material parameters of WBG semiconductors candidates to replace Si in the next generation of power devices. GaN and especially SiC process technologies are by far more mature and, therefore, more attractive from the device manufacturer's perspective, especially for high power and high temperature electronics (HTE). GaN can offer better high-frequency and high-voltage performances, but the lack of good quality bulk substrates is a

Table 16.2 Physical properties of various semiconductors for power devices

Material	Eg(eV) 300 K	μ_a (cm^2/Vs)	μ_p (cm^2/Vs)	v_{sat} (cm/s)	E_c (V/cm)	ε_r
Si	1.12	1450	450	10^7	3×10^5	11.7
GaAs	1.4	8500	400	2×10^7	4×10^5	12.9
3C–SiC	2.3	1000	45	2.5×10^7	2×10^6	9.6
6H–SiC	2.9	415	90	2×10^7	2.5×10^6	9.7
4H–SiC	3.2	950	115	2×10^7	3×10^6	10
GaN	3.39	1000	35	2×10^7	5×10^6	8.9
GaP	2.26	250	150		10^7	11.1
Diamond	5.6	2200	1800	3×10^7	5.6×10^7	5.7

Reproduced from [14]

disadvantage for vertical devices. Nevertheless, GaN presents a lower thermal conductivity than SiC. At present, SiC is considered to have the best trade-off between properties and commercial maturity with considerable potential for both HTE and high power devices. However, the industrial interest for GaN power devices is increasing recently. For this reason, SiC and GaN are the more attractive candidates to replace Si in these applications. In fact, some SiC devices, such as Schottky diodes, are already competing in the semiconductor market with Si power diodes. On the other hand, GaN allows forming hetero-junctions (In-AlGaN alloys) and can be grown either on SiC or Si substrates. Currently, it is a sort of competition between SiC and GaN in a battle of performance versus cost. Nevertheless, scientific and industrial actors agree in considering that both will find their respective application fields with a tremendous potential market.

SiC power devices include high-voltage and high-temperature diodes, junction controlled devices (like JFETs), MOSFETs and MESFETs. Those based on GaN include diodes, HEMTs and MOSFETs; and advanced research on novel devices concerning low-losses digital switches based on SiC and GaN is also of main concern. These novel devices represent a real breakthrough in power devices. Furthermore, the development of modelling and electro-thermal characterization tools for these power devices, and the design of their packaging, drivers and controllers need a great research effort and they represent a world-class innovation.

16.17 SiC Power Devices [14]

Si has long been the dominant semiconductor material for high-voltage applications. The situation has changed due to the significant achievements in SiC bulk material growth, and in SiC process technology. The progress in SiC wafers quality is reflected in the achievement of very low micropipe density (0.75 cm^{-2} for a 75 mm wafer), which provides the basis for a high fabricating process yield of large area SiC power devices. 100 mm SiC wafers are already in the market and it is expected that 150 mm SiC wafers will be available in a near future. It has been shown that the active regions of SiC-based devices overlying micropipe defects in the substrate fail to operate. Therefore, one of the major concerns has been reducing the micropipe density in SiC substrates, which has been driven by the phenomenological understanding of the mechanisms that are responsible for pipe formation during SiC crystal growth. The understanding of such mechanisms has led the identification and removal of most of these defects. However, other defects, such as basal plane dislocations, are still under investigation causing poor reliability in bipolar devices. In conclusion, the most challenging approach is to grow low-doped epitaxial layers while maintaining features such as purity, surface smoothness and growth rate (epilayer thickness). Besides, minority carrier lifetime in thick epilayers appears to be long enough for conductivity modulation, as inferred from high-voltage (19 kV) reported diodes.

Many high-voltage experimental SiC-based two terminal rectifiers and three-terminal switches have been demonstrated. 4H–SiC unipolar devices are expected to replace Si bipolar rectifiers in the 600–6500 V range, and power switches higher than 1.2 kV in the future. Generally speaking, there are three types of power rectifiers:

- Schottky Barrier Diodes (SBD) with extremely high switching speed and low on-state losses, but lower blocking voltage and high leakage current;
- PiN diodes with high-voltage operation and low leakage current, but showing reverse recovery charging during switching; and
- Junction Barrier Schottky (JBS) diodes with Schottky like on-state and switching characteristics, and PiN-like off-state characteristics.

Figure 16.8 shows the cross-sections of the three SiC rectifiers. Schottky and PiN diodes are the two basic types of power rectifiers. However, hybrid rectifiers such as JBS rectifiers, which combine the best features of each type, are particularly attractive.

The most remarkable advantage of SiC SBDs is the continuing increase in the blocking voltage and conduction current ratings. They range from the initial 300 V, 10 A and 600 V, 6 A to the actual 600 V, 20 A and 1.2/1.7 kV. With the latest ratings, it is foreseen that these diodes may replace Si bipolar diodes in medium power motor drive modules. Power Factor Correction and High-Voltage Secondary Side Rectification are applications of 600 V SiC SBDs. Besides, it is expected that SBDs can be advantageously applied for blocking voltages up to 3.5 kV. Large area 3.3 kV SBDs have been fabricated with high-temperature operation that are able to supply forward currents in the range of 20 A. In comparison with Si counterpart, a ×10 increase in voltage blocking is possible with the same SiC drift layer thickness. The main difference to ultra fast Si PiN diodes lies on the absence of reverse recovery charge in SBDs. Therefore, SiC SBDs are well suited for high switching speed applications. 1.2 kV SiC SBDs match perfectly as freewheeling diodes with Si IGBTs.

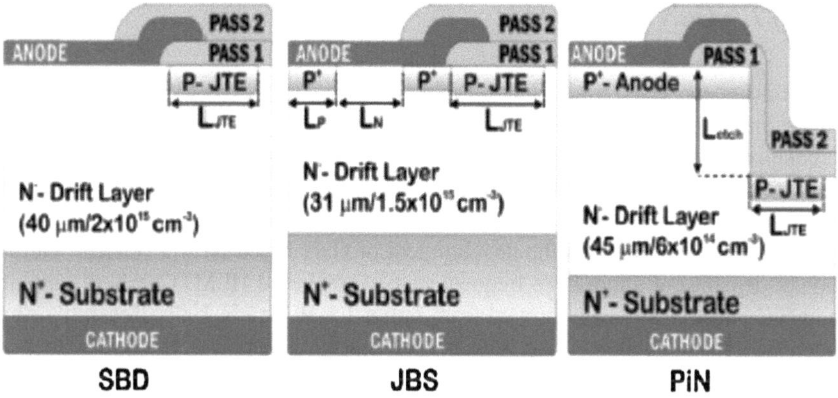

Fig. 16.8 Cross-section of 4H-SiC 3.3 kV Schottky, JBS and PiN diodes. Reproduced from [14]

16.18 GaN Power Devices [14]

GaN is of interest for high-voltage and high-temperature devices due to its remarkable material properties like wide bandgap, large critical electric field, high electron mobility and reasonably good thermal conductivity. In addition, a large conduction band discontinuity between GaN and AlGaN and the presence of polarization fields allows a large two-dimensional (2D) electron gas concentration to be confined. Until recently, because of the lack of electrically conducting GaN substrates, most of the GaN Schottky power diodes reported are either lateral or quasi-vertical. Breakdown voltages of lateral GaN rectifiers on Sapphire substrates could be as high as 9.7 kV, but the forward voltage drop is still high. The interest of these diodes lies on their lower cost when implemented on Si or Sapphire substrates. In fact, with the availability of high-temperature HVPE (Hydride Vapour Phase Epitaxy) GaN substrates, 600 V vertical GaN Schottky diodes are due to be launched to compete with SiC Schottky rectifiers. GaN JBS diodes could further increase the performance of GaN-based power rectifiers in the 600 V–3.3 kV range. In this sense, we are working on the optimization of contact resistance to implanted p-type GaN. We have found that protection during post-implantation annealing is very important to obtain a good uniformity on the contact properties. However, the fact of having much more dispersion, as well as lower contact resistance for some of the samples (in the range of 4×10^{-5} Ωcm^2) for the unprotected samples, makes us to suggest that the contact resistance mechanism is related with the formation of N vacancies on the GaN surface during both the post-implantation and contact annealing.

16.18.1 GaN HEMT [14]

In recent years, GaN High Electron Mobility Transistors (HEMTs) have attracted most attention with remarkable trade-off between specific on-resistance and breakdown voltage. The GaN HEMTs are expected as microwave power devices used for base station of cellular phone and as switching power devices in DC/DC converters. Since the demonstration of the first GaN based HEMT switch, rapid progress has been made in the development of GaN-based HEMT devices. Output power densities at microwave frequencies of GaN based HEMTs on both sapphire and SiC substrates have improved from initial 1.1 W/mm in 1996. Recently, it has been demonstrated impressive AlGaN/GaN microwave power HEMTs with high output power capability, as high as 40 W/mm. A major obstacle has been controlling the trap densities in the bulk and surface of the material affecting the performance of these devices by trapping effects though drain current collapse. To efficiently operate the transistor at high frequency and high voltage, the drain current collapse must be suppressed and the gate-drain breakdown voltage must be improved. Several solutions for the device structure have been proposed including the surface charge-controlled n-GaN-cap structure, the recessed gate and field-modulating plate structure or the passivation of surface states via silicon nitride or other dielectric.

High voltage AlGaN/GaN HEMTs over 1 kV were reported in 2006. In this sense, a high-voltage/low RON AlGaN/GaN HEMT on semi-insulating SiC has been also reported, which exhibits a record of a high-power figure of merit ($\sim 2.3 \times 10^9$ $V^2/W\ cm^2$) and exceeds the 6HSiC theoretical limit. A step further for the most cost-effective and industrially relevant GaN-on-silicon is the removal of the silicon, which limits the amount of power of the GaN-on-Si power HEMTs. Srivastava et al. [15] reported in 2011 a record breakdown voltage for HEMT fabricated on <111> Si by a new local Si substrate removal technology with $V_{BR} = 2.2$ kV for devices with gate-todrain distances of 20 μm (buffer thickness was only 2 μm). This is a remarkable enhancement compared to the reference (on bulk Silicon), which has a saturated $V_{BR} = 0.7$ kV. Furthermore, an extremely high blocking voltage of 8.3 kV has been achieved while maintaining relative low specific on-state resistance of 186 mW cm^2, via-holes through sapphire at the drain electrodes enable very efficient layout of the lateral HFET array as well as better heat dissipation.

It has been also demonstrated a GaN power switch for kW power conversion. The switch shows a speed higher than 2 MHz with rise- and fall-time of less than 25 ns, and turn-on/turn-off switching losses of 11 μJ with a resistive load. Switching at 100 V/11 A and 40 V/23 A was achieved with resistive and inductive loads, respectively.

The GaN HEMT is an intrinsically normally-on device due to the existence of the 2DEG (2-Dimensional Electron Gas). Therefore, a negative gate bias is required to switch the device off. Nevertheless, normally-off devices are preferred in power electronic applications. Several approaches have been developed for converting the

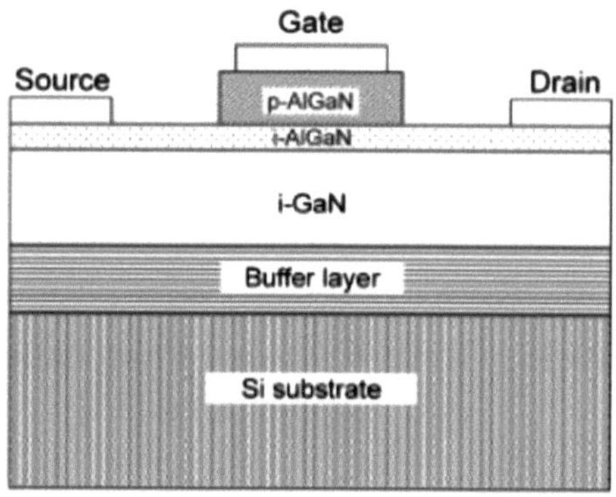

Fig. 16.9 Schematic cross-section of a p-type gate HEMT. Reproduced from [14]

GaN HEMTs from the conventional normally-on mode to the desired normally-off mode:

1. One approach is to employ a recessed-gate structure so that the AlGaN layer under the gate is too thin for inducing a 2DEG, resulting in a low and positive Vth (threshold voltage). The enhancement mode AlGaN HEMT was first reported by Khan et al. in 1996.
2. A second approach is to use fluorine based plasma to dope the semiconductor beneath the gate metal, so that acceptors are formed in this region, effectively depleting the 2DEG.
3. The third approach a p-doped GaN or AlGaN cap layer to deplete the 2DEG underneath (Fig. 16.9). In 2000 Hu et al. proposed an E-mode (enhancement-mode) AlGaN/GaN HEMT with selectively grown pn junction gate.
4. It is also possible to include the MOS-HEMT as a technique for getting an E-HEMT as a structure combining the HEMT 2DEG current capability and the MOS normally-off operation. Recessed MISHEMTs have been also proposed but the objective of the reported devices is not to make the device normally-on but increase the transconductance of MIS-HEMTs or reducing its gate leakage.
5. A combination of the previous techniques with a customized growth of the AlGaN/GaN stack also allows improving the performances of the Emode HEMTs.

16.18.2 New Generation of Power Devices

The new generation of power devices for power converters will be based on Wide Band Gap semiconductors to replace traditional silicon power devices. Currently the highest breakdown voltage capability of the commercial dominant power switch (Si IGBT) is 6.5 kV. In any case, a Silicon-based device could not operate over 200 °C. These inevitable physical limits reduce drastically the efficiency of current power converters, which requires among others, complex and expensive cooling systems.

The use of these new power semiconductor materials will allow increasing the efficiency of the electric energy transformations for a more rational use of electric energy, thus reducing carbon footprint. The most promising WBG semiconductor materials for this new generation of power semiconductor devices are SiC and the GaN.

16.19 Heterogeneous Photocatalysis: Recent Advances and Applications [16]

Semiconductor heterogeneous photocatalysis is a versatile, low-cost and environmentally benign treatment technology for a host of pollutants. These may be of biological, organic and inorganic in origin within water and air. The efficient and successful application of photocatalysis demands that the pollutant, the catalyst and source of illumination are in close proximity or contact with each other. The ability of advanced oxidation technology to remove low levels of persistent organic pollutants as well as micro-organisms in water has been widely demonstrated. This review considers recent developments in the research and application of heterogeneous semiconductor photocatalysis for the treatment of low-level concentrations of pollutants in water and air using titanium dioxide as a "model" semiconductor. The review considers charge transport characteristics on the semiconductor surface, photocatalyst reactor design and organic degradation mechanistic pathways. The effects of photoreactor operating parameters on the photocatalytic process are discussed in addition to mineralization and disinfection kinetics.

16.20 New Semiconductor Materials for Magnetoelectronics at Room Temperature [17]

Most of the semiconductor materials are diamagnetic by nature and therefore cannot take active part in the operation of the magneto electronic devices. In order to enable them to be useful for such devices a recent effort has been made to develop diluted magnetic semiconductors (DMS) in which the small quantity of magnetic ion is introduced into normal semiconductors. The first known such DMS are II–VI and III–V semiconductors diluted with magnetic ions like Mn, Fe, Co, Ni, etc. Most of these DMS exhibit very high electron and hole mobility and thus useful for high speed electronic devices. The recent DMS materials reported are (CdMn)Te, (GaMn)As, (GaMn)Sb, ZnMn(or Co)O, TiMn(or Co)O etc. They have been produced as thin films by MBE and other methods.

Semiconductors and magnetic materials are two very important materials in electronic industries. On combining both the properties and form of both these, new materials is formed in which the performance of devices can be improved. Those

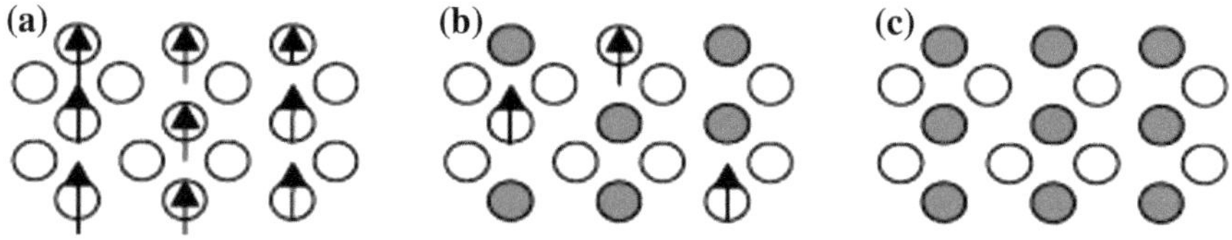

Fig. 16.10 Three types of semiconductors: **a** a magnetic semiconductor, in which a periodic array of magnetic element is present, **b** a diluted magnetic semiconductor, an alloy between nonmagnetic semiconductor and magnetic element and **c** a non-magnetic semiconductor, which contains no magnetic ions. Reproduced with permission from [17]

materials are known as 'diluted magnetic semiconductors' (DMS). Thus the concept of 'magnetoelectronics' has come up i.e. the electronic chips consisting of DMS materials. Diluted magnetic semiconductors (DMS) are expected to play an important role in interdisciplinary materials science and future electronics because charge and spin degrees of freedom accommodated into a single material exhibits interesting magnetic, magneto-optical, magnetoelectronic and other properties.

It is expected that magnetoelectronic chips will be used in quantum computers. An inherent advantage of magnetoelectronics over electronics is the fact that magnet tend to stay magnetized for long. Hence this arises interest in industries to replace the semiconductor-based components of computer with magnetic ones, starting from RAM. The new magnetic RAM will retain data even when the computer is turned off. And most important advantage will be eliminating the time consuming process of 'booting up' information from hard drive to processor like a TV set, all the information would be there. One challenge in realizing magnetic RAM involves addressing individual memory elements, flipping their spins up or down to yield the zero and ones of binary computer logic.

DMS are semiconductors in which a fraction of the component ions are replaced by those of transition metals or rare earths, Fig. 16.10. Most importantly, state of magnetization changes the electronic properties and vice versa through the spin exchange interaction between local magnetic moments and carriers. In order to be practical, magnetoelectronics will need to use semiconductors that maintain their magnetic properties at room temperature. This is a challenge because most magnetic semiconductors lose their magnetic properties at temperatures well below room temperature, and would require expensive and impractical refrigeration in order to work in actual computer.

16.20.1 Materials and Methods [17]

Preparations of these materials are mainly done by (i) bulk crystal growth e.g. Bridgman method, and (ii) thin film growth e.g. (a) molecular beam epitaxy (MBE), (b) pulse laser depositions (PLD) and (c) chemical vapour depositions (CVD). The bulk growth of DMS materials are limited by solid solubility of magnetic ions in the lattice of the semi- conductor crystals. This is because of the fact that the equilibrium growth is controlled by phase diagrams.

In contrast, the thin film growth is non-equilibrium in nature, therefore, by controlling the growth temperature a large concentration of magnetic ions can be introduced in the crystal lattice of the semiconductor. One of the approaches for the integration of semiconductors and magnetic materials is the epitaxial growth.

16.20.2 Alloy of II–VI Semiconductors with Magnetic Materials [17]

Diluted magnetic semiconductors (DMS) are compound of alloy semiconductors containing a large fraction of magnetic ions (Mn^{+2}, Cr^{+2}, Fe^{+2}, Co^{+2}) and are studied mainly on II–VI based materials such as CdTe and ZnSe etc. This is because such +2 magnetic ions are easily incorporated into the host II–VI crystals by replacing group II cations. In such II–VI based DMS such as (CdMn)Se, magneto-optic properties were extensively studied, and optical isolators were recently fabricated. Although this phenomenon makes these DMS relatively easy to prepare in bulk form as well as thin epitaxial layers, II–VI based DMS are difficult to dope to create p- and n-type, which makes the material less attractive for applications.

The magnetic interaction in II–VI DMSs is dominated by the anti-ferromagnetic exchange among the Mn spins, which results in the paramagnetic, antiferromagnetic, or spins glass behaviour of the material. Recently, the II–VI compound semiconductors ZnO, GaN have attracted revival attention since it was found that high quality epitaxial thin film display excitonic ultraviolet laser action at room temperature. In addition, the energy gap of this compound can be extended up to ~4 eV by synthesizing alloy compounds of $Mg_xZn_{1-x}O$. Heavy electron doping ($>10^{21}$ cm^{-3}) was readily achieved in contrast to the other II–VI compound semiconductors.

16.20.3 Alloys of III–V Semiconductors with Ferromagnetic Properties [17]

An approach compatible with the semiconductors used in present day electronics is to make nonmagnetic III–V semiconductors magnetic, and even ferromagnetic, by introducing a high concentration of magnetic ions. The III–V semiconductors such as GaAs are already in use in a wide variety of electronic equipment in the form of electronic and optoelectronics devices, including cellular phones (microwave transistors), compact disks (semiconductor lasers), and in many other applications.

The major obstacle in making III–V semiconductors magnetic has been the low solubility of magnetic elements (such as Mn) in the compounds. Because the magnetic effects are roughly proportional to the concentration of the magnetic ions, one would not expect a major change in properties with limited solubility of magnetic impurities, of the order of 10^{18} cm^{-3} or less. Using molecular beam epitaxy (MBE), a thin film growth technique in vacuum that allows one to work far from equilibrium, made a breakthrough. When a high concentration of magnetic elements is introduced in excess of the solubility limit, formation of the second phase occurs if conditions are near equilibrium.

16.21 Tunable Left-Handed Metamaterial Based on Electro-Rheological Fluids [18]

Electric permittivity ε and magnetic permeability μ are two fundamental electromagnetic parameters describing the electromagnetic response of continuous medium. Most of the naturally existing substances have positive permittivity and permeability. Electro-rheological fluids (ERF), a kind of smart materials, are suspensions of polarizable dielectric particles dispersed in insulating liquid, which can abruptly change from liquid to solid upon the application of an external electric field (E-field) due to structural change when the particles align themselves along the field direction to form chains. ERF have the characteristic of being anisotropic dielectric, i.e. the dielectric constant increases along the field direction while it decreases in the transverse direction. Moreover, the dependence of the resonance frequency of the SRRs on the substrate permittivity has been studied in some research papers. With the increasing permittivity of the substrate, the resonant frequency of the split ring resonators (SRRs) moves toward lower frequency. Therefore, this property of ERF may be utilized to dynamically control the position of passband of the LHMs by applying an external E-field.

16.21.1 The Left-Handed Dendritic Model [18]

The traditional SRRs are a two-dimensional (2D) anisotropic medium, which is disadvantageous to design the isotropic LHMs. In addition, the majority of research methods on the LHMs and the negative permeability metamaterials are static and passive, which enormously restrict their practical applications. Concerning these deficiencies, the feasibility of dynamically tunable electromagnetic behaviours in negative permeability metamaterials and LHMs has been studied by several methods, such as E-field, optical excitation, defect effect, loading varactor, adjusting the permittivity of the substrate or varying the substrate thickness. However, so far, no study is reported on analyzing the omnidirectional resonance of this dendritic model, which simultaneously has $-\varepsilon$ and $-\mu$, and using Electro rheological fluids (ERF) to realize the tunable passband of this unitary dendritic. In this work, the authors have proposed a kind of dendritic model which is easier to cancel electromagnetic coupling than the traditional SRRs. The simulated results show that this left-handed model has the 2D isotropic property. Furthermore, the effects of electrorheological fluids on the passband of the periodic dendritic array, have been demonstrated.

References

1. Guina, M., Härkönen, A., Korpijärvi, V. M., Leinonen, T., Suomalainen, S.: Semiconductor disk lasers: recent advances in generation of yellow-orange and Mid-IR radiation. Advances in optical technologies. Hindawi Publishing Corporation, New York (2012). Article ID 265010, doi:10.1155/2012/265010
2. Chilla, J., Shu, Q. Z., Zhou, H., Weiss, E., Reed, M., Spinelli, L.: Recent advances in optically pumped semiconductor lasers. In: Lasers and Applications in Science and Engineering. International Society for Optics and Photonics, pp. 645109–645109 (2007)
3. Khlyabich, P.P., Burkhart, B., Rudenko, A.E., Thompson, B.C.: Optimization and simplification of polymer–fullerene solar cells through polymer and active layer design. Polymer **54**(20), 5267–5298 (2013)
4. Lim, J.A., Liu, F., Ferdous, S., Muthukumar, M., Briseno, A.L.: Polymer semiconductor crystals. Mater. Today **13**(5), 14–24 (2010)
5. Liao, L., Duan, X.: Graphene for radio frequency electronics. Mater. Today **15**(7), 328–338 (2012)
6. Irimia-Vladu, M., Głowacki, E.D., Voss, G., Bauer, S., Sariciftci, N.S.: Green and biodegradable electronics. Mater. Today **15**(7), 340–346 (2012)
7. Chiechi, R.C., Havenith, R.W., Hummelen, J.C., Koster, L.J.A., Loi, M.A.: Modern plastic solar cells: materials, mechanisms and modeling. Mater. Today **16**(7), 281–289 (2013)
8. Pan, S., Zhang, Z., Weng, W., Lin, H., Yang, Z., Peng, H.: Miniature wire-shaped solar cells, electrochemical capacitors and lithium-ion batteries. Mater. Today **17**(6), 276–284 (2014)
9. Chortos, A., Bao, Z.: Skin-inspired electronic devices. Mater. Today **17**(7), 321–331 (2014)
10. Ye, M., Wen, X., Wang, M., Iocozzia, J., Zhang, N., Lin, C., Lin, Z.: Recent advances in dye-sensitized solar cells: from photoanodes, sensitizers and electrolytes to counter electrodes. Mater. Today (2014)
11. Sahraei, N., Venkataraj, S., Aberle, A.G., Peters, I.M.: Investigation of the optical absorption of a-Si: H solar cells on micro-and nano-textured surfaces. Energy Procedia **33**, 166–172 (2013)
12. Dahiya, R.S., Adami, A., Collini, C., Lorenzelli, L.: POSFET tactile sensing arrays using CMOS technology. Sens. Actuators, A **202**, 226–232 (2013)
13. Fortunato, E., Barquinha, P., Martins, R.: Oxide semiconductor thin-film transistors: a review of recent advances. Adv. Mater. **24**(22), 2945–2986 (2012)
14. Millan, J., Godignon, P., Perez-Tomas, A.: Wide band gap semiconductor devices for power electronics. Automatika J. Control Measur. Electron. Comput Commun. **53**(2)
15. Srivastava, P., Das, J., Visalli, D., Van Hove, M., Malinowski, P.E., Marcon, D., Lenci, S., Geens, K., Cheng, K., Leys, M., Decoutere, S., Mertens, R.P., Borghs, G.: Record Breakdown Voltage (2200 V) of GaN DHFETs on Si with 2-μm Buffer Thickness by Local Substrate Removal. IEEE Electron Device Lett. **32**(1), 30–32 (2011)
16. Ibhadon, A.O., Fitzpatrick, P.: Heterogeneous photocatalysis: recent advances and applications. Catalysts **3**(1), 189–218 (2013)
17. Kamilla, S.K., Basu, S.: New semiconductor materials for magnetoelectronics at room temperature. Bull. Mater. Sci. **25**(6), 541–543 (2002)
18. Huang, Yong, Zhao, Xiaopeng, Wang, Liansheng, Luo, Chunrong: Tunable left-handed metamaterial based on electrorheological fluids. Prog. Nat. Sci. **18**(7), 907–911 (2008)

Appendix

CYSTAL STRUCTURE*, LATTICE CONSTANT** a AND *(cla)* *** RATIO OF ELEMENTS *at* room *temperature** for cubic structures*** for non-cubic structures* (Values are *in* Å). 1 Å = 10^{-10} m

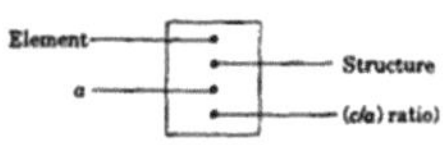

H Hex																	He HCP 3.57 1.633
Li BCC 3.49	Be HCP 2.27 1.581											B Rhom.	C Hex., Cubic 3.57Hex.	N Hex.	O Rhom.	F	Ne FCC 4.46
Na BCC 4.22	Mg HCP 3.21 1.623											Al FCC 4.05	Si DC 5.43	P Complex	S FCO	Cl Tetra.	Ar FCC 5.31
K BCC 5.22	Ca FCC 5.58	Sc HCP 3.31 1.592	Ti HCP 2.95 1.586	V BCC 3.03	Cr BCC 2.88	Mn Cubic	Fe BCC 2.8	Co HCP 2.51 1.621	Ni FCC 3.52	Cu FCC 3.61	Zn HCP 2.66 1.860	Ga FCO	Ge DC 5.66	As Rhom.	Se Hex.	Br Ortho.	Kr FCC 5.64
Rb BCC 5.58	Sr FCC 6.08	Y HCP 3.65 1.569	Zr HCP 3.23 1.594	Nb BCC 3.30	Mo BCC 3.15	Tc HCP 2.74 1.605	Ru HCP 2.71 1.579	Rh FCC 3.80	Pd FCC 3.89	Ag FCC 4.09	Cd HCP 2.98 1.885	In FCT 3.25 1.523	Sn BCT	Sb Rhom.	Te Hex.	I Ortho.	Xe FCC 6.13
Cs BCC 6.04	Ba BCC 5.02	La Hex.	Hf HCP 3.19 1.583	Ta BCC 3.30	W BCC 3.16	Re HCP 2.76 1.615	Os HCP 2.74 1.576	Ir FCC 3.84	Pt FCC 3.92	Au FCC 4.08	Hg Rhom.	Tl HCP 3.46 1.595	Pb FCC 4.95	Bi Rhom.	Po SC 3.34	At	Rn
Fr	Ra	Ac FCC 5.31		Rf	Ha	Unh	Uns	Uno	Une								

Ce FCC 5.16	Pr Hex.	Nd Hex.	Pm	Sm Complex	Eu BCC 4.58	Gd HCP 3.63 1.592	Tb HCP 3.60 1.583	Dy HCP 3.59 1.573	Ho HCP 3.58 1.569	Er HCP 3.56 1.570	Tm HCP 3.56 1.570	Yb FCC 5.48	Lu HCP 3.50 1.585
Th FCC 5.08	Pa Tetra. 3.92 0.826	U Complex. & Ortho.	Np Complex	Pu Complex	Am Hex.	Cm	Bk	Cf	Es	Fm	Md	No	Lr

K.M. Gupta and N. Gupta, *Advanced Semiconducting Materials and Devices*,
Engineering Materials, DOI 10.1007/978-3-319-19758-6

Glossary

Amorphous materials They are non-crystalline solids having no periodic structure at all. SiO_2 glass used as raw material to prepare a single-crystal Si is its example.

Ballistic transport is a phenomenon that occurs in heterojunction devices in which, an electron can pass from one side of the barrier to another side without striking any atom in between.

Bias The voltage applied to a rectifying junction or any other two electrodes of an electronic device.

Biased junction Biasing is a method of connecting *p-n* junction with the battery. It can be of two kinds viz. **1.** Forward biasing, and **2.** Reverse (or backward) biasing. When the positive terminal of the battery is connected on positive side of junction and negative terminal of negative side of junction, it is called *forward-biasing;* while for the connection in just opposite way, the junction is known to be in reverse- biasing.

Binary, ternary and quaternary compounds The binary compounds are made of two-elements such as GaAs, the ternary compounds are made of three-elements such as GaAsP, and quaternary compounds are made of four-elements such as InGaAsP.

Bipolar (solids) are those which conduct current by means of two charge carrying particles of opposite sign (+ and –). For example, the semiconductors are bipolar solids as they conduct current by means of negative electrons and positive holes. These are the solids which conduct current by means of two charge carrying particles of opposite sign (+ and –). For example, the semiconductors conduct current by means of negative charge carrier i.e. electrons and positive charge carriers i.e. holes both, so they are bipolar solids.

Birefringence is a phenomenon in which the material displays double refraction. It so happens when the material/fibre behaves as a birefringent medium due to difference in effective refractive indices and hence phase velocities.

K.M. Gupta and N. Gupta, *Advanced Semiconducting Materials and Devices*, Engineering Materials, DOI 10.1007/978-3-319-19758-6

Birefringence also occurs due to different crystal orientations and anisotropy of fibre cross-section.

Breakdown devices These are the devices whose working depends on avalanche breakdown. Thyristors are such devices.

Breakdown of *p-n* junction is a situation in which the reverse current rises sharply to a very high value. This value can be limited only by the resistance connected in series with the junction.

Brightness It is the power emitted by a source per unit area per unit solid angle.

Bulk negative differential conductivity (BNDC) effect Since the negative conductivity effect of a semiconductor depends on its bulk properties and not on its surface or the junction, therefore this effect is known as BNDC.

Carrier lifetime is the duration until the excess carriers (electrons or holes) do not decay.

Compensation The neutralization of donor impurities by acceptor impurities and vice-versa.

Czochralski (crystal pulling) technique The withdrawal of a crystal from the melt as fast as it grows.

Contact potential (or junction voltage or barrier voltage or potential barrier) is an equilibrium voltage that exists across the depletion region of a *p-n* junction. It cannot induce any net flow of current.

Degenerate *n*-type semiconductors are those in which the electron concentration in conduction band exceeds the effective density of states i.e. $n_e > N(E)_{\text{eff}}$.

Density of state means the population density of electrons in a semiconductor over a certain energy range. It may be expressed as effective density of states N_c located at conduction band edge, or N_v located at valence band edge.

Depletion (or transition) region It is a region of small thickness at the junction in which the rectifying action takes place.

Diac This device is similar to triac, but without the gate terminal. For operation, it breakdowns in either directions.

Diffusion is a flow process in which the carriers move from the regions of high concentration to the regions of low concentration.

Direct band gap semiconductors are those in which the most probable recombination process occurs in such a way that the electrons and holes have the same momentum, and the wave number $k = 0$.

Doping The intentional addition of donor or acceptor impurities to a semiconductor.

Drift velocity v_d. It is the additional velocity acquired by an electron above its normal velocity v_n, when accelerated by an applied electric field.

Effective mass It is the mass of the carriers when they move in an actual crystal on application of external electric field. Effective mass is different from the mass of carriers at rest.

Efficiency of solar cell It is defined as the percentage of incident solar energy that can be converted into useful electrical energy. This efficiency is generally very less, about 10 % only.

Electron hole pair (EHP) These are electron and hole of conduction band, created by the excitation of a valence band electron into the conduction band.

Electronegativity It is the tendency of an atom to attract electrons during bond formation with other atoms.

Electro-optic effect It is the behaviour of a material in which its optical isotropic nature changes to anisotropic nature on application of an electric field. This effect is seen in $LiNbO_3$, $LiTiO_3$ etc.

Epitaxy means an 'arranged layer'. It is a word from Greek literature. The lattice structure of a newly grown layer is an exact extension of the substrate crystal structure of semiconductor. Epitaxy process is one among the sequential processes used in the construction of a microelectronic circuit.

Equilibrium conditions refers to a state in which there is neither any external excitation nor flow of any net current across the *p-n* junction.

Excess carriers are those charge carriers which are in excess of the charges available at thermal equilibrium. The excess carriers can be created by optical excitation, electron bombardment, or other means.

Field-effect transistor (FET) A FET is a three-terminal unipolar solid-state device in which the current is controlled by an electric field.

Forward bias Bias applied to a rectifying junction in the conducting direction.

Hall angle It is the non-vanishing angle made by the electric field in a semiconducting sample, with the axis along which the carriers drift.

Heterojunctions are those *p-n* junctions which are formed between two semiconductors, whose energy band gaps are different. These semiconductors have matched lattice. A heterojunction possesses unique behaviour of 'negative resistivity' and 'ballistic transport'.

Homojunctions are those *p-n* junctions which are formed within a single semiconductor.

Homopolar bonds are those in which the atoms share electrons with each other. Such bonds are also known as covalent bond. Si and Ge are examples of such bonds.

Hyperabrupt junction is a type of linearly graded junction whose voltage sensitivity is very high. This junction is fabricated by epitaxial growth technique.

Intrinsic semiconductor is a perfect crystal having no impurities or lattice defects. Ultra-high pure Si and Ge crystals are intrinsic.

Junction transistor It is a two junction device having their separate regions viz. base, emitter and collector. These regions are separated by base-emitter and base-collector junctions.

Junction The boundary between *p* and *n* semiconductors is known as junction.

Latching It is the ability of a SCR to remain ON, even when the gate current is removed. The SCRs fall in this category, and so are known as latching device.

Light-activated SCR (or LASCR) It is a four-layer device that can be triggered by a pulse of light. This device is useful in n any optoelectronic systems.

Liquidus It is the boundary between a liquid phase and a (solid + liquid) phase. Knowing this boundary in terms of temperature and composition of two alloying elements, is important for zone refining of semiconductors.

Luminescence is the ability of a material by virtue of which it remits visible light on absorption of energy.

Masking It is the process of repeating oxidation-photolithography-doping epitaxial growth metallization a number of times on wafer surface during making of an integrated circuit (IC).

Microelectronics The extreme miniaturization of circuits. In its most extreme form the integration of circuits, junctions, etc. into a single small chip of solid material.

Mobility μ is defined as the proportionality constant between the drift velocity and applied field.

Narrow base diode is such a diode whose neutral region (it is lightly doped) is shorter than the diffusion length of minority carrier. A shorter region structure of diode makes its switching time lower, which is advantageous.

Negative resistance refers to such a behaviour of some materials which shows a decrease in voltage when current is increased through it. It is just opposite to Ohm's law in which $V \propto I$, therefore $V = RI$.

Optical absorption means absorption of electromagnetic (e.m) radiations of light by the semiconducting material. Whether the optical absorption will be less or more, depends on the property of material.

Photoconductivity refers to the phenomenon of increase in conductivity of a semiconductor on account of excess carriers arisen from optical luminescence.

Photoemissivity It refers to the phenomenon of emission of electrons from a metal cathode, when exposed to light or any other radiations.

Photocell A junction whose sensitivity to light is due to the extra minority carriers excited by the radiation.

Photoconductor A semiconductor whose conductivity is sensitive to light due to the creation of conduction electron-hole pairs by absorption of photons.

Photoelectric effect It refers to a phenomenon in which the ejection of electrons from a metal surface takes place, when the metal surface is illuminated by light or any other radiation of suitable frequency (or wavelength). Several devices such as phototube, solar cell, fire alarm etc. work on this effect (principle).

Photons are discrete radiation emitted from the materials in the form of small energy packets. These packets are also called as '*quanta*'. They possess energy, mass and momentum.

***p-n* diode** It is a semiconductor diode made by joining *n*-type and *p*-type extrinsic crystals. By doing so the favourable properties of each type is contributed in the buffer zone (or depletion region). It is a one-way device because it conducts in one direction only.

Power devices The switching devices which are used in power control applications, are called as power devices. They possess the advantages of a rheostat and of a switch.

Recombination is a process of formation of an electron-hole pair (EHP), when an electron in conduction band makes a direct or indirect transition to an empty state (i.e. hole) in valence band.

Rectification is the process of conversion of alternating current (or voltage) into direct current (or voltage). The rectification may be either half-wave type or full-wave type.

Relaxation time is defined as about one half the average time taken between the collisions of conduction electrons with lattice defects phonons etc.

Reverse bias Bias applied to a rectifying junction in the insulating direction.

Schottky effect refers to the phenomenon of lowering of effective work function of a metal, when positive charge force combines with an electric field. The positive charges are induced in the metal when the negative charges are brought near the metal surface.

Seed crystal A piece of single crystal which is used as a nucleus to grow a larger crystal.

Shockley diode This name is often used to represent a two-terminal, four-layer diode which switches a system from its forward-blocking state to the forward-conducting state.

Silicon-controlled switch (SCS) It is a low current SCR having two gate terminals. It can be switched ON or OFF by a pulse applied at either gate.

Solidus It is the boundary between a solid phase and a (solid + liquid) phase. Knowing this boundary in terms of temperature and composition of two alloying elements is important for zone refining of semiconductors.

Steady state conditions are those in which the responses of *p-n* junctions do not depend on time, and the operation of devices is time- independent.

Storage delay time It is the time required for the stored charge to become zero. Thus, it also means the time taken for the junction voltage to become zero.

Thermistor A semiconductor thermometer, which uses the temperature dependence of the carrier densities.

Thyristor It is a solid-state device having two or more p-n junctions, and which can be switched from OFF to ON or vice-versa states. Some devices of this family are silicon-controlled rectifier (SCR), Diacs, and Triacs.

Transient condition means a non-steady state in which the behaviour of p-n junction is a function of time. It is a non-equilibrium state.

Transistor cut-off state is a situation when it does not conduct current. In that case the base current I_B and collector current I_C are zero (i.e. $I_B = 0$, $I_C = 0$) and voltage across the common-emitter V_{CE} and common-collector V_{CC} are equal (i.e. $V_{CE} = V_{CC}$).

Transistor saturation A transistor is known to be in saturation when it carries the maximum current *i.e.* I_c sat $= V_{CC}/R_L$. Here R_L is load resistance. In saturation state, $V_{CE} = 0$.

Transit time device is a high-frequency device that operates at or above the microwave frequencies. They are very useful for converting d-c to microwave a-c signals and in generation of microwave power for various applications.

Translucent materials are those through which the light is transmitted diffusely i.e. it scatters within the material. As a result of this behaviour, the objects are not clearly distinguishable when viewed through a translucent specimen.

Trap It is an energy level that can capture either electrons or holes easily, but not the both.

Triac It is a 3-terminal bi-directional device having two anodes and a gate. It allows flow of current in both directions. A triac can be triggered to conduction state by either of the positive and negative voltages at its anode.

Triggering It is the process of switching of power devices like a p-n-p-n diode, from the forward-blocking state to the forward-conducting state.

Tunnelling (or quantum-mechanical tunnelling) It is a mechanism by which a particle 'penetrates' the potential barrier. The penetration may be of an electron through a barrier of finite height and thickness. This occurrence is important only over a very small dimension, but it is of great importance in conduction of electrons in the solids.

Unijunction transistor (UJT) is a three-terminal *p-n* junction diode having three leads. It differs from FET in that it has no ability to amplify. An UJT exhibits a negative resistance characteristic which makes it useful as oscillator. This device can control a large a-c power with small signal.

Unipolar (solids) are those which conduct current by means of mobile charges of one sign only (+ or −).

Wave number *k* It is the wave function of an electron which is assumed to be moving in plane waveform along *x*-direction. It is expressed as $k = 2\pi/\lambda$, where λ is de-Broglie's wavelength.

Zener diode It is a reverse-biased heavily doped two-terminal *p-n* junction that operates in breakdown region.

Zincblende It is a type of lattice generally found in III–V semiconductor compounds in which the basic structure is diamond cubic (DC), but the arrangement of atoms is different on alternating sites.

Zone refining It is a method of purifying the material from the impurity contents present in it. Zone refining is based on the principle of phase separation. Highly pure single-crystal Si is produced by this method.

References

1. Gupta, K.M., Gupta, N.: Advanced Electrical and Electronics Materials. Scrivener Wiley Publishing, USA (2015)
2. Gupta, K.M.: Engineering Materials Research, Applications and Advances. CRC Press, Taylor and Francis Group, USA (2014)
3. Tyagi, M.S.: Introduction to semiconductor materials and devices. Wiley India Pvt. Ltd., Delhi, Reprint (2008)
4. Streetman, B.G., Banerjee, S.: Solid State Electronic Devices, Prentice Hall International, INC, USA (2000)
5. Gupta, K.M.: Electrical Engineering Materials, 3rd edn. Umesh Publications, Delhi (2006)
6. Millman, J., Halkias, C.C.: Integrated Electronics: Analog and Digital Circuits and Systems, Tata-McGraw Hill Pub. Co. Ltd., New Delhi (1995)
7. Manasrch, O: Semiconductor Heterojunctions and Nanostructures. McGraw-Hill Companies Inc., USA (2005). ISBN 0-07-145228-1
8. Pierret, R.F.: Modular Series on Solid State Devices-Semiconductor Fundamentals, vol. I. Addison-Wesley (1996)
9. Dekker, A.J.: Electrical Engineering Materials. Prentice-Hall of India Private Limited, New Delhi (2009)
10. Bar-Lev, A.: Semiconductor and Electronic Devices. Prentice Hall (1993)

K.M. Gupta and N. Gupta, *Advanced Semiconducting Materials and Devices*,
Engineering Materials, DOI 10.1007/978-3-319-19758-6

MIX
Papier aus verantwortungsvollen Quellen
Paper from responsible sources
FSC® C105338

If you have any concerns about our products,
you can contact us on
ProductSafety@springernature.com

In case Publisher is established outside the EU,
the EU authorized representative is:
Springer Nature Customer Service Center GmbH
Europaplatz 3, 69115 Heidelberg, Germany

Printed by Libri Plureos GmbH
in Hamburg, Germany